D0908308

The Proteins
Third Edition

Volume I

The Proteins

Third Edition

Volume I

Edited by

HANS NEURATH

Department of Biochemistry
University of Washington
Seattle, Washington

ROBERT L. HILL

Department of Biochemistry
Duke University Medical Center
Durham, North Carolina

Assisted by

CAROL-LEIGH BOEDER

Department of Biochemistry
University of Washington
Seattle, Washington

ACADEMIC PRESS *New York San Francisco London 1975*
A Subsidiary of Harcourt Brace Jovanovich, Publishers

ACADEMIC PRESS, INC.
111 Fifth Avenue, New York, New York 10003

United Kingdom Edition published by
ACADEMIC PRESS, INC. (LONDON) LTD.
24/28 Oval Road, London NW1

Library of Congress Cataloging in Publication Data

Neurath, Hans, Date ed.
The proteins.

Includes bibliographies.
1. Proteins. I. Hill, Robert L., Date joint ed. [DNLM: 1. Proteins. QU55 N494p]
QD431.N453 547'.75 74-10195
ISBN 0-12-516301-0

PRINTED IN THE UNITED STATES OF AMERICA

Contents

CHAPTER 3. **Proteins and Sodium Dodecyl Sulfate: Molecular Weight Determination on Polyacrylamide Gels and Related Procedures**

Klaus Weber and Mary Osborn

CHAPTER 4. **Sedimentation Analysis of Proteins**

K. E. Van Holde

CHAPTER 5. **Quaternary Structure of Proteins**

Irving M. Klotz, Dennis W. Darnall, and Neal R. Langerman

CHAPTER 6. Electron Microscopy of Proteins

J. T. Finch

List of Contributors

Numbers in parentheses indicate the pages on which the authors' contributions begin.

Gary K. Ackers, Department of Biochemistry, University of Virginia, Charlottesville, Virginia (1)

Dennis W. Darnall, Department of Chemistry, New Mexico State University, Las Cruces, New Mexico (293)

J. T. Finch, Medical Research Council Laboratory of Molecular Biology, Hills Road, Cambridge, England (413)

Irving M. Klotz, Biochemistry Division, Department of Chemistry, Northwestern University, Evanston, Illinois (293)

Tore Kristiansen, Institute of Biochemistry, University of Uppsala, Uppsala, Sweden (95)

Neal R. Langerman, Department of Biochemistry and Pharmacology, Tufts University School of Medicine, Boston, Massachusetts (293)

Mary Osborn,* The Biological Laboratories, Harvard University, Cambridge, Massachusetts (179)

Jerker Porath, Institute of Biochemistry, University of Uppsala, Uppsala, Sweden (95)

* Present address: Max-Planck-Institut für Biophysikalische Chimie, Göttingen, West Germany.

K. E. Van Holde, Department of Biochemistry and Biophysics, Oregon State University, Corvallis, Oregon (225)

Klaus Weber,* The Biological Laboratories, Harvard University, Cambridge, Massachusetts (179)

* Present address: Max-Planck-Institut für Biophysikalische Chimie, Göttingen, West Germany.

Preface

Some twelve years have passed since the second edition of "The Proteins" was published. In view of the rapid expansion of the field of protein chemistry during this time, we felt that a new edition was both necessary and justified.

The third edition, like its predecessors, will emphasize recent accomplishments that have proved to be particularly important in advancing our understanding of the structure of proteins and the molecular basis for their many biologic functions. This field of research has become so large and diversified that no single author can be expected to cover it comprehensively. We have therefore relied on individuals who are generally recognized as experts in their fields to review specific subjects. Several groups of readers should benefit from the combined efforts of these experts: protein chemists who wish to obtain the particular views of others in the field, advanced students who require more knowledge of protein chemistry than that provided in introductory courses, scientists engaged in biologic and medical research, and others who desire a working knowledge of protein chemistry.

In the selection of topics and their sequential arrangement, we have attempted to place generalities before specifics and to stress, whenever possible, unifying principles. The first three volumes will be devoted to subjects of general interest such as the isolation and characterization of proteins, chemical modification, polypeptide synthesis, sequence analysis, and X-ray crystallography of protein conformation. Individual proteins will be discussed only as illustrations of the basic principles underlying these procedures. The remaining five volumes will deal with specific proteins which are of interest either because of their unusual structural features (e.g., lipoproteins, metalloproteins, glycoproteins, and flavoproteins) or their unique biologic functions (e.g., antibodies, hormones, viruses, and blood-clotting proteins). The organization of this treatise not only gives the reader an opportunity to compare closely related subjects but also provides a broad background against which he can measure the scope and usefulness of current research.

Volume I focuses on procedures for the isolation and characterization of proteins and methods for the establishment of homogeneity. Chapter 1 describes molecular sieve techniques which are of relatively recent origin but have already been widely used both for preparative and analytical purposes. The rigid theoretical treatment which the subject has received in this chapter testifies to the precision that can be attained in a great variety of experimental situations including the determination of molecular size and shape, the study of self-associating protein subunits, and the binding of small ions to proteins. Chapter 2 contains a comprehensive description of the theory and practice of biospecific affinity chromatography. The principle of using protein–ligand or protein–protein interactions as the basis for separation of specific proteins was recognized over sixty years ago but only in the past few years have major improvements in methodology allowed affinity chromatography to be perfected. This technique, largely developed in the laboratory of the authors, now ranks as one of the most reliable and efficient methods for the purification of proteins and promises to eliminate a great deal of the mystery, empiricism, and frustration of previously available isolation procedures. Chapter 3 examines gel electrophoresis of proteins in sodium dodecyl sulfate. This elegant procedure, first introduced in 1967, has already been spectacularly successful in determining the molecular weights of polypeptide chains and characterizing complex protein mixtures. The detailed description of experimental procedures found in this chapter will be extremely useful in preventing those mistakes which so often occur when a relatively simple method is uncritically applied by the uninitiated experimentalist.

Chapter 4 gives an account of sedimentation analysis of proteins. In contrast to the methods treated in the first three chapters, sedimentation analysis is relatively old yet remains the most precise and theoretically sound method for establishing the molecular weights of native proteins. As our knowledge of proteins has advanced, sedimentation techniques have been refined and improved and continue to be employed today not only for the determination of molecular weights but also in the study of the states of aggregation of proteins and protein subunits and the attendant conformational changes. Chapter 5 deals with the subunit structure of proteins. Although it was recognized over fifty years ago that proteins might be composed of subunits, only in the past decade have methods become available which define subunit structure in terms of the number, composition, and amino acid sequence of the constituent polypeptide chains and provide the information necessary for an understanding of the role of subunits in protein function. This complex, far-reaching subject has been approached by the authors in a generalized manner to embrace all noncovalent associations between protein subunits of similar and dissimilar composition. An extensive tabulation of the subunit structure of over 300 proteins (with references) has been included as a guide to the field. Chapter 6 describes the analysis of subunit structure by electron microscopy. Although only recently applied to the study of proteins, electron microscopy has already proved to be of great value in delineating the geometric arrangement of subunits in enzymes, enzyme complexes, muscle proteins, viruses, and other macromolecular assemblies. The utility of this method, recently enhanced by improvements in the design of electron microscopes, advances in specimen preparation, and computational analysis and processing of images, is convincingly demonstrated by the fifty illustrations found in this chapter.

As in previous editions, the author and subject indexes will be found at the end of the volume.

The preparation of this new edition has been a cooperative effort, and we wish to express our gratitude to all those who have aided us in this venture. In particular we would like to thank the authors for their valuable contributions, their good-natured response to editorial suggestions, and their patience. Other friends and colleagues have been of great help in reviewing the manuscripts and we are grateful for the ideas and suggestions they have provided. We have also benefited from the advice and guidance of the staff of Academic Press and wish to thank them for their assistance in the planning and production of these volumes.

HANS NEURATH
ROBERT L. HILL

The Proteins
Third Edition

Volume I

1

Molecular Sieve Methods of Analysis

GARY K. ACKERS

I. GENERAL INTRODUCTION

This article deals with a family of techniques based on the penetration and distribution of protein molecules within porous networks of cross-linked polymers or porous glass materials. These molecular sieve techniques can be conveniently subdivided into two major groups: The first group includes chromatographic transport methods, usually referred to as gel filtration, gel chromatography, gel permeation chromatography, or exclusion chromatography. The second group includes equilibrium methods in which differences between distribution properties of the various molecular species are utilized and measured by direct optical scanning or by batch equilibration procedures. In current practice, molecular sieves are used extensively for protein fractionation, for characterization of molecular size and weight, and for studies of macromolecular interaction, including conformational transitions, subunit interactions, ligand binding, and reaction rates. A list of the principal applications of these techniques is given in Table I. In contemplating a given application it is important for the protein chemist to have an accurate understanding of the fundamental processes that underlie the technique so that experimental design may be optimized and maximum information extracted from the data. The aim of this chapter is to provide a description of relevant principles as well as selected examples of application that will be useful to the protein chemist. The approach is interpretive rather than encyclopedic, and no attempt has been made to review systematically the immense literature that now exists. The objective, rather, has been to describe the major areas of application and to provide a conceptual framework which will be useful to the reader in critical evaluation of the literature. Although the basic principles of gel chromatography and applications to single solute systems

TABLE I

Applications of Molecular Sieve Methods in Protein Studies

A. Preparative applications
  1. Separation of protein constituents
  2. Desalting and dialysis of proteins
B. Analytical applications
  1. Determination of molecular size
  2. Determination of molecular weight
  3. Determination of heterogeneity
  4. Determination of stoichiometries and equilibrium constants for interactions between protein constituents:
     a. Self-associating systems
     b. Isomerizing systems (conformational transitions)
     c. Association between dissimilar subunits
  5. Studies of macromolecule–ligand binding reactions

have been described in an earlier review (Ackers, 1970), some of this material has been reproduced here in condensed form in an attempt to reassess and evaluate certain aspects of the subject in the light of recent developments. A reader already familiar with basic principles and concepts could proceed directly to Section V and subsequent sections dealing with the more recent applications to polydisperse solutes and protein interactions.

A. History and Survey of Applications

The widespread use of molecular sieves for separation and characterization of polymeric materials began in 1959. Since that time a number of techniques have been devised and applied to the broad range of problems outlined in Table I. The earliest applications were aimed at preparative procedures that continue to be of major importance. The rapid development of analytical applications has occurred more recently, largely since 1965.

The initial work leading to these developments was the introduction by Porath and Flodin (1959) of cross-linked dextrans for the chromatographic separation of macromolecules. This technique was called "gel filtration" and continues to be of major importance in protein chemistry. Early observations had suggested the basic principle involved in these separations (Wheaten and Baumann, 1953; Lathe and Ruthven, 1956); however, the extensive development and wide utilization of the technique was dependent upon the commercial production of dextrans cross-linked with epichlorohydrin ("Sephadex"). Successful application of these materials has also provided the stimulus for the development and use of other gel-forming materials including cross-linked polyacrylamide gels and agaroses. For a given experimental problem it is now possible to select a suitable material from a variety of gel-forming substances, depending upon the solvent conditions and porosities desired. This diversity of chromatographic systems has been increasingly exploited in recent years, both for preparative and analytical purposes. Theoretical and experimental studies on the mechanism of molecular sieving processes have also provided information that is useful for the further development of molecular sieve materials.

The extremely rapid growth of analytical gel chromatography has resulted from several lines of research. First was the early realization by biochemical workers using preparative gel chromatography that the technique could also be used for analytical determinations. Since molecular species of differing sizes exhibited different characteristic elution volumes, it was apparent that the method could be used to provide a fundamental characterization of the solute. A calibrated column can be used to determine the molecular radius and to provide an estimate of molecular weight. This realization quickly led to the development of a variety of new techniques and theories for the determination of molecular size and weight (Porath, 1963; Squire, 1964; Ackers, 1964 1967b; Laurent and Killander, 1964; Andrews, 1965). These methods depend upon the interpretation of partition coefficients (or related quantities) in terms of molecular parameters and are described in Section IV. Second, in 1964, Moore introduced the use of cross-linked polystyrene for fractionation of synthetic polymers. Following this important work a number of procedures were introduced for the detection of size heterogeneity and determination of molecular weight distributions. These methods (usually grouped under the heading "gel permeation chromatography") have proved invaluable to synthetic polymer chemists and are reviewed elsewhere (Altgelt and Segal, 1971).

During the last decade there has been increasing interest among biochemists and molecular biologists in the way protein molecules interact to form functional association complexes. Such interactions have been found to play important roles in the regulation of enzyme activities, the stability of viruses, and the formation of antigen–antibody complexes. Other transport techniques, e.g., sedimentation and electrophoresis, had previously been applied to the study of interacting multicomponent protein systems (see Nichol *et al.*, 1964; Cann, 1970), and a substantial theoretical foundation existed for analysis by such methods. Transport experiments on gel columns appeared to offer several distinct advantages for the study of protein interactions. Qualitative effects were initially reported by several workers (Andrews, 1964; Winzor and Scheraga, 1964), and analogies were drawn between the observed elution profiles and corresponding phenomena known to exist in freely migrating transport systems (e.g., sedimentation, electrophoresis). Further advancement depended upon the formulation of an exact theoretical basis for quantitative description of the chromatographic process. Such an analysis was carried out by Ackers and Thompson (1965), who showed that, in the transport equations for chromatography, the elution volumes play the same role as velocity terms play in the classic transport equations for freely migrating systems. Ackers and Thompson applied these theoretical results to an experimental study of the dissociation of human carboxyhemoglobin into α,β dimers and determined the dissociation constant for this reaction under isoelectric conditions. Extension of these methods has constituted a significant line of development since 1965 (Gilbert, 1966a; Ackers, 1967a; Chiancone *et al.*, 1968; Chun *et al.*, 1969a; Henn and Ackers, 1969a,b; Gilbert and Gilbert, 1968; Zimmerman and Ackers, 1971a; Zimmerman *et al.*, 1971; Warshaw and Ackers, 1971). New techniques of "difference chromatography" introduced by Gilbert and associates (Gilbert, 1966b; Gilbert *et al.*, 1972) have also provided a powerful approach to the study of interactions between protein subunits. Applications of molecular sieve methods to the study of protein subunit interactions are described in Sections VI and VII.

Molecular sieve techniques have also been applied to the study of the binding of small molecules by macromolecules such as proteins and nucleic acids. Determination of stoichiometries and binding constants for such systems can be conveniently carried out using techniques that depend upon the equilibrium dialysis properties of molecular sieves. These techniques generally offer high precision and convenience, as compared with other methods used for binding studies (see Section VIII).

From initial studies it became clear that analytical gel chromatography had certain advantages over other transport techniques for the study of protein systems. However, as a transport technique, gel chromatography was initially quite primitive since the amount of information obtainable was severely limited by the experimental necessity of measuring solute zone profiles only after elution from a column. Although monitoring and collecting samples as they are eluted from a column is clearly the desired procedure for preparative separations, a great deal of potentially useful information is lost in the process of analyzing the sample. In order to solve this problem, direct optical column scanning has been developed (Brumbaugh and Ackers, 1968, 1971; Warshaw and Ackers,

1971). This approach permits direct analysis of solute profiles at many stages during a single column experiment, and has placed analytical gel chromatography on the same footing as other transport methods in terms of data acquired during an experiment. Scanning nonelution chromatography appears so promising that it may be expected to replace the traditional elution method for many analytical determinations. This approach is described in Section III,B and applications of the scanning technique are described in subsequent sections.

B. Partitioning of Solutes into Porous Networks

In order to discuss molecular sieving processes and their applications to protein systems of interest, only a few definitions and concepts are necessary. These will be described below.

1. The Partitioning System

The primary process involved in all molecular sieve techniques is the partitioning of solute molecules between the solvent spaces within porous particles (the *stationary phase* in a chromatographic column) and the solvent space exterior to the particles (the *mobile phase* in a chromatographic experiment). A diagrammatic representation of the partitioning system is shown in Fig. 1. The stationary phase usually consists of a gel-forming material that has been allowed to imbibe solvent until

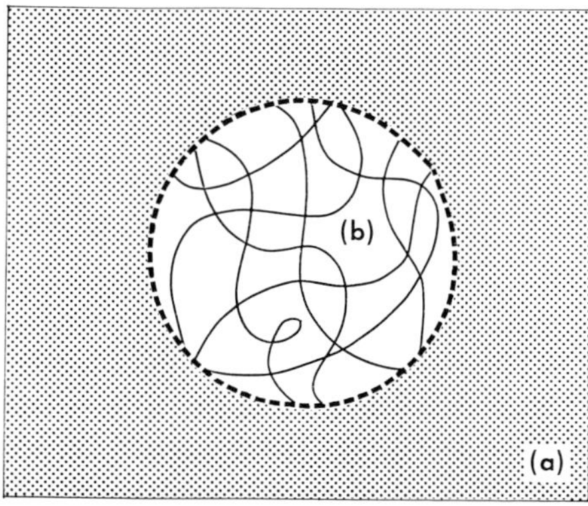

Fig. 1 Diagrammatic representation of partitioning system. The stippled area (a) represents solvent of the mobile phase. A porous gel particle enclosed by the dashed line makes up the stationary phase. It consists of two regions: the internal solvent (b) and gel matrix (solid lines). When solute molecules are introduced into the system, partitioning occurs between regions (a) and (b). Taken from Ackers (1970).

a swelling equilibrium has been achieved. In glass chromatography it consists of a rigid porous glass structure containing solvent within its pores. For any gel partitioning process it is useful to distinguish between three distinct regions within the experimental system: (1) the solvent region exterior to the gel particles which has a volume V_0, termed the *void volume;* (2) solvent within the interior of the gel particles which has an *internal volume* V_i, and is involved in diffusional exchange of solute with the void spaces; and (3) the solid, gel-forming material (the "gel matrix") which has a volume V_g. The total system is the sum of these three regions and occupies a volume V_t:

$$V_t = V_0 + V_i + V_g \tag{1}$$

For any gel-forming material there is a fixed relationship between V_i and V_g, determined by the swelling equilibrium under given parameters of state (temperature, pressure, solvent, ionic strength, etc.). The relationship is conveniently expressed in terms of the solvent regain, S_r. This quantity is the volume of solvent imbibed per unit weight W_g of anhydrous gel-forming material.

$$S_r = V_i/W_g \tag{2}$$

S_r is constant for a given gel material and given environmental conditions. The matrix volume, V_g, is equal to the product of the anhydrous partial specific volume of the gel matrix material, $\bar{V}_g$, and the weight W_g:

$$V_g = \bar{V}_g W_g \tag{3}$$

From Eqs. (2) and (3) the fixed relationship between V_i and V_g can be expressed as:

$$V_i/V_g = S_r/\bar{V}_g \tag{4}$$

The theoretical basis of swelling equilibria in polymer gels has been developed by Flory (1953). In his analysis the equilibrium is treated as a balance between the entropy of swelling and the internal osmotic pressure of the gel. In principle, an externally imposed osmotic pressure produced by a solution of nonpenetrating solute molecules could change the swelling equilibrium and hence the relationship between V_i and V_g. Such osmotic effects have been observed with Sephadex gels using high concentrations of excluded dextran (Edmond *et al.*, 1968). Appreciable changes were observed in the internal volumes of Sephadex beads at dextran concentrations as high as 20 gm/dl. For protein solutions partitioned on most gels at low concentration, the osmotic effects of solute are negligible, but in certain cases it may be desirable to take these effects into account. A thermodynamic analysis of partitioning effects has been developed by Hjertén (1970).

2. *Partition Coefficients*

When solute is introduced into a gel–solvent system it is distributed by diffusion between the solvent regions inside and outside the gel. At equilibrium the distribution is described by a partition isotherm that defines the relationship between the weight of solute Q_i inside the gel and the solute concentration C in the void space exterior to the gel:

$$Q_i = f(C) \tag{5}$$

It is often convenient to formulate the isotherm [Eq. (5)] in terms of equilibrium partition coefficients that provide a thermodynamic description of the system. There are several ways in which partition coefficients can be defined. Two particularly convenient formulations that are widely used in gel chromatography are described below.

a. **Partition Coefficient Referred to Internal Solvent Volume.** This coefficient σ is defined as the amount of solute distributed into the gel per unit internal volume V_i and external concentration C. It is also referred to by the symbol K_D. In terms of this coefficient the isotherm [Eq. (5)] can be written

$$Q_i = \sigma V_i C \tag{6}$$

The dimensionless quantity σ is a measure of the degree of solute penetration within the gel's interior solvent region. The isotherm [Eq. (6)] is applicable to thermodynamically nonideal systems as well as ideal ones. A particularly simple interpretation of σ can be made for the case in which all parts of the system are thermodynamically ideal.

The penetrable volume within the gel, V_p, that is occupied by solute molecules at equilibrium is

$$V_p = Q_i/C_p \tag{7}$$

where C_p is the solute concentration within the region of distribution. Under these conditions, the partition coefficient simply represents the fraction of the internal volume, V_i, that is penetrable by the solute molecule under consideration:

$$\sigma = V_p/V_i \tag{8}$$

This relationship is a good approximation for many systems of interest, especially at low solute concentration where the partition isotherm is found to be nearly linear. However, under more general conditions (e.g., higher solute concentration), the partition coefficient must be written:

$$\sigma = (V_p/V_i)(\gamma_0/\gamma_g) \tag{9}$$

where γ_0 and γ_g are, respectively, the activity coefficients of solute in the void phase and the gel solvent phase. In general a correction term equal to the ratio of these activity coefficients must be applied to the volume ratio V_p/V_i of Eq. (8) in order to satisfy the criterion of thermodynamic equilibrium; σ becomes the volume ratio only in the limit of infinite dilution where $\gamma_0 = \gamma_g$. Nonideality of the isotherm that takes the form of a concentration dependence of σ is to be expected since γ_0 and γ_g will generally have different concentration dependencies. At finite concentrations, σ can be represented by

$$\sigma = \sigma^\circ(1 - gC) \tag{10}$$

where $\sigma^\circ = V_p/V_i$, the limiting value of σ at infinite dilution, and g is the coefficient of concentration dependence. [See Ackers (1970) for a derivation of these relationships.]

b. **Partition Coefficient Referred to Total Volume.** The partition coefficient σ has been defined in Eq. (6) relative to the internal volume of the gel, V_i. It is also possible to define a partition coefficient with respect to the total volume of the gel phase, $V_t - V_0$ (Laurent and Killander, 1964). In this case the partition isotherm can be written

$$Q_i = K_{\text{av}}(V_t - V_0)C_0 \tag{11}$$

where

$$K_{\text{av}} = \frac{V_p}{V_t - V_0}(1 - gC) \tag{12}$$

This partition coefficient, K_{av}, is thus characterized by the same coefficient of concentration dependence (g) as σ [see Eq. (10)]. In the limiting case of infinite dilution, K°_{av} is the volume fraction of the total stationary phase occupied by the solute molecule:

$$K^\circ_{\text{av}} = \frac{V_p}{V_t - V_0} \tag{13}$$

The coefficients K_{av} and σ are related by:

$$\sigma = K_{\text{av}}(1 + \bar{V}_g/S_r) \tag{14}$$

Thus for a given gel system, a constant ratio exists between the two partition coefficients σ and K_{av}. This ratio is independent of the particular values of these coefficients (i.e., it is the same for all molecular species) and also independent of solute concentration. If the partial volume and solvent regain are known, a measured value of one partition coefficient can always be converted to the other. In this article, equations will usually be formulated in terms of σ.

3. The Partition Cross Section

A most useful quantity in both chromatographic transport and equilibrium partitioning experiments is the partition cross section, ξ. This is simply the fraction of a column's cross-sectional area available to the solute:

$$\xi = \alpha + \beta\sigma \tag{15}$$

In Eq. (15), α is the fraction of the column's cross section which is void, and β is the corresponding internal fraction.

The partition cross section enters into the description of molecular sieve phenomena in three important ways.

1. It relates the mean transport velocity, u, of a solute species along the column to the flow rate, F, and total cross-sectional area, A.

$$u = F/\xi A \tag{16}$$

2. It relates the distribution volume (or elution volume) to the distance coordinate in a column transport experiment

$$V = A\xi x \tag{17}$$

or

$$V = A\xi l$$

where l is column length. [Note that Eq. (17) follows from Eq. (16) and from the fact that $V = Ft$, where t is time.]

3. It relates the concentration of solute within the total column (C') to the bulk concentration (C) of solute added to the column (e.g., that which exists within the void volume and penetrable regions of the gel):

$$C' = \xi C \tag{18}$$

Equation (18) is of fundamental importance both to the theory of transport and the analysis of optical scanning experiments. A brief

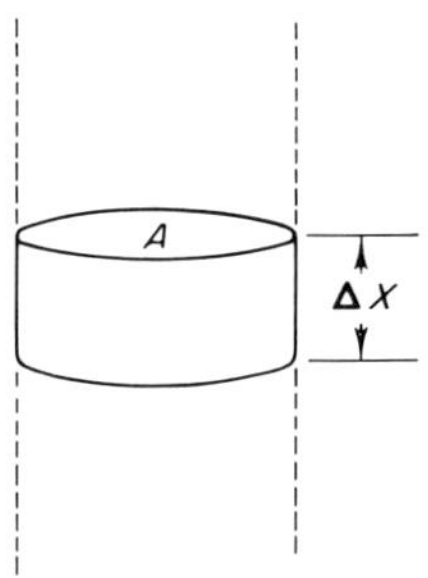

Fig. 2 Small segment of column.

derivation follows. Suppose we consider a small section of column into which beads of gel have been packed (Fig. 2). The slice of column has a height Δx and a cross-sectional area A so that the volume $V_t = A\ \Delta x$. This volume can also be represented as a sum of volumes pertaining to the three regions defined by Eq. (1).

$$V_t = \alpha A\ \Delta x + \beta A\ \Delta x + \gamma A\ \Delta x \tag{19}$$

Here α and β are defined as in Eq. (15) and γ pertains similarly to the gel matrix region. At equilibrium, the volume within the slice occupied by solution at concentration C is given by

$$\begin{aligned} V_p &= V_0 + \sigma V_i \\ &= \alpha A\ \Delta x + \sigma\beta A\ \Delta x \\ &= (\alpha + \beta\sigma)A\ \Delta x \end{aligned} \tag{20}$$

Then,

$$V_p = \xi A\ \Delta x \tag{21}$$

The total mass Q_T of solute distributed within the slice is

$$Q_T = V_p C = \xi A\ \Delta x C \tag{22}$$

The concentration C' of solute within the slice is

$$C' = \frac{Q_T}{V_T} = \frac{\xi A\ \Delta x C}{A\ \Delta x} = \xi C$$

which is the same as Eq. (18). The concentration C in the free solution frame of reference is greater than the concentration C' in the total column frame of reference by the factor $1/\xi$. This transformation between frames of reference for the expression of concentration is important for the analysis of both equilibrium scanning experiments and nonequilibrium transport experiments (see Sections II,B and III,C).

C. Molecular Sieve Materials

With porous materials now available it is possible to separate and study molecules ranging in diameter from only a few angstroms to several hundred angstroms, or in molecular weight from several hundred daltons to several hundred million daltons. The physical and chemical properties of these materials vary greatly. However, they share a common set of basic requirements: (1) It is essential that the gel-forming materials have a strong affinity for the desired solvent in order to swell with a high solvent regain. The equilibrium structure of gels consists of a three-dimensional network formed by the association of long polymeric chains. The structure is swollen by imbibed solvent to an equilibrium limit determined by the nature of the matrix material, the solvent, and the environmental conditions of temperature, pressure, etc. Hydrophilic gels usually contain polar groups such as hydroxyl groups, whereas hydrophobic materials contain aromatic and other less-polar groups. Solvent affinity is not a necessary requirement for porous glass materials since the rigidity and porosity conferred upon them in their initial formation is maintained regardless of the presence of solvent. (2) The porous material must have a low affinity for the solute molecules of interest in order to minimize the adsorptive or charge interactions in analytical applications (for separation purposes, such side effects are not always undesirable and may actually be helpful). (3) The three-dimensional network formed by the swollen gel must have sufficient rigidity to withstand being packed into a column and subjected to the flow of solvent under pressure. In addition, it should be sufficiently rigid so that osmotic swelling and contraction will not produce significant errors. (4) For experiments involving direct optical scanning of gel columns, the material should have a minimum of chromophoric groups in the wavelength region used for detection of solute. A number of materials that satisfy these criteria have been employed for chromatographic use. Specific details of their preparation and chemical structure have been described elsewhere (Ackers, 1970) and will not be repeated here. However a brief review will be given of the principal porous materials in order to indicate the ranges of molecular size and weight in which they have been effectively employed (see Ch. 2 for another discussion of materials). For most protein systems it is possible to find a gel or porous glass material that, under a desired set of ionic and pH conditions, is free of detectable adsorptive interactions and has a desired porosity.

1. Cross-Linked Dextrans

The most extensively used gel-forming materials are cross-linked dextrans, first introduced by Porath and Flodin (1959). These materials are commercially available in bead form under the trade name "Sephadex." The dextran itself is a soluble polysaccharide of glucose and is produced during the growth of the microorganism *Leuconostoc mesenteroides.* The glucose residues are predominantly in α-1,6-glycosidic linkage and there are three hydroxyl groups per glucose unit which can be cross-linked between chains with epichlorohydrin, forming glyceryl ether bonds. These gels are hydrophilic and their water regain is determined by the relative percentage of epichlorohydrin reacted and by the molecular weight of the starting dextran material.

TABLE II

Properties of Commercial Dextran Gels (Sephadex)[a,b]

Type	Particle size[c] (dry; in μm)	Water regain (ml/gm)	Gel bed (ml/gm)	Approximate separation range: Peptides and glob. proteins	Approximate separation range: Dextran fractions
G-10	40–120	1.0 ± 0.1	2–3	Up to 700	Up to 700
G-15	40–120	1.5 ± 0.1	2.5–3.5	Up to 1,500	Up to 1,500
G-25, coarse	100–300				
G-25, medium	50–150	2.5 ± 0.2	4–6	1,000–5,000	100–5,000
G-25, fine	20–80				
G-50, coarse	100–300				
G-50, medium	50–150	5.0 ± 0.3	9–11	1,000–30,000	500–10,000
G-50, fine	20–80				
G-75	40–120	7.5 ± 0.5	12–15	3,000–70,000	1,000–50,000
G-100	40–120	10.0 ± 1.0	15–20	4,000–150,000	1,000–100,000
G-150	40–120	15.0 ± 1.5	20–30	5,000–400,000	1,000–150,000
G-200	40–120	20.0 ± 2.0	30–40	5,000–800,000	1,000–200,000

[a] Taken from Ackers (1970).

[b] Specifications provided by the manufacturer: Pharmacia Fine Chemicals, Uppsala, Sweden.

[c] All porosities are also manufactured as beads in the particle size—Superfine (10–40 μm diameter).

In moderate-to-high salt concentrations the Sephadex gels are relatively free of adsorptive interactions with large molecules. However, a high affinity for small aromatic compounds has been noted by many workers (Janson, 1967; Eaker and Porath, 1967; Determann and Walter, 1968). Evidence has been presented which indicates that the sites of interaction are the ether cross-linkages (Determann and Walter, 1968). Table II summarizes the types and general specifications of commercial dextrans.

2. *Polyacrylamide*

Polymerization of acrylamide by the use of a bifunctional agent such as *N,N′*-methylenebisacrylamide leads to the formation of a cross-linked hydrophilic network. The polymerization is carried out in aqueous solutions and the resulting hydrophilic gel can be granulated and used for chromatographic purposes (Hjertén and Mosbach, 1962). Polyacrylamide gels are generally similar in range of porosity to the cross-linked dextrans. The acrylamide gels most widely used for chromatographic purposes are the commercially produced spherical beads of Bio-Gel P. The types available and their corresponding specifications are listed in

TABLE III

Bio-Gel P Types and Specifications[a]

Type	Particle size (wet mesh) (μm)	Approximate "exclusion limit" (molecular weight)	Hydrated bed volume (ml/gm dry gel)	Water regain (gm water/gm dry gel)
Bio-Gel P-2	50–100	1,600	3.8	1.6
Bio-Gel P-2	100–200	1,600	3.8	1.6
Bio-Gel P-4	50–150	3,600	6.1	2.6
Bio-Gel P-6	50–150	4,600	7.4	3.2
Bio-Gel P-10	50–150	10,000	12.0	5.1
Bio-Gel P-20	50–150	20,000	13.0	5.4
Bio-Gel P-30	50–150	30,000	14.0	6.2
Bio-Gel P-60	50–150	60,000	18.0	6.8
Bio-Gel P-100	50–150	100,000	22.0	7.5
Bio-Gel P-150	50–150	150,000	27.0	9.0
Bio-Gel P-200	50–150	200,000	47.0	13.5
Bio-Gel P-300	50–150	300,000	70.0	22.0

[a] Taken from Ackers (1970).

Table III. Under many experimental conditions polyacrylamide gels have been found to be more inert than dextran gels with respect to adsorptive interactions, particularly at low ionic strength. For the most demanding analytical studies it is desirable to prepare the acrylamide gels from reagents freshly purified in one's own laboratory.

3. *Agar and Agarose*

Agar is a mixture of two polysaccharide components (Araki, 1956). The main component is a linear polymer of D-galactose and 3,6-anhydro-L-galactose, termed agarose. The second component is also a galactose polymer, agaropectin, which contains a substantial number of carboxyl and sulfate groups. Since these groups can produce ion exchange effects within the gel, the agarose component is much more desirable for analytical chromatography. Early applications of granulated agar (Polson, 1961; Steere and Ackers, 1962a) indicated a wide range of use for chromatographic separation of components and for determination of molecular size (Steere and Ackers, 1962b) and weight (Andrews, 1962). Procedures for separating agar into its components have been developed by Hjertén (1962) and Russell *et al.* (1964). The formation of agarose beads can be effectively carried out by the procedures described by

TABLE IV

Properties of Commercial Agarose Gels according to the Manufacturer[a]

Type	Supplier	Average particle size (μm)	Approximate fractionation range for proteins ($\overline{MW}$)
Sag 2	Seravac[b]	70–140 (crushed)	50×10^4–1.5×10^8
Sag 4	Mann	—	20×10^4–15×10^6
Sag 6	—	—	5×10^4–2×10^6
Sag 8	—	—	2.5×10^4–70×10^4
Sag 10	—	—	1×10^4–25×10^4
Sepharose 2B	Pharmacia	60–300 (beads)	8×10^4–20×10^6[c]
Sepharose 4B	—	30–200 (beads)	1×10^4–3×10^6[c]
Bio-Gel A-150 m	Bio-Rad	50–100 (beads) 100–200 (beads)	1×10^6–1.5×10^8
Bio-Gel A-50 m	—	—	10×10^4–50×10^4
Bio-Gel A-15 m	—	—	4×10^4–15×10^6
Bio-Gel A-5 m	—	—	1×10^4–5×10^6
Bio-Gel A-1.5 m	—	—	1×10^4–1.5×10^6
Bio-Gel A-0.5 m	—	—	1×10^4–50×10^5

[a] Taken from Ackers (1970).
[b] Seravac Laboratories (PTY). Lts., Holyport Maidenhead, Berkshire, England. Also Mann Research Laboratories, Inc., New York, under the name "Ago-gel."
[c] Refer to dextran fractions.

Hjertén (1964) or Bengtsson and Philipson (1964). Commercial agarose products are listed in Table IV.

4. *Porous Glass*

A procedure has been developed (Haller, 1965) for making rigid, high-silica glass with a network of interconnected pores. The pores within these materials appear to consist of tortuous channels with roughly circular cross sections and varying diameters. A very narrow distribution of pore diameters can be achieved, covering a range from 170 to 1700 Å. These materials have been used for studies on the mechanism of separation (Haller, 1968). Commercially produced versions of the porous glass media are listed in Table V. Structural studies of porous glass materials have been carried out by Barral and Cain (1968). The special advantage of porous glass columns is that they can be cleaned with acid or sterilized by autoclaving without disturbing the packing. Also, they form an extremely rigid column that does not change volume when pressure is applied. This permits considerable variability in the useful

TABLE V

Fractionation Range of Commercially Available Porous Glass[a,b]

Type	Approximate pore size (Å)
Bio-Glass-200	200
Bio-Glass-500	500
Bio-Glass-1000	1000
Bio-Glass-1500	1500
Bio-Glass-2500	2500

[a] Taken from Ackers (1970).
[b] Specifications provided by the manufacturer: Bio-Rad Laboratories, Richmond, California.

flow rates that can be attained. Their main disadvantage lies in a relatively high adsorptive interaction with most proteins.

5. *Polystyrene Gels*

The cross-linking of polystyrene with divinyl benzene can be carried out to produce gels of varying porosity, depending upon the solvent used in the polymerization of styrene with divinyl benzene (Moore,

TABLE VI

Fractionation Range of Different Types of Styragel[a,b]

Type (Å)	Approximate fractionation range for vinyl polymers (MW)	Approximate "exclusion limit" (MW_{lim})
60	800	1,600
100	2,000	4,000
400	8,000	16,000
1×10^3	20,000	40,000
5×10^3	100,000	200,000
10×10^3	200,000	400,000
30×10^3	600,000	1,200,000
1×10^5	2,000,000	4,000,000
3×10^5	6,000,000	12,000,000
5×10^5	10,000,000	20,000,000
10×10^5	20,000,000	40,000,000

[a] Taken from Ackers (1970).
[b] Specifications provided by the manufacturer: Waters Associates, Inc., Framingham, Massachusetts.

1964). These gels are swollen by nonpolar solvents such as toluene, methylene chloride, and dimethylformamide. The commercially available polystyrene beads (Styragel) are listed in Table VI along with some of their properties.

6. *Other Materials*

Other gel materials that may be potentially useful for gel chromatography are cross-linked products of locust bean gum, polyvinyl alcohol, sorbitol, cellulose, starch, gelatin treated with tannic acid, silica gel beads, and elastin fibers.

Modified dextrans have been widely used for separation in nonaqueous systems (Nyström and Sjövall, 1965). Commercial preparations of methylated Sephadex have lipophilic properties and can be used with a variety of nonaqueous solvents.

D. Concept of Porosity

In discussions of molecular sieving phenomena the term "porosity" is frequently used. This term sometimes denotes a critical molecular size limit for penetration into the gel network. Such usage suggests an "all-or-none" type of penetration mechanism instead of the continuous gradation of exclusion properties actually observed. Thinking of porosity as denoting a *range* of molecular sizes that can be accommodated by a given network is more useful. The "porosity" of the network then is defined by the curve relating molecular size to partition coefficient. As will be seen in Section IV, two parameters are generally required to define the state of the gel with regard to its partitioning properties. Adoption of a precise and universally acceptable means of specifying porosity would avoid confusion on this point. The specification of two molecular radius values and corresponding partition coefficients would suffice. The molecular radii corresponding to partition coefficients of 0.1 and 0.9, for example, would fully characterize most gels with respect to partitioning properties of all molecules.

II. CHROMATOGRAPHIC TRANSPORT METHODS

In protein chemistry, preparative and analytical gel filtration is most commonly performed in a chromatographic system in which particles of porous materials serve as the stationary phase. A pulse of solute is allowed to flow with the mobile phase, and diffusional distribution occurs between solvent of the two phases. The large molecules spend a small fraction of their time within the stationary phase and consequently move more rapidly through the column. The limiting condition for diffusional exchange is determined by the equilibrium partition coefficients defined in the previous section. However, in transport experiments, a nonequilibrium perturbation is always present so that a steady state is achieved with respect to solute partitioning, and true equilibrium does not exist. Nonequilibrium effects of diffusion within the column as well as effects of nonuniform flow around the gel particles also contribute to solute transport. The exact shape and position of solute zones depend upon the initial configuration of the solute sample and on all interactions between solute and column. The general aspects of chromatographic transport on gel columns as well as the basic theory behind different types of chromatographic experiments will be described in the following section.

A. Elution Chromatography

The most common experimental technique is the elution experiment in which a concentration–volume profile is measured for the sample emerging at the bottom of the column. The column usually consists of a cylindrical tube into which the swollen gel particles or beads have been packed. The vertical tube is fitted with a suitable arrangement of inlet and outlet devices which retain the gel bed while permitting the flow of solvent through the column. Sometimes provision is made for pumping solvent in both directions (upward and downward) through the vertical column and for controlled introduction of a solute sample. As the solute is eluted from the bottom of the column, its concentration is monitored as a function of the volume of solvent that has passed through the column after introduction of the sample at the top. For chromatography in aqueous solvents, the detector is usually a spectrophotometric device; for organic solvents, it is usually a refractometer. The output signal from the detector is commonly fed into a strip chart recorder and measured as a function of time. If the rate of flow is known, the correlation between concentration and volume is established. For precise analytical determinations and subsequent automated computational procedures, the output may be fed into a digital data acquisition device that interfaces the experiment with a computer. Many systems based on this general scheme have been employed experimentally using various specific components, some of which are manufactured commercially.

In the column experiment, the essential results are the solute's characteristic elution volume and the amount of spreading of solute zone that has taken place. These aspects will be described here briefly for several of the most useful types of experiments. The theoretical basis of the observed behavior of solute zones will then be discussed. For detailed derivations, the reader is referred to the original papers.

1. Characteristic Elution Volumes

There are two different methods of applying the solute to the column:

***a.* Small-Zone Experiments.** If a solution containing a macromolecular component is introduced at the top of the column in a sample of very small volume compared to the bed volume, V_t, and is followed by a flow of solvent, the effluent volume, V_e, of solvent which passes through the column between the introduction of the sample and the subsequent emergence of its maximum concentration can be very closely approximated by the equation

$$V_e = V_0 + \sigma V_i \tag{23}$$

The constant of proportionality, σ, can be shown to be identical to the partition coefficient described previously (Section I,B,2) and satisfies the following conditions:

$\sigma = 0$ for a molecular species totally excluded from the gel phase
$\sigma = 1$ for a molecular species that diffuses freely with no restrictions within the gel network
$\sigma < 1$ for a molecule of intermediate size in the absence of specific interaction (e.g., adsorption, ion exchange)
$\sigma > 1$ when specific interaction effects sufficiently retard the elution velocity within the column

From Eq. (23) and the above conditions it is seen again that the *larger* molecular species are eluted *first* in chromatography of this type (Fig. 3). An equivalent expression for the elution position may be written in terms of the partition coefficient K_{av}.

$$V_e = V_0 + K_{av}(V_t - V_0) \tag{24}$$

Equations (23) and (24) provide the most common means of evaluating the partition coefficients. The volumes V_0 and V_i are determined from separate experiments using solute species having partition coefficients of zero (totally excluded) and unity (nonexcluded), respectively.

***b.* Large-Zone Experiments.** If the sample is introduced in a volume S which is large in relation to the bed volume of the column, and if the

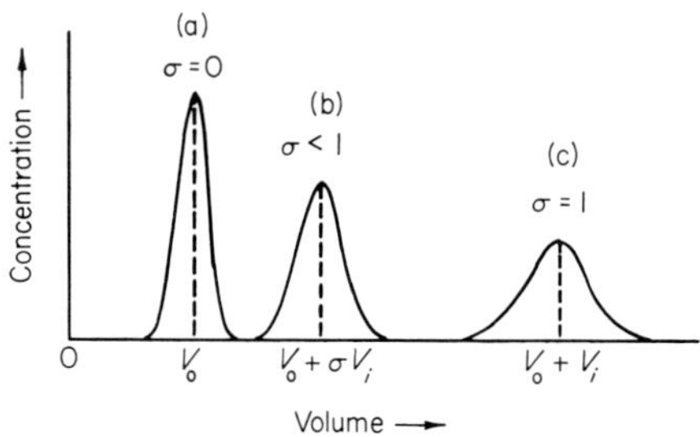

Fig. 3 Schematic elution diagram illustrating three types of solute behavior: (a) total exclusion; (b) a penetrant molecule; and (c) a small, totally nonexcluded molecule. Taken from Ackers (1970).

effluent concentration of the solute species is measured as a function of the volume V that has flowed through the column, a "plateau region" of constant solute concentration (equal to that initially applied) is found. This plateau is delimited by leading and trailing boundaries which are diffuse, i.e., they do not possess sharp edges but have been broadened by the axial dispersion effects mentioned above. For each of these diffuse boundaries an *equivalent sharp boundary* may be determined from which the elution volume may be calculated. The procedure is shown diagrammatically in Fig. 4 for a trailing boundary. Consider an arbitrary reference position V_r chosen within the plateau region of the elution diagram (Fig. 4). The equivalent boundary elution position $\overline{V}'$ is then chosen in such a way that the mass of solute $C_0(\overline{V}' - V_r)$ represented by the area under the resulting "idealized" diagram equals the true mass represented by the area under the experimentally deter-

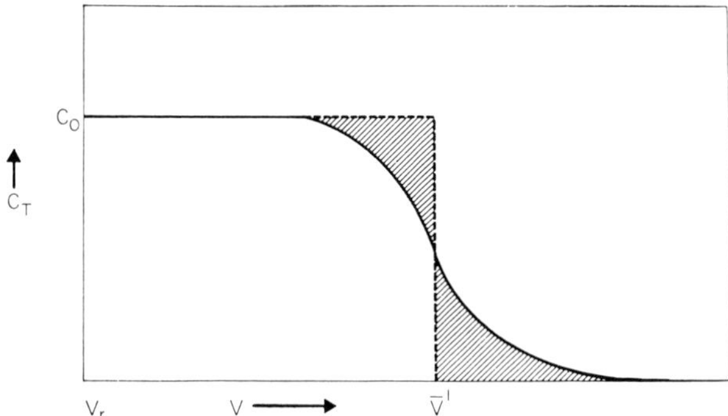

Fig. 4 Sharp boundary equivalent to the trailing elution boundary of a large zone. Taken from Ackers (1970).

mined elution curve (solid line, Fig. 4). This equality can be expressed as

$$C_0(\bar{V}' - V_r) = \int_0^{C_0} (V' - V_r)\, dC \tag{25}$$

By evaluating the terms in Eq. (25), it is seen that

$$\bar{V}' = \frac{1}{C_0} \int_0^{C_0} V'\, dC \tag{26}$$

The equivalent boundary position (centroid volume) $\overline{V'}$ is independent of the particular reference position (V_r) chosen. Its value is obtained by graphical or numerical integration of the elution diagram. When this procedure is carried out, it is found that $\overline{V'}$ satisfies the relationship:

$$\bar{V}' = \bar{V} + S = V_0 + \sigma V_i + S \tag{27}$$

In this experiment V is defined as zero when the leading boundary enters the column. A similar calculation for the leading boundary yields a centroid volume $\overline{V}$ that is equal to the right-hand side of Eq. (23). The parameter σ is found to have the same numerical value for both the small-zone peak position determination and the large-zone "plateau experiment." The plateau experiment is particularly important in the study of interacting solute species (see Sections VI, VII, and VIII).

B. Theory of Column Operation

Theoretical analyses of chromatographic transport processes provide a rational basis for the observable behavior of gel partitioning systems. Theoretical treatment makes it possible to predict behavior and optimum conditions for separation or analysis. An understanding of the mechanics of gel chromatography for single solute systems is also a prerequisite for extending molecular sieve methods to multicomponent systems, including those in which chemical reactions are superimposed on the transport behavior.

A number of approaches to the theory of chromatographic behavior exist and have been reviewed elsewhere (Ackers, 1970; Altgelt, 1967). Although many processes may play limited roles in determining the behavior of macromolecules on gel columns, the essential features of column behavior can best be described as a consequence of equilibrium solute partitioning onto which a steady-state nonequilibrium perturbation is superimposed. The mean rate of solute transport along the column is determined by equilibrium partitioning properties, whereas the nonequilibrium perturbation is caused by lengthwise spreading (axial dispersion) as the sample moves down the column (Halvorson and Ackers, 1971).

1. Equation of Continuity

For the exchange of solute between mobile and stationary phases in a chromatographic process, it is useful to have continuity equations that express the conservation of mass for each of the components undergoing the exchange process. The basic continuity equation for transport of solute can be written as follows for a column of cross-sectional area A:

$$\frac{\partial C}{\partial t} + \frac{F}{\xi A}\frac{\partial C}{\partial x} = L\frac{\partial^2 C}{\partial x^2} \tag{28}$$

where L is the *coefficient of axial dispersion* (with units of cm^2/sec) and x is the distance from the top of the column. The flow equation for total solute is:

$$J = \frac{FC}{\xi A} - L\frac{\partial C}{\partial x} \tag{29}$$

where J is the rate of solute transport per unit of the cross-sectional area occupied by solute (ξA). Equation (29) follows directly from Eq. (28) and the general continuity equation

$$\frac{\partial C}{\partial t} = -\frac{\partial J}{\partial x} \tag{30}$$

Equation (29) has the form of Fick's first law of diffusion for a moving frame of reference. The reference frame is located at the position of the "average solute molecule" (e.g., the peak of a profile within the column) and moves with a velocity of $F/\xi A$. If we consider transport relative to a moving frame of reference, the new position coordinate is $\phi = x - (Ft/\xi A)$. Since $F(\partial C/\partial V) = \partial C/\partial t$, Eq. (28) becomes

$$\frac{\partial C}{\partial V} = L_v\frac{\partial^2 C}{\partial \phi^2} \tag{31}$$

where $L_v = L/F$. This coefficient of axial dispersion (L_v) is the quantity determined directly in experiments. The coefficient L can subsequently be calculated since F is known.

2. Solution of the Continuity Equation for Different Solute Profiles

The two types of experiments described in the previous section require different solutions of Eq. (31). In the "small-zone" experiment, the solute is applied as an instantaneous pulse (amount s). The solution of Eq.

(31) in this case is best formulated with the Dirac delta function $\delta(x)$. For the boundary conditions

$$C(x, 0) = \delta(x) \cdot s \qquad x \geq 0$$
$$C(0, V) = 0 \qquad V > 0$$

the solution of Eq. (31) is

$$C = (s/2\xi A(\pi L_v V)^{1/2}) \exp\{-\phi^2/4L_v V\} \tag{32}$$

Substituting for ϕ and noting that $x = \bar{V}/\xi A$ at the bottom of the column, we have

$$C = (s/2\xi A(\pi L_v V)^{1/2}) \exp\{-(\bar{V} - V)^2/4\xi^2 A^2 L_v V\} \tag{33}$$

It is apparent that, although C is a Gaussian function with respect to ϕ (and hence x) at fixed V, it is clearly non-Gaussian with respect to V at constant x (equal to the column length in an elution experiment). Whereas the solute distribution along the column at any instant is Gaussian, the distribution with respect to time is non-Gaussian as a zone moves past any point along the column. The physical basis of this effect is simply the prolonged time required for dispersion of the latter part of the solute zone. The phenomenon is thus an "end effect" of elution from a column of finite length. These characteristics have been studied by Halvorson and Ackers (1971) and are illustrated in Fig. 5 where the linear tranformation of Eq. (33) is compared with corresponding curves representing the Gaussian function. The experimental points clearly fall on the non-Gaussian curve representing Eq. (33).

In the second type of experiment ("large-zone" experiment), solute of concentration $C = C_0$ is applied in a volume large enough to produce a plateau ($C = C_0$) in the elution profile. The leading and trailing edges of the profile have separate solutions. For the leading edge,

$$C(x, 0) = 0$$
$$C(0, V) = C_0$$

the solution is[1]

$$C(x, V) = \frac{C_0}{2} \operatorname{erfc}\left[\frac{\phi}{(4L_v V)^{1/2}}\right] \tag{34}$$

[1] The error function complement is defined by

$$\operatorname{erfc}(x) = \frac{2}{(\pi)^{1/2}} \int_x^\infty e^{-t^2}\, dt = 1 - \frac{2}{(\pi)^{1/2}} \int_0^x e^{-t^2}\, dt$$

The inverse error function complement is defined as $\operatorname{inverfc}(y) = x$, where $y = \operatorname{erfc}(x)$.

or, at fixed x:

$$C(V) = \frac{C_0}{2} \operatorname{erfc} \left[\frac{\bar{V} - V}{(4\xi^2 A^2 L_v V)^{1/2}} \right]$$

For the trailing edge,

$$C(x, 0) = C_0$$
$$C(0, V) = 0$$

the solution is

$$C(V) = \frac{C_0}{2} \operatorname{erfc} \left[\frac{(V - \bar{V})}{(4\xi^2 A^2 L_v V)^{1/2}} \right] \tag{35}$$

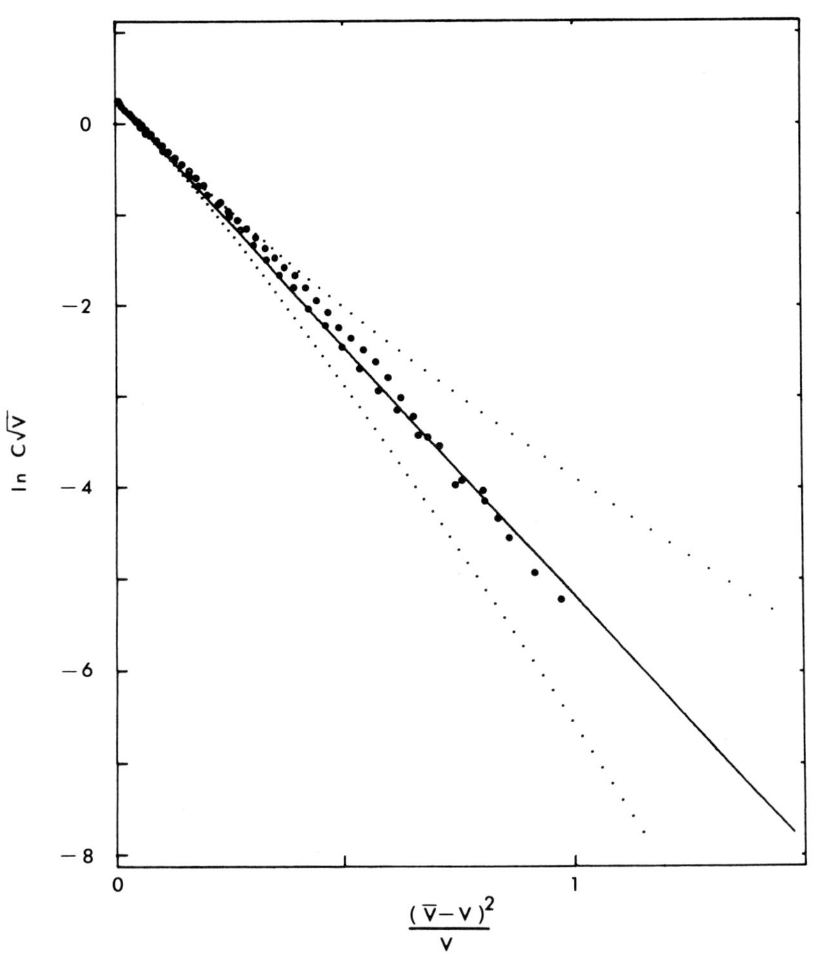

Fig. 5 Linear transformation of Eq. (33): (●) applied to data; (···) best Gaussian fit; and (—) Eq. (33). The slope $(-1/4\xi^2A^2L_v)$ and intercept ln $[s/2\xi A(\pi L_v)^{1/2}]$ are used to determine L_v. Taken from Halvorson and Ackers (1971).

It is apparent from these relations that the concentration profile of an elution experiment is not a simple error function complement of the volume, since the V coordinate appears in a complex manner in both numerator and denominator of the argument of the error function complement. This deviation represents the same "end effect" as that mentioned above in the case of the small-zone experiment. It can be seen from the above solutions to the continuity equation that the two types of experiments provide independent means for the determination of column parameters and the verification of theory.

3. *Determination of the Axial Dispersion Coefficient*

By use of Eqs. (33)–(35) the axial dispersion coefficient L_v (and subsequently L) can be determined from experimental data. Such determinations are of interest for two reasons: (1) Experimental study of the variation of L_v with different chromatographic parameters (such as flow rate, porosity, and particle size) provides a means of testing theories of gel chromatography. (2) Precise knowledge of zone-spreading as a function of system parameters establishes a rational basis for the problem of maximizing the resolution of zones in preparative separations. Halvorson and Ackers (1971) have described various methods of estimating L_v. The simplest is the height–area method for calculation from the small-zone experiment. When $V = \overline{V}$, the exponential term of Eq. (33) is unity and the equation becomes:

$$C_{\max} = s/2\xi A(\pi L_v V)^{1/2}$$

or (36)

$$L_v = \left(\frac{s}{2\xi A C_{\max}}\right)^2 \frac{1}{\pi V}$$

The total profile area, $s = \int_0^\infty C\, dV$, equals the mass of sample initially applied, and the product (ξA) is calculated as $\overline{V}/V_t$. Such experimentally determined coefficients are listed in Table VII for several sets of conditions.

4. *Interpretation of the Axial Dispersion Coefficient*

The axial dispersion coefficient arises from several kinds of processes superimposed on the basic solute partitioning. The major contributing factors are: (1) local nonuniform flow of the mobile phase due to finite gel particle size, size inhomogeneity, and nonuniform packing of the particles; (2) diffusion of solute along the axis of the column in both the mobile and stationary phases; (3) nonequilibrium between the mobile and stationary phases of the column with respect to diffusional

TABLE VII

Axial Dispersion Coefficients (L_v) for Small-Zone Experiments on Sephadex G-100 Columns

Sample	Column	Flow rate (ml/hr)	L_v
Glycylglycine	A	12.85	0.0230
	B	4.64	0.0205
	B	9.59	0.0212
	B	14.47	0.0254
Cytochrome *c*	A	12.08	0.0798
	B	9.49	0.0665
	B	13.93	0.1482
	B	14.11	0.0533
Myoglobin	A	12.08	0.1113
	B	9.82	0.0742
	B	14.11	0.1219

exchange of solute. Formulation of exact equations to describe the above effects is not feasible. However, considerable progress has been made by using approximate theories to describe these three basic processes that account for zone-spreading.

For partition coefficients greater than 0.3, the following relationship has been found to hold (Halvorson and Ackers, 1971):

$$L_v = L_p + \frac{\xi D}{F} + \frac{qd^2F}{\xi^3A^2D} \tag{37}$$

The terms on the right describe effects of the three factors mentioned above. D is free diffusion coefficient of the solute, d is gel particle diameter, and q is a geometric factor. L_p and qd^2 are constants to be determined. Rearrangement of Eq. (37) gives a linear relation between known terms from which the two constants may be determined.

$$L_v - \frac{\xi D}{F} = L_p + \frac{qd^2F}{\xi^3A^2D} \tag{38}$$

Values on the left side of Eq. (38) are plotted against corresponding values of F. The resulting linear plot yields L_p as ordinate intercept and qd^2 as slope, and known values of ξ, A, and D. Figure 6 shows data plotted in this fashion for the small-zone experiments of Halvorson and Ackers (1971) and Table VIII contains the values of L_p and qd^2 for the columns used in these experiments. Reasonable estimates of d indicate that q is the order of 1 (i.e., $0.3 < q < 3$).

The general behavior of L_v with flow rate is shown in Fig. 7 for

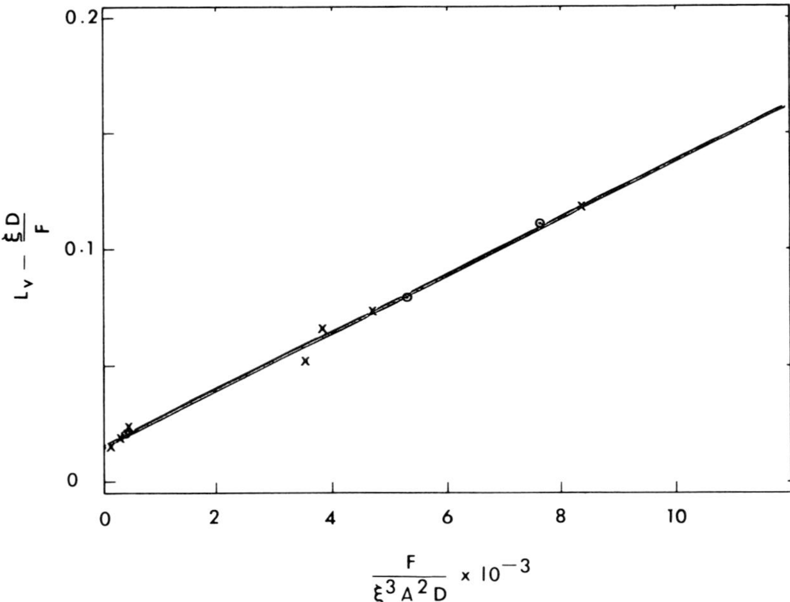

Fig. 6 Analysis of L_v based on Eq. (38) for data from small-zone experiments: (⊙) column A; (×) column B. The upper line is a regression plot for data of column A and the lower is for column B. The slope of the regression line is qd^2, and the intercept is L_p. Taken from Halvorson and Ackers (1971).

columns of Sephadex G-100 and G-75. It should be noted that axial dispersion of a solute zone as a function of flow rate exhibits a minimum since the second term of Eq. (37) dominates at low flow rates and the last term at moderate-to-high flow rates. At moderate flow rates, axial dispersion increases with increasing molecular size (decreasing σ, ξ, and D).

TABLE VIII

Characteristic Axial Dispersion Parameters for Sephadex Columns

Gel	Column	L_p (cm^{-1})	$qd^2 \times 10^5$ (cm^2)
Sephadex G-100[a]	100-A	0.0158	1.23
	100-B	0.0148	1.23
Sephadex G-75	75-A	0.0384	1.57
	75-B	0.0340	1.57

[a] Data for the Sephadex G-100 columns shown in Fig. 6.

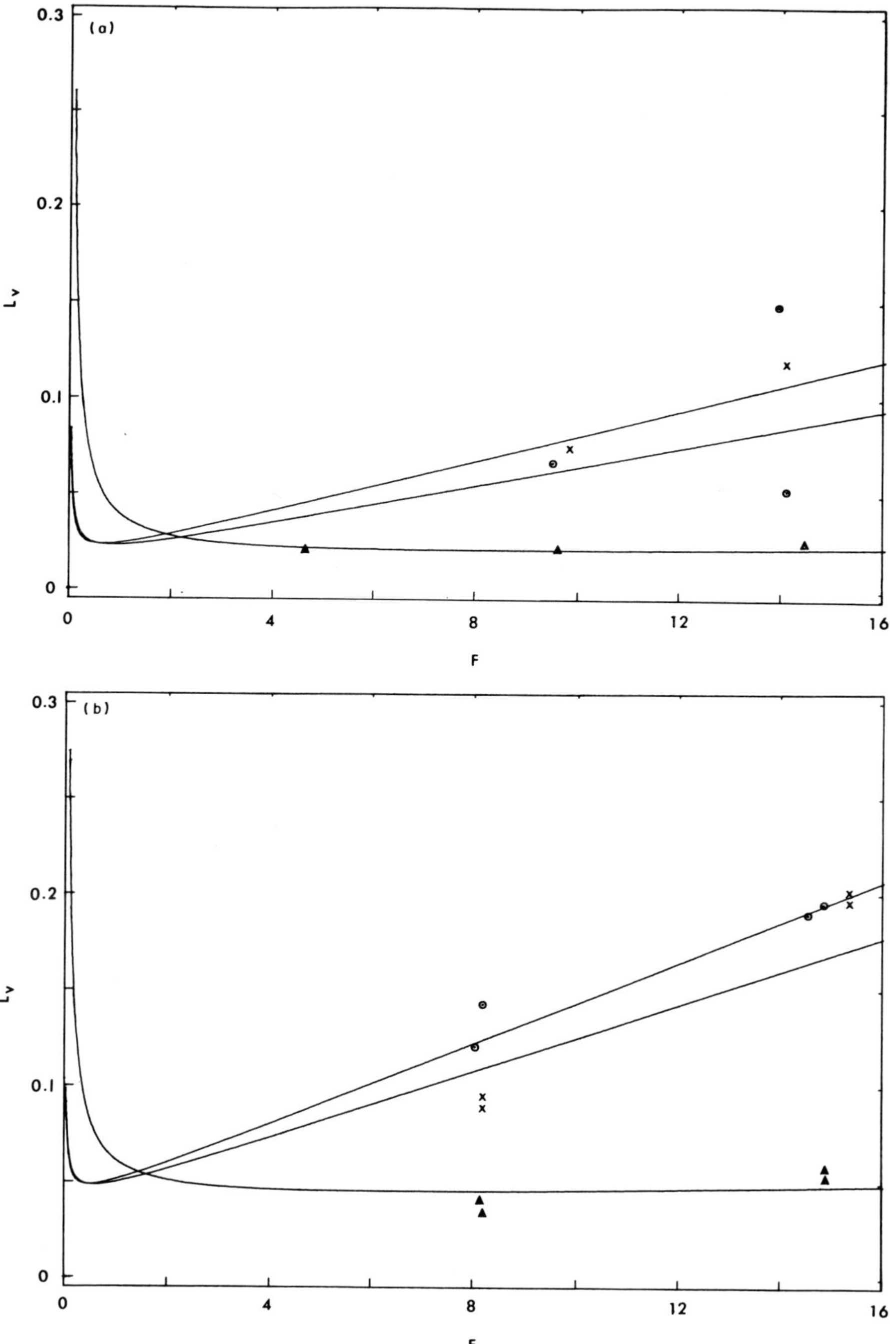

Fig. 7 Behavior of L_v as a function of flow rate on columns of (a) Sephadex G-100, and (b) Sephadex G-75: (▲) glycylglycine; (⊙) cytochrome *c*; (×) myoglobin. Taken from Halvorson and Ackers (1971).

It is important to distinguish between the absolute amount of zone-spreading observed after a sample has been eluted from the column and the rate of spreading during the time the sample is on the column. A small molecular species that spends more time on the column prior to its elution may exhibit a larger amount of absolute dispersion than a large molecule, even though the larger molecule spreads at a greater rate (larger L_v).

5. *Effect of Concentration Dependence*

Additional complexity is encountered when solutions of the continuity equation include the linear concentration dependence of the partition coefficient as described by Eq. (10). A solution for linear isotherms of this type has been obtained by Houghton (1963). The complete continuity equation is a complicated nonlinear equation for which analytic solutions do not exist in the general case. However, a solution can be obtained by limiting considerations to the small concentration dependencies of σ which exist in gel chromatography of most single-component solutes. Under conditions that justify the approximation $\xi^{-1}(2\sigma^\circ gl'C) \ll 1$, the solution of this equation for a solute band of finite width is

$$\frac{C}{C_0} = \frac{\exp(r)[\mathrm{erf}(p+h) - \mathrm{erf}(q+h)]}{1 - \mathrm{erf}(p) + \exp(r)[\mathrm{erf}(p+h) - \mathrm{erf}(q+h)] + \exp(m)[1 + \mathrm{erf}(q)]} \tag{39}$$

where

$$p = \frac{\phi + s/2}{2(L_v V)^{1/2}}, \quad q = \frac{\phi - s/2}{2(L_v V)^{1/2}}$$

$$r = \frac{-\sigma^\circ g C_0}{2L_v}\left[\xi + \frac{s}{2}\frac{-\sigma^\circ g C_0 V}{\xi^2}\right]$$

$$h = \frac{-\sigma^\circ g C_0 V}{\xi^2 (L_v V)^{1/2}}, \quad m = \frac{-F\sigma^\circ g C_0 s}{L_v \xi^2}$$

The effect of the concentration dependence of σ on the shapes of solute bands as they are developed within the column is to produce asymmetric leading and trailing boundaries. Thus, for a single solute species there is a boundary sharpening effect on the trailing edge and a boundary spreading effect on the leading edge of the zone. Experimental observations of these effects in gel chromatography have been described by Winzor and Scheraga (1963). For high concentrations of ovalbumin on columns of Sephadex G-100 the effect can be particularly pronounced. Experimental results obtained by Winzor and Nichol (1965) for this system are shown in Figs. 8 and 9. Figure 8 also shows another con-

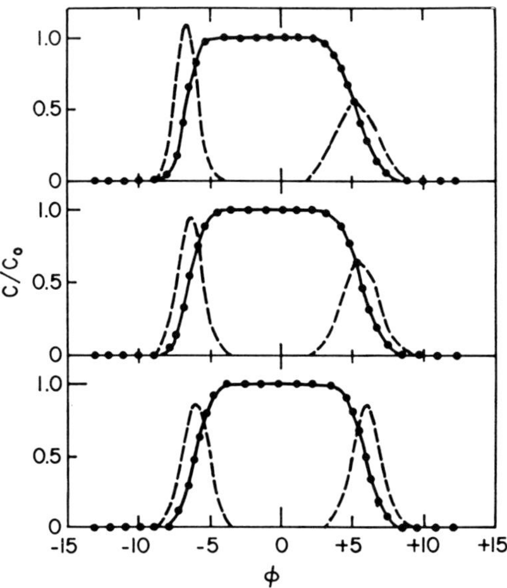

Fig. 8 Effects of concentration dependence of partition coefficient of ovalbumin solutions on Sephadex G-100. Plateau concentrations are, from top to bottom, 12.0 mg/ml, 7.0 mg/ml, and 1.6 mg/ml. Taken from Winzor and Nichol (1965).

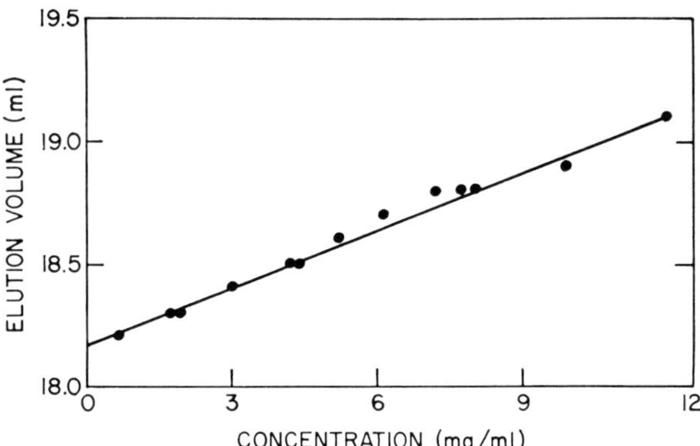

Fig. 9 Concentration dependence of centroid elution volume of ovalbumin solutions chromatographed on Sephadex G-100 in phosphate buffer (pH 6.8, ionic strength 0.1). Data are obtained from centroid boundary positions of large-zone experiments. These positions correspond to partition coefficients for the plateau concentrations of the respective zones. Taken from Winzor and Nichol (1965).

sequence of Eq. (39), namely, that the (normalized) boundary shapes depend upon the initial (plateau) concentration of solute applied.

C. Column Scanning Chromatography

It is evident from a comparison of elution profiles with solutions of the continuity equations that only a small fraction of the quantitative information obtainable from a given experiment is utilized by conventional procedures of solute zone analysis. The measurement of profiles at only a single point of the distance coordinate (corresponding to the bottom of the column) leads to a two-dimensional elution profile. However, the solutions of equations describing the solute zone are three-dimensional surfaces with concentration, distance, and volume or time as variables. By restricting measurements to a single value of the distance coordinate, only a single curve within the surface is sampled, corresponding to the intersection of the surface with a fixed plane at the distance coordinate chosen (Ackers, 1967a). This procedure might be compared to a sedimentation velocity experiment in which only a single picture or scan is taken after the solute boundary has been allowed to sediment toward the bottom of the cell. In 1967 it was proposed that direct optical scanning of gel columns could be utilized in order to derive more information (Ackers, 1968a). With the development of optical column scanning procedures (Brumbaugh and Ackers, 1968) it is now possible to make precise determinations of solute profiles at many intervals during a single experiment. Since several studies (Brumbaugh and Ackers, 1968, 1971; Warshaw and Ackers, 1971) have indicated that for many analytical applications of gel chromatography to protein systems the desired experimental information can best be obtained by direct optical scanning of the solute within the column, the basic principles of this new approach will be reviewed here. For specific details of technique and instrumentation, the reader should consult the article by Brumbaugh and Ackers (1968).

1. Column Scanning Systems

The primary information provided by an optical scanning system is the concentration profile of solute within the gel column. For column transport experiments the profiles are determined at a series of times corresponding to increments of volume flow. A scanning system that has been successfully used for this purpose is shown diagrammatically in Fig. 10. In this system the column is passed through a beam of horizontally collimated monochromatic light by means of a drive system.

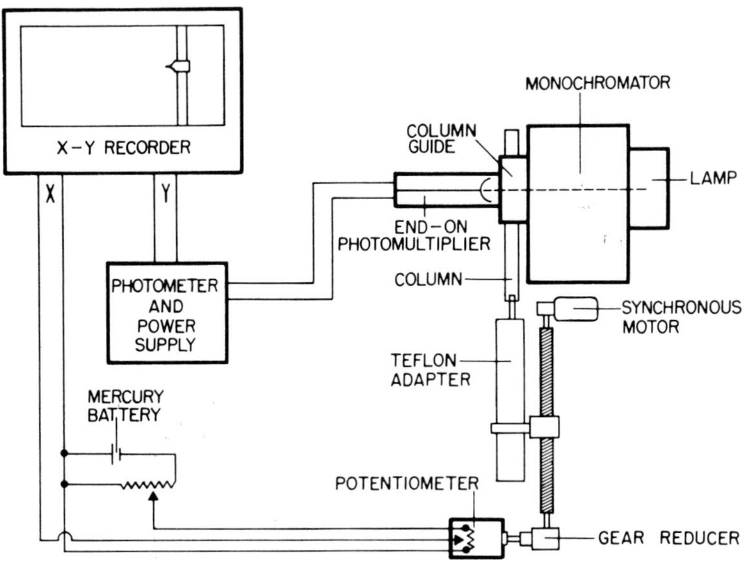

Fig. 10 Schematic diagram of a chromatographic column scanning system. Taken from Brumbaugh and Ackers (1968).

The drive system provides a signal denoting the position of the column with respect to the light beam. This signal is fed into the x axis of an xy plotter. The photomultiplier output is amplified and the logarithm of this signal is plotted to give linear absorbance on the y axis of the recorder. The signals may also be fed into a digital data recording system or interfaced with a computer.

The essential design features in the spectrophotometric system are the close positioning of the column to the end-on photomultiplier and the horizontal collimation of the light beam. It is also essential that the monochromator system have very low stray light characteristics in the wavelength regions of interest, since the internal light-scattering of the gel produces a high baseline absorbance and the protein must be monitored accurately above the baseline. When the gel itself contains no chromophoric groups with significant absorption bands in the wavelength region employed, these high apparent absorbances are attributable to internal scattering of light within the gel (Brumbaugh and Ackers, 1968).

2. *Determination of Solute Concentration within the Column*

The validity of results obtained from column scanning procedures of course depends upon the reliable and accurate determination of protein concentration within the column. Because of the high internal light-scattering by the column bed material, it is of primary importance to determine whether a Beer's law relationship applies to the solute distributed within the gel. The determination of solute concentration depends upon such a known relationship. If deviations are found to occur it is still possible to calibrate the system for concentration against apparent absorbance. However, when appropriate quality is achieved in the spectrophotometric system, it is possible to demonstrate strict adherence of most protein gel partitioning systems to a linear Beer's law plot (Brumbaugh and Ackers, 1968). In studies carried out to date it has been found that, to a very high degree of approximation, the internal scattering of the gel has no effect on the measured absorbance of solute molecules partitioned into the gel. All types of column materials should not, however, be expected to exhibit this "ideal" behavior. In addition to light-scattering effects on absorbance, spectral changes could result from interactions between gel and solute. The possibility of wavelength shifts and band broadening have been tested with proteins in a number of gels by comparing the spectrum within the column with the corresponding spectrum obtained at the same path length in the free solution above the gel.

3. *Small-Zone Experiments*

A useful procedure for the determination of partition coefficients involves the scanning of a zone within the column in order to measure peak position as a function of volume flow. In this procedure a small sample (0.1 ml or less) of protein solution is applied to a small column (e.g., 5 ml bed volume), allowed to enter the column, and followed by a fresh buffer. The column is allowed to flow at a constant rate (e.g., 3–4 ml/hr) for a desired period and then is scanned. The scanning procedure is repeated at a series of volume increments and measurements are made of volume flow at each increment. The information obtained in this type of experiment is the distance coordinate, x, of the solute zone's peak position as a function of the volume V, of liquid passed through the column. Some multiple scans of a small solute zone are shown in Fig. 11, and typical plots of elution volume vs. peak position are given in Fig. 12. The slopes of these plots can be used to calculate the partition coefficient, σ, using the relationship

$$dV/dx = \alpha + \beta\sigma \tag{40}$$

One small-zone experiment is thus equivalent to a set of conventional small zone elution experiments corresponding in number to the number of scans taken. The multiple determinations of σ from a single experi-

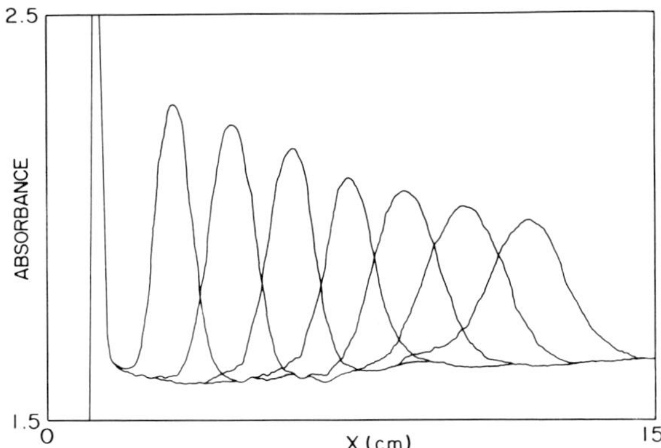

Fig. 11 Multiple scans taken during passage of a small zone (originally 0.1 ml) of potassium chromate through a Sephadex G-100 column. Taken from Brumbaugh and Ackers (1968).

ment may be analyzed statistically. Representative values obtained from the data for several proteins are listed in Table IX. The three experiments shown in Fig. 12 are equivalent to twenty-seven elution experiments but required only about three hours. The efficiency of data acquisition can be improved considerably beyond this by the use of smaller columns and faster scanning systems.

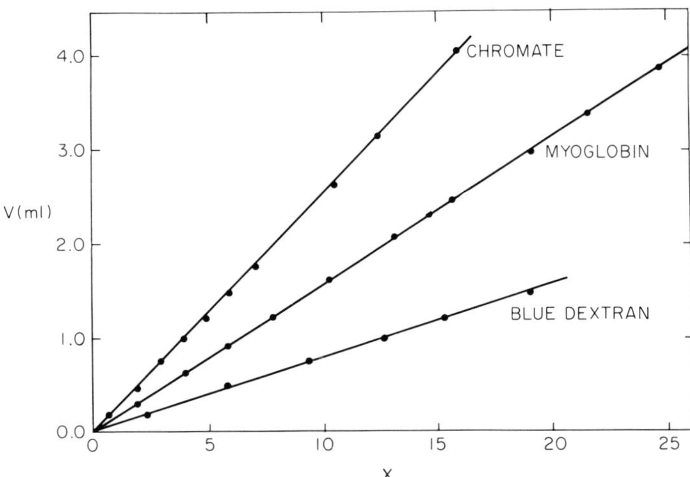

Fig. 12 Peak position plotted against volume flow for small-zone experiments. The abscissa represents peak position and the ordinate gives corresponding values of volume passed through the column. The slopes of these plots can be used to calculate the partition coefficient for myoglobin. Taken from Brumbaugh and Ackers (1968).

TABLE IX

Partition Coefficients for Proteins on Sephadex G-100 Superfine Gels[a]

Species	Partition coefficients (σ)		
	Centroid	Saturation	Small-zone
γ-Globulin[b]	0	0	0
Ovalbumin	0.162	0.160 ± 0.003[d]	0.158 ± 0.004[e]
		0.155 ± 0.006	0.156 ± 0.004[f]
Myoglobin	0.448	0.441 ± 0.008	0.447 ± 0.005
Cytochrome *c*	0.538	0.542 ± 0.004	0.539 ± 0.003[e]
			0.540 ± 0.003[f]
Glycylglycine[c]	0.998	0.998	—
Chromate	—	—	1

[a] Taken from Ackers (1970).

[b] The value of zero used for the partition coefficient of γ-globulin was verified by measurements with two other excluded molecules, thyroglobulin and Blue Dextran (Pharmacia).

[c] The value of unity for the partition coefficient of glycylglycine was used in the calculation of partition coefficients of the three protein molecules.

[d] Duplicate sets of determinations were made on two columns using ovalbumin solutions of differing concentration.

[e] Ovalbumin and cytochrome *c* run individually on the same column.

[f] Ovalbumin and cytochrome *c* run in mixture. Peak concentrations in both cases were 0.037 mg/ml.

4. *Large-Zone Experiments*

Useful information can also be obtained by means of integral boundary experiments in which the solute zone is large enough to establish a concentration plateau. In such experiments the rate of movement of the boundary centroid is measured with respect to volume flow. In addition, the scanning system may be used as a single-point monitor by observing absorbance at some level in the column, thus eliminating the necessity of flow cells. A leading boundary is shown in Fig. 13. This trace was obtained by holding the column at a fixed position and recording absorbance vs. time (volume). Here the absorbance scale span is 2–3 Å. Experiments of this type were performed simultaneously with the saturation experiments for the protein molecules listed in Table IX by driving the column to the lower limit of the scanner and monitoring at that point during the process of saturation. The centroid volume $\overline{V}$ was then related to the void volume, V_0, internal volume, V_i, and partition coefficient, σ, by Eq. (27).

Adsorption of protein to some gels is known to occur under conditions of low ionic strength. Column scanning provides a convenient means of

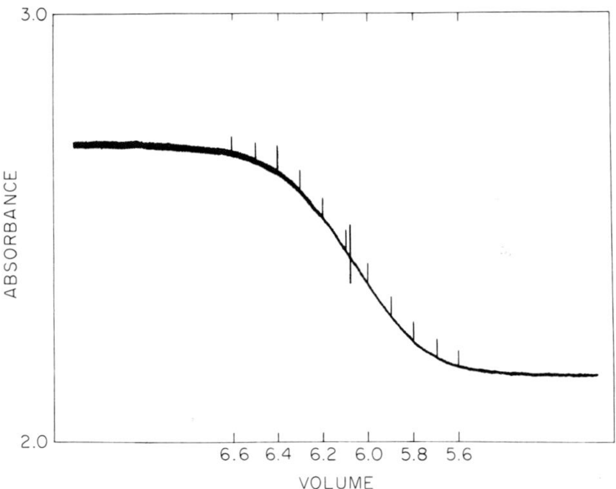

Fig. 13 Leading edge monitored at a single point within the column. Observation of the leading edge in a saturation experiment may be used to determine partition coefficients by a procedure analogous to that used for flow cells. This plot was obtained by monitoring absorbance against time at a fixed distance, *x*, within the column during saturation by a solution of cytochrome *c*. The absorbance scale is 2–3 Å. Taken from Brumbaugh and Ackers (1968).

detecting such interactions. When adsorption occurs or contamination (e.g., from tubing) is present, these effects may be observed as baseline elevations. In addition, the scanning approach appears to be ideally suited to studies of interacting components (cf. Sections VI, VII, and VIII).

5. *Variation of α and β within the Column*

Scanning gel chromatography reveals the nonuniformities in column packing which are invariably present. The main consequence of a variation of the parameters α and β with the distance x is a nonconstant solute velocity along the column since $dx/dt = F/\alpha + \beta\sigma$. For elution chromatography, the value of σ determined from Eq. (23) is a weighted average of $\sigma(x)$ with respect to $\beta(x)$ over the length l of the column.

$$\sigma = \frac{\int_0^l \beta(x)\sigma(x)\,dx}{\int_0^l \beta(x)\,dx} \tag{41}$$

A significant feature of the elution experiment is that, although different molecular species may be partitioned differently at various points within the column, the integrated average value of the partition coefficient (represented by elution volume) is taken over the same path for all molecular species. This is not the case for thin-layer chromatography, a fact that inherently limits the accuracy of this method for analytical determinations.

D. Thin-Layer Gel Chromatography

Migration of proteins on thin layers of porous gels was first investigated by Johansson and Rymo (1962). Using a modification of the standard thin-layer chromatography technique, they were able to separate serum proteins on a variety of Sephadex gels (G-25, G-50, G-75, G-200). The method consists of preparing a thin (0.5 mm) gel–solvent layer on a glass plate from a slurry of swollen gel particles. The particles must be small for good resolution, e.g., Superfine grades of Sephadex, and must be applied very evenly. (A variety of spreaders are manufactured commercially for this purpose.) With the plate either in a vertical or tilted position, solvent is fed through the bed, usually by means of a filter-paper wick. The flow rate can be controlled by the angle of inclination. Determann and Michel (1965) have used a "sandwich" arrangement of a gel layer between double glass plates. After the thin-layer plate has been formed and equilibrated with buffer, a series of protein solutions may be applied in spots at a starting line, as in paper chromatography. After development with solvent, the distances of migration are measured. Detection is usually by a staining procedure with a dye such as Amido Black (Johansson and Rymo, 1962). The migration distances of proteins are then correlated with their molecular weights on empirical grounds (Andrews, 1964; Morris and Morris, 1964; Determann and Michel, 1965). Figure 14 shows such a correlation for proteins covering a wide range of molecular weights (Andrews, 1964). A marker protein of known molecular weight is used as a standard of reference and the molecular weights of other proteins are calculated from the ratios of their migration distances to that of the marker. Determann has applied the technique to the study of size differences in autolysates of pepsin and has demonstrated a concentration-dependent apparent molecular weight of isozymes of lactate dehydrogenase (Determann, 1967a).

Partition coefficients may also be determined from thin-layer chromatograms. If F is not a function of t, and α and β do not vary with

x or y, then the mean distance x_m migrated by a molecular species in time $(t = V/F)$ is related to the partition coefficient σ by

$$1/x_m = (\alpha/V) + (\beta/V)\sigma \tag{42}$$

The excluded molecule $(\sigma = 0)$ migrates a distance $x_0 = V/\alpha$, and the included molecule $(\sigma = 1)$ migrates a distance $x_i = V/(\alpha + \beta)$. Therefore the partition coefficient can generally be calculated as follows:

$$\sigma = \frac{(1/x_m) - (1/x_0)}{(1/x_i) - (1/x_0)} \tag{43}$$

The main advantage of this technique appears to be the convenience with which large numbers of samples can be run simultaneously. However, the reproducibility of relative migration distances for different proteins leaves much to be desired (Determann, 1967a,b). This is probably due to variations in the packing of gel particles, resulting in variations of flow rates in the different parts of the gel bed, and to the dependence of α upon both x and y. The migration distances measured after time t are integrated averages over different paths for the different molecules. The problem of reproducibility is not encountered in single-column elution experiments.

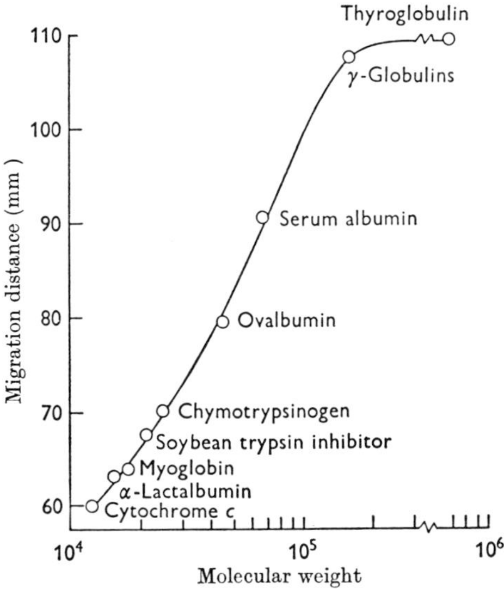

Fig. 14 Thin-layer gel chromatography of proteins on Sephadex G-100. Migration distances are plotted against the logarithm of molecular weight. Taken from Andrews (1964).

III. EQUILIBRIUM PARTITIONING METHODS

In addition to the chromatographic uses of molecular sieves described in the previous section, a number of useful procedures have been developed in which no chromatographic separations occur but only static equilibrium properties are utilized. These procedures are described below.

A. Batch Partitioning

A number of investigations have involved the determination of partition coefficients by static equilibrium experiments (Ackers, 1964; Fasella *et al.*, 1965; Stone and Metzger, 1968; Brumbaugh and Ackers, 1968; Warshaw and Ackers, 1971). In the simplest procedure, a known weight of gel-forming material is swollen in excess solvent in a volumetric container. Then a measured volume of solution containing a quantity Q_T of the molecular species of interest is added, making up a total volume, V_t, for the system. The contents are then stirred or shaken and equilibration is allowed to take place. After a period of time, the gel is allowed to settle and an aliquot of the supernatant liquid is removed and its concentration C_0 assayed. The apparent total volume V'_p occupied by the solute is then calculated as the ratio Q_T/C_0. The experiment is first carried out with a large molecule ($\sigma = 0$) for the determination of V_0, the volume of liquid exterior to the gel. Then the internal volume V_i is determined either with a small molecule ($\sigma = 1$) or from the known partial volume and water regain of the gel. After these values have been determined the partition coefficient for the molecule of interest is calculated from Q_T and C_0 by the relationship

$$\sigma = \frac{Q_T/C_0 - V_0}{V_i} \tag{44}$$

In the use of this static method, particular care must be taken to ensure thorough washing of the swollen gel particles. Since the gel-forming material has large exclusion limits but comprises only a small percent of the weight of the gel phase, release of soluble material into the exterior spaces upon swelling can result in apparently anomalous partition coefficients. In addition to removal of materials that might have substantial exclusion properties, it is desirable to wash out small molecules that might react chemically with the protein or absorb light and thus interfere with the spectrophotometric assay. It is desirable to pack the gel particle into a glass tube forming a small column for washing. After washing with several volumes of the desired buffer, the gel bed can be transferred to a volumetric flask and the partitioning experiment carried out as described above.

B. Optical Scanning of Saturated Columns

For a column saturated with solution of concentration C, absorbance measurements can be used to determine the partition coefficient, σ, by comparing the measured absorbance, A_b, of the solute at any point within the column bed with the corresponding absorbance, A_a, of the free solution above the column bed. The Beer's law relationships for the measurements are

$$\begin{aligned} A_a &= \epsilon l C \\ A_b &= \epsilon l C' \end{aligned} \tag{45}$$

where ϵ and l are, respectively, the extinction coefficient and path length. The concentrations C and C' refer to the bulk solution and to the total column frames of reference, respectively. Since $C' = \xi C$ [see Eq. (18)], the ratio A_b/A_a becomes

$$A_b/A_a = \xi = \alpha + \beta\sigma \tag{46}$$

Defining $P(x) = A_b/A_a$ at a point x within the column and noting that when the column is scanned the parameters are all functions of x, we can write

$$P(x) = \alpha(x) + \beta(x)\sigma(x) \tag{47}$$

The ratio, $P(x)$, of solute absorbance in the gel (corrected for baseline absorbance) to absorbance above the gel is thus a measure of solute partitioning and can be used to determine $\sigma(x)$. In order to do this it is necessary to evaluate the functions $\alpha(x)$ and $\beta(x)$. The first of these can be evaluated using a molecular species large enough to be totally excluded from the gel particles. Then, for all values of x, $\sigma = 0$ and the experimentally determined $P(x) = \alpha(x)$. Subsequently, the function $\beta(x)$ is determined using a small species so that $\sigma = 1$ for all values of x. In this case $\beta(x) = P(x) - \alpha(x)$. Representative column scans taken before and after saturation are shown in Fig. 15. The lower traces are baseline scans of a Sephadex G-100 Superfine column; the upper scans were taken after this column had been saturated with a solution of ovalbumin. The off-scale peak in the left-hand portion of the plot is an opaque porous polyethylene disk at the top of the gel bed. The scale for the baseline scan is 0–1 Å above the gel and 1.5–2.5 Å inside the gel, while the scans of the saturated column are on scales of 1–2 Å above the gel and 1.5–2.5 Å inside the gel. Subtracting the absorbance of the

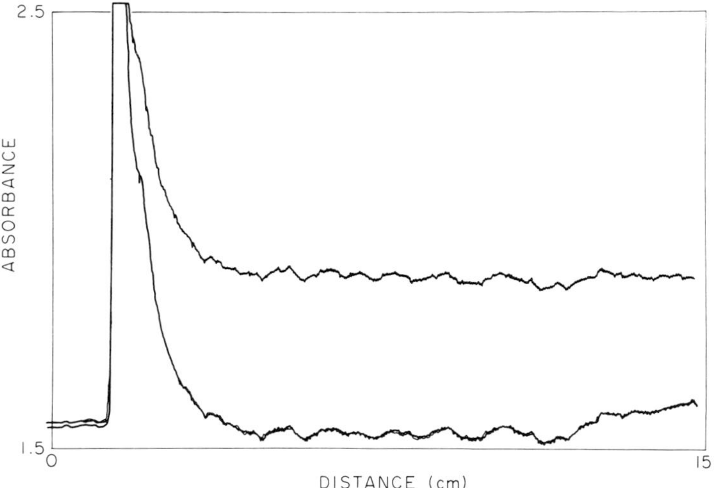

Fig. 15 Representative scan of a saturated column of Sephadex G-100 (Superfine). Recorded plots show the column baseline (low traces) and the absorbance of the column after saturation with ovalbumin. Scans were taken at 220 nm. The plots are divided into two sections by the off-scale peak, which represents an opaque porous disk at the top of the gel bed. The section to the *left* of this peak is the solution above the column bed. Above the gel bed (*left* of disk) the lower trace (buffer) is on a scale of 0–1 Å, and the upper trace (ovalbumin) is on a scale of 1–2 Å. Recorder span was 1 absorbance unit. For both scans of the gel bed (*right* of disk) the scale was 1.5–2.5 Å. Taken from Brumbaugh and Ackers (1968).

baseline gives values for absorbance of the sample at each point in the column. Table X lists the partition coefficients of various proteins on a series of saturated Sephadex gels. Values of α and β vary between different columns packed with the same gel and sometimes vary within the same column. If the gel has been thoroughly washed before the column is packed, $\beta(x)/(1 - \alpha(x))$ will be constant throughout the length of the column although slight variations in $\alpha(x)$ will occur as a result of variations in the packing of gel particles. This effect is shown in Fig. 15 by the variation in distance between the (bottom) baseline trace and (upper) saturation trace for a Sephadex G-100 column saturated with ovalbumin. Variations in particle packing can be corrected by measuring $\alpha(x)$, using a protein which is totally excluded from the gel. This procedure will yield a constant value of $\sigma(x)$ for a given molecular species. However, if care is not taken in packing the column, all three functions, $\alpha(x)$, $\beta(x)$, and $\sigma(x)$, will vary with the distance because small dextran particles will be washed out of the Sephadex upon swelling and may still be present in the slurry with which the column is packed. These particles appear to establish a gradient within

TABLE X

Partition Coefficients of Proteins Obtained from a Stacked Gel Column[a]

Molecule	σ_{200}	σ_{150}	σ_{100}	σ_{75}
RNase[b]				
Sigma	0.7622 (23)[c] ± .0025[d]	0.7165 (48) ± .0016	0.6086 (62) ± .0006	0.5023 (55) ± .0009
Mann	0.7089 (26) ± .0026	0.6814 (48) ± .0013	0.5759 (63) ± .0005	0.4891 (58) ± .0008
Cytochrome *c*	0.7329 (22) ± .0018	0.6983 (48) ± .0012	0.5795 (62) ± .0006	0.4788 (58) ± .0009
Myoglobin	0.6888 (24) ± .0029	0.6260 (48) ± .0024	0.5221 (62) ± .0011	0.3878 (57) ± .0008
Ovalbumin[b]	[e]	[e]	0.3063 (62) ± .0010	0.1709 (58) ± .0003
Ovalbumin[b]	0.5185 (24) ± .0013	0.4397 (48) ± .0010	0.3039 (63) ± .0006	0.1728 (58) ± .0007
Serum albumin	0.3979 (24) ± .0045	0.3159 (48) ± .0027	0.1794 (63) ± .0014	0.0529 (57) ± .0008
Aldolase	0.2818 (22) ± .0020	0.2123 (48) ± .0014	0.1007 (63) ± .0008	0.0[f]

[a] Taken from Warshaw and Ackers (1971).
[b] Two different commercial preparations of ribonuclease were used.
[c] Numbers in parentheses are the number of data points used in each analysis of partition coefficient and associated error.
[d] Error values listed are standard errors of the mean.
[e] Data are for two runs with the same commercial ovalbumin preparation, one done six days after the other. G-200 and G-150 points of the first run were lost due to a baseline perturbation during the measurement.
[f] Aldolase was used to calculate α, the void volume cross section, on G-75.

the column as they are washed down through the bed and trapped. They then contribute to the exclusion properties of the column, resulting in variations of $\beta(x)/(1 - \alpha(x))$ and $\sigma(x)$ as well. It is also possible to observe variations in these quantities if the gel bed has been subjected to high pressures or to packing forces producing distortion of the gel particles. With careful packing it is possible to eliminate almost all such variations over a column length suitable for the measurements.

IV. DETERMINATION OF MOLECULAR SIZE AND WEIGHT

Calibrated molecular sieve columns have been used extensively in recent years for determinations of molecular size and weight. These determinations can be carried out rapidly with very little equipment and at very small expense. The procedures employed vary greatly in accuracy and reliability, but in general they compare quite advantageously with other methods of molecular size determination (e.g., diffusion, viscosity, light-scattering, electron microscopy). The determination of molecular weight, however, always depends upon a correlation between the molecular radius and the molecular weight. This correlation is usually subject to greater uncertainty than is the determination of the molecular radius itself. Under special circumstances, e.g., when the protein molecules are in a completely denatured state, the correlation is extremely good and the molecular weights can be determined with correspondingly greater accuracy. There are, in all cases, certain precautions that must be taken in the intelligent application of these techniques to a given experimental problem. Several of these procedures as well as their limitations will be discussed in this section. For a more thorough discussion of column calibration procedures, the reader is referred to a previous review (Ackers, 1970).

A. Treatment of Molecular Sieve Data

The experimental procedure most commonly employed for molecular size and weight determination is the small-zone elution experiment, in which the primary quantity measured is the peak elution volume, V_e. Data from such experiments have been represented in many different ways by various investigators. Frequently, workers have normalized the elution volumes with respect to total column volume or void volume and expressed their results in terms of V_e/V_t or V_e/V_0, respectively. However, there is very little to be gained by this procedure since packing of the gel particles always varies significantly from one column to another. On the other hand, the calculation of partition coefficients from elution volumes provides an expression of the fundamental property of the system which is invariant from one column to the next for a given gel. Ideally, all published data should be expressed in terms of partition coefficients. A specification of the gel preparative procedure (or lot number of a commercial preparation) and conditions of the experiment will then enable other workers to reproduce the experimental results as precisely as possible.

A major reason for the representation of data in terms of partition coefficients is that it facilitates the estimation of errors involved in the determination of molecular size or weight. A modest error in the determination of elution volume could have a devastating effect on the accuracy of a molecular size or weight determination if the molecular species is eluted close to the void volume, whereas the same error could have a relatively minor effect if the molecule is eluted toward the end of the column. In relating experimental data to molecular parameters, it is sometimes desirable to perform numerical calculations directly on the primary data (e.g., on elution volumes obtained from a given column) in order to eliminate errors of data reduction in the calculation of partition coefficients.

B. Interpretation of the Partition Coefficient

The validity of molecular size or weight determinations depends upon (1) the accuracy with which the partition coefficient or other relevant parameters can be determined and (2) the reliability of the theory that relates these measured parameters to molecular properties. With methods presently available the partition coefficients can be reproducibly determined to an accuracy of three digits. Since their values range between zero and unity, accuracy is greatest for the larger partition coefficients. With regard to the second point there are two questions that must be answered: (a) Which properties of a molecule determine its partition coefficient with respect to a given porous material? and (b) What relationships exist between the molecular properties of the solute and the measured partition coefficient? At the present time these questions cannot be answered as rigorously as one would like on the basis of experimental evidence. Nevertheless, considerable progress has been made toward a practical understanding of these matters.

Under most experimental conditions the measured parameter σ (or K_{av}) corresponds to the distribution of solute within the gel matrix at thermodynamic equilibrium. This coefficient then represents the steric constraints that lead to exclusion of the solute molecule from a fraction of the gel's internal solvent region. Although charge interactions and surface adsorption, when present, play some role, the partition coefficient may be viewed primarily as a result of steric exclusion effects. The major factors that determine the steric constraints are molecular size and shape and the structure of the gel matrix.

In many respects the ideal porous material would be one with the "all-or-none" molecular sieving effect illustrated in Fig. 16 (solid line). For a gel of given "porosity," all molecules larger than a critical size would have partition coefficients equal to zero, while all smaller molecules would have partition coefficients of unity. This would provide for maximum efficiency in the resolution of molecular species, both for

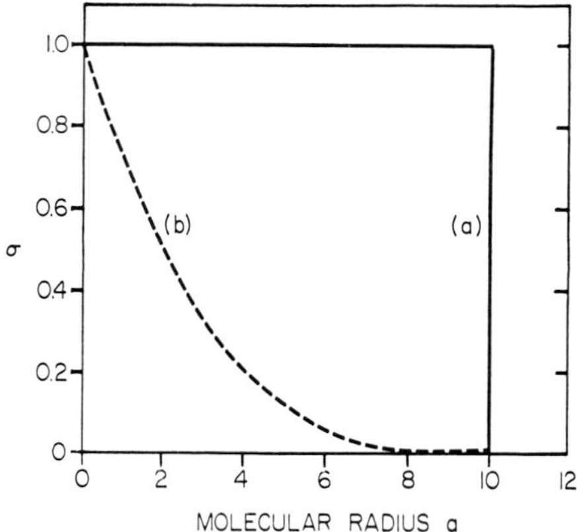

Fig. 16 Variation of partition coefficient with molecular size for (a) "all-or-none" partitioning and (b) conical pores. The curves represent the same limit ($r = 10$) for total exclusion. Taken from Ackers (1970).

preparative and analytical purposes. However, the experimental curves that describe the partitioning of molecules on porous materials always exhibit a more gradual exclusion behavior and are characterized by a nonlinear form, such as that shown in Fig. 16 (dashed line).

There are a number of reasons for the curve having this general shape. For structural models involving any conceivable pore shape or arrangement of constraining surfaces, the curve must decrease continuously with increasing molecular size. For example, consider equilibrium partitioning within uniform cylindrical pores oriented in such a way that all pores are totally accessible to penetration by molecules in the surrounding solvent regions. Even such a structure as this would not exhibit an "all-or-none" type of sieving, and the partition coefficient would be $(1 - a/r)^2$. Similarly, for conically shaped pores, the function becomes

$$\sigma = (1 - a/r)^3 \qquad (48)$$

C. Calibration of Columns

For a given column, the most straightforward way of establishing the relationship between the partition coefficient and a molecular size parameter is to calibrate the column with a known molecular species in a completely empirical fashion. The resulting calibration curve can then be used to determine the molecular size parameter of an unknown molecular species (Steere and Ackers, 1962b). For molecules of similar shape and density, a direct correlation of the partition coefficient with the molecular weight becomes meaningful (Andrews, 1962, 1964; Ackers, 1964; Laurent and Killander, 1964; Siegel and Monty, 1966). In order to minimize the number of calibration standards that must be used to establish a reliable calibration curve, it is useful to have a general equation expressing the relation between the molecular size or weight and the partition coefficient. Equations have been proposed for column calibration based on empirical curve-fitting, on various geometric models of the gel, or on statistical assumptions. These calibration functions have been reviewed elsewhere in detail (Ackers, 1970). Only the three most commonly used procedures will be described here.

1. *Logarithmic Plots*

Nonlinear curves can frequently be straightened out over some region by plotting the variables as various powers of logarithms of each other. This device was first employed by Granath and Flodin (1961) for the treatment of gel chromatographic data and has subsequently been used by many other workers (Andrews, 1962, 1964, 1965; Whitaker, 1963; Leach and O'Shea, 1965). A logarithmic relationship found empirically to obtain for many gel systems can be written in terms of the partition coefficient:

$$\sigma = -A \log M + B \tag{49}$$

where A and B are empirical constants and M is molecular weight. For spherical molecules or random coils the molecular weight can be expressed in terms of some power, p, of molecular radius a

$$M = Ka^p \tag{50}$$

and Eq. (49) becomes

$$\sigma = -A' \log a + B' \tag{51}$$

where $A' = -pA$ and $B' = (B - A \log K)$.

A very thorough investigation of these relationships has been carried out by Andrews in a series of studies (Andrews, 1962, 1964, 1965) aimed at evaluating the general validity of direct molecular weight determinations using gel chromatography. A representative calibration curve that correlates the elution volumes with the molecular weights of various proteins is shown in Fig. 17. This correlation can be derived from Eqs. (21) and (49), giving

$$V_e = A'' \log M + B'' \tag{52}$$

where $A'' = V_i A$ and $B'' = (BV_i + V_0)$. It can be seen (Fig. 17) that this relationship holds only over the central portion of the curve. Highly

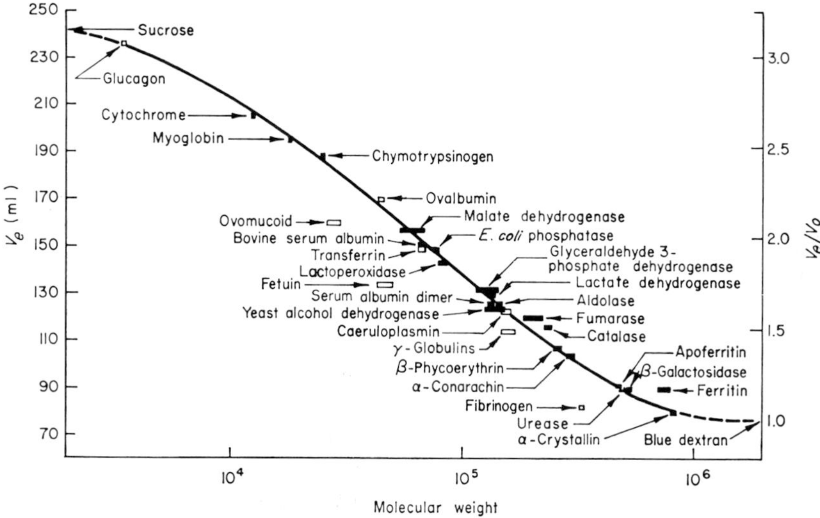

Fig. 17 Logarithmic calibration plot for proteins of different molecular weight on Sephadex G-200. Taken from Andrews (1965).

asymmetric molecules and those containing carbohydrate deviate significantly from the curve generated from elution volumes of "normal" proteins. Furthermore, the behavior of all molecular species could be better correlated by a curve relating the molecular weight to the diffusion coefficient. Thus Eq. (49) is a more general representation of the data than Eq. (50), provided a is taken as the equivalent hydrodynamic radius (Stokes radius). The conclusion that elution volumes are best correlated with molecular size rather than molecular weight has been reached by a number of workers (Steere and Ackers, 1962a,b; Laurent and Killander, 1964; Ackers, 1964; Siegel and Monty, 1965; Giddings *et al.*, 1968). A careful investigation by Ward and Arnott (1965) of four glycoproteins indicated large deviations from the molecular weight calibration curves obtained with unconjugated globular proteins. Since this effect was observed both for gels of dextran (Sephadex) and polyacrylamide (Bio-Gel), it was attributed to the different partial specific volumes of the two groups of proteins rather than to surface adsorption or charge effects. These studies indicate that the logarithmic calibration procedure is subject to uncertainties if the calibrating standards and the unknown molecules have different shapes or densities. For proteins this difficulty can largely be circumvented if chromatography is carried out in the presence of denaturing agents so that all denatured molecules assume the conformation of "random coils" (see Section IV,E).

2. *Cylindrical Rod Model*

By assuming the gel network to be made up of rigid cylindrical rods, Laurent and Killander (1964) were able to obtain an expression for the partition coefficient referred to the total volume of gel.

$$K_{\text{av}} = \exp[-\pi \mathcal{L}(a + r)^2] \tag{53}$$

This equation is an application of the expression derived by Ogston (1958) for the fractional volume available to spheres within a random collection of cylinders. The same relationship was derived by Giddings *et al.* (1968) on the basis of a statistical–mechanical approach and by Chun *et al.* (1969a) using a moment-generating function. In Eq. (53), r is the radius of the obstructing rod and $\mathcal{L}$ is the concentration of rods within the system. Since the polymer chains of a gel network are not really cylindrical rods, these parameters can be viewed as adjustable constants which provide a good correlation between molecular radius and partition coefficient (Fig. 18). An extensive investigation of this relationship for the partitioning within polyacrylamide gels as a function of gel composition has been carried out by Fawcett and Morris (1966). In this study the apparent rod concentration, $\mathcal{L}$, was found to be a linear function of monomer concentration for a constant degree of cross-linking,

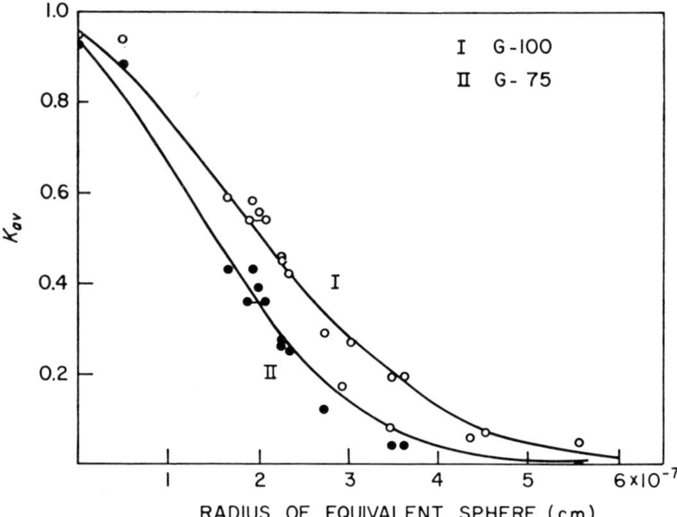

Fig. 18 Correlation of partition coefficient values (K_{av}) with molecular radius by means of the cylindrical rod model [Eq. (53)]. Data are for molecules partitioned into Sephadex G-100 (curve I) and G-75 (curve II). Taken from Laurent and Killander (1964).

while the apparent rod diameter r was constant. Furthermore, it was found that $\mathcal{L}$ increased to a maximum with an increasing degree of cross-linking at constant monomer concentration, and then dropped sharply. The apparent rod diameter increased to a plateau under these conditions. This increase of r was interpreted as a lateral aggregation of the gel strands (Fawcett and Morris, 1966).

3. *Random Distribution of Penetrable Volume Elements*

In order to circumvent the necessity of postulating specific geometric shapes for the pores within gel partitioning systems, a statistical calibrating function was proposed (Ackers, 1967b). This function is based on three assumptions: (a) the microregions within the porous network are heterogeneous in size, (b) the individual region may be characterized by the radius, a, of the largest molecule that it can accommodate, and (c) the frequency distribution of sizes follows a normal curve. The differential fraction of the internal volume penetrable by a molecule of radius a then can be represented by a Gaussian probability curve. The total fractional volume within the gel that can be occupied by a molecular species of radius a is the partition coefficient, approximated by the error function complement of the Gaussian distribution:

$$\sigma = \operatorname{erfc}\left[\frac{a - a_0}{b_0}\right] = 1 - \frac{2}{\pi^{1/2}} \int_0^{(a-a_0)/b_0} e^{-a^2}\, da \tag{54}$$

The constant a_0 is the position of the maximum value of the distribution, and b_0 is a measure of the standard deviation. These parameters, a_0 and b_0, are calibration constants for a given gel. Solving Eq. (54) for the molecular radius, a linear relation is predicted between a and the inverse error function complement (erfc^{-1}) of σ.

$$a = a_0 + b_0 \operatorname{erfc}^{-1} \sigma \tag{55}$$

Representative plots showing this correlation are given in Fig. 19 for a variety of column materials and molecular species. The two calibration constants a_0 and b_0 appear as the ordinate intercept and slope of each plot, respectively.

From these examples it is evident that the characteristic relationship between partition coefficient and molecular size can be represented by equations based on a variety of different models. The models are equivalent in the sense that they can be fitted to the experimental curve through adjustable constants. This is reminiscent of the fact that molecules can be represented by a variety of geometric structures (spheres, rods, coils, etc.) that are hydrodynamically equivalent. A consequence of both the

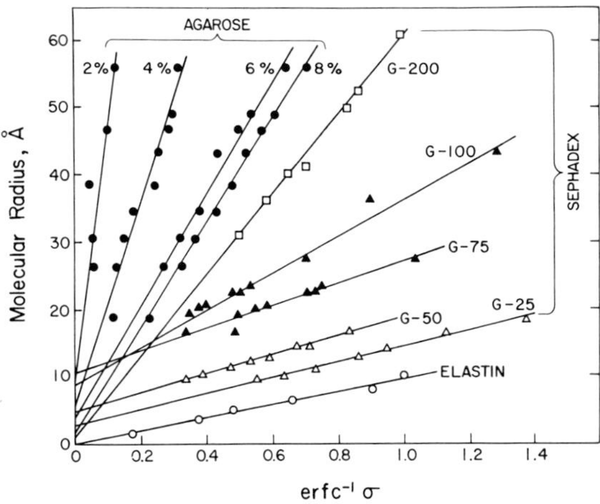

Fig. 19 Correlation of partition coefficient σ with molecular radius according to Eq. (55), assuming random distribution of penetrable volume elements. Taken from Ackers (1967b).

geometric pore model and the statistical models is that, generally, two adjustable constants must be used to characterize the state of a porous network with respect to its solute partitioning properties.

D. Combination of Chromatographic Results with Other Kinds of Data

Combination of chromatographic data with sedimentation, viscosity, or light-scattering measurements can be used to determine an accurate molecular weight or to estimate the degree of molecular asymmetry, provided the molecular weight is already known. In most cases the diffusion coefficient, D, is calculated from the molecular radius, a, by means of the Einstein relation (see Gosting, 1956)

$$D = RT/N6\pi\eta a \tag{56}$$

yielding the standard value of $D_{20,\mathrm{w}}$ corrected to water at 20°. Here R is the gas constant, N is Avogadro's number, and η is the viscosity. The diffusion coefficient can be combined with the sedimentation coefficient, $s_{20,\mathrm{w}}$, and the partial specific volume, $\bar{v}$, to yield the anhydrous molecular weight by means of the Svedberg equation

$$M = RTs_{20,\mathrm{w}}/D_{20,\mathrm{w}}(1 - \bar{V}\rho) \tag{57}$$

where ρ is the density of water at 20°. Similar calculations can be performed by combining viscosity measurements with the diffusion coefficient.

If the molecular weight is known, the frictional ratio f/f_0 may be calculated from the molecular radius, a. The frictional coefficient f is related to a by Stokes' law

$$f = 6\pi\eta a \tag{58}$$

while the frictional coefficient for an equivalent anhydrous spherical molecule is given by

$$f_0 = 6\pi\eta(3M\bar{v}/4\pi N)^{1/3} \tag{59}$$

The frictional ratio then is

$$f/f_0 = a(3M\bar{v}/4\pi N)^{-1/3} \tag{60}$$

This ratio provides a measure of both molecular asymmetry and hydration. If the latter can be estimated, then the axial ratio for an equivalent ellipsoid of revolution can be calculated (Siegel and Monty, 1966).

In some cases a protein may not be pure enough to be studied except by activity measurements. In such cases the partition coefficient, measured by activity assay on a calibrated column, can be used to calculate an upper limit for the hydrated molecular weight (Ackers, 1964). By combining the molecular radius, a, with Avogadro's number, N, the molecular weight, M_{max}, is

$$M_{\text{max}} = 4\pi Na^3/3\bar{v} \tag{61}$$

The partial specific volume must be known or reasonably estimated. The true anhydrous molecular weight will generally be less than this calculated value M_{max} by a large percentage. Based on an assumed hydration of 0.35 gm of water/gm anhydrous protein, an approximate correction of 1.46 can be applied.

E. Molecular Weight Determinations with Denatured Proteins

The inherent difficulties of direct molecular weight determination of compact globular proteins by gel chromatography stem largely from differences in molecular shape and density of the various molecular species employed. These difficulties can be largely overcome by carrying out the determinations under conditions that cause denaturation of the proteins. The most effective applications of this approach to date are those in which a high concentration of guanidine hydrochloride (e.g., 6 M) is present in solution. This approach was first used by Small and co-workers (Small *et al.*, 1963; Cebra and Small, 1967) who studied immunoglobulin derivatives in the presence of 5 M guanidine hydrochloride on Sephadex G-200 columns. They found a linear correlation between the square root of the molecular weight and the square root of the partition coefficient. The method seems to have been rediscovered by Davison (1968) who used 6% agarose equilibrated with 6 M guanidine hydrochloride containing mercaptoethanol and a denaturing agent. In this study the log–molecular weight plot [Eq. (52)] was found to hold approximately for a series of polypeptide chains ranging in molecular weight from 2000 to 100,000.

A much more extensive inquiry into the accuracy and reliability of this technique has recently been made by Fish *et al.* (1969). These workers investigated the behavior of polypeptide chains on columns of 6% agarose equilibrated with 6 M guanidine hydrochloride, 0.1 M mercaptoethanol. In addition, they attempted to use the relation between molecular weight and radius of gyration of randomly coiled polypeptide chains in order to test the validity of column calibrating equations such as Eqs. (53) and (54). They derived a relationship between $M^{0.527}$ and the radius of gyration based on the Flory–Fox parameter obtained by Tanford *et al.* (1967). Some representative data are shown in Fig. 20. Good correlation between molecular weight and partition coefficient can also be obtained when proteins are denatured in sodium dodecyl sulfate (Fish *et al.*, 1970; see also Ch. 3). It may be expected that these methods will be widely used in the future. They are especially valuable

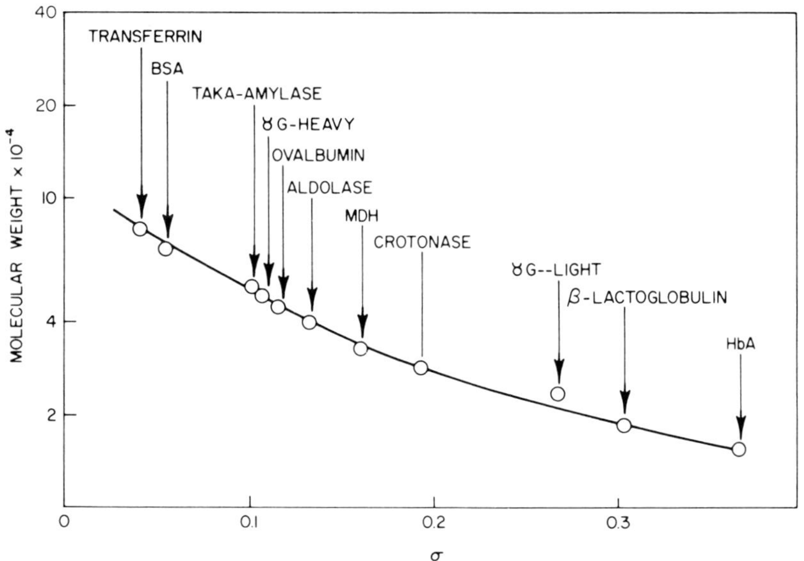

Fig. 20 Correlation between molecular weights and partition coefficients for proteins chromatographed on 6% agarose columns in 6 M guanidine HCl, 0.1 M mercaptoethanol. BSA is bovine serum albumin, γG is gamma globulin, MDH is malate dehydrogenase, and HbA is hemoglobin A. Taken from Fish *et al.* (1969).

for determining the number and identity of subunit polypeptide chains in a protein. It appears that an accuracy of at least 10% can be achieved routinely in the determination of molecular weights by this approach.

In addition to studies carried out under conditions of complete denaturation, the chromatographic behavior of partially denatured proteins may be studied as a guide to the changes in molecular size and shape when these properties are perturbed by denaturing solvents.

F. Evaluation of Conformational Changes in Chemically Modified Proteins

Because partition coefficients are sensitive to molecular size and shape, they can be used to detect conformational changes brought about by chemical modification of proteins. Such studies have been carried out by Habeeb (1966) who found that succinylated bovine serum albumin gave two components with Stokes radii of 7.75 and 11.2 nm; acetylated albumin showed two components with radii of 5.5 and 8.4 nm. Nitroguanylated, guanylated, and amidinated albumin had radii of 4.5, 3.78, and 3.88, respectively, whereas untreated bovine serum albumin had a single Stokes radius of 3.7 nm. The use of gel chromatography to study conformational changes may be expanded in the future to include studies of changes associated with allosteric transitions.

V. MOLECULAR SIEVE PROPERTIES OF POLYDISPERSE SOLUTES

A. Introduction

Porous gel partitioning has proved to be one of the most powerful and promising new approaches in the study of protein subunit interactions. Many recent studies have shown that subunit interactions play an important role in the structural organization of biologically functional protein complexes and in the regulation of their biochemical activities. In many instances, protein subunits are known to undergo association reactions, but the degree of reaction cannot be measured spectroscopically. In these cases an average property related to molecular size or weight can be employed to measure the degree of reaction at various well-defined equilibrium states of the system. The average property is usually a weight-average partition coefficient determined by molecular sieve methods. (Application of these methods to associating systems will be described in Section VI.) Molecular sieve methods have also been used effectively to study "mixed association" reactions between two or more different molecular species. (These applications are described in Section VII.) In the present section the basic concepts applicable to polydisperse solutes of all kinds will be discussed.

On operational grounds the transport behavior of multicomponent systems can be divided into two categories. The first category includes solute systems for which separation of individual species is possible by the transport technique at hand, i.e., all systems of noninteracting components and those in which interaction is so slow that at least partial separation occurs during the transport experiment. The second category includes those systems in which the components cannot be separated because the interactions between species are rapid in comparison with the rates of their separation. Under these circumstances average behavior is observed. Appropriately, there are two fundamental approaches that can be taken in the analysis of multicomponent systems by transport methods. The first includes methods of analysis based on the macroscopic separation characteristics of the species. This approach generally depends upon the shapes and positions of solute zones. The second approach involves analyses based on appropriate average equilibrium properties; the experiments are designed in such a way that separation properties of the components may be ignored. As will be seen, both approaches can be effectively applied in a complementary fashion.

In this chapter the primary emphasis will be placed on the analysis of multicomponent systems in which the various species undergo chemical reactions. But first the formulation of partition coefficients for multicomponent systems and some procedures for their experimental determination will be considered.

B. Partition Coefficient for Total Solute

For a polydisperse system containing various molecular species, the equilibrium partition coefficient for total solute is a weight-average of the partition coefficients $\bar{\sigma}_w$ of the individual species (Ackers and Thompson, 1965).

$$\bar{\sigma}_w = \sum_j C_j \sigma_j \Big/ \sum_j C_j \tag{62}$$

Here C_j is the concentration of species j exterior to the gel, and σ_j is the corresponding partition coefficient. This relationship can be seen in the following way. For each species, the partition isotherm can be written as

$$Q_j = V_i \sigma_j C_j \tag{63}$$

The total solute Q_T partitioned into the stationary phase is

$$Q_T = \sum_j Q_j = V_i \sum_j \sigma_j C_j \tag{64}$$

and the partition coefficient for total solute is equal to the ratio:

$$Q_T / V_i C_T = \sum_j \sigma_j C_j \Big/ \sum_j C_j = \bar{\sigma}_w \tag{65}$$

In some systems it is necessary to consider subclasses of the species j. For example, if aggregating subunits exist in a variety of isomeric forms representing different geometric arrangements of subunits, then each j-meric class (containing j subunits) may be considered to consist of m_j isomers, each with a partition coefficient σ_{ji} and concentration C_{ji}. The partition coefficient for total solute is then

$$\bar{\sigma}_w = \frac{\Sigma_j C_j (\Sigma_{i=1}^{m_j} \sigma_{ji} C_{ji} / \Sigma_{i=1}^{m_j} C_{ji})}{\Sigma_j C_j} \tag{66}$$

where $C_j = \Sigma_{i=1}^{m_j} C_{ji}$ is the total concentration of species belonging to class j (Ackers, 1967a).

It can be seen from Eq. (65) that for noninteracting systems the weight-average partition coefficient will be independent of total solute concentration (except for the linear concentration dependence of individual partition coefficients) since a change in total concentration will change all values of C_j proportionately. However, if interactions are present between species, the various equilibria will be shifted according to the law of mass action and $\bar{\sigma}_w$ will exhibit a marked dependence upon C_t. The resulting dissociation curve of $\bar{\sigma}_w$ vs. concentration may then be used to resolve the stoichiometry and equilibrium constants for the interactions between species, as described in the next section.

C. Experimental Determination of Weight-Average Partition Coefficients

There are essentially three experimental methods for determining the parameter $\bar{\sigma}_w$.

1. Direct Optical Scanning of Saturated Columns

The column saturation scanning method described in Section III,B can be effectively employed to measure solute partitioning in polydisperse systems. For these systems, the weight-average partition coefficient at any distance coordinate, x, within the column is given by

$$\bar{\sigma}_w = \frac{P(x) - \alpha'(x)}{\beta'(x)} \tag{67}$$

With this method a very high degree of accuracy can be achieved in the determination of $\bar{\sigma}_w$ since several hundred data points can be obtained from a single scan of the column (Warshaw and Ackers, 1971).

2. *Integral Boundary Method*

If a large-zone experiment is carried out on a multicomponent system, the weight-average partition coefficient may be determined by measurement of the centroid positions of the boundaries on either side of the plateau (Ackers and Thompson, 1965; Ackers, 1967a). The fundamental quantity of interest within the plateau region is the fraction $\bar{\xi}$ of the total column into which partitioning occurs.

$$\bar{\xi} = \alpha + \beta\bar{\sigma}_w = \sum_j C_j\xi_j \Big/ \sum_j C_j \tag{68}$$

In order to determine $\bar{\sigma}_w$ experimentally from the elution diagram, it is useful to identify a particular value of the volume coordinate V', i.e., $V' = \bar{\xi}l$ and hence $V' = V_0 + \bar{\sigma}_w V_i$. The desired parameter $\bar{\sigma}_w$ can then be determined from this particular volume coordinate V' (knowing V_0 and V_i) in the same way that a partition coefficient is determined for a single component. The volume coordinate that satisfies this requirement is the centroid volume $\bar{V}'$ defined by Eq. (26) (Ackers, 1970).

It is obvious from the foregoing that the centroid elution volume is a weight-average of the elution volume V_j which the components would have if present individually.

$$V_j = V_0 + \sigma_j V_i \tag{69}$$

Solving this expression for σ_j and substituting into Eq. (62) leads directly to the weight-average relationship.

$$\bar{V} = \sum_j V_j C_j \Big/ \sum_j C_j \tag{70}$$

3. *Equilibrium Distribution Method*

The equilibrium distribution method described in Section III can be directly applied to the determination of weight-average partition coefficients. From Eqs. (44) and (65) it is immediately seen that the coefficient measured by this technique is in fact the weight-average $\bar{\sigma}_w$. Of the three methods for determining $\bar{\sigma}_w$, this method generally provides the least accurate results.

D. Constituent Parameters and Frames of Reference

The theory of transport phenomena for multicomponent solute systems is greatly simplified by the use of *constituent parameters.* Since its introduction (Tiselius, 1930), the concept of constituent quantities has been used frequently in the analysis of interacting systems in transport experiments of the freely migrating type, such as electrophoresis and sedimentation. Application to transport systems involving a space-filling stationary phase such as chromatography or gel electrophoresis requires careful consideration of frames of reference. In fact the main difference between the two kinds of transport systems lies in the frames of reference necessary for the expression of concentration and certain other constituent parameters.

General transport equations formulated previously for freely migrating systems (see Nichol *et al.*, 1964) are applicable to gel chromatography only if the total column is used as a frame of reference. For elution chromatography, however, it is most useful to formulate the transport equations in terms of directly accessible experimental quantities, i.e., elution volume and bulk-solution concentration. Such a formulation was first introduced by Ackers and Thompson (1965) who showed that chromatographic elution volumes play the same role as velocity terms play in classic transport equations for freely migrating systems. This important transformation provides the basis for direct applications of existing transport equations to elution experiments with gel chromatography (Gilbert, 1966a; Zimmerman and Ackers, 1971a; Zimmerman *et al.*, 1971).

Consider a solution containing two components, A and B, and suppose they interact to form a complex, C. Then the constituent concentrations of the two components in this solution are usually defined as

$$\begin{aligned}\bar{C}_{\mathrm{A}} &= C_{\mathrm{A}} + C_{\mathrm{C}} \\ \bar{C}_{\mathrm{B}} &= C_{\mathrm{B}} + C_{\mathrm{C}}\end{aligned} \tag{71}$$

The constituent concentration $\overline{C}_{\mathrm{A}}$ of component A is merely the total concentration of A in all of its forms (free A and A combined with B), and the same is true of B. These concentrations are expressed in the bulk solution frame of reference (i.e., mass/moles of solute per unit volume *of solution*) appropriate to samples that may be added to and eluted from a chromatographic column. This concentration frame pertains to the solution within the void volume and other local regions accessible to solute. However, in order to relate solute concentration and other system parameters within the column it is more useful to consider the constituent concentrations (primed quantities) defined with respect to the total column frame of reference.

$$\begin{aligned} \bar{C}_A' &= C_A' + C_C' = \xi_A C_A + \xi_C C_C \\ \bar{C}_B' &= C_B' + C_C' = \xi_B C_B + \xi_C C_C \end{aligned} \tag{72}$$

The primed quantities are mass/moles of solute *per unit total volume within the column.* These are related to the bulk concentrations (unprimed) through the partition cross sections, as shown in Section I, yielding the relations on the right of Eq. (72). The *constituent velocity* $\bar{u}_A'$ is also defined in the total column frame in such a way that the product $\bar{u}_A'\bar{C}_A'$ equals the total flux, expressed as the sum of fluxes of all the species[2]:

$$\begin{aligned} \bar{u}_A &= (1/\bar{C}_A')(C_A'u_A + C_C'u_C) \\ \bar{u}_B &= (1/\bar{C}_B')(C_B'u_B + C_C'u_C) \end{aligned} \tag{73}$$

In this frame, the constituent velocity is a weight-average of species velocities with respect to C' concentrations (but *not* with respect to C concentrations) in contrast to transport systems of the freely migrating type. The species velocities are

$$u_A = F/\xi_A, \qquad u_B = F/\xi_B, \qquad u_C = F/\xi_C \tag{74}$$

Substituting the relations for species velocities into Eq. (73), the following relations between the various constituent quantities are obtained:

$$\bar{u}_A = F\bar{C}_A/\bar{C}_A', \qquad \bar{u}_B = F\bar{C}_B/\bar{C}_B', \qquad \bar{u}_C = F\bar{C}_C/\bar{C}_C' \tag{75}$$

The *constituent partition cross section* $\bar{\xi}_A$ is defined in such a way that it relates the constituent velocity to flow rate in the same way as the species velocities are related to flow rate by Eq. (74).

$$\bar{u}_A = F/\bar{\xi}_A, \qquad \bar{u}_B = F/\bar{\xi}_B, \qquad \bar{u}_C = F/\bar{\xi}_C \tag{76}$$

From Eqs. (75) and (76), one obtains

$$\bar{\xi}_A = \bar{C}_A'/\bar{C}_A = (C_A\xi_A + C_C\xi_C)/(C_A + C_C) \tag{77}$$

Thus it is seen that $\bar{\xi}$ is a weight-average of species partition cross sections *with respect to bulk concentrations* C (but not with respect to C' terms).

The *constituent distribution* (*or elution*) *volume* is similarly defined:

[2] In freely migrating systems, the constituent mobilities are $\bar{u}_A = (1/\bar{C}_A)(C_A u_A + C_C u_C)$, etc.

$$\bar{V}_{\mathrm{A}} = \bar{\xi}_{\mathrm{A}} l = \frac{\xi_{\mathrm{A}} C_{\mathrm{A}} l + \xi_{\mathrm{C}} C_{\mathrm{C}} l}{C_{\mathrm{A}} + C_{\mathrm{C}}} \tag{78}$$

Since the species elution volumes are

$$V_{\mathrm{A}} = \xi_{\mathrm{A}} l, \qquad V_{\mathrm{B}} = \xi_{\mathrm{B}} l, \qquad V_{\mathrm{C}} = \xi_{\mathrm{C}} l$$

we see that $\overline{V}_{\mathrm{A}}$ is also the same type of average as in Eq. (77):

$$\bar{V}_{\mathrm{A}} = \frac{V_{\mathrm{A}} C_{\mathrm{A}} + V_{\mathrm{C}} C_{\mathrm{C}}}{C_{\mathrm{A}} + C_{\mathrm{C}}} \qquad \text{or} \qquad \bar{V}_{\mathrm{A}} \bar{C}_{\mathrm{A}} = V_{\mathrm{A}} C_{\mathrm{A}} + V_{\mathrm{C}} C_{\mathrm{C}} \tag{79}$$

For n species the constituent elution volume is

$$\bar{V} = \sum_{j=1}^{n} V_j C_j \Big/ \sum_{j=1}^{n} C_j \tag{80}$$

i.e., it is a weight-average of species volumes with respect to bulk concentrations C (Ackers and Thompson, 1965). Again it should be noted that this quantity is *not* a weight-average with respect to the total concentrations, C', within the column. Thus in order to apply general transport equations of freely migrating systems to gel chromatography, the velocities must be replaced by elution volumes (or partition cross sections) and constituent concentrations within the column must be replaced by free solution values of constituent concentrations.

VI. SELF-ASSOCIATION OF PROTEIN SUBUNITS

In this section equilibrium partitioning experiments in porous media and related transport experiments on gel columns are presented as a means of studying subunit association equilibria. Reactions of the mixed association type ($A + B \rightleftharpoons C$) are discussed in Section VII, and ligand-binding reactions are described in Section VIII.

A. Partition Isotherm for Associating Systems

Consider a self-associating system comprised of j subunits in equilibrium.

$$\begin{aligned} A_1 + A_1 &\rightleftharpoons A_2 \\ A_2 + A_1 &\rightleftharpoons A_3 \\ A_{j-1} + A_1 &\rightleftharpoons A_j \end{aligned}$$

At constant temperature and pressure the formation of each j-mer can be characterized by an equilibrium constant:

$$K_j = C_j/C_1{}^j \tag{81}$$

where C_j and C_1 represent the constituent concentrations of j-mer and monomer, respectively. The total concentration of solute, with association up to n-mer, is:

$$C_T = \sum_{j=1}^{n} C_j = \sum_{j=1}^{n} K_j C_1{}^j \tag{82}$$

When such a system is subjected to a partitioning experiment, the amount of solute (per unit column length) distributed into the gel phase at equilibrium is

$$Q_T = \sum_{j=1}^{n} \sigma_j K_j C_1{}^j \tag{83}$$

The parametric Eqs. (82) and (83) define the partition isotherm for total solute under conditions of equilibrium. The corresponding partition coefficient for total solute is the weight-average defined for all multicomponent systems by Eq. (65). From Eqs. (65) and (81) we see that the coefficient can be expressed as

$$\bar{\sigma}_w = \Sigma \sigma_j K_j C_1{}^j / \Sigma C_j \tag{84}$$

When the total concentration C_T is changed, the various equilibria between species are shifted according to the law of mass action so that the average degree of aggregation increases with increasing solute concentration. Consequently the value of $\bar{\sigma}_w$ decreases with increasing concentration, since σ_j decreases with increasing molecular size.

B. Chromatographic Transport of Associating Systems

1. *Small-Zone Experiments*

A number of studies have been carried out in which small samples of an interacting protein system were chromatographed on a gel column and an apparent molecular weight calculated from the peak elution position (see, for instance, Andrews, 1964; Sullivan and Riggs, 1967). Although this type of experiment is a useful qualitative means of detecting interaction, there is no theoretical justification for a quantitative interpretation of such an experiment in terms of dissociation constants. As a result of continuous dilution caused by axial dispersion, the peak concentration of the zone moves with a decreasing velocity down the column. The apparent partition coefficient is then a function of column length as well as of the equilibrium constants and initial concentration

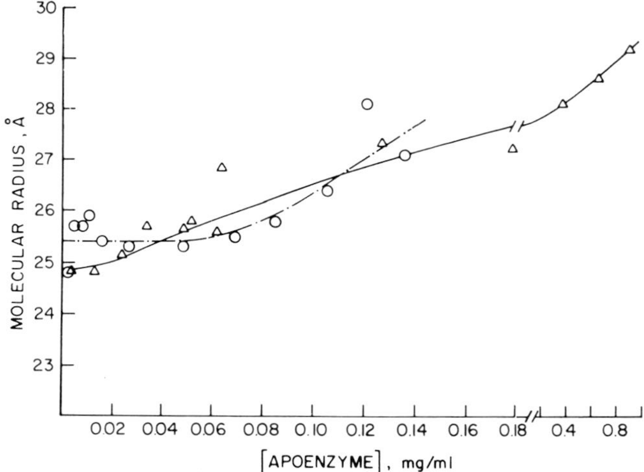

Fig. 21 Apparent molecular radius as a function of concentration of the apoenzyme of D-amino acid oxidase. Protein concentration was determined by optical density at 220 nm; molecular radius was calculated according to Eq. (55). ○—○, small-zone experiments on Sephadex G-75; △—△, integral boundary experiments on Sephadex G-100. Taken from Henn and Ackers (1969a).

of applied sample. A recent investigation (Zimmerman and Ackers, 1971b) has shown that, if this procedure is used to estimate apparent equilibrium constants, the errors may be great indeed, and quantitative interpretations of such curves should be avoided.

Small-zone experiments are useful for the determination of minimum subunit size and molecular weight. In the limit of infinite dilution the measured apparent partition coefficient must correspond to the completely dissociated species. This approach has been used for estimation of the subunit molecular weight of L-glutamate dehydrogenase (Rogers *et al.*, 1965; Andrews, 1965), D-amino acid oxidase (Henn and Ackers, 1969a), and β-lactoglobulin A (Andrews, 1964) (Fig. 21).

2. *Large-Zone Experiments*

If the volume of sample applied to the column is large enough to create a plateau region throughout the duration of the experiment, then the centroids of leading and trailing boundaries move at velocities which depend only upon the parameter C. The centroid elution position can be used to determine the weight-average partition coefficient $\bar{\sigma}_w$ pertaining to each plateau concentration C_0 according to Eq. (70). The theory of these integral boundary experiments has been developed for interacting systems by Ackers and Thompson (1965), Ackers (1967a), and Chiancone *et al.* (1968). The information that can be obtained from large-zone experiments with interacting systems depends upon the different kinds of analyses described below.

a. **Analysis of Characteristic Boundary Shapes.** When a chemically reacting system is subjected to a transport experiment there are two kinds of processes that operate to produce dispersion of the boundaries between solution and solvent. The first of these is the normal diffusion or axial dispersion also present in the transport of noninteracting systems. The second is the dispersion (or contraction) of boundaries which arises from the chemical reactions. The latter effect is the result of the tendency of the various molecular species to move at different velocities because of differences in molecular size, counteracted by their tendency to move with the same velocity because of molecular association. The shapes of boundaries therefore depend upon the rates of equilibration between species (reactants and products) and upon the differences in their respective transport coefficients.

The two processes of boundary dispersion described above do not operate independently of each other and therefore a complete analysis of the system behavior can be obtained only by solution of the complete continuity equation. Unfortunately, analytical solutions are not possible for the transport behavior of interacting systems (which are formally analogous to single-component systems exhibiting nonlinear concentration dependence of transport coefficients). In spite of this difficulty, considerable progress has been made toward understanding the way in which boundary shapes are influenced by chemical reactions in systems where the equilibrium can be considered to be established instantaneously (i.e., rapidly in comparison with the length of time required for the transport experiment under consideration). This problem was originally investigated by Gilbert in an elegant series of papers (Gilbert, 1955, 1959, 1963; Gilbert and Jenkins, 1959). The behavior of free boundaries (i.e., those in which no stationary phase is present) was described theoretically by these authors for the idealized situation in which diffusion is ignored. In spite of the inherent limitations imposed by this idealization, the Gilbert theory has been notably successful in predicting the correct qualitative features of reaction boundaries. The corresponding theory for transport behavior in the presence of a stationary phase has been developed more recently (Ackers and Thompson, 1965; Ackers, 1967a; Chiancone *et al.*, 1968; Winzor *et al.*, 1967) and has been applied to a number of experimental systems. In modern practice the use of idealized theories is largely replaced by complete solutions to the continuity equations obtained by computer simulation. This approach will be described below.

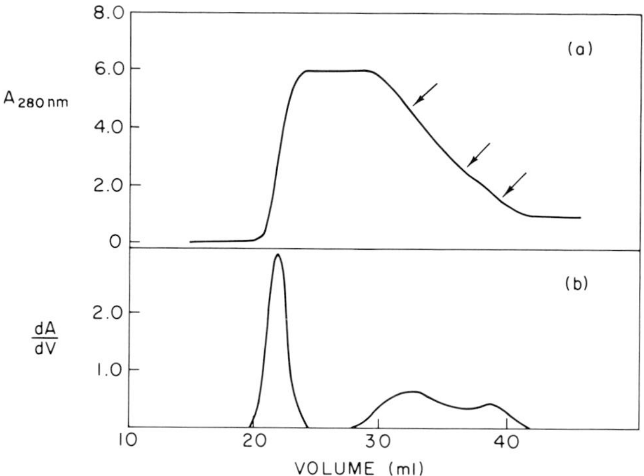

Fig. 22 Elution profile of α-chymotrypsin (3.8 mg/ml) chromatographed on Sephadex G-100. (a) Concentration–volume profile showing boundary-sharpening effect on leading edge (*left*) and boundary spreading of trailing edge (*right*). Arrows indicate inflection points. (b) First derivative curves of absorbance vs. volume for the elution profile shown in (a). Taken from Winzor and Scheraga (1963).

The general qualitative features exhibited in gel chromatography of associating systems were demonstrated by Winzor and Scheraga (1963), who showed that a boundary-sharpening was present on the leading edge of the solute zone and a corresponding boundary-spreading occurred on the trailing side. This effect, shown in Fig. 22, arises from the tendency of the larger molecules to move faster within the column than the smaller ones. However, on the leading boundary, the molecules that move ahead of those in the plateau find themselves in a region of lower concentration which promotes their dissociation, leading to decreased velocity. Thus the boundary-sharpening effect is in continuous operation at the leading edge. These effects produce a continuous spreading of the trailing boundary.

Winzor and Scheraga (1963) also demonstrated that the qualitative features predicted by the Gilbert theory for the behavior of reaction boundaries were analogous to those of chromatography. The gradient of concentration across the trailing boundary of α-chymotrypsin in low ionic strength exhibited two maxima and a single minimum above a certain critical plateau concentration (Fig. 23). This is analogous to the behavior of a monomer–n-mer associating system in sedimentation experiments when n is greater than 2. In contrast, the gradient for chymotrypsin under conditions of dimerization ($n = 2$) exhibited only a single maximum and no minimum. Furthermore, the elution volume

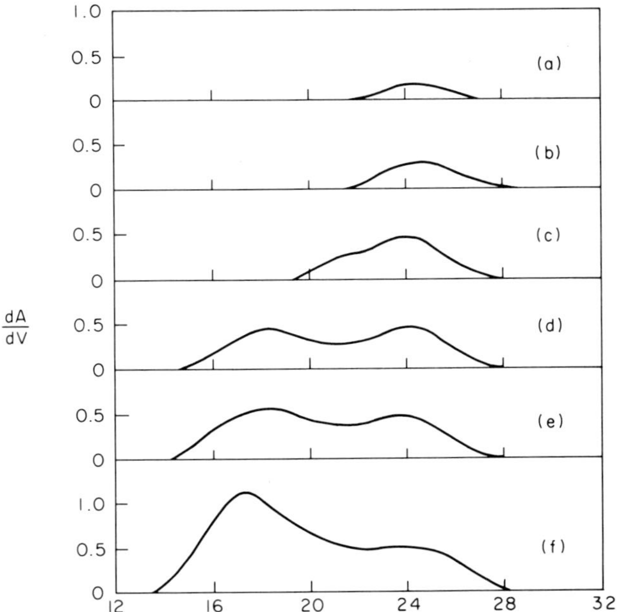

Fig. 23 Derivative curves of the trailing edge of the protein zone in the chromatography of α-chymotrypsin on Sephadex G-100. Concentrations (mg/ml) are (a) 0.6, (b) 1.2, (c) 1.4, (d) 2.8, (e) 3.4, and (f) 5.0. Taken from Winzor and Scheraga (1963).

of the leading boundary centroid could be correlated linearly with the independently determined weight-average molecular weight (Winzor and Scheraga, 1964). This empirical correlation procedure was used to determine the minimum subunit molecular weight of bovine thrombin.

The quantitative theory for these effects has been formulated in terms of asymptotic equations, ignoring axial dispersion (Ackers and Thompson, 1965). More complete analyses have recently been made, and numerical solutions to the continuity equations (which include the effects of axial dispersion) were obtained (Zimmerman and Ackers, 1971a,b; Zimmerman *et al.*, 1971). The continuity equation for total solute written in the total column frame of reference is

$$J' = u_T C_T' - L_T(\partial C_T'/\partial x) \tag{85}$$

In this equation J' is the total solute flux (mass transported along the distance coordinate x per unit time per unit of cross section), and C_T' is total mass concentration. The total flux is seen as the sum of contributions from a nondispersive transport term ($u_T C_T'$) and a dispersion term $[-L_T(\partial C_T'/\partial x)]$. The parameter L_T is a dispersion coefficient that, for many processes, is simply a diffusion coefficient. For countercurrent distribution, an "analogue" diffusion coefficient is defined which depends upon the partition coefficient. In gel chromatography the dispersion coefficient arises from the combination of processes described in Section II.

For a multicomponent system the coefficients u_T and L_T are appropriate averages of the corresponding coefficients for each of the species. The velocity u_T characterizing total solute is a weight-average of species velocities u_j with respect to the mass concentrations C_j'

$$u_T = \Sigma u_j C_j' / \Sigma C_j' \tag{86}$$

The velocity term for each species is

$$u_j = F/\xi_j \tag{87}$$

The other phenomenological parameter of Eq. (85), L_T, is a dispersion coefficient that represents the gradient average of the dispersion coefficients L_j.

$$L_T = \frac{\Sigma L_j(\partial C_j/\partial x)}{\Sigma \partial C_j/\partial x} \tag{88}$$

The axial dispersion coefficients of the individual species are in turn given by

$$L_j = FL_p + \xi_j D_j + \frac{qd^2F^2}{\xi_j^3 A^2 D_j} \tag{89}$$

In this equation the quantities with j subscripts refer to a given molecular species, while the other quantities refer to behavior of all species on a given column at specified flow rate F. The constant L_p is the dispersive contribution resulting from flow perturbation, D_j is the free diffusion coefficient for species j, A is the cross-sectional area of the column, and qd^2 is a term relating the gel particle diameter and geometry to the average solute equilibration time between the mobile and stationary phases of the column. It is evident from Eq. (37) that axial dispersion coefficients of molecular species may be varied experimentally in many different ways through variation of flow rate, gel particle size, column cross-sectional area, or gel porosity (expressed through the parameter ξ). In gel chromatography the values of the dispersion coefficients usually increase with an increasing degree of polymerization, whereas in sedimentation velocity the diffusion coefficients vary in the opposite way (Halvorson and Ackers, 1971). Numerical solutions of Eqs. (85)–(89) have been obtained for several cases of subunit association (Zimmerman and Ackers, 1971a,b; Zimmerman *et al.*, 1971) using the elegant computer simulation method of Cox (1969). The parameters investigated by simulation include development time on the column,

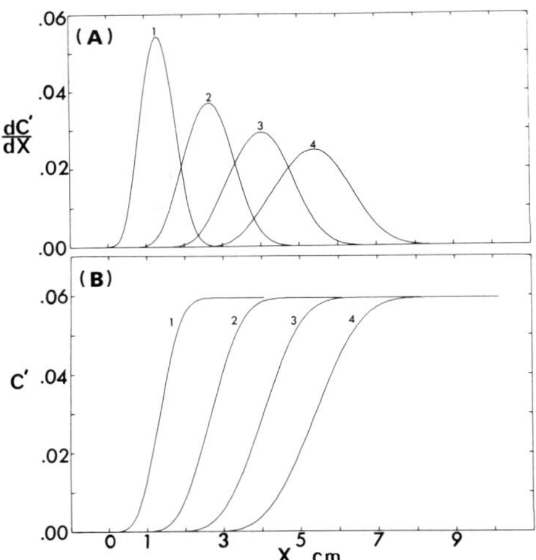

Fig. 24 Development of reaction boundary with time. Theoretical profiles calculated for a rapidly reversible dimerizing solute at 50% dissociation, being chromatographed on a Sephadex G-100 column (regular bead size). The monomer has a molecular radius of 18.9 Å, corresponding to an approximate molecular weight of 17,000. The column cross-sectional area is 1 cm². Flow rate is 9.6 ml/hr. The solute is initially present as a step function ($x = 0$, $t = 0$) and the loading concentration (corresponding to solute concentration of the mobile phase within the plateau) is 0.100 mg/ml. Times are (1) 5 min, (2) 10 min, (3) 15 min, and (4) 20 min. Upper curves (A) represent gradients of concentration profiles shown in lower curves (B). Taken from Zimmerman and Ackers (1971b).

flow rate, degree of molecular association, gel porosity, and gel particle size. It was found that a given self-associating solute can be expected to yield appreciably different boundary profiles on gels of different porosities and bead sizes, or on the same column at different flow rates. Some representative profiles are shown in Figs. 24 and 25.

Results of the simulations indicate that axial dispersion contributes substantially to the shape of solute profiles. Its effect on boundary shapes was evaluated by comparing the simulated profiles with quantitative predictions of the corresponding asymptotic equations in which axial dispersion is assumed to be absent. Substantial differences between the equilibrium constants and centroid parameters predicted from the asymptotic equations and those of the simulated profiles were found. (For details of the asymptotic solutions, see Ackers, 1970.) The sensitivity of boundary shapes to changes in those column parameters that can be experimentally manipulated makes gel chromatography, in principle, a uniquely powerful tool for distinguishing one mode of self-association from another. Although simulated chromatography of an incorrect model

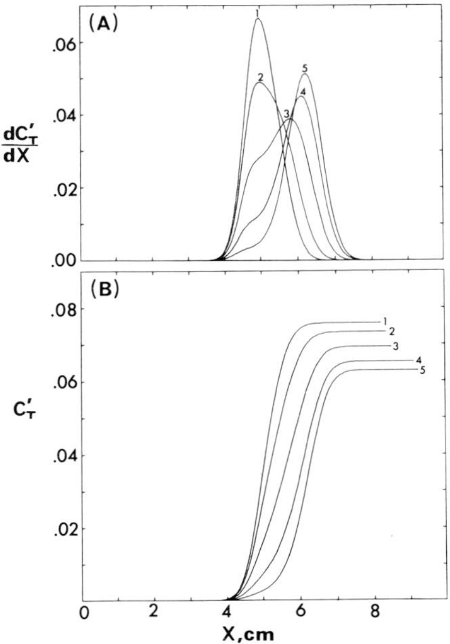

Fig. 25 Simulated gel chromatography of rapidly equilibrating monomer–tetramer systems on Sephadex G-200R. Monomer molecular weight = 17,000. Loading concentration = 0.1 mg/ml. Sharp initial boundary at $x = 0$. Column cross section = 1.0 cm^2. Flow rate = 1.2 ml/hr; time = 192 min. Numbers (1–5) denote the percentage of monomer by weight, i.e., 90, 75, 50, 25, and 10%, respectively. Curves (A) are derivatives of concentration profiles (B). Taken from Zimmerman *et al.* (1971).

of a polymerizing solute can sometimes reproduce the behavior of the solute under a given set of experimental conditions, it is most unlikely that the same incorrect model can be made to generate the different boundary shapes obtained at various flow rates in gels of different porosities and bead sizes.

***b.* Analysis of Partition Coefficient Averages.** In order to obtain reliable stoichiometries and equilibrium constants from integral boundary experiments, use can be made of the weight-average partition coefficient. The determination of this parameter is rigorously exact (Section V,C). The dependence of $\bar{\sigma}_w$ upon plateau concentration is determined by the equilibrium constants for the reactions and the partition coefficients of the individual associating molecular species. The species partition coefficients are functions of gel porosity as well as of molecular size in a linearly independent fashion (Section IV,B). There are, therefore, two kinds of experimental variations that can be introduced in order to determine the nature of a given reaction system. In elution chromatography, as the plateau concentration C_0 is varied, the equilibria are shifted so that the concentration terms, C_j, pertaining to the individual species are altered [Eq. (84)] while the corresponding partition coefficient terms, σ_j, remain constant. Alternatively, the plateau concentration can be maintained constant while the porosity of the gel is varied (i.e., $\bar{\sigma}_w$ is measured for a series of gels saturated at the same plateau concentration, C_0). In this case the partition coefficient terms change while the C_j terms remain constant.

The simplest variation that can be introduced is that of the plateau concentration, C_0. The dissociation curve obtained for a given column (i.e., $\bar{\sigma}_w$ as a function of C_0) can then be compared with models representing various reaction stoichiometries and equilibrium constants until the best fit is obtained. For most studies (except at very low protein concentration) it is necessary to make corrections for the linear concentration dependency of individual partition coefficients. If it is assumed that g is the same for all species, substitution of Eq. (10) into Eq. (62) leads to a linear dependence of $\bar{\sigma}_w$ on C_T (Ackers, 1967a). This procedure has been carried out for a number of systems including human hemoglobin (Ackers and Thompson, 1965; Chiancone *et al.*, 1968), α-chymotrypsin (Ackers, 1967a), L-glutamate dehydrogenase (Chun *et al.*, 1969a), and D-amino acid oxidase (Henn and Ackers, 1969a,b). Critical evaluations of this approach have been made by Gilbert (1967) and by Chiancone *et al.* (1968). Usually a large amount of accurate experimental data is required in order to virtually eliminate all ambiguities in the possible models that can be fit to the data.

In a very careful study of human oxyhemoglobin, Chiancone *et al.* (1968) could not distinguish unequivocally between dissociation models involving monomer ⇌ dimer ⇌ tetramer or dimer ⇌ tetramer reactions (Fig. 26), although they could eliminate models involving trimer. Furthermore, they ascertained that dimer–tetramer association predominated in the concentration range studied (down to 0.009 gm/dl) and determined an accurate value for the dimer–tetramer equilibrium constant. The use of nonlinear least-squares parameter-fitting procedures made it possible for data to be critically tested against a variety of models. Each model required a different solution of Eq. (84).

With the development of direct column scanning it is now possible to determine accurate $\bar{\sigma}_w$ values conveniently and rapidly on column beds stacked with gels of different porosities. By use of the saturation technique the $\bar{\sigma}_w$ values can be determined in a single experiment for all of the porosities present in the column bed (Warshaw and Ackers,

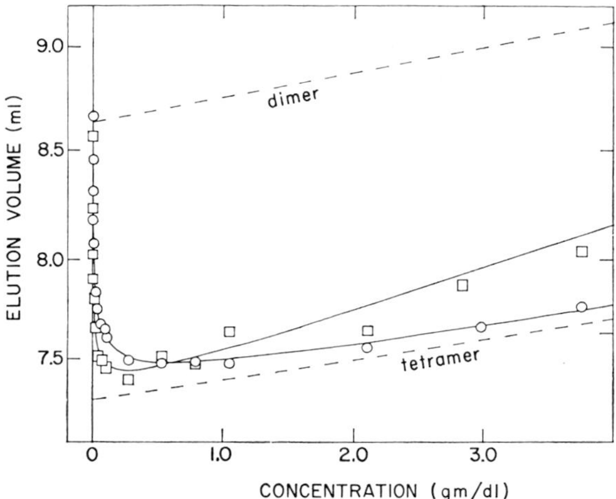

Fig. 26 Elution volume as a function of human oxyhemoglobin concentration for a 50 × 0.8-cm column of Bio-Gel P-100 at 2.5°–3.0°, equilibrated with 0.1 *M* Na^+ (0.09 as chloride, plus phosphate), pH 7.00 at 20°. (○) Integral boundary formed with solvent; (□) finite difference boundary; and (—) theoretical curves. Taken from Chiancone *et al.* (1968).

1971). It has been proposed (Ackers, 1968b) that variations in porosity as well as concentration can be used to determine the number of components present in the reaction mixture (corresponding to C_0) and the weight fraction of all species present. If M different reaction mixtures are each chromatographed on N different gels of different porosities (provided that M and N are at least as great as k, the number of species), then the MN experimental parameters of the type

$$\theta = \bar{\sigma}_w C_0 \tag{90}$$

define a k-dimensional linear vector space, and the calculated rank of the matrix of experimental observations

$$\begin{vmatrix} \theta_{11} & \theta_{12} & \cdots & \theta_{1N} \\ \vdots & \vdots & & \vdots \\ \theta_{M1} & \theta_{M2} & \cdots & \theta_{MN} \end{vmatrix} \tag{91}$$

is equal to k, the number of components present. In addition, Eq. (90) together with Eq. (81) can be solved for stoichiometries and equilibrium constants. An alternative method of analysis, proposed by Winzor *et al.* (1967), is based on a combination of the z-average elution volume and the centroid elution volume. This method involves evaluation of a series of iterated integrals and is based on the idealized expression for volume within the reaction boundary, neglecting axial dispersion. The method has been successfully used to determine combining ratios and equilibrium constants for the self-association of α-chymotrypsin.

3. *Differential Methods*

An alternative approach to the integral methods described above are methods that make use of small differences in the plateau concentration C_0 (finite difference boundaries) or slight differences in the partition coefficient of samples run in series on a column (layering technique). Both of these approaches have been developed by Gilbert and co-workers (Gilbert, 1966a,b; Chiancone *et al.*, 1968; Gilbert and Gilbert, 1968).

***a.* Difference Boundaries.** A solution at concentration C_0 is applied to a gel column until a plateau is established in the effluent. Subsequently the same solute is added at a slightly different concentration, $C_0 + \Delta C_0$, until a second plateau is established. A finite difference boundary will then be established between the two plateau concentrations. For such a boundary, the elution volume, V_Δ, of the equivalent boundary position is

$$V_\Delta = \Delta(C_0\bar{V})/\Delta C_0 \tag{92}$$

where $\bar{V}$ is the centroid elution volume of the boundary corresponding to the plateau concentration, C_0. In the limit as ΔC_0 approaches zero, the finite difference boundary approaches a true differential boundary so that

$$\lim_{\Delta C_0 \to 0} V_\Delta = V = d(C_0\bar{V})/dC_0 \tag{93}$$

This limiting volume V reflects the "diffusion-free" profile of the reaction boundary (Chiancone *et al.*, 1968), which is defined (Ackers, 1967a) in terms of partition cross sections by

$$(V - S)/x = \Sigma jK_j\xi_jC_1^{j-1}/\Sigma jK_jC_1^{j-1} \tag{94}$$

The differential boundary pertaining to a particular concentration C_0 can be determined by extrapolating measurements on finite difference boundaries to zero concentration difference. In their study of hemoglobin dissociation, Chiancone *et al.* (1968) carried out finite difference boundary measurements ($\Delta C_0 = 0.1C_0$) and compared models of the reactions to the experimental data (Fig. 26). It was found that the finite difference boundary determinations were less accurate than the corresponding integral boundary measurements. Nevertheless, this appears to be a useful approach and may be employed to obtain a different kind of average property than the weight-average. Under appropriate conditions, a differential boundary will theoretically split into a number of boundaries which can be related to the individual partition coefficients (σ_j) and concentration terms (C_j) of a reaction mixture (Jenkins, 1965).

b. **Layering Technique with Constant Plateau Concentration.** A second differential technique closely related to that described above is based on the differential rate of migration of two boundaries for solute zones of identical plateau concentration (Gilbert, 1966b). If a solution containing one species is layered over another in the column, discontinuity will arise at the interface between the solutions whenever the solute moves at different rates within the two plateau regions. If the second sample applied to the column is more highly aggregated than the first, it will tend to move faster and overtake the first, producing a "hump" in the concentration profile of the region of interface between the two. If the opposite condition obtains, a "trough" will result between the two zones. Gilbert (1966b) has shown that the area of this discontinuity, $\bar{V}C_0$, is related to the difference in the degree of dissociation, f', of the two solutes. For two closely related solutes, both exhibiting a monomer–n-mer association:

$$\Delta\bar{V} = (f_2' - f_1')(V_1 - V_n) \tag{95}$$

where f_1' and f_2' are the degrees of dissociation of total solute in the first and second samples applied to the column. The volume terms V_1 and V_n are elution volumes of the monomer and n-mer, respectively. This technique is especially useful for the investigation of the dissociation behavior of closely related proteins which have only slight differences in amino acid composition (e.g., hemoglobins). The application of column scanning affords an additional measurement of the differences in degree of association, since the concentrations "seen" by the scanning system in the plateau region (of constant mobile phase concentration C_0) will differ if molecular species occupy different fractions of the cross-sectional area of the column.

4. *Temperature Dependence Studies*

Study of interacting protein systems can be carried out at various temperatures in order to obtain a thermodynamic characterization of the reactions. Entropy and enthalpy changes for the reaction can be calculated from the measured temperature dependence of the equilibrium constant. A study of the D-amino acid oxidase apoenzyme (Henn and Ackers, 1969a,b) revealed a sharp transition in the dimerization constant K_D over a narrow temperature range (12°–14°) and a molar

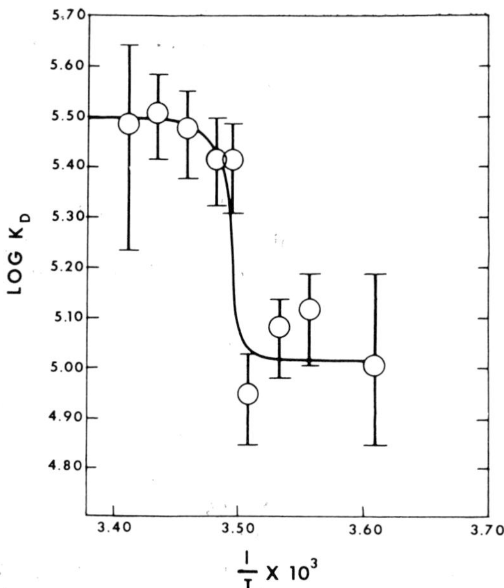

Fig. 27 Van't Hoff plot for the dimerization of the D-amino acid oxidase apoenzyme at temperatures of 4°–20°. The protein undergoes a sharp thermal transition in the range of 12°–14°. Taken from Henn and Ackers (1969b).

enthalpy change of 72 kcal (Fig. 27). The corresponding change in molar entropy over the transition region was 230 eu, whereas above and below this temperature range the change was approximately 25 eu. These findings are consistent with the discovery by Massey *et al.* (1966) of a sharp thermal transition in the catalytic and physical properties of the holoenzyme, presumably due to a large conformational transition ("melting").

5. *Combination with Other Methods*

Combination of analytical molecular sieve chromatography with other techniques is useful in the study of interacting multicomponent systems. Since partition coefficients are sensitive to molecular shape as well as size, the combination of weight-average partition coefficient data with weight-average molecular weight data can yield information about the mode of subunit aggregation. In a study of the association of L-glutamate dehydrogenase subunits (Chun *et al.*, 1969a), theoretical values of $\bar{\sigma}_w$ were calculated as a function of concentration for compact aggregation models (in which all species were assumed to be spherical) and for linear aggregation models. These coefficients were calculated from

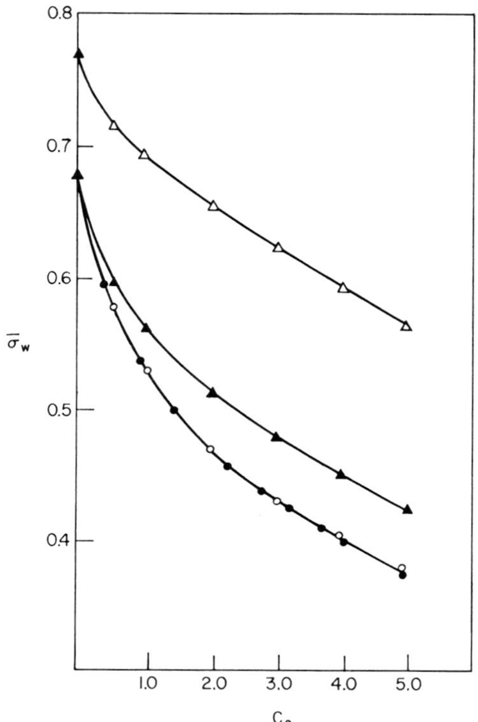

Fig. 28 The weight-average partition coefficients of bovine liver L-glutamate dehydrogenase, generated from various calibration models as a function of protein concentration. (△) Association model based on the relationship between σ_j and log M_i (log model).

$$\bar{\sigma}_w = \sigma_1 - A \sum_i f_j \ln j$$

(▲) Association based on a spherical model (sphere model) where $a_1 = 54.5$ Å; $a_0 = 9.1$ Å; and $b_0 = 157.5$ Å. (○) Indefinite linear association model from the weight fractions

$$\bar{\sigma}_w = \sum_j \sigma_j f_j$$

where $\sigma_j = \text{erfc}[(f/f_0)^j a_1 j - a_0/b_0]$; $a_1 = 54.5$ Å; $a_0 = 9.1$ Å; and $b_0 = 157.5$ Å. (●) Experimentally obtained curve in 0.2 M sodium phosphate buffer, 10^{-3} M EDTA (pH 7.0) at 25°. The column gel was composed of Sepharose 4B. Taken from Chun *et al.* (1969b).

weight fractions of species and equilibrium constants obtained from measurements of weight-average molecular weight. The resulting plots are shown in Fig. 28. The experimental points coincide with the theoretical curve for linear aggregation (lower curve, Fig. 28). The middle curve is the theoretical plot for compact aggregation, and the upper curve is the predicted dissociation curve based on direct molecular weight calibration of the (agarose) column and on the assumed logarithmic relation between molecular weight and partition coefficient. Failure of the logarithmic calibration is attributable in part to the fact that the "monomeric subunit" of this enzyme has a hexameric chain structure which contains voids. Because of these voids the relation between molecular weight and size is quite different from that of the compact globular proteins used for the calibration of the column.

VII. ASSOCIATION BETWEEN DISSIMILAR SUBUNITS

A. Introduction

Frequently it is of interest to study reversible association reactions of the type

$$A + B \rightleftharpoons C$$

where A and B are dissimilar subunits and C is the association complex. Such complexes are known to be involved in immunochemical systems (antigen–antibody or hapten combinations), in inducer–repressor interactions, in the association of regulatory and catalytic subunits of enzymes, and in many others.

Several procedures have been developed for the analysis of such systems by elution chromatography (Nichol and Winzor, 1964; Nichol *et al.*, 1967; Gilbert and Kellett, 1971). In the latter study the analysis was based on the different spectral properties of the constituents, ovalbumin and myoglobin (Gilbert and Kellett, 1971). This approach permits the experimenter to resolve profiles for the complex as well as for the individual constituents. We will consider a system in which a 1:1 complex is formed between the two species A and B.

B. Analysis of Elution Profiles

In general the chromatography of such a reversible interacting system will give rise to an elution profile having the features shown in Fig. 29.

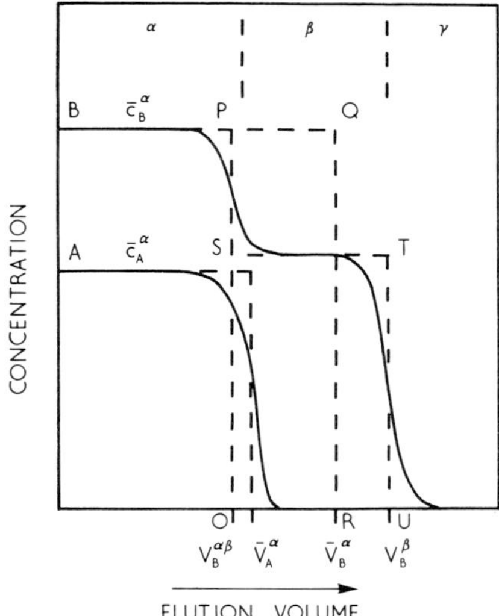

Fig. **29** Trailing elution profile for the association between dissimilar species $A + B \rightleftharpoons C$. See the text for details. Taken from Gilbert and Kellett (1971).

There are two plateaus in the profile of one constituent (B) and a single plateau in the other (A). Using the notation of Longsworth (1959), the profile can be divided into three regions, α, β, and γ. We will denote elution volumes and molar concentrations by $V_x{}^y$ and $C_x{}^y$ where $x =$ A, B, C and $y = \alpha, \beta, \gamma$. The two elution volumes pertaining to the trailing boundary of the B component are $V_B{}^{\alpha\beta}$ and $V_B{}^\beta$ and are defined by the equivalent sharp boundaries *OP* and *UT*. The vertical line *RQ* defines the centroid for the total B component over the entire boundary, denoted $\bar{V}_B{}^\alpha$ and corresponding to the concentration $\bar{C}_B{}^\alpha$ of B in the α phase, i.e., the area under the dashed lines (*BQR*) equals the total amount of B component under the solid curve B of Fig. 29. This area *BQR* is also equal to the area defined by *BPSTU* (the line *ST* denotes the plateau within the trailing boundary of the B component, and the vertical lines *PS* and *TU* denote centroids of the two boundaries of the B component). Consequently, the areas *OPQR* and *OSTU* must be equal. In the notation this equality is written

$$(V_B{}^{\alpha\beta} - \bar{V}_B{}^\alpha)\bar{C}_B{}^\alpha = (V_B{}^{\alpha\beta} - V_B{}^\beta)C_B{}^\beta \tag{96}$$

The elution volume $V_B{}^{\alpha\beta}$ of the $\alpha\beta$ boundary is therefore given by the expression

$$V_B{}^{\alpha\beta} = \frac{\bar{V}_B{}^\alpha \bar{C}_B{}^\alpha - V_B{}^\beta C_B{}^\beta}{\bar{C}_B{}^\alpha - C_B{}^\beta} \tag{97}$$

This equation is analogous to the corresponding equations for electrophoresis (Longsworth, 1959) and sedimentation (Hersh and Schachman, 1955) in which elution volumes have replaced velocities.

Substituting the definitions of constituent concentrations and elution volumes into Eq. (96) leads to the relationship

$$C_A^\alpha(V_A^\alpha - V_B^\alpha) = \bar{C}_A^\alpha(\bar{V}_A^\alpha - \bar{V}_B^\alpha) = \bar{C}_B^\alpha(V_B^{\alpha\beta} - V_B^\alpha) + C_B^\beta(V_B^{\alpha\beta} - V_B^\beta) \quad (98)$$

In this expression all the parameters except V_B^α and V_A^α are experimentally measurable from the elution profile. The two remaining terms can be estimated by independent experiments. In order to do this, however, it is necessary to know their concentration dependence.

The equilibrium constant for the dissociation of complex C can be written in terms of the α-phase concentrations as

$$K = \frac{C_A^\alpha C_B^\alpha}{C_C^\alpha} = \frac{C_A^\alpha(\bar{C}_B^\alpha - \bar{C}_A^\alpha - C_A^\alpha)}{(\bar{C}_A^\alpha - C_A^\alpha)} \quad (99)$$

The expression on the right provides a means of calculating K without having to know the elution volume of complex C. Once K has been

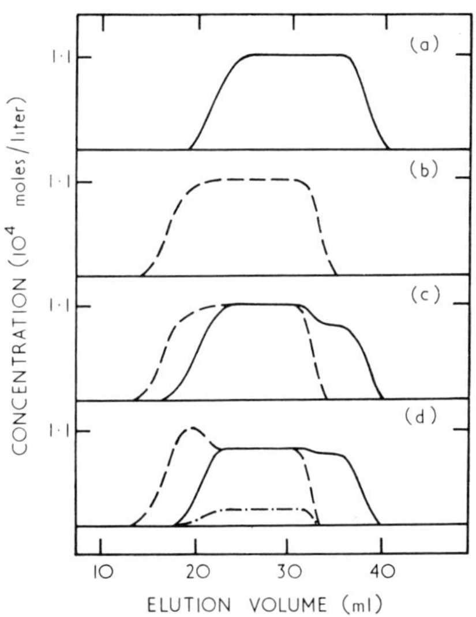

Fig. 30 Elution profiles for an interacting system of ovalbumin and sperm whale myoglobin in a 1:1 mixture. (a) Free myoglobin, 1.1×10^{-4} *M*; (b) free ovalbumin, 1.1×10^{-4} *M*; (c) constituent concentrations in the mixture of (a) and (b); and (d) subspecies concentrations in the mixture. ——, Myoglobin; - - - - -, ovalbumin; -·-·-·-·-, complex. Taken from Gilbert and Kellett (1971).

determined, the concentration of the complex—and subsequently of free A and B—can be determined at any point across the profile from the measured constituent concentrations of A and B by the relationship:

$$C_C^2 - C_C(K + \bar{C}_B + \bar{C}_A) + \bar{C}_A\bar{C}_B = 0 \quad (100)$$

The procedure outlined here was used by Gilbert and Kellett (1971) to study the interaction of ovalbumin with sperm whale myoglobin on a column of Sephadex G-100. The profiles obtained for this system are shown in Fig. 30. Measurements were made at 280 nm for ovalbumin and at 409 nm for metmyoglobin. In order to obtain appropriate values for $V_A{}^a$ and $V_B{}^a$, studies of the concentration dependence of these parameters were carried out. By varying pH, ionic strength, and concentration it could be shown that the principal difference in g values is attributable to electrostatic effects, provided that the effects are assumed to be independent and additive for mixtures of species. Consequently, when myoglobin is at its isoelectric point, the expressions used are

$$V_m = (V_m)_0[1 - g_m(\bar{C}_m + \bar{C}_{ov})] \quad (101)$$

where V_m and $(V_m)_0$ are the myoglobin elution volume and its value at zero concentration, respectively. Similarly, for ovalbumin,

$$V_{ov} = (V_{ov})_0[1 - g_m\bar{C}_m - g_{ov}\bar{C}_{ov}] \quad (102)$$

and for the complex

$$V_C = (V_C)_0[1 - g_m\bar{C}_m - g_{ov}\bar{C}_{ov}] \quad (103)$$

The equilibrium constants obtained using this approach to the concentration-dependence problem are shown in Table XI.

TABLE XI

Calculation of the Equilibrium Constants (K) for the Interaction of Ovalbumin and Myoglobin[a]

$\bar{c}_{ov}{}^{\alpha} \times 10^4$	$\bar{c}M^{\alpha} \times 10^4$	$cM^{\beta} \times 10^4$	$\bar{V}_{ov}{}^{\alpha}$	$V_m{}^{\alpha\beta}$	$V_m{}^{\beta}$	$K \times 10^4$	$(V_C)_0$
1.125	1.131	0.894	16.58	16.43	21.88	4.3	13.1
2.247	1.134	0.163	17.62	17.28	21.86	5.7	10.4
3.261	1.110	0.287	18.83	18.12	21.84	4.8	10.7
2.536	2.275	1.274	17.35	17.11	21.90	4.9	11.1
3.261	3.379	1.730	17.48	17.14	21.93	5.4	10.0

[a] The mean value of K is $5.0 \pm 0.5 \times 10^{-4}$ moles/liter and the mean value of $(V_C)_0$ is 11.1 ± 1.2 ml. Concentration is in moles/liter; elution volume is in ml.

VIII. BINDING OF SMALL MOLECULES TO PROTEINS

A. Introduction

Several procedures have been developed in recent years which utilize gel partitioning systems for the study of ligand binding. Of the four methods to be described in this section, the first two are essentially equilibrium dialysis procedures, while the others depend upon separation of components during a transport experiment. Even the transport techniques, however, depend upon equilibrium properties, and analysis of boundary shapes is necessary for their application. The methods generally offer considerable convenience and precision, compared with alternative procedures.

Before describing the various techniques, a few definitions will be presented concerning the type of molecular system to be studied and the experimental quantities of interest. The system is represented by the general scheme for reaction equilibria involving the successive binding of a small ligand species, L, to a macromolecular component, P.

$$\begin{aligned} P + L &\rightleftharpoons PL \\ PL + L &\rightleftharpoons PL_2 \\ &\vdots \\ PL_{i-1} + L &\rightleftharpoons PL_i \\ &\vdots \\ PL_{n-1} + L &\rightleftharpoons PL_n \end{aligned} \tag{104}$$

Successive binding constants are represented by

$$K_i = [PL_i]/[PL_{i-1}][L] \tag{105}$$

The experimental quantity required for determination of successive binding constants is the binding ratio, r, as a function of the free ligand concentration [L]:

$$r = ([L_0] - [L])/[P_0] \tag{106}$$

In Eq. (106), $[P_0]$ is the total (constituent) concentration of macromolecule, defined by the relationship

$$[P_0] = \sum_{i=0}^{n} [PL_i] \tag{107}$$

The quantity $[L_0]$ is the total (constituent) concentration of ligand.

$$[L_0] = [L] + \sum_{i=1}^{n} i[PL_i] \tag{108}$$

Although a complete discussion of binding ratios in terms of molecular models is beyond the scope of this chapter, consideration of the simplest case will indicate the types of conditions under which binding ratios should be determined. If all of the binding sites are identical and have identical intrinsic binding constants, K, then

$$K_i = ((n - i + 1)/i)K$$

The Scatchard equation relates the experimental quantities r and [L] to the stoichiometry, n, and binding constant, K, for a simple system of this type.

$$r/[L] = K(n - r) \tag{109}$$

Binding constants can, in principle, be determined from measurement of the binding ratio r as a function of free ligand concentration [L]. Such data are usually analyzed by plotting $r/[L]$ vs. r. When Eq. (109) is applicable, a straight line with a slope of $-K$ and an abscissa-intercept of n is obtained. If a nonlinear relation is found, more complex forms of analysis must be attempted which are beyond the scope of the present discussion (see Klotz and Hunston, 1971). It is important to note that binding ratios of extremely high precision are required for unequivocal resolution of all but the simplest cases (Weber, 1965). The remainder of this section will deal with ways of accurately determining the binding ratio for protein systems in which ligand species are reversibly bound.

B. The Batch Method

The equilibrium dialysis properties of porous gel networks can be utilized in a batch procedure similar to that described earlier for the determination of partition coefficients (see Section III,A). The method is simple, straightforward, and requires very little equipment. Although this technique is not as precise as the others to be described, it does provide a rapid alternative to conventional equilibrium dialysis.

This technique requires the use of a porous gel that will totally exclude the protein. Experiments are usually performed in replicate with constant quantities of gel-forming material weighed into volumetric containers. The gel should be previously swollen, washed, and then redried in order to remove soluble material that can interfere with determinations. A sufficient volume of solvent is added to ensure maximum swelling, and the gels are allowed to set until this has occurred. Subsequently a solution containing known amounts of protein and ligand is added and the total volume brought to some constant value. Two of the determinations are carried out with protein or ligand alone in order to determine distribution volumes for these species. In each case a quantity Q_T of the molecular species of interest (either ligand or protein) is added and the system stirred or shaken until equilibration has taken place (only a few minutes are usually required). After equilibration, an aliquot of the supernatant liquid is removed and its concentration, C_0, is assayed. The apparent total volume occupied by the solute within the container is calculated as the ratio Q_T/C_0. When the protein alone is assayed, this ratio is equal to V_0, the void volume. In the case of ligand alone (or any other small molecule), the ratio is equal to $V_0 + V_i$ so that V_i is also determined (this volume can alternatively be determined from the partial volume and solvent regain of the gel). For each of the samples to which mixtures have been added, the ratio of total ligand Q_T added to ligand concentration $[L_0]$ measured in the supernatant provides a determination of the *constituent distribution volume*, $\bar{V}_L$, defined by

$$\bar{V}_L = \frac{1}{[L_0]}\left\{[L]V_L + \sum_{i=1}^{n} i[PL_i]V_{PL_i}\right\} \tag{110}$$

Here V_L is the distribution volume for free ligand ($V_L = V_0 + V_i$), and V_{PL_i} is the corresponding volume occupied by complex PL_i. Since all complexes are excluded (all $V_{PL_i} = V_0$), Eq. (110) can be simplified by combining it with Eq. (108) to yield

$$\bar{V}_L = \frac{1}{[L_0]}\{[L](V_0 + V_i) + V_0([L_0] - [L])\} \tag{111}$$

and

$$\bar{V}_L = V_0 + \frac{[L]}{[L_0]} V_i \tag{112}$$

Solving for the free ligand concentration,

$$[L] = \frac{\bar{V}_L - V_0}{V_i}[L_0] \tag{113}$$

It is immediately apparent that the constituent distribution volume $\bar{V}_L$ is equal to the ratio $Q_T/[L_0]$ since the total ligand is the sum of that distributed inside and outside the gel:

$$\begin{aligned} Q_T &= Q_0 + Q_i \\ &= V_0[L_0] + V_i[L] \end{aligned} \tag{114}$$

and consequently

$$\frac{Q_T}{[L_0]} = \frac{V_0[L_0] + V_i[L]}{[L_0]} = V_0 + V_i\frac{[L]}{[L_0]} \tag{115}$$

which is identical to the right side of Eq. (112). Thus the free ligand concentration can be obtained from Eq. (113) by determining the ratio $Q_T/[L_0]$ for the totally excluded (protein) and nonexcluded (ligand) species and for the mixtures of protein and ligand. The binding ratios can then be calculated by Eq. (106).

When ligand also binds reversibly to the gel, the apparent quantity $(V_0 + V_i)$, determined with ligand alone, will not be independent of the amount of ligand added to the system. Corrections must be made by first determining the ligand-gel binding isotherm in a series of control experiments with different amounts of ligand. Q_b, the mass of ligand bound to the gel, must also be added to Eq. (114). Consequently, when the experiment is carried out with ligand alone, $[L_0] = [L]$ and

$$\frac{Q_T}{[L]} = V_0 + V_i + \frac{Q_b}{[L]} \tag{116}$$

In general Q_b will be a monotonically increasing function of free ligand concentration [L], and values of $Q_T/[L]$ will approach $(V_0 + V_i)$ in the limit of $[L] = 0$. The isotherm Q_b is determined from

$$Q_b([L]) = [L]\left\{\frac{Q_T}{[L]} - (V_0 + V_i)\right\} \tag{117}$$

In many cases it may be desirable to determine $(V_0 + V_i)$ by independent experiments with another small, noninteracting species. A term $Q_L/[L_0]$ must then be added to Eq. (110). The determined ratio $Q_T/[L_0]$ is an apparent constituent volume, V_i, which is equal to $V_L + Q_L/[L_0]$. The relationship corresponding to Eq. (113) by which [L] is calculated then becomes

$$[L] + \frac{Q_b([L])}{V_i} = \frac{[L_0]}{V_i}(\bar{V}_L' - V_0) \tag{118}$$

Once the isotherm $Q_L([L])$ has been determined empirically, the left-hand side of Eq. (118) becomes a known function from which the free ligand concentration [L] may be calculated for each measured value of $\bar{V}_L'$.

C. The Brumbaugh–Ackers Method

A much more precise equilibrium dialysis technique utilizes direct optical column scanning and measures ligand binding at many points within a single column (Brumbaugh and Ackers, 1971). Although this approach requires more sophisticated and expensive instrumentation, the advantages of speed and precision certainly warrant the additional investment.

1. Principle

This method is based on the direct optical scanning of a small chromatographic column that has been saturated with solution containing the desired mixture of ligand and macromolecule. Usually conditions are arranged so that the macromolecule, P, and all complexes, PL_i, are totally excluded from the gel. The system is again similar to an equilibrium dialysis experiment in which the mobile phase (void spaces) of the column corresponds to the "inner" compartment, containing macromolecular species, complexes, and unbound ligand. The solvent space of the stationary phase (internal volume) acts as the "outer" compartment containing only free ligand. When such an equilibrated column is optically scanned, the measurement of absorbance yields, at each point within the column, a determination of the binding ratio. Since the absorbance measurements are recorded digitally at several hundred points along the column, the experiment is equivalent to a large collection of replicate equilibrium dialysis experiments. Appropriate statistical analysis can be used to evaluate the accuracy of the results. In addition to providing high precision, the method is simple and convenient and requires only small quantities of material.

2. The Equilibrium Saturation Experiment

In the equilibrium saturation experiment (previously described in Section III,B), the column is first equilibrated with buffer and optically scanned at the desired wavelength to establish a baseline. Subsequently the column is completely saturated with a solution chosen with respect to the solute species of interest. After equilibration has taken place throughout all penetrable volume elements of the column, a second scan is performed in order to determine the apparent solute concentration as a function of distance along the column axis. There are seven regions of interest in the scan, shown schematically in Fig. 31: (1) the air space above the column, (2) a spike at the interface between air and the top of the column, (3) a region of air space within the column, (4) a meniscus at the interface between air and solution above the column bed, (5) liquid above the gel bed, (6) a spike corresponding to the porous disk that separates the column bed from inflowing solution above it, and (7) the column bed itself. A comparison of absorbance values (after baseline subtraction) within region 5 (constant throughout the region)

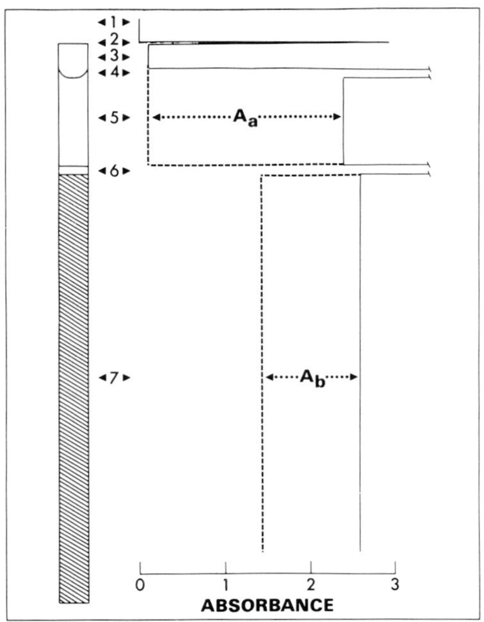

Fig. 31 Schematic diagram of a column and an idealized scan. Regions of interest are (1) air above the column, (2) spike corresponding to the air–column interface, (3) air inside the column, (4) meniscus at the air–liquid interface, (5) liquid above gel bed, (6) porous disk at top of gel bed, and (7) gel bed. Dashed lines represent baseline absorbance of the buffer-saturated column. Solid line represents absorbance of column saturated with solution. A_a is absorbance of solution above the column, and A_b is corresponding absorbance within the column bed. Taken from Brumbaugh and Ackers (1971).

and any point within region 7 yields a measure of the *partition cross section* ξ pertaining to that point in region 7:

$$A_b/A_a = \xi$$

where A_b is absorbance at a point within the gel bed, and A_a is absorbance in free solution above the column bed.

The scanning chromatograph provides data in the form of several hundred absorbance values corresponding to positions along the column axis. Calculations from a scan are best performed by computer, and the data are collected in digital form (e.g., on punched paper tape) for this purpose. The absorbance measurements corresponding to region 5 are first averaged, yielding a single value of A_a for comparison with each value of A_b within region 7. The values of A_b are determined by subtraction of corresponding baseline absorbances at each point from the absorbances at each point measured in the saturation scan. Subsequently the ratio A_b/A_a is calculated at each point to obtain corresponding values of ξ.

3. *Procedure for Binding Studies*

In order to calculate binding ratios from measured values of ξ, it is necessary to know α and β at each point within the column. The saturation experiment described above is first carried out using a totally excluded solute species (e.g., tobacco mosaic virus) for which $\sigma = 0$. The values of ξ derived from these scans thus represent the void fraction α as a function of distance along the column. Corresponding values of $\alpha + \beta$ are subsequently obtained by carrying out the experiment with a totally nonexcluded solute (e.g., glycylglycine, scanned at 200 nm) for which $\sigma = 1$. The partition cross sections obtained in this way are equal to $\alpha + \beta$ at each point within the column, and β is determined at each point by subtraction. Once the values of α and β are known, corresponding saturation experiments are carried out with mixtures of ligand and protein. Each mixture is prepared from stock solutions by weight so that the total concentrations, $[P_0]$ and $[L_0]$, are accurately known. From scans of the saturated column, the free ligand concentration and the binding ratio are calculated at each point within the column bed as described below.

4. *Calculation of Free Ligand Concentration and Binding Ratio*

Since the total concentrations of both ligand $[L_0]$ and macromolecule $[P_0]$ with which the column has been saturated are known, it is evident from Eq. (106) that the binding ratio can be determined if the free ligand concentration [L] can be found. The calculation of [L] follows from the measurement of A_a and A_b at each point within the column, according to the relationships described below. The absorbance per unit path length in the solution above the column bed is written in terms of species concentrations and extinction coefficients (ϵ):

$$A_a = \epsilon_P[P] + \epsilon_{PL}[PL] + \epsilon_{PL_2}[PL_2] + \cdots + \epsilon_{PL_n}[PL_n] + \epsilon_L[L] \quad (119)$$

The corresponding absorbance at any point within the column bed is

$$A_b = \epsilon_P\xi_P[P] + \epsilon_{PL}\xi_{PL}[PL] + \epsilon_{PL_2}\xi_{PL_2}[PL_2] + \cdots + \epsilon_{PL_n}\xi_{PL_n}[PL_n] + \epsilon_L\xi_L[L] \quad (120)$$

Since all macromolecular species are excluded,

$$\xi_P = \xi_{PL} = \xi_{PL_2} = \cdots = \xi_{PL_n} = \alpha \quad (121)$$

and since $\xi_L = \alpha + \beta$,

$$A_b = \alpha A_a - \epsilon_L \beta [L] \quad (122)$$

Hence the free ligand concentration is given by

$$[L] = \frac{A_b - \alpha A_a}{\epsilon_L \beta} \quad (123)$$

Since the extinction coefficient of free ligand, ϵ_L, can be obtained independently, the determination of α, β, A_a, and A_b in this experiment enables one to calculate the free ligand concentration regardless of the complexity of the system. The calculation is performed at many points within the column, yielding constant values of [L] and consequently of r.

5. *Method of Correction for Binding to the Gel*

If the ligand also binds to the gel it is necessary to determine the concentration dependence of this binding in order to calculate the free ligand concentration. The ligand–gel binding isotherm is determined by measuring the apparent partition cross section $\xi_L' = A_b/A_a$ of the column saturated with different ligand concentrations in the absence of protein. In these measurements $A_a = \epsilon_L[L]$ and $A_b = \epsilon_L \xi_L [L] + \epsilon_L Q_L$, where Q_L is the apparent concentration of ligand bound to the gel. At each point within the column, this quantity can be determined as

$$Q_L = [L](\xi_L' - \xi_L) \quad (124)$$

When ξ_L' is measured at a series of different saturating concentrations [L] of ligand, the binding isotherm $Q_L([L])$ is determined. Note that a comparison of ξ_L' with ξ_L provides a test for binding to the gel. If $\xi_L' = \xi_L$, then $Q_L = 0$ and the absence of binding is established. When binding is present, ξ_L' will be greater than ξ_L and will approach ξ_L in the limit of zero ligand concentration. Accuracy of the binding isotherm depends upon a proper determination of ξ_L. For Sephadex gels the use of glycylglycine is recommended since it has been shown to occupy the same distribution volume as water within the gel (i.e., $\sigma = 1$). If the ligand binds to the gel, Eq. (116) must include an $\epsilon_L Q_L$ term with the result

$$[L] + \frac{Q_L([L])}{\beta} = \frac{A_b - \alpha A_a}{\epsilon_L \beta} \quad (125)$$

Once $Q_L([L])$ has been determined, the left-hand side of Eq. (125) becomes a known function from which the free ligand concentration [L] may be calculated for each measured value of A_b/A_a. Equation (125) defines a completely general method for determination of [L]. The form of the function Q_L will depend upon the particular system studied and

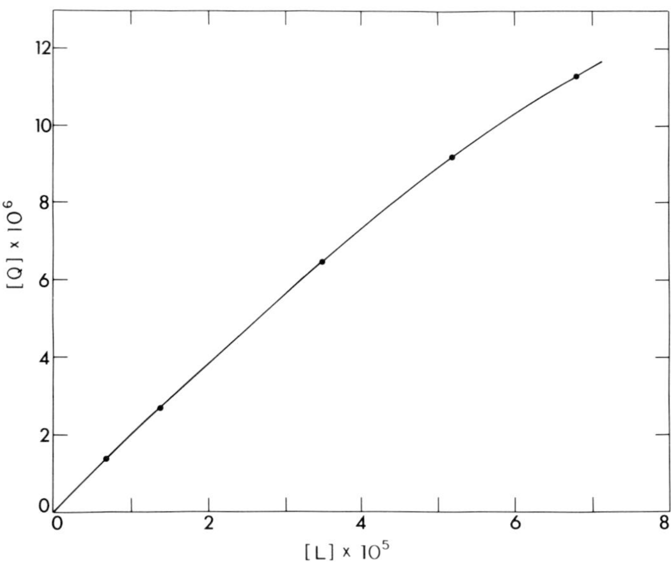

Fig. 32 Isotherm for the binding of methyl orange to Sephadex G-75. The ordinate represents the concentration of methyl orange bound to the gel and the abscissa is the corresponding free ligand concentration. Data are averages over a 30-point segment of the gel bed. Taken from Brumbaugh and Ackers (1971).

must be determined experimentally. An example of a binding isotherm is shown in Fig. 32, representing the binding of methyl orange to Sephadex G-75. Data of this kind can usually be fitted to a polynomial function of low degree which provides a convenient way of expressing Q_L for use in Eq. (125). A second-degree polynomial is frequently adequate for this purpose.

$$Q_L = a_0 + a_1[L] + a_2[L]^2 \tag{126}$$

From data of the type shown in Fig. 32, the constants a_0, a_1, and a_2 are determined by polynomial regression.

Representative results obtained by use of this method are shown in Table XII for the binding of methyl orange to bovine serum albumin. The standard deviations pertain to final calculated values of [L] after correction for the binding isotherm and provide a measure of the experimental precision obtainable. In cases in which such correction is not necessary, the precision can be as much as an order of magnitude higher. A careful selection of gels is therefore warranted. For studies of binding with aromatic ligand species, it is recommended that polyacrylamide gels be used instead of Sephadex gels.

TABLE XII

Determination of Bovine Serum Albumin–Methyl Orange Binding by the Equilibrium Saturation Method[a]

Expt.	$[P_0] \times 10^5$	$[L_0] \times 10^5$	$[L] \times 10^5$	S.D. $\times 10^5$	No. of points	r
1	1.4412	7.6552	4.128	0.0176	138	2.448
2	0.7146	7.6839	5.483	0.0314	245	3.080
3	0.1320	7.7077	7.151	0.0086	156	4.216
4	1.5503	5.1448	2.658	0.0358	249	1.604
5	0.7670	5.0953	3.531	0.0237	240	2.039
6	0.3779	5.1702	4.266	0.0188	246	2.392
7	0.0607	5.1693	4.973	0.0122	237	3.224
8	1.4927	1.0398	0.510	0.0089	241	0.355
9	0.7409	1.0517	0.713	0.0066	228	0.457
10	0.1476	1.0456	0.970	0.0080	204	0.510

[a] Taken from Brumbaugh and Ackers (1971).

D. The Hummel–Dreyer Method

A simple method in terms of equipment required is the Hummel–Dreyer technique carried out by elution chromatography (Hummel and Dreyer, 1962). Basically the only equipment needed is a chromatographic column, a small amount of gel, and a spectrophotometric monitoring device. For many applications to enzyme systems the last of these is either used in conjunction with a sensitive biologic assay procedure or replaced altogether by a biologic assay. Activity assays permit the study of enzyme systems in impure form and also enable the concentration range to be extended down to the nanomole range. Because of these advantages and the ready accessibility of necessary equipment in any biochemical laboratory, the Hummel–Dreyer elution method has been used extensively in protein studies.

1. Principle

A gel column is equilibrated with a solution containing the ligand at a desired concentration [L]. A small sample of protein solution in which the total ligand concentration equals that of the column-saturating solution is then added to the column. If the protein binds ligand, the solvent of this sample will be depleted with respect to ligand. When the sample is chromatographed on the column, the protein will be separated from the ligand-depleted solvent, moving ahead with its bound ligand. The resulting elution profile will exhibit a peak in ligand concentration above the ligand-saturation baseline. This peak represents the excess ligand bound. A corresponding trough will follow, representing the depletion that resulted when the bound ligand was carried ahead with the protein. These effects are illustrated in Fig. 33. In principle, the area of the trough and peak are equal. However, measurement of the trough area is usually simpler and more reliable. From the area of the trough and the known total amount of protein applied, the binding ratio is determined, corresponding to the ligand saturation level used.

Since maximum separation between ligand and macromolecule is desired, a gel should be chosen from which the macromolecular species is totally excluded. For most protein systems Sephadex G-25 or G-50 will be satisfactory, but if ligand–gel binding occurs, an appropriate polyacrylamide gel should be used instead (e.g., Bio-Gel P-50). The column should be jacketed so that the temperature may be regulated by an external thermostated bath. It is desirable that this bath also have space for thermal equilibration of samples prior to their addition to the column. A reservoir containing buffer or the desired saturation solution of ligand allows flow into the column from the top.

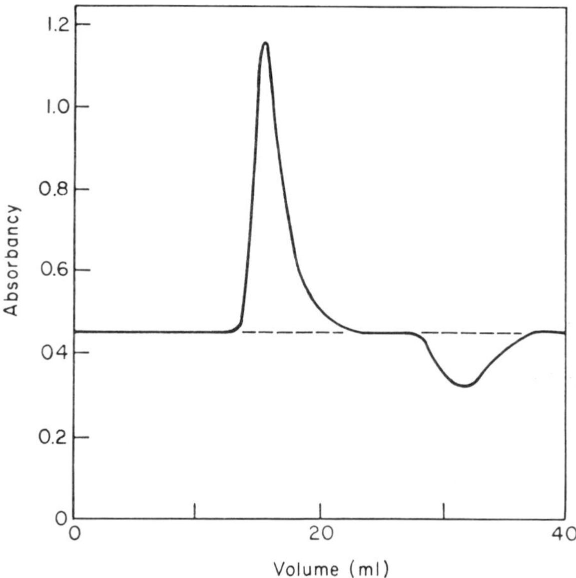

Fig. 33 Binding of 2′-cytidylic acid and ribonuclease as detected by the Hummel–Dreyer procedure on a Sephadex G-25 column equilibrated with 0.1 *M* acetate buffer, pH 5.3. Absorbency measurements at 285 nm (ordinate) indicate a positive peak (*left*) and a negative trough (*right*) with respect to the (horizontal) baseline absorbancy. Taken from Hummel and Dreyer (1962).

After the column has been equilibrated at the desired temperature, it is saturated with 1–2 volumes of solution containing ligand of the desired free ligand concentration [L]. The column has reached saturation when the concentration of ligand emerging from the bottom of the column is equal to that introduced at the top. If there is any doubt, these concentrations should be checked spectrophotometrically and compared. It is usually sufficient to record effluent ligand concentrations until a constant value is reached.

A small sample containing a mixture of ligand and protein is then prepared from a stock solution so that (1) the concentration of protein in the sample is known, and (2) the total concentration of ligand in the sample is equilibrated in a thermostated bath at the same temperature as the column to which it will be applied.

After the sample has been equilibrated, it is carefully applied to the top of the column and then eluted by applying the original saturating ligand solution. The elution profile is determined either by continuous recording of absorbance or by graphical plotting of the concentrations of collected fractions vs. time (or volume). It is essential that the elution profile exhibit a distinct peak and trough, and that a *plateau of constant ligand concentration exist between them* equal to the saturating level applied. If this plateau is not present, it is likely that the binding equilibria are slow and the experiment must be repeated at a lower flow rate. If no plateau can be established, the experiment cannot provide a valid determination of the binding ratio and another method must be tried.

2. *Calculations of Amount Bound and Binding Ratio*

The immediate object of the experiment described above is to determine the amount of ligand bound by the macromolecular sample applied to the column. Although in principle one can use either the peak or trough area to determine this binding ratio, the trough is usually the more reliable and convenient. This is especially true if concentrations of ligand are monitored spectrophotometrically, since there is no interference due to absorption or light-scattering by the protein. Determination of the binding ratio is carried out as follows.

***a.* Automatically Recorded Elution Profiles.** From the strip-chart recording of absorbance vs. time, the area of the trough is first determined (in square inches) by graphical integration or planimetry. The number of μmoles of ligand bound by the protein sample is then calculated from the formula

$$\mu\text{moles bound} = \frac{\text{area of trough}}{ab(\epsilon_L \times 10^{-3})} \tag{127}$$

In Eq. (127), a is the recorder pen displacement corresponding to unit absorbance, and b is the chart movement (in inches) corresponding to a 1-ml increment of column effluent.

***b.* Concentrations Measured from Collected Fractions.** From data obtained as a series of absorbances corresponding to collected fractions, the amount of ligand is calculated as

$$\mu\text{moles bound} = \frac{\Sigma_i(\Delta A_i)(\text{ml}_i)}{\epsilon_L \times 10^{-3}} \tag{128}$$

In Eq. (128), A_i is the difference between the absorbance of fraction i and the baseline absorbance (i.e., that of saturating ligand solution). The volume of fraction i is ml_i, and ϵ_L is the molar extinction coefficient of the ligand. The summation is carried out for all fractions within the trough region of the elution profile. For other types of ligand concentration determinations (e.g., biologic activity assay or radioactive assay), a procedure similar to that of Eq. (128) is used.

Once the number of moles of bound ligand has been determined, the binding ratio r is calculated as the number of μmoles of protein in the sample applied.

3. *Variation of Conditions*

Because the binding ratio is *only* a function of [L], it is not useful to carry out a series of experiments in which only the amount of protein in the sample is varied without changing the saturating level of ligand. Therefore it would appear necessary to carry out the tedious process of saturating the column with solution at each desired free ligand concentration [L]. In order to minimize such labor it is possible to use only two different concentrations [L] and still determine K and n for a system with n identical binding sites. Substitution of the defining relation for r (Eq. (106)) into Eq. (109) yields, after rearrangement,

$$\frac{1}{[\text{L}_0] - [\text{L}]} = \frac{1}{[\text{P}_0]}\left\{\frac{1}{n} + \frac{1}{nK([\text{L}_0] - [\text{L}])}\right\} \tag{129}$$

Therefore a plot of the reciprocal of ligand bound vs. the reciprocal of protein concentration should yield a straight line having a slope of

$$\frac{1}{n}\left\{1 + \frac{1}{K([\text{L}_0] - [\text{L}])}\right\}$$

If two such plots are constructed from data obtained at different saturating levels [L], the two unknowns, n and K, can be calculated from the two slopes. In principle, then, only one protein concentration $[P_0]$ plus two saturating levels of [L] need be used. However, it is desirable to obtain data from several $[P_0]$ values in order to estimate the precision of the analysis.

E. The Large-Zone Elution Method

Elution profiles from large-zone experiments carried out with mixtures of ligand and protein have been used to determine binding ratios (Cooper and Wood, 1968; Wood and Cooper, 1970; Nichol *et al.*, 1971). The method is similar to that described in Section VII for the study of interactions between dissimilar protein subunits. The experiment is illustrated in Fig. 34 (Nichol *et al.*, 1971) for the binding of methyl orange to bovine serum albumin on a column of Sephadex G-50 (the gel again being chosen to exclude protein and complexes).

After the column has been equilibrated with buffer (not containing ligand), a large sample of the protein–ligand mixture is applied in sufficient volume to establish a series of plateaus (Fig. 34). First to emerge (at the void volume) is a boundary of pure protein from which ligand has been "stripped." The plateau region extending to the right then represents the constituent concentration, $[P_0]$, of protein in the applied sample. This is followed by a region that contains, in addition to free protein, the protein–ligand complexes (between 50–75 ml in the elution diagram), followed by a region of pure ligand (80–115 ml). The plateau

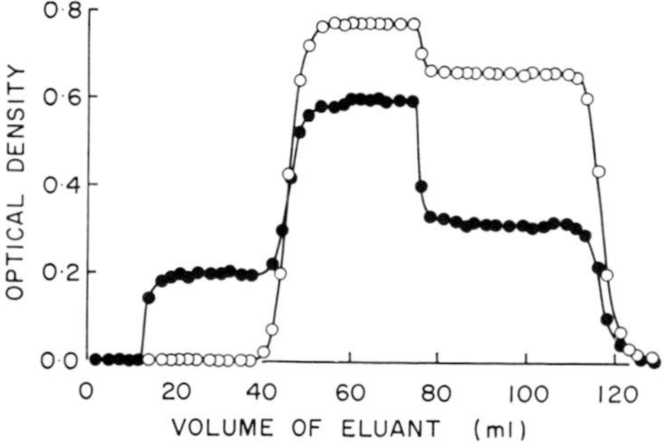

Fig. 34 An elution profile of a frontal analysis experiment performed at 4° with a mixture containing total concentrations of 1.45×10^{-2} gm/liter of methyl orange and 0.26 gm/liter of bovine serum albumin in phosphate buffer, pH 5.68. The volume of mixture loaded onto a 1 × 38-cm column of Sephadex G-50 was 60 ml. Readings at 495 nm (○) and 280 nm (●). Taken from Nichol *et al.* (1971).

concentration within this part of the profile is equal to the free ligand concentration [L] in the original mixture. Once the free ligand concentration has been determined, the binding ratio may be calculated by Eq. (106) in the usual way, since values of $[P_0]$ and $[L_0]$ in the initial mixture are already known.

ACKNOWLEDGMENTS

The author is grateful to the United States Public Health Service, National Institute of General Medical Sciences, for research grants in support of the work described in this article. The author also thanks the John Simon Guggenheim Memorial Foundation for a Fellowship, and the Department of Biochemistry, University of Birmingham, Birmingham, England, for hospitality and use of facilities during the writing of this article.

REFERENCES

Ackers, G. K. (1964). *Biochemistry* **3**, 723.
Ackers, G. K. (1967a). *J. Biol. Chem.* **242**, 3026.
Ackers, G. K. (1967b). *J. Biol. Chem.* **242**, 3237.
Ackers, G. K. (1968a). *Proc. Int. Congr. Biochem., 7th, 1967* Vol. **5**, p. 1008.
Ackers, G. K. (1968b). *J. Biol. Chem.* **243**, 2056.
Ackers, G. K. (1970). *Advan. Protein Chem.* **24**, 343.
Ackers, G. K., and Thompson, T. E. (1965). *Proc. Nat. Acad. Sci. U. S.* **53**, 342.
Altgelt, K. H. (1967). *Advan. Chromatogr.* **4**, 3.
Altgelt, K. H., and Segal, L. (1971). "Gel Permeation Chromatography." Dekker, New York.
Andrews, P. (1962). *Nature (London)* **196**, 36.
Andrews, P. (1964). *Biochem. J.* **91**, 222.
Andrews, P. (1965). *Biochem. J.* **96**, 595.
Araki, C. (1956). *Bull. Chem. Soc. Jap.* **29**, 543.
Barral, E. M., and Cain, J. H. (1968). *J. Polym. Sci., Part C* **21**, 253.
Bengtsson, S., and Philipson, L. (1964). *Biochim. Biophys. Acta* **79**, 399.
Brumbaugh, E. E., and Ackers, G. K. (1968). *J. Biol. Chem.* **243**, 6315.
Brumbaugh, E. E., and Ackers, G. K. (1971). *Anal. Biochem.* **41**, 543.
Cann, J. R. (1970). "Interacting Macromolecules." Academic Press, New York.
Cebra, J. J., and Small, P. A. (1967). *Biochemistry* **6**, 503.
Chiancone, E., Gilbert, L. M., Gilbert, G. A., and Kellett, G. L. (1968). *J. Biol. Chem.* **243**, 1212.
Chun, P. W., Kim, S. J., Stanley, C. A., and Ackers, G. K. (1969a). *Biochemistry* **8**, 1625.
Chun, P. W., Thornby, J. I., and Saw, J. G. (1969b). *Biophys. J.* **9**, 163.
Cooper, P. F., and Wood, G. C. (1968). *J. Pharm. Pharmacol.* **20**, 1505.
Cox, D. J. (1969). *Arch. Biochem. Biophys.* **129**, 106.
Davison, P. F. (1968). *Science* **161**, 906.

Determann, H. (1967a). "Gel Chromatography." Springer-Verlag, Berlin and New York.
Determann, H. (1967b). *Protides Biol. Fluids, Proc. Colloq.*, p. 563.
Determann, H., and Michel, W. (1965). *Z. Anal. Chem.* **212**, 211.
Determann, H., and Walter, I. (1968). *Nature (London)* **219**, 604.
Eaker, D., and Porath, J. (1967). *Separ. Sci.* **2**, 507.
Edmond, E., Farquhar, S., Dunstone, J. R., and Ogston, A. G. (1968). *Biochem. J.* **108**, 755.
Fasella, P., Hammes, G. G., and Schimmel, P. R. (1965). *Biochim. Biophys. Acta* **103**, 708.
Fawcett, J. S., and Morris, C. J. O. R. (1966). *Separ. Sci.* **1**, 9.
Fish, W. W., Mann, K. G., and Tanford, C. (1969). *J. Biol. Chem.* **244**, 4989.
Fish, W. W., Reynolds, J. A., and Tanford, C. (1970). *J. Biol. Chem.* **245**, 5166.
Flory, P. J. (1953). "Principles of Polymer Chemistry." Cornell Univ. Press, Ithaca, New York.
Giddings, J. C., Kucera, E., Russell, C. P., and Myers, M. N. (1968). *J. Phys. Chem.* **72**, 4397.
Gilbert, G. A. (1955). *Discuss. Faraday Soc.* **20**, 68.
Gilbert, G. A. (1959). *Proc. Roy. Soc., Ser. A* **250**, 377.
Gilbert, G. A. (1963). *Proc. Roy. Soc., Ser. A* **276**, 354.
Gilbert, G. A. (1966a). *Nature (London)* **210**, 299.
Gilbert, G. A. (1966b). *Nature (London)* **212**, 296.
Gilbert, G. A. (1967). *Anal. Chim. Acta* **38**, 275.
Gilbert, G. A., and Jenkins, R. C. L. (1959). *Proc. Roy. Soc., Ser. A* **253**, 420.
Gilbert, G. A., and Kellett, G. L. (1971). *J. Biol. Chem.* **246**, 6079.
Gilbert, G. A., Gilbert, L. M., Owens, C. E., and Shawky, N. A. (1972). *Nature (London)* **236**, 110.
Gilbert, L. M., and Gilbert, G. A. (1968). *Fed. Eur. Biochem. Soc. Symp.* **6**, 73.
Gosting, L. J. (1956). *Advan. Protein Chem.* **11**, 429.
Granath, K. A., and Flodin, P. (1961). *Makromol. Chem.* **48**, 160.
Habeeb, A. F. S. A. (1966). *Biochim. Biophys. Acta* **121**, 21.
Haller, W. (1965). *J. Chem. Phys.* **42**, 686.
Haller, W. (1968). *J. Chromatogr.* **32**, 676.
Halvorson, H. R., and Ackers, G. K. (1971). *J. Polym. Sci.* **9**, 245.
Henn, S. W., and Ackers, G. K. (1969a). *J. Biol. Chem.* **244**, 465.
Henn, S. W., and Ackers, G. K. (1969b). *Biochemistry* **8**, 3829.
Hersh, R. T., and Schachman, H. K. (1955). *J. Amer. Chem. Soc.* **77**, 5228.
Hjertén, S. (1962). *Biochim. Biophys. Acta* **62**, 445.
Hjertén, S. (1964). *Biochim. Biophys. Acta* **79**, 393.
Hjertén, S. (1970). *J. Chromatogr.* **50**, 189.
Hjertén, S., and Mosbach, R. (1962). *Anal. Biochem.* **3**, 109.
Houghton, G. (1963). *J. Phys. Chem.* **67**, 84.
Hummel, J. P., and Dreyer, W. J. (1962). *Biochim. Biophys. Acta* **63**, 530.
Janson, J.-C. (1967). *J. Chromatogr.* **28**, 12.
Jenkins, R. C. L. (1965). *J. Phys. Chem.* **69**, 3785.
Johansson, B. G., and Rymo, L. (1962). *Acta Chem. Scand.* **16**, 2067.
Klotz, I. M., and Hunston, D. L. (1971). *Biochemistry* **10**, 3065.
Lathe, G. H., and Ruthven, C. R. J. (1956). *Biochem. J.* **62**, 665.
Laurent, T. C., and Killander, J. (1964). *J. Chromatogr.* **14**, 317.
Leach, A. A., and O'Shea, P. C. (1965). *J. Chromatogr.* **17**, 245.

Longsworth, L. G. (1959). *In* "Electrophoresis" (M. Bier, ed.), Vol. 1, p. 91. Academic Press, New York.
Massey, V., Curti, B., and Ganther, H. (1966). *J. Biol. Chem.* **241,** 2347.
Moore, J. C. (1964). *J. Polym. Sci.* **2,** 835.
Morris, C. J. O. R., and Morris, P. (1964). "Separation Methods in Biochemistry." Wiley (Interscience), New York.
Nichol, L. W., and Winzor, D. J. (1964). *J. Phys. Chem.* **68,** 2455.
Nichol, L. W., Bethune, J. L., Kegeles, G., and Hess, E. L. (1964). *In* "The Proteins" (H. Neurath, ed.), 2nd ed., Vol. 2, p. 305. Academic Press, New York.
Nichol, L. W., Ogston, A. G., and Winzor, D. J. (1967). *J. Phys. Chem.* **71,** 726.
Nichol, L. W., Jackson, W. J. H., and Smith, G. D. (1971). *Arch. Biochem. Biophys.* **144,** 438.
Nyström, E., and Sjövall, J. (1965). *J. Chromatogr.* **17,** 574.
Ogston, A. G. (1958). *Trans. Faraday Soc.* **54,** 1754.
Polson, A. (1961). *Biochim. Biophys. Acta* **50,** 565.
Porath, J. (1963). *Pure Appl. Chem.* **6,** 233.
Porath, J., and Flodin, P. (1959). *Nature (London)* **183,** 1657.
Rogers, K. S., Hellerman, L., and Thompson, T. E. (1965). *J. Biol. Chem.* **240,** 198.
Russell, B., Mead, T. H., and Polson, A. (1964). *Biochim. Biophys. Acta* **86,** 169.
Siegel, L. M., and Monty, K. J. (1965). *Biochem. Biophys. Res. Commun.* **19,** 494.
Siegel, L. M., and Monty, K. J. (1966). *Biochim. Biophys. Acta* **112,** 346.
Small, P. A., Kehn, J. E., and Lamn, M. E. (1963). *Science* **142,** 393.
Squire, P. G. (1964). *Arch. Biochem. Biophys.* **107,** 471.
Steere, R. L., and Ackers, G. K. (1962a). *Nature (London)* **194,** 114.
Steere, R. L., and Ackers, G. K. (1962b). *Nature (London)* **196,** 475.
Stone, M. J., and Metzger, H. (1968). *J. Biol. Chem.* **243,** 5049.
Sullivan, B., and Riggs, A. (1967). *Biochim. Biophys. Acta* **140,** 274.
Tanford, C., Kawahara, K., and Lapanje, S. (1967). *J. Amer. Chem. Soc.* **89,** 729.
Tiselius, A. (1930). *Nova Acta Regiae Soc. Sci. Upsal.* **4,** 1.
Ward, D. N., and Arnott, M. S. (1965). *Anal. Biochem.* **12,** 296.
Warshaw, H. S., and Ackers, G. K. (1971). *Anal. Biochem.* **42,** 405.
Weber, G. (1965). *In* "Molecular Biophysics" (B. Pullman and M. Weissbluth, eds.), p. 369. Academic Press, New York.
Wheaten, R. M., and Baumann, W. C. (1953). *Ann. N. Y. Acad. Sci.* **57,** 159.
Whitaker, J. R. (1963). *Anal. Chem.* **35,** 1950.
Winzor, D. J., and Nichol, L. W. (1965). *Biochim. Biophys. Acta* **104,** 1.
Winzor, D. J., and Scheraga, H. A. (1963). *Biochemistry* **2,** 1263.
Winzor, D. J., and Scheraga, H. A. (1964). *J. Phys. Chem.* **68,** 338.
Winzor, D. J., Loke, J. P., and Nichol, L. W. (1967). *J. Phys. Chem.* **71,** 4492.
Wood, G. C., and Cooper, P. F. (1970). *Chromatogr. Rev.* **12,** 88.
Zimmerman, J. K., and Ackers, G. K. (1971a). *J. Biol. Chem.* **246,** 1078.
Zimmerman, J. K., and Ackers, G. K. (1971b). *J. Biol. Chem.* **246,** 7289.
Zimmerman, J. K., Cox, D. J., and Ackers, G. K. (1971). *J. Biol. Chem.* **246,** 4242.

2

Biospecific Affinity Chromatography and Related Methods

JERKER PORATH AND TORE KRISTIANSEN

I. INTRODUCTION

The extremely rapid expansion of protein chemistry during the last fifty years has been due to the development of very effective methods of protein analysis and separation. Among the separation procedures, a group of methods based on the molecular affinities found in biologic systems has become increasingly important. A historical review shows, however, that the idea of separating substances on the basis of association–dissociation equilibria of naturally occurring macromolecular complexes is by no means new: it is almost as old as the knowledge of the existence of these complexes. Some significant events in the development of separation methods based on the biospecific formation of complexes are listed in Table I. One may wonder why the inherently most effective method—namely, biospecific adsorption chromatography

TABLE I

Some Historical Events Pertaining to Biospecific Adsorption and Bioaffinity Chromatography

Event	Reference
Specific desorption of trypsin from charcoal	Hedin (1907a,b)
Specific adsorption of an enzyme (amylase) to its insoluble substrate (starch)	Starkenstein (1910)
Purification of antibodies by adsorption on cellulose to which antigens had been attached by diazocoupling	Campbell *et al.* (1951)
Production of a biospecific adsorbent by covalent attachment of substrate analogs to an insoluble carrier (cellulose)	Lerman (1953)
Introduction of modified ion exchangers as specific adsorbents	Grubhofer and Schleith (1954)
Introduction of Sephadex (cross-linked dextran) as a molecular sieve, later to be used for immobilized enzymes and bioadsorbents	Porath and Flodin (1959)
Preparation of an immobilized enzyme (trypsin) to be used in columns	Bar-Eli and Katchalski (1960)
Introduction of coenzymes (flavins) as ligands	Arsenis and McCormick (1964)
Introduction of antibodies on an insoluble support for specific detection of radioactive antigens	Gurvich and Drizlikh (1964)
Introduction of cyanogen bromide for coupling of proteins and other ligands containing amino groups	Axén *et al.* (1967)
Introduction of agarose as a matrix for immobilized proteins and peptides	Porath *et al.* (1967)
Introduction of the concept of "affinity chromatography" and demonstration of the importance of an "arm" (spacer) between matrix and ligand	Cuatrecasas *et al.* (1968)

or affinity chromatography—has not fully proven itself before, since the major breakthrough was achieved only two or three years ago. The reason for this, in our opinion, has been a lack of interest in systematically improving an initially promising but imperfect technique. Research workers have been content with results produced by rather unsatisfactory methods. Instead of exerting the effort clearly required to develop a general method approaching the ideal, the individual investigator has devised methods that might be moderately useful to him but of dubious value to others. For many years progress has therefore been slow indeed.

The development of biospecific adsorption chemistry to some extent coincides with the growth of another field of biopolymer chemistry, that of immobilized enzymes. (Progress in the former field has, however, more rapidly reached the level of development that allows its general application in research and technology.) The subject of immobilized enzymes has been treated in several reviews by Katchalski and co-workers (Silman and Katchalski, 1966; Katchalski *et al.*, 1971; Goldman *et al.*, 1971) and a review containing a very extensive list of references has been published by Melrose (1971). We shall discuss immobilized enzymes only in the context of biospecific adsorption for separation purposes.

Biospecific adsorption in various forms may be used in many areas of biochemistry in order to isolate and study nucleic acids, membranes, particulate cellular elements, etc. We shall discuss these applications only briefly and instead focus our attention on protein chemistry, especially the purification of enzymes and the isolation of antigens and antibodies. Review articles on biospecific adsorption and related topics have recently been published by Kato (1969), Jerina and Cuatrecasas (1970), Nezlin (1970), Cuatrecasas (1970a,b,c, 1971a,b), Cuatrecasas and Anfinsen (1971a,b), Feinstein (1971a), Friedberg (1971), Pihar (1971), Melrose (1971), Reiner and Walch (1971), Chudzik and Koj (1972), Kocemba-Sliwowska (1972), Weetall (1972), Lang (1972), Orth and Brümmer (1972), and Porath and Sundberg (1970, 1972a).

II. BASIC PRINCIPLES

The adsorption capacity of gels and microporous materials such as glass and active charcoal is without exception conditioned by the presence of "adsorption active" groups of one kind or another, usually matrix-fixed ions or ionogenic groups. Active charcoal, for example, adsorbs

TABLE II

Principal Interactions in Different Forms of Liquid Chromatography

Main separation parameter	Nonspecific affinity chromatography	Specific affinity chromatography: Degree of complementarity — Low	Specific affinity chromatography: Degree of complementarity — High
Molecular size	Permeation chromatography	Chromatography based on ion trapping, chelate formation, and molecular adduct formation	Biospecific affinity chromatography
Molecular size (+ ionic charge)	Ionic exclusion chromatography		
Ionic charge (+ molecular size)	Ionic adsorption chromatography		
Nonionic interactions:	"Adsorption chromatography"		
Hydrogen bond formation			
Hydrophobic interaction	Hydrophobic affinity chromatography		
Charge transfer	Charge transfer chromatography		
Other non-Coulombic interactions			

unsaturated compounds, particularly aromatic compounds and hydrophobic substances such as hydrocarbons but also, although more weakly, hydrophilic compounds such as carbohydrates. Other compounds are adsorbed through ionic interactions. The same kinds of forces operate in biospecific adsorption. The difference is that the forces involved in the latter are coordinated because of steric requirements, resulting in a higher or lower degree of specificity. Commonly used ion exchangers occupy an intermediate position because adsorption is mainly conditioned by ionic forces, i.e., by the attraction between ions or ionic groups of opposite charge.

The matrix of the carrier polymer itself contributes somewhat to adsorption effects of various kinds by the formation of hydrogen bonds, hydrophobic interactions, charge transfer phenomena, van der Waals-London forces, etc. With biospecific adsorbents these undesirable interactions between matrices and solutes can never be avoided but it is imperative to repress them as much as possible in order to minimize nonspecific adsorption. Table II shows schematically the principal interactions involved in different forms of liquid chromatography.

In analytical chemistry, selective reagents are used which form complexes of characteristic solubility and color. For separation purposes it is obviously important to develop an adsorbent containing a complex with one such characteristic property and to join it to the matrix. A classical example is Skogseid's adsorbent for potassium ions (Skogseid, 1948, 1952). This dipicrylamine polymer has a very high specificity. Gel matrices, the supports most often used for adsorbents devised for biologic substances, usually possess no exclusive biospecificity themselves, although there are exceptions which will be treated later. Adsorbents for isolating well-defined substance classes or special substances may be made to order by fixing complex-forming ligands to the matrix. A particularly high degree of specificity may be attained if adsorption is based on cooperative forces (each of them not necessarily very strong) along and between contact surfaces with a high degree of steric fit. The course of adsorptions may generally be described as shown below:

$$\text{Ⓜ}\!-\!L + A \rightleftarrows \text{Ⓜ}\!-\!L\cdots A$$

where adsorbate A is selectively bound to the matrix M via the ligand L, and L constitutes a specific adsorption center. It is evident that A and L may exchange positions. Immobilized A is an adsorbent for the substance forming the ligand L. More complicated complexes are also possible. Adsorbate A may, for instance, in turn specifically bind a substance B:

$$\text{(M)}-\text{L}\cdots\text{A}+\text{B} \rightleftarrows \text{(M)}-\text{L}\cdots\text{A}\cdots\text{B} \rightleftarrows \text{(M)}-\text{L}+\text{A}\cdots\text{B}$$

$$\text{(M)}-\text{L}\cdots\text{A}\cdots\text{B} \rightleftarrows \text{(M)}-\text{L}\cdots\text{A}+\text{B} \rightleftarrows \text{(M)}-\text{L}+\text{A}+\text{B}$$

We shall return to this kind of adsorption-desorption phenomena in Section VI.

III. NOMENCLATURE

The question of nomenclature has almost always been the subject of disagreement among chromatography experts. The very word "chromatography" is or was a misnomer when it was applied to column separations based on adsorption of noncolored substances. The term has since been extended to include partition chromatography. Column separation according to the principle of molecular sieving (gel filtration, see Ch. 1) was then also included despite the fact that in this case a completely different, new principle was being applied. All forms of chromatography except the one mentioned last and "zone precipitation" in various forms (Porath, 1962; Swanljung, 1971) are based on affinity. "Affinity chromatography" therefore should be considered as the opposite of molecular sieving.

As already mentioned, affinity chromatography of biopolymers may be based on interactions of other kinds which are less specific than ionic adsorption. For example, "charge transfer chromatography" has been suggested but has not yet been systematically studied. It seems likely to us that aromatic and hydrophobic adsorption as well as "hydrogen bond adsorption" will be best exploited in conjunction with simultaneously operating ionic interactions. Fractionation of lipophilic proteins by mixed hydrophobic–ionic interactions has been described by Yon (1972). Such mixed interactions no doubt play an important role in bioaffinity chromatography, but if the stereospecific factor is lacking there will be a much lower degree of selectivity.

A possible term for chromatography based on sterically governed affinity would be *"stereospecific affinity chromatography."* The disadvantage of this term is the difficulty of establishing the influence of steric factors in each separate case. When there is no doubt, we propose that this term, abbreviated "SAC," be used. In biochemical separation methodology, the concept of affinity chromatography has a narrow meaning, namely, that of chromatography based on biospecific adsorption or stereospecific adsorption of biologic significance. It will be for the IUPAC Nomenclature Commission to decide how to solve the nomenclature problem. Because the concept of affinity chromatography is so well established, we shall use this term here, but to stress its special character we propose the specification "*biospecific affinity chromatography*" which may be shortened to bioaffinity chromatography or "BAC."

As a rule, the ligand will have to be covalently bound to the matrix by way of a third substance. We suggest that this substance be called a "*connector substance*" and that for convenience "*connector*" be used for the group involved in the attachment as well as for the group created in the attachment reaction. In order to nullify or depress steric hindrance by the matrix it will often be necessary to increase the distance of a ligand from the matrix by introducing a spacer as an arm between the ligand and the matrix. The expression "*spacer*" could also be used in a dual sense without risking a misunderstanding, i.e., both for the spacer substance and for the resulting interlinking group. The spacer substance can contain none, one, or two reactive connector groups. For example:

(Matrix)$-O-CH_2-CH(OH)\sim\sim CH(OH)-CH_2-O-$ Ligand
Spacer

$CH_2-CH(-O-)\sim\sim\sim\sim CH-CH_2(-O-)$
Connector group   Connector group

Spacer (substance): $H_2N-(CH_2)_n-COOH$

(Matrix)$(-O-)_2C=N-(CH_2)_6-CO-NH-$Ligand
Connector   Spacer   Connector

In the second example, the connecting substance serves as a spacer and as a double connector.

IV. MATRICES (CARRIERS)

A. Background

In Uppsala ten years ago when we considered starting intensified studies on biospecific adsorption methods along the lines drawn up by

Engelhardt (1924), Campbell *et al.* (1951), Isliker (1953), Lerman (1953), Gurvich *et al.* (1959), Nezlin (1960), Sehon (1963), Pressman and Grossberg (1968), and others, we encountered two weak points common to the techniques used up to then. The first was that all carrier substances contained charged groups or other adsorption centers and many carriers had undesirable physical properties or showed other defects. The second weak point concerned the method of fixing the ligands to the matrices. With some ligands, large quantities of unwanted substituents were formed, and with others, unstable products. Together these negative effects severely retarded the general acceptance of biospecific adsorption techniques, however attractive these methods appeared in principle. The molecular sieves that had been introduced in the fifties (Sephadex) seemed to us to offer obvious advantages compared with cellulose and the organic polymers that had been so extensively used to immobilize enzymes and other proteins. Even though we were well aware of the value of the contributions by Manecke (1962) and others, we also realized that the present means and methods of biospecific adsorption had limitations. The comparatively low permeability of cross-linked dextrans such as the Sephadex gels limited their use as matrices for biospecific adsorbents. The agarose gels developed by Hjertén (1961, 1971) for molecular sieve chromatography and electrophoresis seemed to be more suitable with respect to rigidity and permeability but certain weaknesses were nevertheless obvious in these materials. In the following sections we shall discuss the requirements that should be fulfilled by an ideal matrix and briefly review how different matrix materials meet these requirements. Finally we shall discuss at length agar and agarose gels and describe our improvement of agar gels, being well aware of the fact that our description is in part based on subjective judgment.

B. Chemical and Physical Properties

The ideal matrix fulfills the following requirements.

1. Insolubility

If the matrix is not completely insoluble, the solubility should of course be as low as possible, and it should not be difficult to remove dissolved products from the substances to be purified. If the separated products are to be used clinically, maybe even injected, the substances dissolved from the matrix should not be immunogenic. Immunogenicity is to be expected from leaking protein–matrix conjugates, which means that any leakage at all is virtually unacceptable. Even a stable linkage achieved by a proven coupling technique can be jeopardized by a leaky matrix.

2. *Permeability*

To allow complexes to be easily and quickly formed and broken, different regions in the gel should be accessible both to the substances that are to be chemically bound to the matrix (the ligands) and to the substances to be adsorbed to them. As a rule a high capacity presupposes extensive penetration. When the particle weight exceeds 10^8–10^9 daltons it would probably be best to localize the adsorption to the phase boundaries between the gel particle and free fluid phase.

3. *Rigidity and Mechanical Stability*

The ideal matrix is mechanically rigid and resistant to possible pressures (ceramic materials approach the ideal in this respect). The matrix should not be ductile or elastic.

4. *Formation of Particles of Suitable Size and Shape*

The matrix material must exist in a form allowing rapid diffusion of the substance to be adsorbed. In cases where chromatography is used, uniformity and optimal size are required in order to increase the number of theoretical plates for column beds. The size of the particles should not exceed 200 μm and not fall below 5 μm (exceptions may be made under unusual conditions). For a specified bed, especially for linear chromatography, the size difference should be as small as possible. By using spherical particles it is possible to move closer to the lower limit of approximately 5 μm.

5. *Hydrophilicity*

The hydrophilic nature of the matrix, usually due to hydroxyl groups, seems to be a prerequisite for high permeability as well as for low adsorption of proteins. It is often also a requirement that the matrix swell in water and form a gel that is rigid and permeable at the same time.

6. *Chemical Stability*

The ideal gel is chemically stable in all media and at all conceivable working temperatures. Matrices intended for biospecific adsorption should not decompose in the temperature range of 0°–70° nor in the pH range of 2–12. Heat resistance is important whenever sterilization by autoclaving is necessary. Resistance to more drastic conditions will occasionally be required during chemical modifications (see below).

7. *Chemical Modification*

It must be possible to modify the matrix by the introduction of reactive groups to which the ligands may be coupled. The activation must be carried out under controlled conditions that allow the molecular network of the matrix to remain intact or at least permeable. All hydroxylic matrices may, in principle, be activated in a controlled manner. Sometimes it is necessary to carry out the activation reactions in an organic phase. The matrix *must* be hydrophilic and should be organophilic, a requirement that is sometimes difficult to meet.

8. *Freedom from Nonspecific Adsorption Centers*

The matrix must not contain ionogenic groups before or after activation or other active groups that can cause nonspecific adsorption.

9. *Resistance to Microorganisms and Enzymes*

Some gel-forming biopolymers (e.g., starch) are not suitable as matrix substances because of their propensity for microbial degradation. The materials now commonly used as separation media, including agar and organic polymers, are stable except under extraordinary conditions. Sephadex may be attacked by dextranase-producing bacteria or by molds in certain buffer systems if no preservative is used.

C. Evaluation of Matrices

The mechanical properties of cross-linked proteins, polyamino acids, and polypeptides make them unsuitable for column experiments unless these substances are mixed with some supporting polymer. However, they can have certain advantages in batch procedures (see Section VI,B). Porous glass and other ceramic materials have several desirable properties (see Table III). However, in our opinion, the strong nonspecific adsorption is a serious disadvantage as is the sensitivity toward alkali. No type of glass available today can compete in this respect with the other materials we suggest.

A guide to matrix choice is compiled in Table III. It should be under-

TABLE III

Guide to Matrix Choice

Matrix	Flow properties	Leakage at pH 7, 25°	Permeability for high molecular weight solutes	Nonspecific adsorption capacity	Chemical stability at		Swelling properties in organic solvents	Density of potentially reactive groups
					pH > 12	pH < 3		
Porous glass	Excellent	Negligible	Very high	Very high	Very low	Very high (except in HF)	—	Low
Polyacrylamide	Good	Low	High	Low	Low	Moderate	?	High
Cellulose	Fair	Moderate–low	Moderate (variable)	Moderate–high	Very high	Moderate	Poor	High
Sephadex	Good	Low	Moderate	Low	Very high	Moderate	Poor	High
Agarose	Good	Moderate	Very high	Low	Very high	Low	Good	Moderate
Glyceryl-bridged, desulfated agarose	Good	Negligible	Very high	Very low	Very high	Moderate	Very good[a]	Moderate
Glyceryl-bridged, hydroxylated agarose	Good	Negligible	Very high	Very low	Low[b]	Moderate	Very good	High
Divinyl sulfone-cross-linked agarose	Excellent	Negligible	Very high	Very low	Low	Moderate	Very good	Moderate

[a] This is valid on the assumption that the primary swelling has taken place in aqueous solution.
[b] Hydroxylation with polyphenols yields products that are easily oxidized at high pH.

stood that the evaluation of the properties of the various matrix substances of necessity is somewhat biased by the opinions of the authors.

1. Cellulose

Cellulose often shows considerable adsorption and permeability that is not easy to control. Biospecific adsorbents based on cellulose often deviate more strongly from the behavior expected from the degree of substitution than do most other adsorbents. Substitution will therefore seem to occur in regions that are sterically unfavorable for the formation of high polymer complexes. Cellulose forms uneven beds with comparatively poor flow rates. Even so, suitably prepared cellulose can be packed to give much narrower zones of nonadsorbed material than any other matrix. Although cellulose beds may be attacked by bacteria and fungi, we have had very good experiences with the stability of cellulose beds under various conditions.

2. Ethylene–Maleic Acid Copolymer

This cross-linked copolymer of ethylene and maleic acid amide contains an unacceptable number of carboxyl groups and better polymers than this are now available. It cannot be denied, however, that very impressive work has been carried out with this type of polymer by Levin and Katchalski (1968), Centeno and Sehon (1971), Fritz *et al.* (1968a), and others.

3. Polyamides, Polyesters, and Polyvinyl Alcohol

Polyamides and polyesters have many excellent properties. A weakness is the limited chemical stability in alkali. Free carboxyl groups are formed by alkaline hydrolysis and also in strongly acid media. However, the hydrogels introduced by Wichterle and Lim (1960) may possess many of the desirable matrix properties. Brown *et al.* (1971) have recently described new types of polymethacrylates for immobilization of proteins and for coupling of amino acids and peptides. The grafting of polymers for covalent binding should also be mentioned (Catt *et al.*, 1967; Hoffman *et al.*, 1972). Cross-linked polyvinyl alcohol is not suitable as a matrix since it forms loose gels usually giving high nonspecific adsorption.

4. Cross-Linked Dextran

Like cross-linked polyacrylamide, cross-linked dextran (Sephadex) has many outstanding qualities. Unfortunately, gels with high permeability become far too soft for column procedures. Compared with agar gels, Sephadex has a somewhat higher stability toward acids. Sephadex has more hydroxyl groups per weight unit of matrix than agarose and hence has a greater propensity for substitution. In other respects agar gels are superior to Sephadex as matrices. Agar and agarose seem to have molecular gel structures especially well suited for biospecific adsorbents. Crude agar shows considerably more adsorption than agarose. Agar as well as agarose can be converted to a sol in boiling water, but even at room temperature solubility can sometimes be disturbing. Resistance toward acids is somewhat lower than that of Sephadex and cellulose. Cross-linked desulfated agarose is at present the most suitable matrix for biospecific adsorption. We shall therefore discuss this material in detail.

5. Agar and Agar Derivatives

Araki (1956) separated naturally occurring agar into two fractions, one weakly charged almost neutral fraction, and one more strongly charged acidic fraction. The former is called agarose, the latter agaropectin. He showed that agarose is a linear galactan with alternating repeating units of 1,4-linked 3,6-anhydro-α-L-galactose and 1,3-linked β-D-galactose (Fig. 1).

The agar polysaccharides, particularly the anhydro form of L-galactose, have unique structures and are very different from the polysaccharides found in animals, microorganisms, and terrestrial plants. Accordingly, it is not surprising that no agar-splitting enzymes are found in the bacteria contaminating laboratories or industrial premises. Microbial attack on beds of agar is extremely unlikely, neither is there any great possibility of agar-depolymerizing enzymes occurring in biologic preparations. Agarases have only been found in certain microorganisms. For these reasons the biologic stability of agar and agar derivatives may be considered entirely satisfactory. Agar has been fractionated by different methods to give products with good gel-forming properties and low electrical charge. Although these fractionated products are also termed agarose, they are not homogeneous since agar contains an indefinite number of galactans of differing charge.

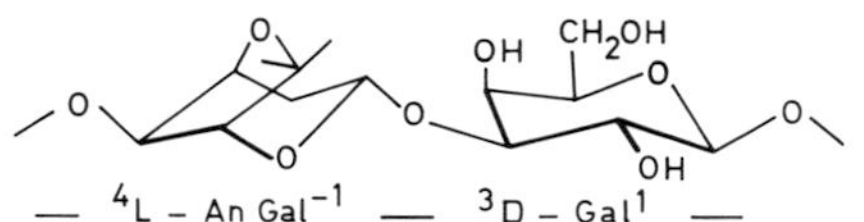

Fig. 1 Structure of the disaccharide unit in agarose.

Apart from Araki's contribution, a more profound knowledge of agar chemistry has been provided by many groups, primarily those of Rees (1969), Duckworth and Yaphe (1971), and Izumi (1971). Agar contains two types of charged groups: carboxyl and sulfate. The carboxyl groups appear as 4,5-ketal-linked pyruvic acid, i.e., carboxyethylidene substituents, and are therefore stable in alkali. The sulfate is half-ester bound in various positions. Since the distribution of sulfate and carboxyethylidene groups is uneven, it is possible to obtain fractions strongly enriched in pyruvic acid. The proportions of different galactans in the agar is furthermore dependent upon the kind of seaweed from which the basic product has been prepared. The polysaccharide chain is so stable in reducing alkaline media that the agarose may be heated to well above 90° without depolymerization. By this treatment the majority of sulfate groups are removed with the partial formation of anhydrogalactose (Rees, 1961, 1963). Alkali stability is important for the preparation of derivatives, which requires alkaline conditions. The ketal-linked pyruvic acid may be cleaved in acid solution. Unfortunately, simultaneous cleavage of the polysaccharide chain occurs, probably at the 1,3-galactoside linkages, and the gel is dissolved. Acid lability is in fact the weakness of agar and agar derivatives. Agar gels may be exposed to aqueous solutions of pH 2–2.5 at room temperature, but the contact time must be very short (less than an hour) in order to minimize hydrolysis. Such a contact is sometimes necessary for the desorption of biospecifically adsorbed material (see Section VII,B,*4*). Even when great care is taken during this operation, an increase in sugar content is observed in the eluate and washing solutions.

Rees (1961, 1963, 1969) has shown that the agar galactans, like proteins, have secondary and tertiary structures. The polymer chains in the agarose gels fold into double helices held together by hydrogen bonds. Chain segments of the random-coil type probably alternate with compact helical structures in localized areas of the gel. One can thus imagine the formation of large cavities enclosed by helical regions conferring high permeability and high gel strength. Hydrogen bonds and hydrophobic interactions probably also contribute to the mechanical stability. Agar forms gels at temperatures below 40°–45°. Gels of varying mechanical strength are formed depending upon the chain length and the content of anhydrogalactose and sulfate. As a rule, gels with a matrix content lower than 1% are too soft for chromatography. In order to increase the flow through agarose beds the particles should be spherical (Hjertén, 1964). The gel particles seem to have a "memory": when they are lyophilized and then swelled in water they regain their previous form, water content, size, and permeability.

It should be borne in mind that *mechanical leakage* may occur in any kind of small-particle bed. Fine or ultrafine particles may pass through the supporting filter and contaminate the solution in suspended form. Such particulate gel matter can easily be removed by centrifugation or filtration.

During washing, agar and agarose gels continuously release carbohydrate. This leakage increases with temperature and eventually the gel is transformed into a sol. This disadvantage may be overcome by cross-linking the gel (Schell and Ghetie, 1968; Porath *et al.*, 1971). We effect cross-linking either by means of bisepoxides

$$\sim\!OH + \underset{\diagdown O \diagup}{H_2C-CH}\sim\!\sim\underset{\diagdown O \diagup}{HC-CH_2} + HO\!\sim \xrightarrow{NaOH}$$

$$\sim\!O-CH_2-\underset{OH}{CH}\sim\!\sim\underset{OH}{HC}-CH_2-O\!\sim$$

or by means of epihalohydrins or 2,3-dihalopropanol in strong alkali in the presence of sodium borohydride:

$$\sim\!OH + CH_2Br-CHBr-CH_2OH + HO\!\sim \xrightarrow{NaOH} \sim\!O-CH_2-\underset{OH}{CH}-CH_2-O\!\sim$$

The cross-linked gel is then autoclaved in a reducing medium in strong alkali at 120° for a couple of hours (Porath *et al.*, 1971). The sulfate groups are almost completely removed by this procedure and gel which is not cross-linked is dissolved. The form and mechanical properties of the agar or agarose beads are retained. As a result of the removal of sulfate groups, the adsorption tendency is strongly reduced so that cross-linked and desulfated crude agar may surpass commercial agarose preparations as a gel matrix. If agarose is used as a starting material, a cross-linked gel with excellent properties is obtained that is well suited as a matrix for biospecific adsorbents. If exceptional demands are made for low adsorption, the remaining carboxyl groups may be reduced with $LiAlH_4$ in dioxane (Låås, 1972). The permeability is hardly changed by cross-linking. We have found that exclusion limits as measured with blue dextran are only very slightly reduced.

Agar gels with a 2% matrix substance in bead form (Sepharose 2B, or epichlorohydrin cross-linked agar, or agarose) provide beds with acceptable flow properties for small-scale preparations. Even low pressures compress the bed and cause a reduced flow rate. We have found that considerably more rigid gels can be obtained by cross-linking with divinyl sulfone:

$$\text{≷}-OH + CH_2{=}CH-SO_2-CH{=}CH_2 + HO-\text{≷} \xrightarrow[Na_2CO_3]{NaOH \text{ or}}$$

$$\text{≷}-O-CH_2-CH_2-SO_2-CH_2-CH_2-O-\text{≷}$$

Epoxide cross-linked agarose gels may be made firmer by additional treatment with divinyl sulfone.

Divinyl sulfone cross-linked gels have mechanical properties approaching the ideal for gels having a concentration of 2% or above. We have also prepared a fully acceptable gel with a lower matrix content. However, rather unexpectedly, the permeability is not increased very much.

Theoretically, according to Fig. 1, three hydroxyl groups per disaccharide unit would be free for substitution. We mentioned that certain groups are blocked with sulfate and are converted by nucleophilic attack with hydroxyl groups to anhydro sugars. The 1,4-linkages are to a certain extent blocked by carboxyethylidene groups. It appears that the degree of substitution obtained with most reagents is rather low and after cross-linking it decreases still further. To compensate for this, in cases where high substitution is desirable, we have prepared agar derivatives with an increased number of hydroxyl groups (Porath and Sundberg, 1972b). This has been achieved by substituting hydroxyl-containing compounds, e.g., sorbitol and phloroglucinol, using epichlorohydrin, for example:

$$\text{Ⓜ}-OH + CH_2\underset{O}{\diagdown\!\diagup}CH-CH_2Cl \longrightarrow \text{Ⓜ}-O-CH_2-CH\underset{O}{\diagdown\!\diagup}CH_2$$

$$\text{Ⓜ}-O-CH_2-CH\underset{O}{\diagdown\!\diagup}CH_2 + HO-C_6H_3(OH)_2 \longrightarrow \text{Ⓜ}-O-CH_2-\underset{OH}{CH}-CH_2-O-C_6H_3(OH)_2$$

This method may also be used in order to increase the capacity of other matrices containing hydroxyl and amino groups.

V. METHODS FOR ATTACHMENT OF LIGANDS

A. Survey of Reactions

Ligands can be directly coupled to the matrix with certain methods. In some cases it is suitable or necessary to attach the ligands indirectly by means of a spacer substance. The spacer is first attached to the matrix, then the ligand is attached to the end of the spacer. Spacers were often used in earlier work. Their importance was first suggested by Dennis (1968) and more strongly emphasized by Cuatrecasas *et al.* (1968). Since the matrix, as previously discussed, should not contain adsorption-reactive groups, interest has been focused on methods for coupling ligands to hydroxyl groups. However, certain substances cannot be coupled satisfactorily by these methods without the introduction of other reactive groups into the matrix as an initial step in the preparation of the adsorbent. Since the terminal groups of these substituents (e.g., carboxyl, carbonyl, sulfhydryl, etc.) are active or activatable, these substances can at the same time act as spacers. A bifunctional substance that forms a spacer and a bilateral connector can also be used.

When preparing biospecific adsorbents it is as important to utilize efficient coupling methods as it is to choose a suitable matrix. The following requirements should ideally be satisfied:

1. The matrix should allow the introduction of reactive substituents by way of hydroxyl, carbonyl, or amino groups already present in the matrix.

2. Apart from the ligand, no adsorbing groups should be introduced in the subsequent coupling of spacers or ligands.

3. The linkages formed in the consecutive steps should be stable in all media to which the adsorbent will be subjected during adsorption, desorption, and regeneration.

4. The degree of substitution should exceed 100 μmoles amino acid or 100 μmoles of a low molecular weight compound containing amino groups or 50 mg protein/gm dry carrier.

A very large number of reactions have been utilized for coupling ligands to suitable matrices. Silman and Katchalski (1966) have reviewed work carried out before 1966. Inman and Dintzis (1969) have listed a series of methods for preparing derivatives of polyacrylamide. It is, however, doubtful whether any of these adsorbents fulfills the above requirements completely. In the following section we shall briefly discuss different coupling methods (see Table IV for references) and then in greater detail present the cyanogen bromide method which we have developed in an attempt to satisfy the above requirements.

1. Reactions with Hydroxyl Groups

Hydroxyl groups are attacked by strong electrophiles. For example, acyl chlorides react as shown below:

$$\text{Ⓜ}{-}\mathrm{OH} + \mathrm{Cl{-}\underset{\underset{\displaystyle O}{\|}}{C}{-}R{-}X} \longrightarrow \text{Ⓜ}{-}\mathrm{O{-}\underset{\underset{\displaystyle O}{\|}}{C}{-}R{-}X}$$

TABLE IV

Methods of Attaching Ligands to Specific Groups on the Matrix

Group on matrix	Method	Reference
Hydroxyl	Cyanogen halide	Axén *et al.* (1967, 1971), Axén and Ernback (1971), Axén and Vretblad (1971a), Cuatrecasas *et al.* (1968)
	Halotriazines (cyanuric chlorides)	Kay and Lilly (1970), Kay and Crook (1967)
	Epoxide (oxirane), halohydrin	Porath and Sundberg (1972b)
	Bromoacetyl bromide	Jagendorf *et al.* (1963)
	Divinyl sulfone	Porath and Sundberg (1972b)
Amino	Amide formation using carbodiimides as condensing agents	Weliky *et al.* (1969), Weetall and Weliky (1964)
	Isonitrile	Axén *et al.* (1971), Axén and Vretblad (1971b)
	Cyanogen halide	Axén *et al.* (1967, 1971), Axén and Ernback (1971), Axén and Vretblad (1971a), Cuatrecasas *et al.* (1968)
	Condensation with glutaraldehyde	Weston and Avrameas (1971)
	Conversion of aromatic amines to diazonion salts and subsequent coupling	Barker *et al.* (1970), Grubhofer and Schleith (1954), Cebra *et al.* (1961), Bar-Eli and Katchalski (1963), Kursanow and Solodkow (1943), Gurvich (1957)
	Isothiocyanate	Barker *et al.* (1970), Axén and Porath (1964), Kent and Slade (1960), Manecke and Günzel (1967)
	Halotriazines	Kay and Lilly (1970), Kay and Crook (1967)
Carboxyl	Acid anhydride, acid chloride	Levin *et al.* (1964), Isliker (1953, 1957), Axén *et al.* (1971), Weliky *et al.* (1969), Weetall and Weliky (1964)
	Azide	Barker *et al.* (1970), Micheel and Evers (1949), Mitz and Summaria (1961)
Carbonyl	Isonitrile	Axén *et al.* (1971)
	Condensation to Schiff's base and subsequent reduction	Sanderson and Wilson (1971), Goldstein *et al.* (1970)
Sulfhydryl	Mercuric chloride	Eldjarn and Jellum (1963)
	Homocysteine lactone	Cuatrecasas (1970a)

where X is a reactive group such as —COCl and R is the spacer. One disadvantage of this method is that negatively charged carboxyls are introduced and cannot be easily blocked without destroying the ligand. In addition, the linkages are relatively unstable at low pH.

In alkaline solution, hydroxyl groups are attacked by substances containing groups suitable for our purpose: X—CH_2—CHOH— where X is a halogen, CH_2—CH—, and CH_2=CH—SO_2—. Very strong linkages

$$\underset{\diagdown\ O\ \diagup}{CH_2—CH—}$$

are formed with epoxides (oxiranes), halohydrins, and cyanogen halides. The use of bifunctional compounds leads to complications such as cross-linking and multipoint attachment of proteins and other biopolymers and a more or less pronounced cross-linking of the matrix itself. For this reason it is important to keep the reactions under control with respect to pH, temperature, and time.

2. *Reactions with Amino Groups*

Aliphatic amines react with the hydroxyl group reagents mentioned above and also with several others. Considering the requirements for a good adsorbent, we think that *aliphatic* amines in the matrix should in principle be avoided since they impart very definite ion exchange characteristics in the pH range of many adsorption–desorption processes. In cases where the formation of a matrix amino group is unavoidable, one of the methods given by Inman and Dintzis (1969) or Cuatrecasas (1970c) can sometimes be used to advantage.

Aromatic amino groups are rather reactive and as a rule less disturbing than aliphatic amines as ionogenic adsorption centers. However, because of their hydrophobic nature and tendency to form charge transfer complexes, aromatic substituents can contribute some unspecific adsorption. Ligands carrying carboxyl groups can condense with aromatic amines in the matrix, with carbodiimides, Woodward's reagent, isonitriles, etc. The amino groups can be converted to diazonium ions allowing coupling with aromatic amines, phenols, and many heterocyclic substances. Diazo coupling is the method that was most often used in the past for preparing immunosorbents and was applied by Lerman in his now classical work on specific adsorbents for tyrosinase (Lerman, 1953). Unfortunately, products are formed which fall short of the high stability called for by immobilized proteins. The reason for this is probably that coupling not only occurs at phenol and imidazole nuclei in tyrosine and histidine residues, respectively, but also at α-terminal and ϵ-amino groups. These side reactions, "Nietzski reactions," lead to the formation of triazene linkages, —N=N—NH—, which are less stable and are eventually hydrolyzed. A slight leakage can often be detected with sensitive methods. If these side reactions could be avoided and if high yields could be obtained, coupling via diazonium salts would be an excellent complement to more recent methods.

Glutaraldehyde has been used by Avrameas and Ternynck (1969) and Onkelinx and co-workers (1969) for condensing proteins and attaching proteins to erythrocytes, and by Weston and Avrameas (1971) and Ternynck and Avrameas (1972) for coupling proteins to polyacrylamides. Since glutaraldehyde should only react with amino but not hydroxyl groups, one can assume that this substance should be a suitable connector for proteins and amines to amino aryl ethers of argarose, Sephadex, and cellulose. Condensation with glutaraldehyde can take place in acid solution. This applies to the isocyanide method as well (see below).

3. *Reactions with Carboxyl Groups*

Mitz and Summaria (1961) modified CM-cellulose by introducing chloride and azide groups and have coupled enzymes to these derivatives with partial retention of enzymatic activity. Other coupling methods have been used for immobilization of enzymes and preparation of adsorbents, examples being amide formation with carbodiimides and the isocyanide method. Since complete blocking is difficult or impossible to achieve, the product will function as an ion exchanger. However, since the coupling reaction can be carried out in acid solution these methods are sometimes indispensable. The immobilization of pepsin affords an example (Vretblad and Axén, 1971). In this context we should also mention the copolymer of maleic anhydride and ethylene which Katchalski and co-workers developed and which has been used extensively and with great success by Fritz, Werle, and others.

4. *Reactions with Carbonyl Groups*

Ligand fixation can take place by reacting substances containing amino groups with keto or preferably aldehyde groups formed in the matrix by periodate oxidation (Smith reaction) or by treatment with dimethylsulfoxide. The imines formed (Schiff bases) can be stabilized by reduction with sodium borohydride. The isonitrile method, a variant of Ugi's four-component condensation, is a very versatile method which requires the presence of an isonitrile, a keto compound, an amine, and an anion, for example, carboxylate. Any one of these groups may be matrix-bound if the others are present in solution. The reaction can be suitably carried out in slightly acid or neutral solution. Unfortunately, with aldehyde-containing matrices the yields of reaction products have not been very high.

5. Reactions with Sulfhydryl Groups

Cuatrecasas (1970a) has prepared sulfhydryl agarose from ω-aminoalkyl agarose by reaction with homocysteine thiolactone using a method analogous to the method of Benesch and Benesch (1956) for the thiolation of proteins. Thiol agarose reacts with sulfhydryl compounds by forming disulfide bridges.

Eldjarn and Jellum (1963) have shown that, in solution, sulfhydryl compounds, including proteins, can be chemically adsorbed (chemisorbed) to Sephadex containing covalently bound mercurichloride groups.

B. Cyanogen Bromide Method

1. General Principles

During our systematic experiments aimed at developing methods of immobilizing proteins, we discovered a very useful side reaction while attempting to couple amines via cyanamide (Axén *et al.*, 1967; Porath, 1968). If Sephadex or cellulose is treated with cyanogen iodide or bromide in alkaline solution, the polysaccharide becomes reactive and capable of binding large quantities of proteins, peptides, or amino acids. The method is generally useful for coupling primary and secondary amines and can also be used indirectly for coupling other classes of substances, for example, carbohydrates. We have been very successful in applying this method to agarose, cross-linked agarose, and hydroxylated agarose. Since cyanogen bromide coupling at present occupies such an important position, we consider a more detailed discussion of this method justified.

It was apparent at an early stage in our experiments that several reactions were occurring simultaneously and that these could be differentially influenced by the alkalinity of the solution. Preliminary studies by infrared absorption showed the existence of —CO— as well as —C=N— structures which suggested the formation of iminocarbonates and carbamates. This suggestion was later confirmed by Axén and Vretblad (1971b) and by Kågedahl and Åkerström (1970), who also found evidence of the presence of isourea. From these results and from the collected experience in this field, we propose the following reaction mechanisms for activation:

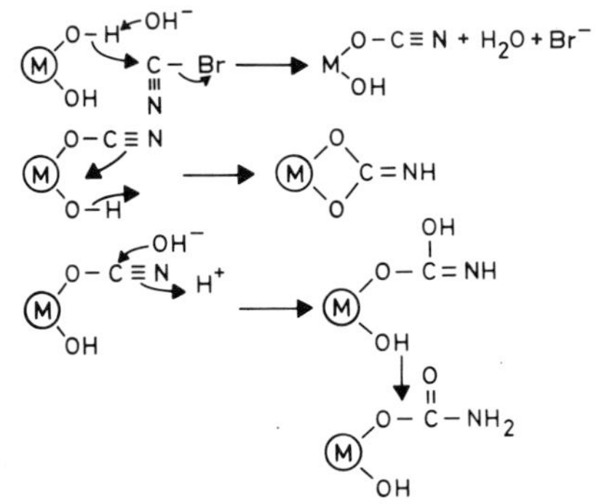

(a)

(b)

(c)

The proposed cyanate intermediate is too reactive and short-lived to be demonstrated with the methods used up until now. The attacking hydroxyl can be so positioned as to form a cyclic iminocarbonate with high bond energy (b). The hydroxyl group can also be situated in another part of the matrix so that a cross-link can be formed. The carbamate group (c) is inert and should not cause adsorption. The work of Axén and Vretblad (1971b) shows that carbamate nitrogen can constitute up to 50% of the total nitrogen content. The activated gel will be hydrolyzed in acid and strongly alkaline solution, and in neutral solution a slow deactivation occurs. At pH 9 the coupling capacity of activated Sephadex is reduced by 40% in 24 hr. Activated Sephadex can be lyophilized or converted into a stable form by controlled dehydration. Activated stable Sepharose is commercially available (Pharmacia Fine Chemicals).

From the extent of ammonia formation and from IR absorption measurements of the product, the following coupling reactions are indicated:

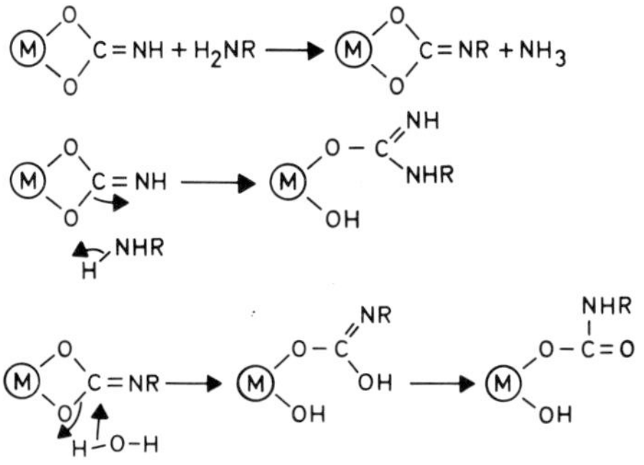

Very low but significant coupling occurs at pH 4.5 (glycylleucine) in acetate buffer. The yield increases with increasing pH. It is low even at pH 7 but then rapidly increases until pH 11–11.5 is reached. Beyond this point there is a rapid decrease as the solution becomes extremely alkaline.

It is, of course, extremely important that the connector-to-matrix as well as the connector-to-ligand bonds are stable. This was apparent from earlier experiments but was more rigorously shown to be true by the following experiments (Kristiansen *et al.*, 1969). [^{14}C]Sorbitol with an activity of 2.22×10^7 dpm was added to a solution of cyanogen bromide, activated at pH 11, and then coupled to aminoethyl cellulose at pH 8.5. Equal parts of the product were transferred to two chromatography columns (pH 7.0). Successive equilibrations in steps of one unit were carried out against buffer of increasing pH. After exposure for 24 hr, the columns were flushed and the effluent assayed for radioactivity. Similar experiments were performed at successively decreasing pH values. No radioactivity was observed in the pH range of 2–12. After 4 days of standing at room temperature in 0.1 *N* HCl and 0.1 *N* NaOH, 1.8% and 42.7%, respectively, of the initial radioactivity was in the solution. This result can be wholly or partially explained by assuming that the matrix was destroyed under these drastic conditions. The experiment shows that the coupling between a polyalcohol and a substance with aliphatically bound amine is extraordinarily stable. On the other hand, it has not been shown that secondary or heterocyclic amines form such stable linkages. The question has been brought to the fore by Tesser *et al.* (1972) who reported that Sepharose-bound cyclic 3′,5′-adenosine monophosphate was not stable in solutions above pH 5. Tesser *et al.* state that "such a cleavage, if it were a general phenomenon characteristic of Sepharose derivatives prepared with cyanogen bromide, would place serious limitations on the use of affinity chromatography for isolating pico- and nanomolar amounts of proteins. . . ."

The increased cross-linking that occurs after activation at higher pH has been investigated by Axén and Ernback (1971) in a study of chymotrypsin immobilized on Sephadex. It was shown in this study as well as in earlier and later work that a high degree of substitution does not in itself guarantee an immobilized protein of high quality. High permeability is an important requirement for a biospecific adsorbent when the ligand or the adsorbate is a protein or another large molecule. The relative adsorption capacity (i.e., adsorption calculated on the basis of bound protein) can be dangerously low if both ligand and adsorbate are high polymers. We believe that, in the commercial promotion of adsorbents, the degree of substitution as a merit in itself has been overemphasized. Average substitution but retention of high activity is preferred. Activation at low pH (9–10) is sometimes preferable as an initial step in the coupling of proteins. Lower protein concentrations give higher relative adsorption capacities, indicating that a more extensive substitution results in the coupling of ligands to sterically hindered or less accessible groups in the matrix. Consequently, a dense substitution of immobilized protein results in adsorbents that are heterogeneous with respect to affinity (Kasche, 1971).

The cyanogen bromide method can be used for attaching both ligands and spacers to agar and other matrices. Attachment of the spacer to the matrix may be followed by the fixation of the ligand to a terminal reactive or activatable group on the spacer.

$$\left.\begin{matrix}-O\\-O\end{matrix}\right\rangle C{=}NH + H_2N-(CH_2)_n-NH_2 \longrightarrow$$

$$\left.\begin{matrix}-O\\-O\end{matrix}\right\rangle C{=}N-(CH_2)_n-NH_2$$

$$\left.\begin{matrix}-O\\-O\end{matrix}\right\rangle C{=}N-(CH_2)_n-NH_2 + HOOC-R \xrightarrow{\text{carbodiimide}}$$

$$\left.\begin{matrix}-O\\-O\end{matrix}\right\rangle C{=}N-(CH_2)_n-NH-COR$$

The spacer can also be introduced into the "adsorption-active" molecule to form a spacer–ligand complex. In fact this is the preferable route since the introduction of additional ionogenic groups on the matrix is thus avoided. For example, Dudai and co-workers (1972) substituted an ε-aminocaproyl group in phenyltrimethyl ammonium and coupled this derivative to Sepharose, thus obtaining a specific adsorbent for acetylcholinesterase.

Many variations of this method have been used to substitute terminal amino groups in coenzymes and low molecular weight enzyme inhibitors (see Section VII,A).

Evidence up to now suggests that the spacer, when needed, should be linear or almost linear. The required length may vary individually with the adsorbate molecule but need hardly ever exceed 20 Å. Immobilized proteins generally work well without spacer substances. Cuatrecasas (1970c) has used alkane chains besides peptides and polyamines.

In earlier work we used *p*-aminophenyloxy-2-hydroxypropyl for biotin-Sephadex (Porath, 1967) and for immobilized enzymes (Axén and Porath, 1964). Spacer substances such as these are especially suitable when a ligand is to be introduced selectively into a matrix carrying long-chain substituents.

2. *Suggestions for Practical Procedures*

One of the advantages of the cyanogen bromide method of fixing ligands to hydroxyl-containing matrices is its reproducibility. Activation may suitably be carried out in the pH range of 10.5–11.5. The cyanogen bromide concentration is not critical, but it strongly affects the capacity and permeability of the gel product. Since a general optimization is not possible, conditions should be adjusted for the problem at hand. Extensive adjustments of activation parameters are rarely needed. Below we suggest a few variations that will be described in more detail. Coupling may be carried out over a wide pH range, but the yield will improve with increasing pH (<12). However, a high degree of substitution is not necessarily advantageous. Matrix-bound proteins often become adsorption-inactive at high substitution density. Low molecular weight ligands bound to strongly activated gels will also become imbedded because of the intensive cross-linking in gel regions unavailable to proteins. The resulting decrease in capacity per amount of ligand should not be underrated and, in our opinion, sufficient attention may not have been paid to this fact so far. The drawback of having to use somewhat larger gel beds associated with a gentler activation should not be overemphasized.

Our suggestions given below for activation, coupling, reactivation, and storage are based on our own experience as well as on discussions with and suggestions from Dr. Rolf Axén and Dr. Kåre Aspberg.

a. **The Activation Step.** i. *Highly activated 1%–4% agarose gels.* To 100 ml of a sedimented desulfated agar or agarose gel X% cross-linked ($X = 1$ to 4) are added at 5°–10° 25X ml of a chilled 5 M phosphate buffer (pH 12) prepared by mixing 3.33 M K_3PO_4 and 1.67 M K_2HPO_4 (Porath *et al.*, 1971). The suspension is diluted to 200 ml with distilled water and stirred continuously. Then 10X gm of fresh cyanogen bromide (BrCN) are added at 5°–10° over a period of about 6 min. After an additional 10 min of reaction time, the gel is filtered on a Büchner funnel and washed with distilled water at 0° until the pH is neutral.

ii. *Highly activated 4%–8% agarose gels.* Conditions are the same as above except the agar gel is first washed with 2 M buffer and then diluted with 2 M buffer to a consistency suitable for stirring (this volume should preferably be less than 100 ml).

iii. *Intermediately activated 1%–8% agarose gels.* The BrCN concentration is reduced to 1/5 of that described in (i).

iv. *Weakly activated 1%–8% agarose gels.* The BrCN concentration is reduced to $\frac{1}{20}$ of that described in (i). The degree of activation will then approximate the one mentioned by Chan and Mawer (1972).

***b.* The Coupling Step.** When extensive substitution is sought, as with low molecular weight ligands, activation alternatives (i) and (ii) are recommended. The choice of solvent and matrix is also important. In our experience, coupling is best carried out in aqueous solution, but if the solubility of the ligand substance in water is low it may be necessary to carry out the coupling reaction in a mixed solvent. Higher yields are obtained with hydroxylated agarose. Sephadex gels afford a higher degree of substitution per gram of matrix substance than do agar gels. For coupling at low pH and in mixed solvents, long coupling times are generally required.

i. *Coupling of low molecular weight ligands.* Immediately after activation the gel (100 ml) is coupled with a saturated solution of the ligand substance at 0°–4°. The temperature may be raised to 25° if necessary. If the ligand substance is very difficult to dissolve in water, one can try optimal proportions of a mixture of water and an organic solvent, e.g., ethanol, ethylene glycol, dimethylsulfoxide, dimethylformamide, etc. The time required for the reaction is at least 24 hr at pH 9.0. Solubility and stability permitting, we suggest that coupling be carried out at pH 10 for 2 hr and that the pH then be adjusted to 9. After addition of a suitable bacteriostatic agent (e.g., sodium azide or toluene) the gel should be stored at 0°–4° with excess ligand.

Excess activated groups may be inactivated by reduction of pH, elevation of temperature, or chemical blocking if the ligand can withstand it. Excess activated groups in cross-linked gels can be inactivated by storage at pH 4–5 for several weeks or by heating to 100° for about 1 hr.

ii. *Coupling of proteins.* At pH values below 7, long coupling times are required for both strongly and weakly activated gels. At high pH there is a risk of too much protein being coupled, giving a relatively low specific activity.

After fixation, proteins not forming specific complexes may very well function as nonspecific ion adsorbers and more or less drastically reduce the selectivity of the adsorbent.

Protein coupling occurs via ϵ-amino groups in lysine residues and terminal α-amino groups (amino-terminal proline also reacts). Proteins having a strong positive charge (e.g., lysozyme and trypsin) are efficiently coupled even in weakly acidic media, whereas proteins without available amino groups are not coupled at all. To obtain a weakly substituted protein gel adsorbent of high specific adsorption capacity, the following coupling method is suggested. Mix 100 ml of activated gel in coupling buffer at 0°–4° with 100 ml of buffer solution containing 1–5 mg protein/ml. Allow the suspension to stand for 24 hr with slow stirring.

Remove the liquid by filtration and wash the gel with buffer (excess protein may be recovered from the washings). Suspend the gel in 100 ml of buffer, saturated with glycine, to block excess activated groups. Add sodium azide to 0.01% or, alternatively, a few drops of toluene to prevent bacterial growth in the buffer. Store in glycine solution for at least 24 hr or until use.

The carboxyl group of glycine is efficiently screened by the basic imino group in the cyclic iminocarbonate (Porath and Fryklund, 1970). Blocking may also be carried out with a weak amine.

Sometimes, despite what has been stated above, it may be appropriate to prepare gels with a high content of bound protein. This applies to proteins of relatively low molecular weight and to gels used for batchwise extraction of large quantities of specifically adsorbed substances as well as to adsorbent beds through which the flow is particularly rapid. It is then possible to raise the protein concentration in the coupling solution ten times or even more, and start from an intermediately activated or strongly activated gel. We do not, however, recommend this as a standard procedure. A high degree of protein substitution is no guarantee of high quality; more often the opposite is true.

VI. ADSORPTION–DESORPTION PROCEDURES

Specific adsorption is used to separate biochemically distinct substances from each other, mostly to isolate one or more components in extracts, body fluids, exudates, etc. Adsorption can also be used to remove undesirable biologic activities from solutions without further fractionation. For instance, extracts may be stabilized by batchwise contact with immobilized proteinase inhibitors. In principle it would be possible to continually remove antinutritional factors from bacteria or other cell cultures by including a biospecific adsorbent in the culture medium. We do not know whether such a technique has already been used. Coagulation factors may be removed by allowing blood plasma to pass through a tube or plastic container coated with complex-forming substances. In short, undesirable reactions can be prevented. We shall not discuss these possible applications further but instead limit ourselves to biospecific adsorption methods for the purification and analysis of proteins. Chromatographic procedures will be discussed separately from batch methods.

A. Chromatographic Methods

1. *Theory*

Biospecific affinity chromatography is in principle no different from other forms of adsorption chromatography, being based on the affinity between the ligand and the adsorbate molecules. High selectivity is often characteristic of this kind of chromatography, but as previously pointed out, exceptionally high affinity may also exist under other conditions. On the other hand, the biospecificity can be very weak. As a matter of fact there is a more or less continuous change from completely nonspecific affinity to the strong interaction between unique antigens and antibodies, thus making it difficult or impossible to define the specificity of bioaffinity chromatography. The technique and theory generally applicable to adsorption chromatography may therefore be applied to biospecific affinity chromatography as well. The adsorption phenomena are based on the interaction between two biopolymers, or between one biopolymer (as ligand or adsorbate) and another substance, usually of low molecular weight. The interaction is of a mixed nature and includes the formation of ionic bonds and hydrogen bonds, hydrophobic interactions, and van der Waals–London forces, all known to be involved in other kinds of adsorption. The only difference is that the forces are exerted along sterically complementary surfaces between the complex-forming components. This is taken into consideration when the conditions for adsorption and desorption are chosen.

As a rule the ligands are fixed within the gel phase. The carrier gel should be permeable enough to allow the substances which are to be separated to penetrate it prior to formation of adsorption complexes. In this manner a high capacity may be obtained.

Occasionally, especially when working with particles and solutes of very large molecular size, it may not be possible to design hydrophilic gels with sufficient permeability and mechanical strength. Then the ligands may be attached to the surface of impermeable particles which have dimensions optimized for high capacity as well as good flow characteristics in chromatographic beds.

In cases where component complexes of the type AB_n or A_mB_n are formed in free solution, different species distribution occurs in the adsorbent because of the heterogeneous microenvironment. A rigorous theoretical treatment is probably impossible.

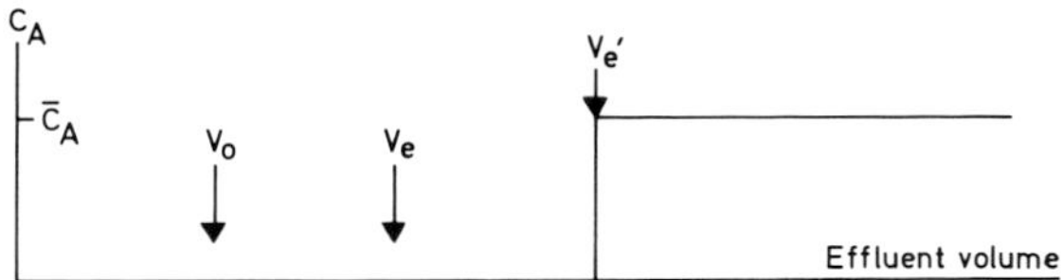

Fig. 2 Principal frontal chromatogram indicating retention due to molecular sieve action ($V_e - V_0$) and adsorption ($V_e' - V_e$).

In order to compare and evaluate experiments performed in different beds of identical or similar adsorbents, we shall deduce a few simple relationships from very simplified postulates.

Consider a gel bed with the total volume of V_t and the void volume V_0 in equilibrium with a buffer containing a displacer substance, D, of concentration C_D, forming complexes with the ligand L. If we apply a solution of buffer in which the substance A has a concentration $\bar{C}_A$, A will appear in the effluent after a volume (V_e') has passed through: in other words, we perform frontal chromatography. An example of a frontal chromatogram is diagrammatically shown in Fig. 2. V_e is the retention in the bed caused by the molecular sieve effect. A sharp front presupposes equilibrium to be instantaneously attained and all adsorption sites to be equal.

We can distinguish between two types of specific displacement:

$$\text{Ⓜ}-L + A \rightleftarrows M-L \cdots A$$
$$\text{I} \quad \text{Ⓜ}-L \cdots A + D \rightleftarrows M-L + A \cdots D$$
$$\text{II} \quad \text{Ⓜ}-L \cdots A + D \rightleftarrows M-L \cdots D + A$$

In case I, a soluble complex between adsorbate and displacer is formed. In case II, the displacer forms an adsorption complex.

With the simple assumptions made above for an adsorption–desorption model, the following equations may be formulated:

$$C_{LA}/C_L C_A = k \tag{1}$$
$$C_{AD}/C_A C_D = k' \tag{2}$$
$$\bar{C}_L = r = C_L + C_{LA} \tag{3}$$
$$\bar{C}_A = C_A + C_{AD} \tag{4}$$
$$\bar{C}_D = C_D \tag{5}$$

where $r = \bar{C}_L$, and $\bar{C}_L$, $\bar{C}_A$, and $\bar{C}_D$ are the total concentration of *available* ligand, substance A, and displacer, respectively. The symbols k and k' are the association or binding constants for the corresponding equilibria. They may be considered as "apparent association constants" or average binding constants. From Eqs. (1)–(5) we obtain

$$C_{LA} = \frac{kr\bar{C}_A}{1 + k\bar{C}_A + k'C_D} \tag{6}$$

For molecular sieving in the gel bed, the following distribution constant has been defined (Laurent and Killander, 1964):

$$K_{av} = \frac{V_e - V_0}{V_t - V_0} \tag{7}$$

V_e for a biospecific adsorbent may be determined under conditions that allow no adsorption to take place. When adsorption occurs, a corresponding distribution constant can be defined:

$$K_{av}' = \frac{V_e' - V_0}{V_t - V_0} \tag{8}$$

From Eqs. (6) and (8) we obtain:

$$\begin{aligned} K_{av}' &= \frac{\bar{C}_A(V_e - V_0) + C_{LA}(V_e - V_0)}{\bar{C}_A(V_t - V_0)} \\ &= K_{av} + \frac{C_{LA}}{\bar{C}_A} K_{av} \\ &= K_{av} + K_{av} \frac{kr}{1 + k\bar{C}_A + k'\bar{C}_D} \\ &= K_{av}\left[1 + \frac{kr}{1 + k\bar{C}_A + k'\bar{C}_D}\right] \end{aligned} \tag{9}$$

Since we are most interested in the influence of biospecific adsorption, we shall concentrate on the second term. It is a dimensionless parameter which we may call the biospecific separation factor, S, for the interacting system in question:

$$S = \frac{V_e' - V_e}{V_t - V_0} = K_{av}' - K_{av} = K_{av}\frac{kr}{1 + k\bar{C}_A + k'\bar{C}_D} \tag{10}$$

We shall first consider two special cases. For $k\bar{C}_A \ll 1 + k'\bar{C}_D$ we obtain

$$S = K_{av}kr \tag{11a}$$

The retention is dependent upon the concentration of the substance A. If a zone is introduced into the bed, it will, under ideal conditions, migrate as a single zone without tailing (in the ideal case, k may be regarded as constant). The retention, which is directly proportional to the effective ligand concentration, need not be particularly high. If affinity is weak, on the other hand, extensive and efficient ("adsorption efficient") ligand substitution is required. When separating substances of nearly identical distribution constants it is always advantageous to work under conditions of linear adsorption. It will then be possible to adjust column dimensions and other parameters so that a number of theoretical plates sufficient to achieve separation will be obtained (Giddings, 1965). This presumably favorable type of biospecific affinity chromatography has as yet hardly been used.

Even in cases where the first condition above [Eq. (11a)] has not been fulfilled, it is possible to adjust the eluant so that conditions for linear chromatography apply. The displacer substance must be of such a concentration that $k\bar{C}_A \ll 1 + k'\bar{C}_D$ (and $\bar{C}_D$ = constant):

$$S = K_{av} \frac{kr}{1 + k'\bar{C}_D} \tag{11b}$$

The position of the front of the zone may be altered by varying $\bar{C}_D$.

Linear biospecific affinity chromatography should become particularly useful for separating complexes of different molecular dimensions (different K_{av}) as well as separating substances with similar affinity for the ligand. In general Eq. (10) may be used as a guide for increasing adsorption capacity and qualitatively adjusting elution conditions. A high capacity is obtained by excluding substances that compete for adsorption sites ($\bar{C}_D = 0$) in which case the capacity is directly proportional to the concentration of sites. It should be noted that the *effective* ligand concentration, r, is lower than the total concentration of ligand substance in the matrix. This is because some of the ligands are damaged and others are unavailable for steric reasons.

Elution and subsequent regeneration often pose the most difficult problems. One must avoid damaging the ligand or the adsorbate while quantitatively displacing the latter from the matrix. In order to make the retention parameter (S) zero or nearly zero, k (and possibly k') must be reduced or $\bar{C}_D$ increased substantially. The former results in nonspecific desorption, the latter in specific desorption. Specific displacement (elution) on nonspecific adsorbents has been reviewed by Pogell (1966).

The binding constant k may be reduced by changing the temperature or the composition of the solvent. Often it is possible to effect elution by increasing the ionic strength or changing the pH, but a drastic change may sometimes result in loss of biologic activities. The constant k may be reduced by sterically modifying biopolymer ligands or adsorbates. This can be effectively achieved using urea, guanidinium salts, or chaotropic ions. These are all substances that tend to break hydrogen bonds or alter the water structure near hydrophobic regions. However, there is always a danger that the components of the adsorption complex may be irreversibly destroyed. Fortunately, a protein ligand is, as a rule, more stable in immobilized form than in free solution, a fact that is well known for enzymes. It is therefore important to adjust the concentration, temperature, and exposure time to allow only minimal conformational changes at the adsorption site during desorption. Hopefully, conformational changes in other parts of the ligand and adsorbate will be reversible under these conditions. The minimal concentration required for elution can be deduced by a preliminary experiment using a gradient of displacing substance in the eluent and then determining concentration and biologic activity. The chromatographic experiment can be carried out by either stepwise or gradient elution, using a specific or nonspecific displacer or eluting medium. Figure 3 diagrammatically shows the types of elution profiles which are obtained.

In connection with the choice of matrix, we pointed out the importance of repressing nonspecific adsorption. In many instances ion exchangers have been used as bases for biospecific adsorbents and successful fractionation has been achieved. The results of Fritz, Werle, and co-workers with EMA-trypsin for the purification of trypsin inhibitor should be mentioned here (see the references in Table V-5). In order to reduce ionic affinity these workers used amphoteric polymers containing carboxyl groups partly blocked by coupling to diamines. It is, however, more practical to avoid this problem altogether.

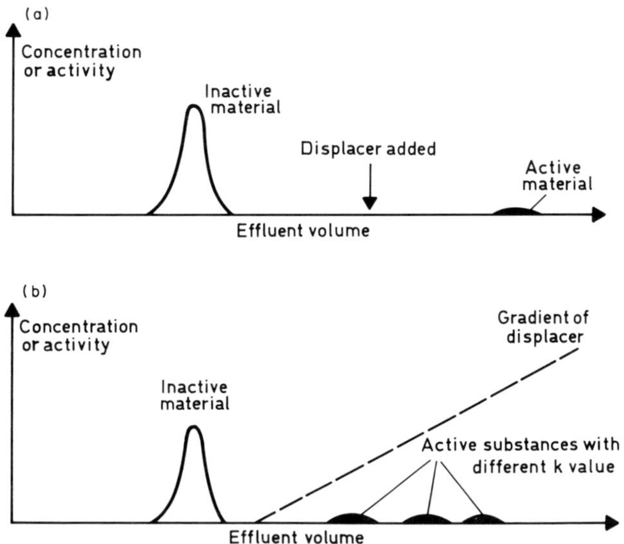

Fig. 3 Ideal biospecific affinity chromatograms. (a) Displacement with fixed concentration of displacer, and (b) gradient displacement.

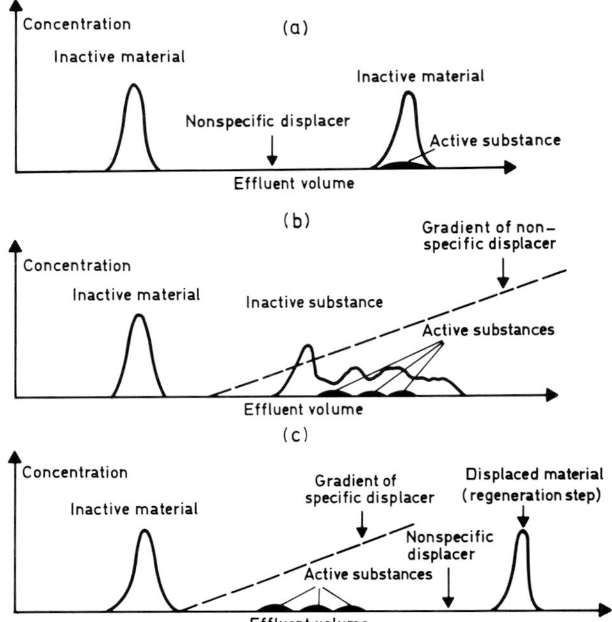

Fig. 4 Nonideal biospecific chromatograms. (a) Nonspecific displacement, (b) nonspecific gradient displacement, and (c) specific gradient displacement.

No matrices available so far are wholly free of charge, not even the best desulfated agarose gels. Desorption can result in a mixture of active components and impurities, as illustrated in Fig. 4a. A considerable increase in the effective separation is obtained with gradient elution (Fig. 4b) or better still with a specific displacer which, if it is devoid of charge, can yield pure active substance (Fig. 4c).

We shall deal with one more typical set of parameters that can be applied in practice under certain conditions. The formation of a ternary complex or a complex of higher order presupposes that the components are placed in a certain order. If the components are L, A_1, and A_2, A_2 cannot be bound directly to L but only to the A_1L complex. If L is the matrix-bound ligand, the following equilibria occur behind the front of A_2 when A_2 (in equilibrium with A_1) is passed through the bed:

$$\mathrm{M{-}L} + \mathrm{A_1} \rightleftharpoons \mathrm{M{-}L \cdots A_1}$$

$$\mathrm{M{-}L \cdots A_1} + \mathrm{A_2} \rightleftharpoons \mathrm{M{-}L \cdots A_1 \cdots A_2} \;\left(\text{or } \mathrm{M{-}L}\begin{matrix} \cdot^{\cdot}\mathrm{A_1} \\ \vdots \\ \cdot_{\cdot}\mathrm{A_2} \end{matrix}\right)$$

Using the same conditions and notation as before, we obtain the following relationships:

$$C_{LA_1}/C_L C_{A_1} = k_1 \tag{12}$$

$$C_{LA_1A_2}/C_{LA_1} C_{LA_2} = k_2 \tag{13}$$

$$\bar{C}_L = C_L + C_{LA_1} + C_{LA_1A_2} \tag{14}$$

$$\bar{C}_{A_1} = C_A \tag{15}$$

$$\bar{C}_{A_2} = C_{A_2} \tag{16}$$

From these equations, we obtain the following expression for the concentration of adsorbed A_2:

$$C_{LA_1A_2} = \frac{k_1 k_2 \bar{C}_{A_1} \bar{C}_{A_2} \bar{C}_L}{1 + k_1 \bar{C}_{A_1} + k_1 k_2 \bar{C}_{A_1} \bar{C}_{A_2}} \tag{17}$$

If we start with a certain concentration $\bar{C}_{A_1}$ in the buffer and successively reduce it, the amount of adsorbed A_2 will decrease in accordance with Eqs. (17) and (18). Eventually one should reach a concentration range where the conditions for linear chromatography are fulfilled. Adsorption is neutralized when $\bar{C}_{A_1} = 0$.

$$S = K_{av} \left[1 + \frac{C_{LA_1A_2}}{\bar{C}_{A_2}} \right] \tag{18}$$

A chromatogram showing the principle of elution with a "negative" concentration gradient is given in Fig. 5.

2. *Practical Considerations*

The routine procedure is as follows: The substance to be isolated is adsorbed and most of the impurities are removed by washing with two or three bed volumes of buffer. The substance is then displaced by a change of medium.

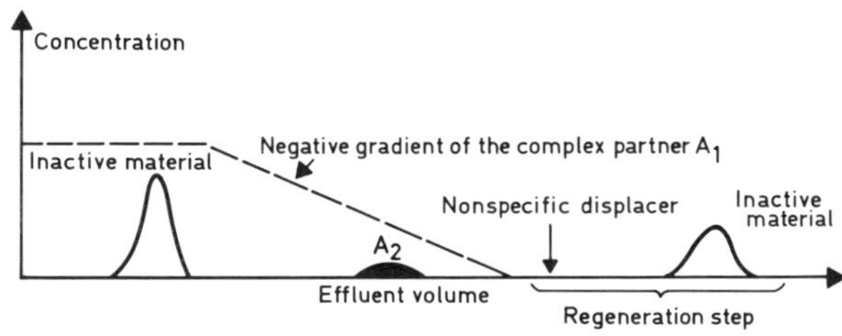

Fig. 5 Selective elution of the partner A_2 from a three-component complex consisting of L, A_1, and A_2 by gradual decrease in concentration of A_1 [Eq. (17)].

For displacement to occur, the binding constant must be sufficiently large [Eq. (6)] and the volume of the medium should not exceed the bed capacity. ($V_e' - V_e$ can be used as an indication of maximum bed capacity.)

We recommend preliminary experiments to determine the bed capacity. Frontal chromatography is carried out in a column bed having a total volume of a few milliliters. The UV absorption of the effluent is measured at a suitable wavelength and small fractions are collected. Activity in the fractions is determined, and the capacity calculated from the retention of activity. Often an approximation from $V_e' - V_0$ or $V_e' - V_t$ is sufficient. In addition, frontal chromatography provides information about the possible presence of several components having the same activities but different retention volumes. Sometimes two activities can be determined simultaneously: for example, trypsin and chymotrypsin can be determined in the eluate from a column of soybean inhibitor gel. The information obtained in the preliminary experiment can be used for planning the main experiment to give maximal purification in a bed of a given volume. A column should be charged with a sample quantity corresponding to approximately half of the retention volume. If several substances are specifically adsorbed at the same time and migrate independently of concentration [Eq. (17)], the sample and bed volume should be adjusted in accordance with the ratio of the S values of the components. A sample volume of 5%–10% of total column volume can be recommended.

The crude extract can often be chromatographed directly on the biospecific adsorbent without any complications. It is possible, as will be shown later, to obtain the desired substance in one single step with a purification factor of several thousand. To approach this ideal, absence of nonspecific adsorption and perfect reversibility of the adsorption–desorption process are essential. Even if these conditions are met, a preliminary fractionation by conventional methods may be advantageous or even necessary in order to dissociate complexes and separate their components.

A preliminary purification may also be required to avoid degradation or irreversible inactivation of the ligand by enzymes or inhibitors present in the crude extract.

B. Batch Procedures

Batch procedures are useful because they can easily be applied on a large scale and it may not be necessary to remove cell debris and fines.

Whenever the binding constant k is large and the substance to be purified is therefore transferred almost quantitatively to the gel phase, batch procedures may be more convenient than chromatography.

Before the substance or substances to be purified can be specifically adsorbed, an initial purification step may be required in order to dissociate and separate the interacting components of naturally occurring complexes containing the desired substances. Such an initial step may consist of precipitation, extraction, or ionic adsorption. Ionic strength, pH, and other important factors should be adjusted to optimize the formation of the adsorption complex. Frequently, biospecific adsorption may be carried out directly on a crude extract, serum, or an exudate.

The adsorbent may be suspended directly in a clear extract, stirred for a while, and then allowed to settle by gravitation or centrifugation. The time required is determined by the speed of transport of the substance from the bulk solution to the gel-solution interphase and its subsequent diffusion into all interior regions which the substance can permeate. Clearly these transport processes are extremely complicated and not amenable to elementary theoretical treatment.

Some orienting experiments were performed by Porath and Sundberg (1970). One gram of soybean trypsin inhibitor–Sepharose was suspended in solutions of trypsin having identical concentrations but different volumes. The activity of trypsin remaining in solution after different periods of time was determined. A set of curves, shown in Fig. 6, was obtained. On the basis of our experience, we suggest as a rough guide a contact time of 20–30 min if the suspension contains 1% adsorbing gel particles 50–100 μm in diameter, if the solute to be

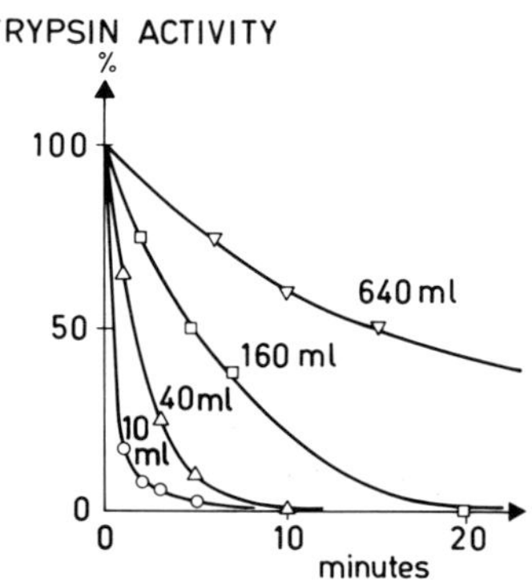

Fig. 6 Rate of adsorption of trypsin onto soybean trypsin inhibitor–agarose particles in suspensions of different dilutions.

adsorbed is in the molecular range of 10,000–100,000 daltons, and if the gel particles consist of agarose (or cross-linked agarose) with a matrix density of 6% or less.

A large molecular size protein or a particle exceeding 10^6 daltons will require a much longer time for the diffusion step if a gel of very high permeability is chosen. Since multipoint adsorption is thus much more likely to occur, an adsorbent of extremely high quality is required. It therefore seems logical to avoid gel permeation and instead let adsorption take place on beads coated with a substance specifically interacting with the particles. Possibly silica gel beads of the kind described by Haynes and Walsh (1969) or specifically designed Sephadex or agarose would be suitable.

To permit adsorption from suspensions of substances forming stable complexes, we have found it useful to enclose the gel particles inside containers with semipermeable walls, permitting free passage of the solute and retention of large particles (Sundberg *et al.*, 1970). In the simplest form, soldered nylon net bags are used. Figure 7 shows the principle permitting simultaneous adsorption of many substances by the use of bags, each containing a specified adsorbent. The bag method can also be used to remove proteolytic enzymes and other substances that may lower the yield of the desired product or otherwise disturb the isolation process.

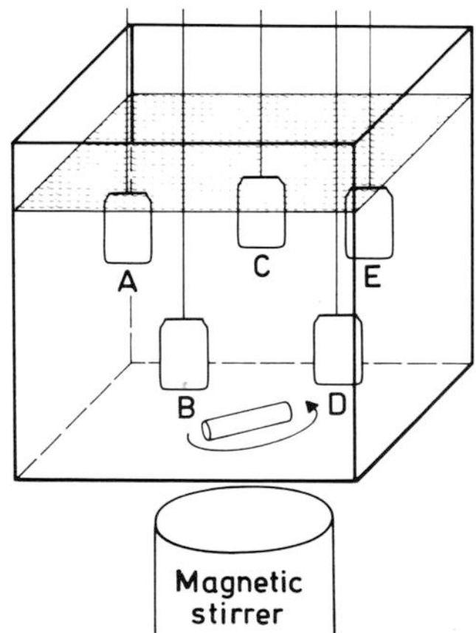

Fig. 7 Diagrammatic illustration of the bag method for simultaneous batchwise specific adsorption of many components. Bags A–E contain biospecific adsorbents directed toward different solutes in the suspension.

VII. APPLICATIONS

A. Enzymes and Enzyme Inhibitors

For the design of a suitable adsorbent, several alternatives are possible and the choice may not be easy. In the case of an enzyme, possibilities exist for preparing ligands with structures similar to (1) substrate(s), (2) product(s), (3) cofactor(s), and (4) modifier(s) or effector(s). In addition, one may consider specific ligands that do not have any immediate relation to the catalytic function of the enzyme such as (5) antibody ligands and (6) ligands derived from substances that happen to respond to group-specific adsorbents because of some particular moiety in the enzyme molecule, e.g., a carbohydrate (lectin-containing adsorbent). Adsorbents that are reactive to SH groups may be considered to be a special case.

Highest specificity is likely to be found in category 5 and next in categories 1 and 2, while an adsorbent belonging to category 6 has a low specificity. This does not mean, however, that an adsorbent in category 5 is *a priori* more suitable than, for example, one in category 3 or 4. Often the opposite may be true simply because cofactors and modifiers are usually more readily available and more stable during handling.

Concerning ligands resembling substrates, it should be borne in mind that the likelihood of strong affinity will increase the closer the structure of the ligand is to that of the substrate. Presumably an alkylated substrate serving as a ligand will most closely meet the requirement of stereochemical fit. There is, however, a risk that the ligand may be converted into a derivative similar to the product, possibly causing a drop in affinity. We have encountered such a case involving certain asparagine gels in contact with asparaginase.

The specificity of enzyme–substrate interaction may sometimes be so high that every possible substitution of the ligand (i.e., that required for attachment) may lead to complete annihilation of affinity, as apparently demonstrated by succinate ligands for succinate dehydrogenase. A better idea might then be to be less ambitious and try a cofactor ligand. A combination of several adsorbents of moderate specificity may sometimes be highly effective (Barker *et al.*, 1972).

Enzymes form complexes with inhibitors or substrates having association constants in the region of 10^4–10^{10} liters/mole. Attachment to a matrix will probably change these constants in some way and to an extent that is not always predictable. The matrix acts as a steric hindrance for complex formation. The connector and the ligand may decrease or increase the ability to associate. These effects are likely to be pronounced for low molecular weight ligands. If the ligand is a protein, however, there is a high probability that the association constant will be the same or very nearly the same for the free and the matrix-bound complex. Unfortunately, we are as yet short of experimental data for judging these affinity changes and their influence on adsorption chromatography. We are only able to give a qualitative evaluation. We shall exemplify pertinent aspects of biospecific affinity chromatography of enzymes by special reference to work on proteolytic enzymes and their inhibitors, since most applications up until now have dealt with such systems and we ourselves have had the most experience in this particular field. This may make the presentation somewhat one-sided, but many observations have general validity.

As will be seen, it is by no means necessary that the association constant for a particular soluble complex be especially large for the corresponding adsorbent complex to be suitable for chromatography. The *N*-ethyl asparagine–asparaginase system, for example, shows comparatively weak interaction in a medium where efficient chromatographic purification is feasible. In spite of this, Sepharose to which D-asparagine has been attached via the spacer substance 1,6-diaminohexane can be used as an efficient adsorbent (Kristiansen *et al.*, 1970). Figure 8 shows a series of tests carried out with and without inhibitors in the eluent. The displacement of the elution peak toward V_e indicates that the enzyme is adsorbed and desorbed specifically. Up until now linear biospecific affinity chromatography of this kind has hardly been used despite the fact that under easily controlled conditions the separation would be very high. Literature references, coupling methods, and spacers for enzymes and enzyme inhibitors are compiled in Table V.

1. *Proteolytic Enzymes*

Under specified conditions, chymotrypsin and trypsin form strong complexes with soybean trypsin inhibitor (STI) and other inhibitors of proteolytic enzymes from plants. In an extensive series of papers, Werle, Hochstrasser, Fritz and their collaborators have shown how immobilized trypsin may be utilized for the purification and isolation of trypsin inhibitors from different kinds of biologic material (for references, see Table V). A copolymer of maleic anhydride and ethylene (EMA) is used as matrix. Unfortunately, the resulting adsorbents have an exceptionally high content of carboxyl groups. Nevertheless, surprisingly good results have been obtained.

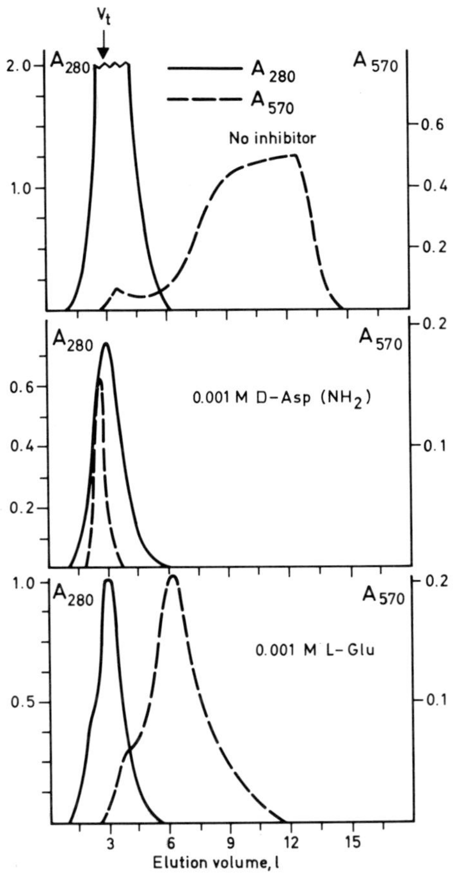

Fig. 8 Chromatograms of an asparaginase preparation obtained from columns of D-Asp-NH_2-Sepharose with and without inhibitors.

The rather high association constant for STI-trypsin is in the range of 10^9–10^{10} liters/mole. The distribution of enzyme between neutral solution and STI-gel therefore favors the gel phase. On the other hand, in slightly acid media chymotrypsin is found in the liquid phase. In strongly acid solution the same is true for trypsin. From this fact it can be inferred that these two enzymes should separate from inactive material and from one another by pH-gradient elution. Figure 9 shows the chromatogram from such an experiment using a crude extract from frozen and dehydrated pancreas. The purification was 200-fold, i.e., corresponding approximately to twice crystallized commercial enzyme. About the same degree of purity was obtained when chymotrypsin was subjected to the same one-step process. After three years of intensive

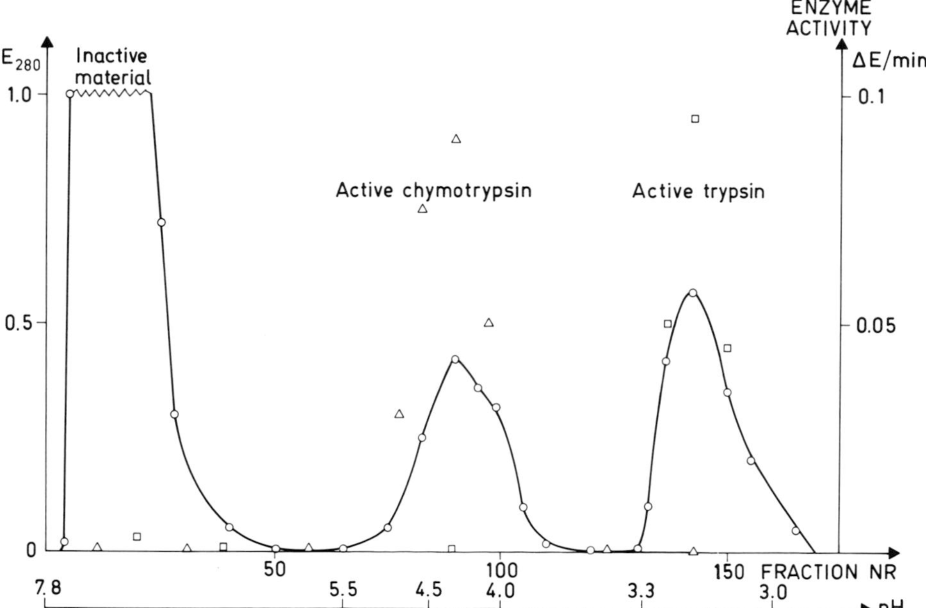

Fig. 9 Biospecific affinity chromatogram of crude pancreatic extract obtained by pH-gradient displacement from soybean trypsin inhibitor–agarose.

use, one of our agarose gels showed a drop in capacity of 33%. The STI-agarose gels are extremely stable with respect to inhibitor activity, but even better adsorbents are now being made from cross-linked desulfated agarose (Porath *et al.*, 1971) and hydroxylated gels (Porath and Sundberg, 1972b). This high stability is really remarkable considering the fact that the inhibitor is hydrolyzed both by trypsin and chymotrypsin, although admittedly to a rather limited extent.

Because of the broad inhibitory specificity of STI, other proteolytic enzymes may contaminate trypsin and chymotrypsin in the gradient-eluted material. In order to avoid contamination by other enzymes, specific displacement chromatography was tried. Chymotrypsin and trypsin were desorbed with tryptamine and benzamidine, respectively, at constant pH and ionic strength. Additional material was subsequently desorbed from the bed by lowering the pH to 3 (Fig. 10). All preparations could be further fractionated by zone electrophoresis. Different forms of chymotrypsin and trypsin as well as some unknown enzymes were obtained. With a similar technique, Kasche (1971) separated chymotrypsin from radiation-induced modified forms and degradation products.

A matrix-bound inhibitor may be used to isolate inactive "chemical

TABLE V

Biospecific Affinity Chromatography of Enzymes and Enzyme Inhibitors

Part A

Oxidoreductases	Matrix	Spacer	Spacer/matrix coupling method	Ligand	Ligand coupling method	Type of desorption	Reference
General[a] (GDH, G6PDH, ICDH, MDH, TDH, ADH, G3PDH)	Cellulose	ε-Aminohexanoic acid;	BrCN	NAD and	Water-soluble carbodiimide	Unspecific gradient elution	Lowe and Dean (1971), Lowe *et al.* (1972), Guilford *et al.* (1972), Ohlsson *et al.* (1972)
		glyglygly-	BrCN	NADP			
	o-Hydroxanilino cellulose	—	—	NADP	Diazo		
General (ADH)	Glass	—	—	NAD	—	—	Weibel *et al.* (1971)
General (G3PDH, LDH)	Agarose (Sepharose 6B)	ε-Aminohexanoic acid	BrCN	NAD^+	DCCI	Specific stepwise	Mosbach *et al.* (1971, 1972)
PK	(Sepharose 4B)	N^6-Aminohexyl AMP	BrCN	—	—	Specific stepwise	Mosbach *et al.* (1971, 1972)
LDH	Agarose (Sepharose 4B)	ε-Aminohexyl	BrCN	Oxamate	Water-soluble carbodiimide	Specific	O'Carra and Barry (1972)
Tyrosine hydroxylase	Agarose (Sepharose 4B)	—	—	3-Iodotyrosine	BrCN	Nonspecific	Poillon (1971)
Xanthinoxidase	Agarose (Sepharose)	—	BrCN[b]	—	—	Specific ($Na_2S_2O_4$)	Edmonson *et al.* (1972)
Xanthinoxidase and guanine dehydrogenase	Agarose (Sepharose)	—	—[c]	—	—	Specific	Baker and Siebeneick (1971)
Tetrahydrofolate dehydrogenase	Aminoethyl agarose	—	BrCN	Methotrexate	Water-soluble carbodiimide	Specific stepwise	Baker and Vermeulen (1970), Kaufman and Pierce (1971), Gauldie and Hillcoat (1972), Chello *et al.* (1972)
	Agarose (Sepharose 4B)	—	BrCN	Various inhibitors	—	—	
	Aminoethyl-Sepharose 4B	—	BrCN	Methotrexate	Water-soluble carbodiimide	Batchwise specific	Poe *et al.* (1972), Newbold and Harding (1971)
	Agarose (Sepharose 4B)	1,6-Hexanediamine	BrCN	Methotrexate	Water-soluble carbodiimide	—	
Tetrahydrofolate-binding proteins	Agarose (Sepharose 6B)	ε-Aminohexyl	BrCN	Folic acid	Water-soluble carbodiimide	Nonspecific	Kaufman and Pierce (1971), Salter *et al.* (1972)
Estradiol-17β-dehydrogenase	Agarose (Sepharose 4B)	ε-Aminohexanoic and ω-aminoundecanoic acid	BrCN	Estrone	DCCI	Specific stepwise	Nicolas *et al.* (1972)

Part B

Hydrolases	Matrix	Spacer	Spacer/matrix coupling method	Ligand	Ligand coupling method	Type of desorption	Reference
α-Chymotrypsin	Agarose (Sepharose)	ε-Aminocaproic acid	BrCN	D-Trp	—	Nonspecific	Cuatrecasas *et al.* (1968)
Chymotrypsin	Agarose (Sepharose 4B and 6B); Sephadex G-100 and G-200; cross-linked polyvinyl alcohol	—	—	Soybean trypsin inhibitor	BrCN, oxirane	Unspecific	Porath and Sundberg (1970), Kasche (1971)
Chymotrypsin	Agarose (Sepharose 4B)	—NH—$(CH_2)_4$—	BrCN	—C_6H_5	—	Nonspecific	Stevenson and Landman (1970)
Chymotrypsin I and α	Agarose (Sepharose 4B)	ε-Aminocaproic acid	BrCN	D-Trp	—	Nonspecific	Rovery and Bianchetta (1972)
Chymotrypsin	Agarose (Sepharose 6B)	—	—	Soybean trypsin inhibitor	BrCN	Specific	Poillon (1971)
Anhydrochymotrypsin	Agarose (Sepharose 2B)	—	—	Lima bean inhibitor	BrCN	Nonspecific	Ako *et al.* (1972)
Trypsin (a)	Agarose (Sepharose)	—	—	Soybean trypsin inhibitor (STI)	BrCN, oxirane	Nonspecific and specific	Porath and Sundberg (1970, 1972a,b)
(General) (b)	Cellulose	*p*-Aminobenzoic acid	Acylation	STI potato inhibitor	Diazo	Batchwise nonspecific	Mosolov and Lushnikova (1970)
(c)	Agarose (Sepharose 4B)	—	—	—	—	—	—
(General) (d)	Cellulose	Aminoethyl	Alkylation	Gly-D-Phe	Alkylation	Nonspecific	Uren (1970)
(e)	Sepharose 2B and 4B	—	—	Ovomucoid STI	BrCN	Nonspecific	Feinstein (1970), N. C. Robinson *et al.* (1971)
(f)	Agarose (Sepharose 4B)	(*p'*-Aminophenoxy)-propoxybenzamidine	—	—	BrCN	Nonspecific	Jameson and Elmore (1971)
(g)	Agarose (Sepharose)	—	—	Arginyl peptides	BrCN	Nonspecific	Kasai and Ischi (1972)
Carboxypeptidase	Agarose (Sepharose)	—	—	D-Ala-L-Arg, L-Tyr, D-Trp	BrCN	Nonspecific	Sokolovsky and Zisapel (1971), Cuatrecasas *et al.* (1968)

(Continued)

TABLE V (*Continued*)

Hydrolases	Matrix	Spacer	Spacer/matrix coupling method	Ligand	Ligand coupling method	Type of desorption	Reference
Carboxypeptidase B	Agarose (Sepharose)	—	—	—	—	—	Akanuma *et al.* (1971)
Carboxypeptidase A and B	Agarose (Sepharose 4B)	ϵ-Aminocaproic acid	BrCN	D-Trp, Soybean trypsin inhibitor	BrCN	Nonspecific	Reeck *et al.* (1971)
Carboxypeptidase A	Porous glass	Amino-, alkyl-, silyl-	—	Gly-D-Phe	Glutaraldehyde	—	P. J. Robinson *et al.* (1971)
Pepsinogen and pepsin	Agarose (Sepharose 4B); CM-cellulose; Sephadex G-200	—	—	Polylysine	BrCN	Nonspecific	Nevaldine and Kassel (1971)
Wheat protease	Sepharose	—	—	Hemoglobin	—	—	Chua and Bushuk (1969)
Papain	Agarose (Sepharose 4B)	—	—	Gly-Gly-Tyr-Bz-Arg	BrCN	Unspecific (H_2O)	Blumberg *et al.* (1970)
Plasminogen	Agarose (Sepharose)	—	—	Lysine	BrCN	Specific	Deutsch and Mertz (1970), Liu and Mertz (1971), Zolton and Mertz (1971), Zolton *et al.* (1972)
	Agarose (Sepharose)	—	—	*n*-Butyl-*p*-aminobenzoate ("Butesin")	BrCN	Specific	
Collagenases	Agarose (Sepharose 4B)	—	BrCN	—	BrCN	Nonspecific	Bauer *et al.* (1971)
Thrombin	Agarose	—NH—$(CH_2)_6$—CO	BrCN	—NH—CH_2—⌬—Cl; —⌬—Cl	—	Specific	Thompson and Davie (1971), Schmer (1972)
	Agarose	—NH—$(CH_2)_6$—CO	BrCN	*p*-Aminobenzamidine	—	Specific	
Kallikrein	Agarose (Sepharose 4B)	—	—	Trasylol	BrCN	Nonspecific	Seki *et al.* (1970), Fritz *et al.* (1969b, 1972a,b)
	CM-cellulose	—	Azide	"g-Inhibitor"	—	Specific	
	CM-cellulose	—	Azide	Soybean trypsin inhibitor		Specific and nonspecific	
	EMA	—	—	Kallikrein inhibitor	—	—	
Asparaginase	Agarose (Sepharose 6B)	—NH—$(CH_2)_6$—	BrCN	Asparagine	Alkylation	Specific	Kristiansen *et al.* (1970)

Asparaginase	Porous glass	—	—	Antibody	Silane	—	Weetall (1970)
Pancreatic lipase	Concanavalin A–agarose	—	—	—	—	Specific	Garner and Smith (1972)
Lipoprotein lipase	Agarose	—	—	—	—	—	Olivecrona and Egelrud (1971)
Acetylcholinesterase	Agarose (Sepharose 2B)	$H_2N-(CH_2)_5CO$ and others	BrCN	$H_2N-C_6H_4-N(CH_3)_3$ and others	Mixed anhydride	Specific and nonspecific	Dudai *et al.* (1972), Rosenberry *et al.* (1971), Kalderon *et al.* (1970)
β-Galactosidase	Agarose (Sepharose 4B); Polyacrylamide (Bio-Gel P300)	$H_2N-C_6H_4-$; 3-aminosuccinyl-3′-aminodipropylamine and others	BrCN	β-D-Galactothiopyranoside		Nonspecific	Tomino and Paigen (1970), Steers *et al.* (1970)
β-D-Xylosidase	Agarose (Sepharose 2B)	*p*-Aminobenzyl-	BrCN	Thio-β-D-xylopyranoside	Alkylation	Specific	Claeyssens *et al.* (1970)
Endopolygalacturonase	Cross-linked pectic acid	—	—	—	—	Specific and nonspecific	Rexová-Benková and Tivensky (1971)
Nuclease (staphylococcal)	Agarose (Sepharose)	—	—[d]	—	BrCN	Nonspecific	Cuatrecasas *et al.* (1968), Cuatrecasas (1970b)
Ribonuclease (tobacco)	Agarose (Sepharose 2B)	—[e]	—	—	BrCN	Specific	Jervis (1972)
	Agarose (Sepharose 2B)	ε-Aminohexanoic acid	BrCN	GMP	Water-soluble carbodiimide	—	—

Part C

Group transfer enzymes	Matrix	Spacer	Spacer/matrix coupling method	Ligand	Ligand coupling method	Type of desorption	Reference
Transaminases (GOT)	Agarose (Sepharose 4B)	Aminododecyl; aminodecyl	BrCN	Pyridoxamine-5′-phosphate	Aldimine reduction	Specific	Collier and Kohlhaw (1970)
Tyrosine aminotransferase	Agarose (Sepharose 4B)	Various	Various	Pyridoxamine-5′-phosphate	Various	Specific	Miller *et al.* (1971)
Aspartate aminotransferase	Agarose (Sepharose 4B)	1,6-Diaminohexane	BrCN	Pyridoxal-5′-phosphate	Aldimine reduction	Specific	Ryan and Fottrell (1972)
Protein kinase (→ cAMP independent form)	Agarose (Sepharose)	ε-Aminocaproic acid	BrCN (spacer-ligand)	cAMP (bound at position 6)	Anhydride of spacer	Specific and nonspecific; irreversible alteration of the enzyme	Wilchek *et al.* (1971)

(Continued)

TABLE V (*Continued*)

Group transfer enzymes	Matrix	Spacer	Spacer/matrix coupling method	Ligand	Ligand coupling method	Type of desorption	Reference
General phosphokinases	Agarose (Sepharose 2B and 4B); Sephadex G-10; Cellulose	—S—CH_2—CH_2—$^+NH_3$; —CH_2CH_2—NH—CO CO—HN—$(CH_2)_5$ | O—C$(Me)_3$; —S—CH_2—CH_2—NH | H_3N^+—$(CH_2)_4$—CO	BrCN	cAMP (bound at position 8)	—	—	Tesser *et al.* (1972)
Phosphofructokinase	Sephadex G-200; cross-linked polyacrylamide	—	—	Cibacron Blue F3G-A	—	Nonspecific and mixed specific-nonspecific	Böhme *et al.* (1972)
Phosvitin-kinase	Cellulose	—	—	Phosphate	—	—	Baggio *et al.* (1970)
Anthranilate-5-, phosphoribosyl-pyrophosphate phosphoribosyl transferase	Agarose (Sepharose)	Succinyl-amido-hexamethylimino-Sepharose	—	Anthranilic acid	—	Nonspecific	Marcus and Balbinder (1972)
Ceramide trihexosidase	Agarose (Sepharose)	—	Succinic acid	Melibiose		Nonspecific (Triton-containing buffer)	Mapes and Sweeley (1972)
Galactosyl-transferase	Agarose (Sepharose 4B)	—	—	α-Lactalbumin	BrCN	Specific (negative gradient)	—
	Agarose (Sepharose 4B)	6-Amino-1-hexanol	—	UDP-*N*-acetyl-glucosamine	BrCN	—	Mawal *et al.* (1971), Barker *et al.* (1972)
Pyruvate kinase	Sephadex G-200	—	—	Cibacron Blue F3G-A	—	Nonspecific	Röschlau and Hess (1972)
Thymidine kinase	Agarose	—	—	—	—	—	Rohde and Lezius (1971)
Nucleoside deoxyribosyl-transferase	Agarose (Sepharose)	Various	BrCN	Nucleosides	—	Nonspecific	Cardinaud and Holguin (1972)
DNA-polymerase	DNA-acrylamide	—	—	—	—	Nonspecific	Cavalieri and Carroll (1970)

	Matrix	Spacer	Spacer/matrix coupling method	Ligand	Ligand coupling method	Type of desorption	Reference
General (DNA-polymerase I and II, RNA-polymerase, exonuclease II, T4-polynucleotide kinase)	DNA 3% Agarose	—	—	—	—	Nonspecific	Brishammar (1972)
Polymerases: RNA-dependent DNA-polymerase	Agarose (Sepharose 4B)	—	—	Antibody	BrCN	—	Schaller *et al.* (1972)
RNA-polymerase	Sephadex G-200	—	—	DNA	—	—	Rickwood (1972)
RNA-polymerases (TMV-infected tobacco leaves)	Agarose (Sepharose 2B)	—	—	RNA	Divinyl sulfone	Specific	Brishammar (1972)
Chorismate mutase	Agarose (Sepharose 4B)	—	BrCN	L-Trp	—	Specific	Sprossler and Lingens (1970)

Part D

Lyases	Matrix	Spacer	Spacer/matrix coupling method	Ligand	Ligand coupling method	Type of desorption	Reference
Aldolase	Agarose (Sepharose 4B)	—	—	Enzyme and enzyme subunit	BrCN (low concentration)	Nonspecific	Chan and Mawer (1972)
Carbonic anhydrase	Sephadex G-150; Agarose (Sepharose 2B, 4B, 6B)	—	—	Sulfanilamide	BrCN	Specific	Porath and Sundberg (1970), Edmonson *et al.* (1972), Falkbring *et al.* (1972)
3-Deoxy-D-arabino-heptulosonate-7-phosphate synthetase	Sepharose	—	—	L-Tyr	—	—	Chan and Takahashi (1969)

Part E

Enzyme inhibitors	Matrix	Spacer	Spacer/matrix coupling method	Ligand	Ligand coupling method	Type of desorption	Reference
Trypsin inhibitors	Agarose (Sepharose 6B)	—	—	Trypsin	—	Nonspecific batchwise	Poillon (1971), Sundberg *et al.* (1970)

(*Continued*)

TABLE V (*Continued*)

Enzyme inhibitors	Matrix	Spacer	Spacer/matrix coupling method	Ligand	Ligand coupling method	Type of desorption	Reference
Trypsin inhibitors	—	—	—	Trypsin	BrCN	Nonspecific	Stewart and Doherty (1971), Chauvet and Acher (1972), Kassel and Harciniszyn (1971)
Trypsin inhibitors	Sephadex G-200	—	—	Trypsin	—	—	Beeley and McCairns (1972)
Serum inhibitors	Agarose (Sepharose)	—	—	Trypsin	BrCN	Nonspecific	Koj *et al.* (1972)
Chicken ovo-inhibitor	Agarose (Sepharose)	—	—	Chymotrypsin	BrCN	—	Feinstein (1971b)
Human sperm plasma	CM-cellulose	—	—	—	—	—	Fink *et al.* (1971, 1972)
Acrosin (boar)	Cellulose	—	—	Soybean trypsin inhibitor	—	—	—
Inhibitor of DNase I	Agarose (Sepharose 4B)	—	—	DNase I	BrCN	Nonspecific	Lindberg and Eriksson (1971)
RNase inhibitor	CM-cellulose	—	—	RNA	—	Nonspecific	Gribnau *et al.* (1970)
Various kinds of proteolytic inhibitors	EMA resin	—	—	Trypsin	—	—	Fritz *et al.* (1966, 1967, 1968a,b, 1969a, 1970a,c, 1971a,b)
	CM-cellulose	—	—	Trypsin	—	—	Fritz *et al.* (1970c, 1972a), Fink *et al.* (1971)
	Aminoalkyl-cellulose	—	—	Trypsin	—	—	Fritz *et al.* (1972a)
	Bio-Gel	—	—	—	—	—	Fritz *et al.* (1970b)

[a] Abbreviations: GDH, L-glutamate dehydrogenase; G6PDH, D-glucose-6-phosphate dehydrogenase; ICDH, *threo*-L_s-isocitrate dehydrogenase; MDH, L-malate dehydrogenase; DCCI, dicyclohexyl carbodiimide; LDH, L-lactate dehydrogenase; TDH, L-threonine dehydrogenase; ADH, alcohol (liver) dehydrogenase; G3PDH, D-glyceraldehyde-3-phosphate dehydrogenase; PK, pyruvate kinase; EMA, ethylene–maleic acid copolymerizate.

[b] [1-*H*-Pyrazolo(3,4*d*)-pyridine-4-ylamine]-1-propyl-6-aminohexanoate.

[c] 9(*p*-Aminoethoxyphenyl)-BrCN guanine.

[d] 3′-(4-Aminophenylphosphoryl)deoxythymidine-5-phosphate.

[e] 5′-(4-Aminophenylphosphoryl)GMP.

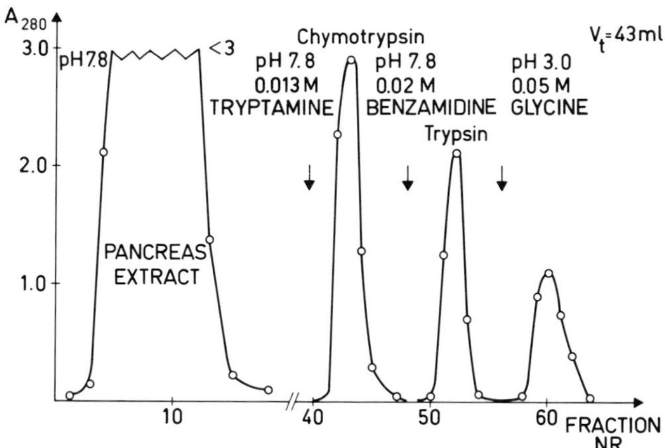

Fig. 10 Isolation of chymotrypsin and trypsin from crude pancreatic extract by one-step specific displacement and nonspecific displacement of residual enzymes.

mutants" from structurally different substances. For instance, Ako *et al.* (1972) have used lima bean inhibitor (LBI) to bind enzymatically inactive anhydrochymotrypsin. Since the serine in the active site of the enzyme has been converted to anhydroserine in the derivative, it is obvious that a covalent bond is not required for adsorption. Anhydrochymotrypsin actually has a greater association constant than chymotrypsin ($\sim 10^{10}$ vs. 7×10^{8} liters/mole) and therefore can easily displace the latter from LBI-Sepharose.

Peanut trypsin inhibitor (PNTI) is an example of an inhibitor that forms complexes not only with the enzyme but also with the corresponding zymogen. Stewart and Doherty (1971) have shown that PNTI can be adsorbed and desorbed under milder conditions from trypsinogen–agarose than from matrix-bound trypsin. They eluted PNTI after applying approximately five total volumes of buffer of constant concentration, a rare example of linear chromatography. The purification is obviously very effective, but published data do not permit an estimation of the purification factor. Separation of enzyme and zymogen should be easy by a combination of chromatography on STI- and PNTI-bound Sepharose. Feinstein (1971b), who has studied "turkey ovomucoid-Sepharose," suggests that the fixation reaction with BrCN blocks the binding site for trypsin but not for chymotrypsin. If this is true, we should be cautious in drawing conclusions about chromatographic properties from the behavior of the enzyme–enzyme inhibitor system in solution. After attachment to the polymer matrix, the specificity of the interaction may in fact increase.

Neurath and collaborators have used biospecific affinity chromatography in their studies of the evolutionary development of trypsins. For this purpose N. C. Robinson *et al.* (1971) prepared "CHOM"-Sepharose with the specific trypsin inhibitor chicken ovomucoid bound to Sepharose 4B. The inhibitor gel may be used for the rapid isolation of chymotrypsin-free trypsin from pancreatic juices of different animal species. An interesting technique employed by Robinson *et al.* (1971) to counteract autolysis was the addition of benzamidine to the equilibrating buffer. Beeley and McCairns (1972) isolated ovomucoid with immobilized trypsin, i.e., they reversed the system used by Robinson *et al.*

Other examples of the purification of protease inhibitors are referred to in Table V. It should be pointed out that there is a risk that the inhibitor might be partially hydrolyzed by the matrix-bound protease and therefore isolated in a partially degraded form.

Chymotrypsin and trypsin have also been fractionated on adsorbents to which low molecular weight ligands have been attached. Cuatrecasas *et al.* (1968) made the interesting discovery that the inhibitor D-tryptophan methyl ester must be extended from the matrix by a spacer in order to give the gel acceptable adsorption properties. A six-carbon chain drastically changes the conditions for the formation of the enzyme–inhibitor adsorption complex. The association constant for this complex is in the region of 10^4 liters/mole. The ligand concentration is about 10 μmoles/ml of gel.

In this connection Stevenson and Landman's (1970) separation of chymotrypsin-like enzymes is of great interest. As ligand and also as spacer they chose 4-phenylbutylamine (PBA). This substance forms a complex with a very low association constant (10^2–10^3 liters/mole). A very effective adsorbent is nevertheless obtained when the amine is coupled to Sepharose by the BrCN method ($\sim$50 μmoles/ml PBA-gel). Since the adsorption seems to be so strong, it is probable that the matrix-bound ligand forms a stronger complex than the free amine. The ligand–spacer–connector contains an interesting constellation of an aromatic ring, an aliphatic hydrophobic group, and a basic group. Blocking the active site of chymotrypsin strongly reduces or abolishes the affinity to the adsorbent, which shows that the strong binding is a result of specific adsorption.

In a very precise study Reeck *et al.* (1971) developed a rapid method for the isolation of carboxypeptidases A and B. In the first step trypsin and chymotrypsin are extracted from pancreatic juice by STI-Sepharose. This is followed by ultrafiltration and another biospecific purification step, this time using Trp-Sepharose. Different carboxypeptidase components are finally separated by ionic adsorption on DE-cellulose. The choice of low molecular weight ligands is not critical.

Sokolovsky and Zisapel (1971) have used adsorption chromatography on D-Ala-L-Arg-Sepharose to show that porcine carboxypeptidase B has the same intrinsic specificity as carboxypeptidase A.

Nevaldine and Kassell's (1971) method of purifying pepsin will serve as an example of specific interaction with a strong ion exchanger. Polylysine coupled to Sepharose 4B was used as adsorbent. In this case both specific and nonspecific adsorption occur, but it is impossible to assess their respective contributions.

2. *Cofactor and Effector-Dependent Enzymes*

An obvious way of using biospecific adsorption for the purification of cofactor and effector-dependent enzymes is to bind the cofactor or effector to a suitable matrix. Pioneer work in this field was carried out by Arsenis and McCormick in the sixties (Arsenis and McCormick, 1964). Although the cellulose adsorbents used by them hardly satisfy present demands, one can still obtain information of value from their documented experience. Of principal interest is the limitation of the method. A holoenzyme saturated with very strongly bound cofactor should have no residual capacity for adsorption to a cofactor gel. Collier and Kohlhaw (1970) have described such a case. Unlike the apoenzyme, glutamic-oxaloacetic transaminase holoenzyme is not adsorbed to an *N′*-(ω-aminoalkyl)-pyridoxamine-5′-phosphate gel. Collier and Kohlhaw therefore suggest incubation of the test solution with the inhibitor *O*-methyl hydroxylamine to release the apoenzyme prior to chromatography in order to increase the yield. We have noticed that no other authors seem to have touched upon this type of problem.

Using different methods, cofactors may be bound to the matrix via a spacer, usually via an amino group if one is available. Pyridoxalphosphate-dependent enzymes may be bound as pyridoxamine-5′-phosphate either in alkylated (Collier and Kohlhaw, 1970) or acylated form (Miller *et al.*, 1971). Cofactors with an adenosine group may be bound as the N^6-alkylated derivative as shown by Mosbach *et al.* (1971). Tesser *et al.* (1972) have synthesized cyclic AMP with different side chains in the 8-position. Two of the side chains contained an amino group for coupling the cofactor to the matrix using BrCN.

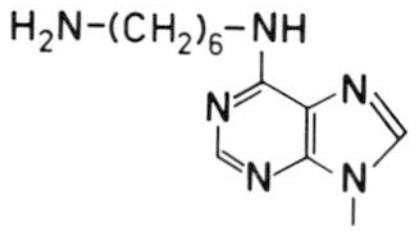

$H_2N{-}(CH_2)_6{-}NH$ (purine structure)

Mosbach derivative

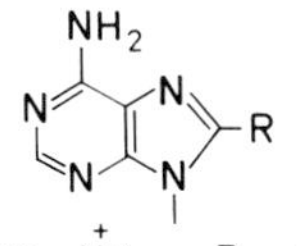

$R = -S-CH_2-CH_2-\overset{+}{N}H_3$ or $R = -S-(CH_2)_2-NH-CO-(CH_2)_5-\overset{+}{N}H_3$

Tesser derivative

Barker *et al.* (1972) prepared still another kind of cofactor derivative in which the amine-spacer is attached to the phosphate moiety of nucleotides. (R is a nucleoside such as adenosine or uridine.)

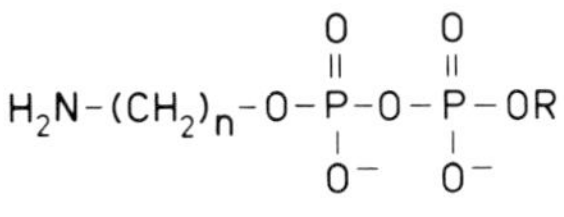

$$H_2N-(CH_2)_n-O-\underset{O^-}{\overset{O}{\overset{\|}{\underset{|}{P}}}}-O-\underset{O^-}{\overset{O}{\overset{\|}{\underset{|}{P}}}}-OR$$

Barker derivative

Lowe and Dean (1971) have pointed out that covalent binding of the adenine group changes the conformation of the coenzyme, which is important for binding to the enzyme. Most NAD(P)-dependent dehydrogenases complex with the coenzyme, causing it to undergo a conformational change.

Among the many interesting observations reported by Lowe and Dean is the fact that NADP-dependent enzymes are adsorbed to NAD-cellulose and, conversely, NAD-dependent enzymes are adsorbed to NADP-cellulose. Glyceraldehyde-3-phosphate dehydrogenase and alcohol dehydrogenase were exceptions. These enzymes were adsorbed neither to NAD- nor to NADP-cellulose.

Mosbach *et al.* (1972) have carried out a comprehensive study of the conditions for the biospecific adsorption and desorption of cofactor-dependent enzymes. Adsorbents have been made with NAD and AMP as ligands. Separation of different enzymes may be carried out by desorption with NAD^+ and NADH in separate steps. The competitive inhibitor salicylate is an effective displacer for separating glyceraldehyde-3-phosphate dehydrogenase adsorbed to AMP-Sepharose.

3. *Other Enzymes*

Many of the principles that have been discussed in the previous section are generally applicable to all enzymes. For instance, glucosidases are adsorbed to glucoside or thioglucoside gels, and acetylcholinesterases are adsorbed to a suitable inhibitor gel. We would like to illustrate biospecific separation of isoenzymes with an example from our own work on carbonic anhydrase which has been carried out over a period of many years in collaboration with Malmström's group (Falkbring *et al.*, 1972).

To obtain suitable adsorbents, sulfanilamide was coupled to Sephadex G-200 and Sepharose 6B. No spacer substance was used. Surprisingly, a considerably higher specific capacity was obtained with the Sephadex adsorbent. An explanation may be that cross-linked dextran contains glyceryl ether groups as "loose ends" on the polysaccharide matrix. These groups may serve as spacers. Adsorbents prepared from cross-linked desulfated agarose with sulfanilamide attached by a long-chain bisoxirane show a high capacity and considerable stability (L. Sundberg and P. Wåhlstrand, unpublished observations). Chromatography of different preparations on sulfanilamide–Sepharose can be carried out in an extremely efficient manner so that pure carbonic anhydrase can be isolated from crude extracts of bacteria or human erythrocytes in a single step.

Figure 11 shows a chromatogram of a hemolysate of human erythrocytes. Hemoglobin and other proteins appear in the eluate while carbonic anhydrase is adsorbed. The two forms of the enzyme, B and C, are displaced with 0.1 *M* NaI and 0.1 *M* KCNO, respectively. The adsorbent has nearly ideal properties, and purification factors exceeding several thousand are attained. The regeneration is quantitative and the

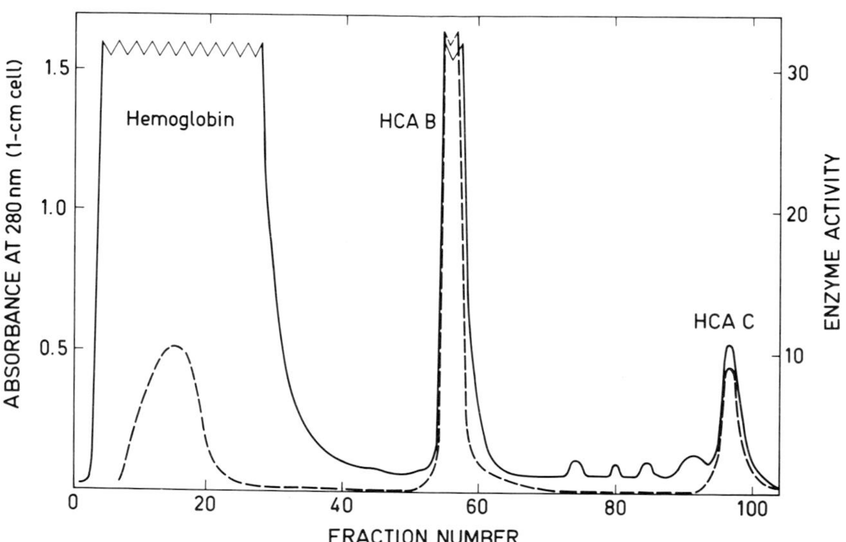

Fig. 11 Biospecific affinity chromatogram showing one-step isolation of the isoenzymes carbonic anhydrase B and C from a hemolysate of human erythrocytes on sulfanilamide-agarose.

chromatographic procedure may be repeated an unlimited number of times.

Similar substrates or substrate derivatives can also be used for the chromatographic purification of polymerases and depolymerases. In some cases it may be suitable to allow the ligand substance to penetrate the activated gel in the usual way and be bound to the carrier polymer. In other cases, particularly when the substrate is of high molecular weight and consists of random coils that considerably hinder permeation, it is possible to include the substrate or the substrate derivative in the matrix gel and then perhaps cross-link the two polymers. This is how, for instance, a DNA- or RNA-agarose gel can be formed. The gel beads are then cross-linked with divinyl sulfone.

Enzymes that directly attack the gel matrix may be considered as special cases. For instance, dextranase is adsorbed to Sephadex under suitable conditions (J.-C. Janson, personal communication). Cross-linked dextran also has affinity for other carbohydrate-splitting enzymes, e.g., amylase. Semenza and Kolinská (1967) used Sephadex G-200 for "enzyme-substrate chromatographic purification" of intestinal sucrose–isomaltase complexes. In this and some other similar cases, we encounter examples of linear chromatographic behavior with $V_e' > V_e$ and often $> V_t$.

Enzymes may be specifically fractionated by making use of characteristics other than their ability to bind substrate derivatives, inhibitors, or effectors. For example, if the enzyme contains the appropriate carbohydrates, immunosorption or adsorption to immobilized lectins can be carried out. Occasionally enzymes with SH groups can be separated on columns containing thiol reagents as ligands (Eldjarn and Jellum, 1963).

The literature contains a number of examples of specific affinity chromatography based on the interaction of enzymes with synthetic ligand substances not occurring in nature. For example, Sephadex G-200 to which Cibacron Blue F3G-A has been coupled is an efficient adsorbent for phosphofructokinase and pyruvate kinase. Böhme *et al.* (1972), who introduced this adsorbent, have discussed possible mechanisms for interaction that may explain the specificity of adsorption. It appears that an accidental similarity to the allosteric inhibitor ATP may be the cause of the specific affinity.

Biospecific adsorption of membrane-bound enzymes, conceivably in organic solvents, is still an almost virgin field. Agar and agarose gels, particularly when cross-linked, can easily be transferred to other media in which they swell (e.g., ethanol or dimethylsulfoxide) with little or no decrease in volume. These adsorbents should therefore be particularly well suited for chromatography in these media when a change of solvent is desirable. Enzymes attacking lipids may occasionally require mixed solvents.

Nicolas *et al.* (1972) have purified placental estradiol-17β dehydrogenase on estrone–Sepharose in a buffer containing 20% glycerol. The tests were carried out on gels of high and low ligand concentration. In the latter case a high ammonium ion concentration (1 *M*) was required for adsorption. Desorption was achieved by lowering the salt concentration. We think that this procedure might be a kind of "salting-out chromatography" (Tiselius, 1948) or a distribution chromatography similar to the one described by Martin and Porter (1951). The adsorption is, however, specific and requires that a ligand with high affinity be bound to the gel.

As a rule, enzyme adsorption–desorption occurs reversibly. There are, however, exceptions, for example, when the enzyme consists of several rather loosely bound subunits.

Cyclic AMP-dependent protein kinase consists of a regulatory unit binding cAMP and a catalytically active unit. When the kinase comes into contact with cAMP-Sepharose, the kinase activity passes through the column, whereas the regulatory activity is adsorbed so strongly that it cannot be eluted. As Wilchek *et al.* (1971) point out, it would be helpful if in this case one were able to remove the regulatory subunit from the matrix by chemical means. The method of biospecific adsorption on cAMP-Sepharose is very suitable for the isolation of the free catalytic unit.

Chan and Mawer's studies on aldolase (1972) may be cited as an example of biospecific affinity chromatography applied to a polymer enzyme, in this case one consisting of four equal subunits. The enzyme is bound by applying a very low BrCN concentration (1–5 mg/ml gel; see Section V,B,*2*) in the activation stage so that a "significant portion of the molecules [are] bound via one subunit." A bed of this matrix-bound aldolase was washed with a solution of 0.1 *M* Tris–HCl (pH 7.5), 8 *M* urea, 5 m*M* 2-mercaptoethanol, and 1 m*M* EDTA. In this way an "aldolase-subunit adsorbent" was obtained which contained 10% of the original activity and retained 25% of the protein content of matrix-bound aldolase. The matrix-bound aldolase could specifically bind additional aldolase subunits. Kinetic studies showed that these matrix-bound subunits are less stable toward alkali and heat treatment, probably because the formation of tetramers counteracts "unfolding."

In R. L. Hill's laboratory, Barker *et al.* (1972) have undertaken a penetrating study of the conditions for biospecific purification of galactosyl transferase. The general reaction involved is the following:

$$\text{Glycosyl}_1\text{-P-P-nucleoside} + \text{glycose}_2 \rightarrow \text{glycosyl}_1\ \text{glycose}_2 + \text{nucleoside-PP}$$

Ligands resembling the nucleoside portion of the substrate and the acceptor portion of the substrate were selected for the preparation of adsorbents. The ligand contained a 6-aminohexyl group as a spacer and thus the *N*-acetylglucosamine agarose had the following structure:

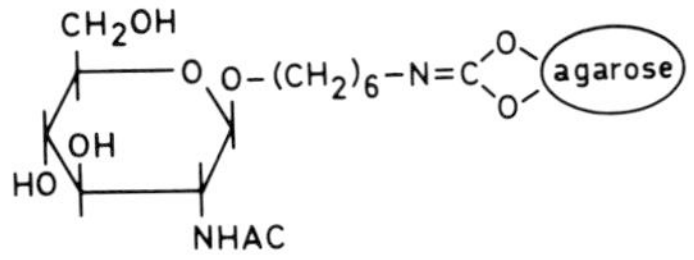

CH_2OH, OH, HO, NHAC, $O-(CH_2)_6-N=C$ (O, O) agarose

Barker *et al.* obtained galactosyl transferase from whey in two consecutive steps, each involving specific affinity chromatography. In the first step they used UDP-Sepharose and in the second, α-lactalbumin–Sepharose. The latter contains the enzyme itself as a portion of the macromolecular ligand. Under optimized conditions about 1000-fold purification was achieved. The product contained three electrophoretically distinct components of different molecular weight but very similar amino acid composition.

The method of Barker *et al.* seems to offer a general procedure for the isolation of glycosyl transferases. In Section VIII we will return to another aspect emerging from this remarkable study.

B. Immunosorbents

1. Soluble Antigens and Antibodies

The pattern that influences parameters for the isolation of antigens and antibodies by specific adsorption is somewhat different from that encountered in the specific sorption of enzymes and inhibitors.

Serum antibody to a given antigen sometimes includes a combination of IgG, IgA, and IgM, ranging in molecular weight from 150,000 for the IgG monomer to 900,000 for pentameric IgM. Within each class of immunoglobulins there is some variation in size and shape of the active sites. Accordingly, antibody molecules may vary considerably in their affinity for the corresponding antigen. Conversely, multivalent antigens, e.g., the blood group antigens, may be extremely polydisperse. The size distribution of antigenically specific blood group fragments extracted from cell membranes is a function of the particular method used for solubilization. In the case of immunosorbents we are therefore generally dealing with a system of reactants having markedly heterogeneous affinity. The immobilization by direct polymerization or by coupling to a matrix of antigens or antibodies will introduce still another affinity distribution of reactive sites. One important consequence of this fact is that the adsorption capacity of a given immunosorbent for the most weakly interacting member of a family of molecules may be exhausted long before the adsorbent is saturated with respect to the member showing the strongest interaction.

The strength of immunocomplexes is usually much higher than that of reversible enzyme–inhibitor complexes and is less dependent upon pH. Consequently, mere removal of unwanted antigens or antibodies by immunosorbents is a lot simpler than purification by adsorption and subsequent desorption. Desorption of minute amounts of antigens and antibodies without substantial loss of activity is still a major problem, particularly when small amounts of material are to be removed from very dilute solutions, a chore only partly eased by recent advances in membrane technology.

Analytical techniques in immunologic chemistry often must be extremely sensitive in order to be useful. If the immunosorbent matrix and the ligand–matrix bond are unstable, leakage will occur and tend to confuse the results. To realize the potential of immunosorption for large-scale preparation of material intended for therapeutic use (maybe even repeated injections into the same individual), it is clearly necessary to suppress leakage of unwanted substances from the adsorbent.

The investigator who wants to select an immunosorption technique likely to solve his particular separation problem is faced with a profusion of matrices, coupling methods, and desorption techniques. Many papers in this field are cheerfully optimistic, and the novice may get the impression, sadly untrue, that all combinations are acceptable and most are highly satisfactory. A number of reviews exclusively or partly concerned with immunosorbents have been published by Silman and Katchalski (1966), Nezlin (1970), Cuatrecasas (1970a,b,c, 1971a,b), Cuatrecasas and Anfinsen (1971a,b), Friedberg (1971), Feinstein (1971a), Pihar (1971), Weetall (1972), and Kocemba-Sliwowska (1972). We hope that a coherent treatment of a few selected methods will be more useful to the reader than scattered comments on a multitude of alternative techniques.

***a.* Directly Cross-Linked Antigens and Antibodies.** As already suggested, application of immunosorbents can be divided into two main sections, depending upon whether recovery of the adsorbate is essential or not. If removal of unwanted antigen or antibody from a sample is the primary goal, direct cross-linking of the complementary protein

TABLE VI

Part A Isolation of Antibody by Means of Directly Polymerized Antigen

Antigen polymerized	Polymerizer	Antibody source	Desorption of antibody	Reference
Human IgG	Glutaraldehyde	Rabbit serum	Glycine-HCl, pH 2.8	Avrameas and Ternynck (1969)
	Glutaraldehyde	Rabbit serum	2.5–5 *M* $MgCl_2$	Avrameas and Ternynck (1969)
	Glutaraldehyde	Rabbit serum	2 *M* NaI	Avrameas and Ternynck (1969)
	Glutaraldehyde	Human serum	Glycine-HCl, pH 2.8	Sordi and Piazzi (1970)
Murine IgG	Glutaraldehyde	Rabbit serum	Glycine-HCl, pH 2.8	Stanislawski and Coeur-Joly (1972)
Rabbit IgM	Ethyl chloroformate	Human serum	Glycine-HCl, pH 2.2	Avrameas and Ternynck (1967b)
Human serum albumin (HSA)	Glutaraldehyde	Rabbit serum	Glycine-HCl, pH 2.8	Avrameas and Ternynck (1969)
	Glutaraldehyde	Rabbit serum	2.5–5 *M* $MgCl_2$	Avrameas and Ternynck (1969)
	Glutaraldehyde	Rabbit serum	2 *M* NaI	Avrameas and Ternynck (1969)
	Ethyl chloroformate	Rabbit serum	Glycine-HCl, pH 2.2	Avrameas and Ternynck (1966)
	Glutaraldehyde	Horse serum	—	Varró and Bátory (1972)
Bovine serum albumin (BSA)	Glutaraldehyde	Horse serum	Glycine-HCl, pH 2.8	Sordi and Piazzi (1970)
Sheep serum	Glutaraldehyde	Horse serum	Glycine-HCl, pH 2.8	Sordi and Piazzi (1970)
Bence Jones K	Glutaraldehyde	Horse serum	Glycine-HCl, pH 2.8	Avrameas and Ternynck (1969)
Bence Jones L	Glutaraldehyde	Horse serum	Glycine-HCl, pH 2.2	Avrameas and Ternynck (1969)
	Ethyl chloroformate	Horse serum	Glycine-HCl, pH 2.2	Avrameas and Ternynck (1967b)
Transferrin	Ethyl chloroformate	Horse serum	Glycine-HCl, pH 2.2	Avrameas and Ternynck (1967b)
Ragweed antigen E	Ethyl chloroformate	Human serum	2 *M* KCl, 2 *M* NaCl	Goldstein *et al.* (1969)
	Ethyl chloroformate	Human serum	2 *M* KI, 2 *M* KSCN	Tannenbaum and Goodfriend (1971)
Australia antigen	Ethyl chloroformate	Human serum	3 *M* NaI	Lehmann *et al.* (1971)
Blood group substance A	L-Leucine-*N*-carboxyanhydride	Human serum	Acetate, pH 3.62	Kaplan and Kabat (1966)
	L-Leucine-*N*-carboxyanhydride	Human serum	*N*-acetyl-D-galactosamine	Kaplan and Kabat (1966)

Blood group substance B	L-Leucine-*N*-carboxyanhydride	Human serum	D-galactose	Kaplan and Kabat (1966)
Gliadin	Ethyl chloroformate	Human serum	2 *M* KCl, 2 *M* NaCl	Goldstein *et al.* (1969)
E. coli lipopolysaccharide	Glutaraldehyde	—	Glycine-HCl, pH 2.2	Eskenazy (1970)
Bacillus subtilis	Tetrazotized benzidine	Human serum	Phosphate-HCl, pH 2.3	Weetall (1967)
Serratia marcescens	Tetrazotized benzidine	Human serum	Phosphate-HCl, pH 2.3	Weetall (1967)
Concanavalin A	Glutaraldehyde	Human serum	—	Avrameas (1970b)

Part B Isolation of Antigens by Directly Polymerized Antiserum

Antigen	Polymerizer	Antiserum polymerized	Desorption of antigen	Reference
Human IgG	Glutaraldehyde	Rabbit	Glycine-HCl, pH 2.8	Avrameas and Ternynck (1969)
	Glutaraldehyde	Rabbit	2.5–5 *M* $MgCl_2$	Avrameas and Ternynck (1969)
	Glutaraldehyde	Rabbit	2–5 *M* NaI	Avrameas and Ternynck (1969)
	Ethyl chloroformate	Rabbit	Glycine-HCl, pH 2.2	Avrameas and Ternynck (1967b)
	Ethyl chloroformate	Rabbit	5 *M* KI in Tris, pH 8.2	Avrameas and Ternynck (1967a)
Human serum albumin	Ethyl chloroformate	Rabbit, IgG fraction	5 *M* KI in Tris, pH 8.2	Avrameas and Ternynck (1967a)
	Ethyl chloroformate	Rabbit	5 *M* KI in Tris, pH 8.2	Avrameas and Ternynck (1967a)
Human normal serum	Ethyl chloroformate	Goat	—	Radermecker and Goodfriend (1969)
Human allergic serum	Ethyl chloroformate	Goat	—	Radermecker and Goodfriend (1969)
Human IgE	Ethyl chloroformate	Rabbit	Glycine-HCl, pH 2.2	Ito *et al.* (1969)
	Ethyl chloroformate	Rabbit	2 *M* $MgCl_2$ in Tris, pH 7.5	Ito *et al.* (1969)
	Ethyl chloroformate	Rabbit	2 *M* NaCl in phosphate, pH 7.2	Ito *et al.* (1969)
Hepatitis antigen	Ethyl chloroformate	Human	—	Rapp *et al.* (1972)
Human myeloma IgG	Ethyl chloroformate	Human anti-myeloma	—	Rivat *et al.* (1971)

to form an insoluble antigen or antibody affords a simple and efficient method. The mechanical properties of insolubilized proteins obtained in this way make them more suitable for batch separation than for column procedures. Satisfactory flow properties during column operation will be restricted to beds of fairly coarse particles with a low surface-to-weight ratio, allowing only slight penetration of adsorbate and resulting in low efficiency. Still, if protein is freely available, the simplicity of the method amply compensates for this imperfection. Furthermore, flow properties may be improved by the addition of beads of polyacrylamide, Sephadex, or agarose. The two most useful agents for direct polymerization of proteins appear to be ethylchloroformate and glutaraldehyde. A brief discussion of their relative merits in the preparation of immunosorbents is given in the key paper by Avrameas and Ternynck (1969). These authors reported use of proteins, polymerized by glutaraldehyde, in columns of 1–2 cm diameter and 2–4 cm height, operated at flow rates of serum of 30–40 ml/hr. Desorption was performed with glycine-HCl (pH 2.8) of concentrations up to 5 *M* NaI or $MgCl_2$. Antiprotein antibodies were recovered in a yield of 60%–98% by weight and claimed to be 70%–100% precipitable by the homologous antigen. Eskenazy (1970) reported the insolubilization of lipopolysaccharide antigen by means of glutaraldehyde and the successful isolation of the homologous antibody. Selected references to methods involving direct polymerization are given in Table VI. A general discussion of desorption techniques will be found in Section VII,B,*4*.

***b.* Methods Involving Inert Carriers.** We will consider only adsorbents consisting of a ligand covalently coupled to the carrier. A few claims of success have been reported for carriers containing ligands that were either physically adsorbed or mechanically imbedded in a polymer lattice. We feel that the inherent instability of these adsorbents, especially toward strong desorbing agents, makes them unreliable as immunosorbents. In general, the use of these inert carriers is justified only for particulate material or when a satisfactory coupling procedure for the ligand in question is not available.

The choice of a matrix for an immunosorbent is governed by the demand for a fairly open structure allowing penetration of immunoglobulins ranging in molecular weight from 150,000 to 900,000, either for coupling as ligands or for adsorption to coupled antigen ligands. In addition, the matrix should be resistant to fairly drastic desorption conditions, e.g., pH < 3.5 or high concentrations of chaotropic ions often required to break immune complexes.

Cellulose lends itself readily to derivatization preceding various coupling reactions and has been much used in the past because of the reasonably high surface-to-volume ratio of swollen cellulose. However, cellulose suffers from several drawbacks. We have found that it has a tendency to continuously release long carbohydrate chains with ligands coupled to them, even at physiological pH and salt concentration. (This was also suspected by Rickwood (1972) but in a different context.) Furthermore, cellulose contains an abundance of charged groups, leading to substantial nonspecific adsorption, and its resistance to chaotropic ions is not satisfactory. Still, immunosorbents based on bromoacetyl–cellulose ("BAC") have been quite useful (Robbins *et al.*, 1967; Clerici and Schechter, 1970; Kostner and Holasek, 1970; Schenkein *et al.*, 1971). Polyacrylamide beads, while readily reacting to form derivatives and admirably inert toward strong desorbing agents, are largely unsuitable because of insufficient porosity to start with and even further shrinkage during the activation reaction.

An interesting technique recently introduced by Ternynck and Avrameas (1972) involves the coupling of antigens and antibodies to polyacrylamide beads activated with glutaraldehyde.

Several reports of the use of chemically substituted porous glass beads as immunosorbents have appeared. This matrix is potentially attractive because of its inertness in acid solution and its favorable mechanical properties. However, due to the high charge density of glass, it is afflicted with considerable nonspecific adsorption, creating problems both in desorption and regeneration of the adsorbent.

The consistently good results reported by Weetall (1970, 1972) have been difficult to duplicate, at least in our laboratory. For a tabulation of the physical and chemical properties of other matrix substances currently used, the reader is referred to Table III.

2. *Agarose Derivatives as Immunosorbents*

Since the somewhat accidental discovery of the BrCN coupling method, agarose beads have been widely used for bioaffinity chromatography of enzymes and inhibitors. Their acceptance as a carrier for immunosorbents has been slower, largely due to the traditional commitment of several leading research groups to other types of matrices. We never quite shared the satisfaction with agarose beads expressed by others in some early papers. Several of our colleagues complained of a small though embarrassing leakage of soluble agarose–protein conjugates, even under mild conditions. Resistance toward chaotropic ions, especially thiocyanate, was low. (This fact may have passed unnoticed by some workers primarily concerned with the immunologic rather than the chemical purity of desorbed material.) In an attempt to overcome these drawbacks, cross-linked types of agar derivatives were developed (Porath *et al.*, 1971). The principal advantages of the new type of gel as an immunosorbent matrix are the following: It can be autoclaved. In most media of biologic interest, continuous leakage of carbohydrate is extremely low, although not completely absent. It can be desorbed with strong solutions of chaotropic ions, e.g., I^-, SCN^-, and CCl_3COO^-, without dissolving. We strongly advise prospective users to consider this kind of gel. Awaiting its commercial availability, the investigator can easily convert conventional agarose beads into specific immunosorbents. Table VII gives several examples of immunosorbents prepared by activation of conventional agarose beads with BrCN.

When coupling immunoglobulins to a matrix, it is essential to avoid

TABLE VII

Immunosorbents Based on Agarose Beads Activated with BrCN[a]

Adsorbate	Adsorbent	Reference
Human IgE	Sheep anti-IgE-Sepharose 2B	Bennich and Johansson (1971)
Human brain-specific antibodies	Rabbit anti-brain extract-Sepharose 4B	Tripatzis *et al.* (1971)
Human brain-specific α_2-glycoprotein	Rabbit anti-brain extract-Sepharose 4B	Warecka *et al.* (1972)
Human, dog, and rabbit α-fetoprotein	Horse anti-α-fetoprotein-Sepharose 6B	Nishi and Hirai (1972)
Antibodies to human myeloma proteins, Bence Jones proteins, Waldenström's macroglobulins	Nonspecifically purified antigens	Mannik and Stage (1971)
Rabbit, goat, and horse antibodies to *Naja nigricollis* α_1-toxin	α_1-Toxin-Sepharose 4B	Détrait and Boquet (1972)
Rabbit antibodies to sperm whale myoglobin	Metmyoglobin-Sepharose 4B	Boegman and Crumpton (1970)
Glucagon from pig ileum	Rabbit anti-glucagon-Sepharose 4B	Murphy *et al.* (1971)
Rabbit antibodies to human FSH	HCG-Sepharose 4B	Sato and Cargille (1972)
Monkey growth hormone	Specifically purified sheep antibodies to human placental lactogen, coupled to Sepharose 4B	Guyda and Friesen (1971)

[a] The following abbreviations are used: FSH = follicle-stimulating hormone; HCG = human chorionic gonadotropin.

multipoint attachment of the antibody molecule by way of ϵ-amino groups. Such a linkage will tend to place the antibody adjacent to the carbohydrate chain of the matrix and make it difficult for the antigen to reach the antibody. As discussed by Cuatrecasas and Anfinsen (1971b), coupling to BrCN-activated carbohydrates seems to occur primarily by way of unprotonated amino groups. A reduction in pH to about 6.5 will enhance protonization and reduce the probability of multipoint coupling of the same antibody molecule. Activation with a larger amount of BrCN will compensate for the ensuing reduction of coupling yield. The penalty will be a marked increase in nonspecific adsorption due to the introduction of charged groups. We have considered the opposite approach, i.e., low level activation (0.1 gm BrCN/gm dry gel), the idea being to increase the distance between active matrix coupling sites so that multipoint attachment will be minimized. The advantage would be a substantial decrease in nonspecific adsorption. Actual experiments with the coupling of horse antihuman lymphocyte globulin (ALG) have shown that activation of ECD-Sepharose 2B with 1 gm BrCN/gm dry gel and coupling at pH 8.6 lead to a high coupling yield (100 mg/gm gel) but also to an adsorbent of poor capacity for lymphocyte membrane antigen. Low level activation (0.1 gm BrCN/gm gel) and coupling at pH 8.6 give a yield of only 3–4 mg/gm gel. Although still unacceptably low, adsorbent capacity is higher than that of the previous adsorbent. However, high level activation (10–20 gm BrCN/gm gel) and coupling at pH 6.5 give yields exceeding 200 mg ALG/gm ECD-Sepharose 2B, representing 80%–100% of protein added to 50 ml sedimented gel.

3. *Immunoselective Fractionation of Cells*

The presence of immunologically active determinants on the surface of cells forms the basis of specific separation methods. Lymphocytes from both immunized and nonimmunized animals generally carry immunoglobulins on their surfaces (see Nossal and Ada, 1971), and numerous papers in recent years have reported enrichment of lymphocytes according to their surface receptors. Antibody-coated (Evans *et al.*, 1969) and antigen-coated (Mage *et al.*, 1969) polyurethane foam have been used to separate guinea pig erythrocytes and antibody plaque-forming cells, respectively. Abdou and Richter (1969) used antigen-coated glass beads to fractionate antigen-reactive cells from rabbit bone marrow.

Polyacrylamide beads carrying covalently coupled haptens were used by Truffa-Bachi and Wofsy (1970) and Wofsy *et al.* (1971). A series of papers on cellular immunosorbents was published by Wigzell's group: selected references are Wigzell (1970, 1971) and Wigzell and Andersson (1971). Antigen cross-linked with ethylchloroformate has been applied by Tallberg and collaborators in the form of slides that are able to selectively bind lymphocytes carrying complementary antibodies (Tallberg *et al.,* 1971). A very interesting technique for fractionation of lymphoid cells differing in Ig receptors was recently introduced by Edelman's group (Rutishauser *et al.,* 1972). This involves adsorption of cells onto antigen-coated nylon fibers, and the authors claim that the technique allows easier quantitation of bound cells than procedures based on beaded column adsorbents.

4. *Desorption from Immunosorbents*

The dissolution of antigen–antibody complexes formed on adsorbents and the subsequent recovery of either antigen or antibody in high yield and pure form is a tall proposition indeed. The problem is partly caused by the nonexistence of conditions for complete desorption with absolute conservation of the original antigen or antibody properties. Besides, the very powerful selectivity of biospecific sorption techniques often leads to the desorption of minute amounts of material of extremely high specific activity, and the absence of stabilizing molecular species so nicely eliminated from the starting material may induce conformational changes that are only partly reversible. Further losses are inevitably suffered during concentration and dialysis of the desorbed material which is frequently very dilute. The investigator who is not free to work with plenty of complementary antigen and antibody is therefore urged not to be discouraged by final recoveries in the 20%–30% range. Several reports claiming antigen and antibody recoveries of 90%–100% describe uncomplicated systems of little scientific interest, apparently chosen to demonstrate the superiority of a particular adsorbent, coupling method, or desorbing agent.

***a.* Specific Desorption.** Gentle desorption of antibodies at nearly neutral pH can sometimes be achieved with a hapten of low molecular weight that is easily removed by dialysis of the desorbed material. For instance, antibodies to polysaccharide or proteoglycan antigens may be released from the adsorbent by high concentrations of the terminal monosaccharide. An example is the desorption by *N*-acetyl-D-galactosamine of antibody to blood group substance A from an adsorbent obtained by the cross-linking of the A substance with *N*-carboxy-L-leucine anhydride (Kaplan and Kabat, 1966). It may be necessary to work at fairly high ionic strength in order to prevent nonspecific readsorption of released antibody onto charged groups on the matrix.

***b.* Nonspecific Desorption.** i. *Adjustment of pH.* Innumerable papers could be quoted on the breaking of antigen–antibody complexes at low pH (<3.5). However, many protein antigens are wholly or partially denatured at this level of acidity.

ii. *Unfolding agents.* Guanidine hydrochloride and urea have been used with some success as disrupting agents (Bata *et al.*, 1964).

iii. *Ionogenic complexing agents.* Svensson *et al.* (1970) took advantage of the ability of borate to form complexes with certain carbohydrates (Weigel, 1963) to dissociate polysaccharides and their antibodies. Other agents complexing with hydroxyl groups (Weigel, 1963) may conceivably be used as desorbers. Due to the reversibility of these complexes, simultaneous complexing with the matrix itself is not a serious objection.

iv. *Detergents.* Crumpton and Parkhouse (1972) studied the ability of some ionic and nonionic detergents to inhibit the formation of antigen–antibody complexes. Strongly charged sodium dodecyl sulfate and Sarkosyl L had a marked effect at a concentration of 0.5%, whereas weakly charged sodium deoxycholate and the neutral detergents Tween 80, Triton X-100, Nonidet P-40, and Brij 58 had next to no effect. Unfortunately, sodium dodecyl sulfate has a reputation for denaturing many proteins and caution is necessary (see Ch. 3).

v. *Chaotropic ions.* This term, coined by Hamaguchi and Geiduschek (1962) in a paper dealing with the denaturation of DNA by various salts, denotes ionic species modifying the structure of water by breaking hydrophobic bonds. The effect of chaotropic ions has been the subject of several papers, chiefly from the groups of Dandliker and Hatefi: selected references are Hatefi and Hanstein (1969) and Levison *et al.* (1970). The subject has been reviewed by Dandliker and de Saussure (1971). Some chaotropic ions appear to be useful for desorption from immunosorbents. In order of decreasing efficiency, the principal chaotropic ions can be listed as follows:

$$CCl_3COO^- > SCN^- > ClO_4^- > I^- > Cl^-$$

However, they are by no means innocuous and have to be used with care. Several authors have reported desorption of antibodies by chaotropic ions from immunosorbents prepared by the coupling of antigens to BrCN-activated agarose beads. To our surprise, few authors have expressed concern about the substantial dissolution of the agarose matrix itself which occurs with 1–4 *M* concentrations of chaotropic ions above the efficiency level of chloride. Some protection against dissolution is provided by the inadvertent cross-linking due to the BrCN method, but it does not protect against the action of thiocyanate and trichloroacetate. We are therefore inclined to be skeptical about the quantitative data presented in earlier papers and tend to regard some of the good results as fortuitous. ECD-Agarose, when properly prepared, has good but not absolute resistance even to 3 *M* trichloroacetate at pH 7.4. Immunoglobulins seem to vary in their proneness to irreversible denaturation by chaotropic ions, even at neutral pH. Dandliker and de Saussure (1968) reported that 19%–96% of precipitable rabbit antibody to bovine and serum albumin and ovalbumin could be recovered when 2–4 *M* SCN^- was used for desorption. Antibody constituted 55%–86% of the eluted material, indicating considerable nonspecific adsorption. Bennich and Johansson (1971) used 3.5 *M* SCN^- to desorb IgE from a Sepharose-based immunosorbent. Recovery of IgE was comparable to that obtained by acid dissociation.

We have treated human hyperimmune anti-A serum with various chaotropic ions at different concentrations. Hemagglutinating activity was completely and irreversibly lost by incubation for 2 hr with 3 *M* SCN^- at room temperature. Under the same conditions ClO_4^- and I^- caused 30%–50% inactivation. Avrameas and Ternynck (1967a) found no inactivation of human anti-A or anti-B serum by 2 *M* I^- after incubation for 24 hr. However, in 5 *M* I^-, hemagglutinating activity was completely destroyed after 1 hr. With 3 *M* CCl_3COO^- we found only very slight inactivation of human anti-A serum after 2 hr, whereas only 10% of adsorbed anti-A activity was recovered from an A-substance–agarose column by a 0–3 *M* gradient of CCl_3COO^-. On the other hand, incubation of anti-A with 2 *M* CF_3COO^-, which is more weakly chaotropic than CCl_3COO^-, caused no loss of activity after 12 hr at room temperature. Gradient elution with 0–3 *M* CF_3COO^- led to more than 50% recovery of anti-A activity (T. Kristiansen, unpublished experiments). It appears that concentrations of haloacetates above 1.5 *M* are rarely needed to desorb ECD-agarose immunosorbents completely. Chaotropic properties of haloacetates have been studied by Hanstein *et al.* (1971). Some antigens seem to be more vulnerable than antibodies to chaotropic ions. Edgington (1971) obtained evidence that membrane antigens may be more liable to irreversible injury by chaotropic ions than the corresponding antibodies. He found that the D antigen of erythrocyte membranes was partially inactivated by 1.5 *M* SCN^- or 2.0 *M* I^- and destroyed altogether by higher concentrations. In our laboratory we have had similar experiences with human lymphocyte membrane antigens. Pending further evidence, our tentative recommendation for nonspecific desorption of immunosorbents is a 0–1.5 *M* gradient of CF_3COO^- or CCl_3COO^- at near neutral pH and low temperature.

5. *Immunoassay Methods Employing Biospecific Adsorbents*

By use of immobilized antigens or antibodies participating in systems where labeled and unlabeled adsorbate molecules compete in solution for their common countersites on the adsorbent, nanogram amounts of antibodies and antigens can be detected and quantitated. The label may be a fluorochrome, an enzyme, or a radioactive isotope.

Several methods have been described for coupling enzymes to antigens and antibodies. The complexes partially retained enzymatic and immunologic activities (see Avrameas, 1968, 1970a) which were used to detect and characterize the corresponding antibodies and antigens. This idea was adapted to immunologic quantitation by Engvall and Perlmann (1971). Antiglobulin was coupled to BrCN-activated cellulose, and alkaline phosphatase was conjugated to the corresponding globulin with glutaraldehyde. Competition for antiglobulin sites on the adsorbent between a known amount of enzyme–globulin conjugate and an unknown amount of globulin will be reflected in the variable enzymatic activity of the adsorbent, which can be determined with a suitable substrate. A similar technique was used by Avrameas and Guilbert (1971a,b).

In a long series of papers, Wide and collaborators have described applications of radioimmunoassays based on the coupling of antigens and antibodies to Sephadex with BrCN. According to the RIST (Radio-Immuno-Sorbent-Test) technique first described by Wide and Porath (1966) and Wide *et al.* (1967a), anti-IgE is coupled to the matrix. Standard curves are obtained by measurement of the radioactivity of the adsorbent after interaction with mixtures of ^{125}I-labeled and unlabeled IgE in different proportions. The RAST (Radio-Allergo-Sorbent-Test) technique (Wide *et al.*, 1967b) involves coupling a specific allergen to the matrix. The specific IgE corresponding to this allergen will be selectively adsorbed from the IgE pool by the allergen–matrix conjugate. After washing the adsorbent–IgE complex, labeled anti-IgE is added and will attach to the IgE already bound. The three-membered complex is then counted in a gamma counter. Variations of these procedures have been successfully applied to hormone determination, for instance the diagnosis of pregnancy by assay of chorionic gonadotropin (Wide, 1969a), serum vitamin B_{12} (Wide and Killander, 1971), and β_2-microglobulin (Evrin *et al.*, 1971). General aspects of the method have been discussed by Wide (1969b, 1972).

C. Other Systems

1. Lectins

Lectins (phythemagglutinins) are substances present in plants, capable of agglutinating red blood cells. Their biologic role is not known, but a characteristic property of lectins is their ability to bind various carbohydrates by forming molecular complexes. Glucose, galactose, fucose, and others are bound to different lectins, often with high stereospecificity (Tobiška, 1964). For example, a lectin present in *Dolichus biflorus* reacts with substances containing terminal 2-acetamido-2-deoxy-D-galactopyranose (Etzler and Kabat, 1970). It is therefore natural to expect that lectin-based adsorbents will gain in importance in biospecific affinity chromatography.

Agrawal and Goldstein (1965) have shown how concanavalin can be adsorbed to Sephadex G-200 after preliminary purification of a jack bean extract by precipitation with ammonium sulfate. The chromatography was performed in 1 *M* NaCl in order to reduce nonspecific adsorption, and 0.1 *M* glucose was used as a specific desorber. In a publication of basic importance in lectin chemistry, Poretz and Goldstein (1970) mapped the saccharide binding site in concanavalin A and investigated the forces involved in complex formation. A diagrammatic representation of the binding site for polar saccharides is shown in Fig. 12. The binding constant for methyl-LD-mannopyranoside was determined to be 2.06×10^4 liters/mole and for methyl-L-α-glucopyranoside 5.4×10^3 liters/mole (Poretz and Goldstein, 1970).

Aspberg and Porath (1970) made an adsorbent that is group-specific for glycoproteins by coupling concanavalin A to agarose by the BrCN method. Human serum proteins were divided chromatographically into two fractions, the principal one containing mainly albumin, γ-globulin, and some other components of low carbohydrate content. The second fraction, desorbed with L-α-mannoside, showed a considerably higher content of α- and β-globulins and, upon electrophoresis at pH 8.6, revealed rapidly migrating γ-globulins.

Lloyd (1970) has prepared concanavalin–agarose in a similar way and tested its adsorption capacity for carbohydrates and blood group substances.

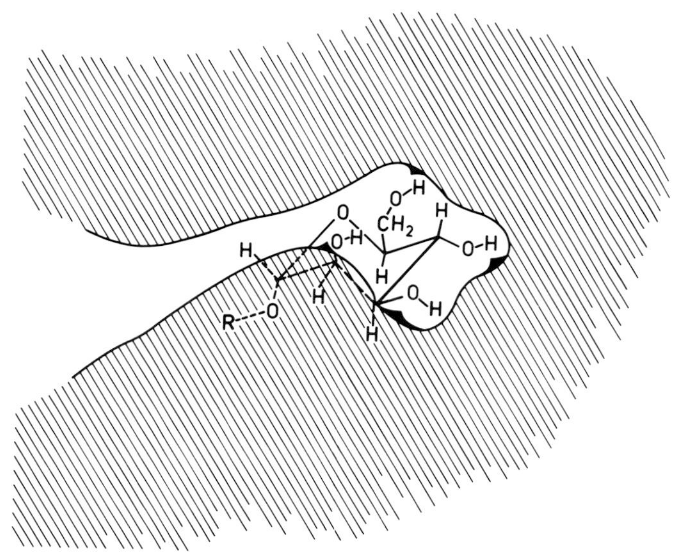

Fig. 12 Schematic representation of the carbohydrate-combining site of concanavalin A. Redrawn from Poretz and Goldstein (1970).

A lectin from *Vicia cracca* is an example of a lectin adsorbing directly to the Sephadex matrix (Aspberg *et al.*, 1968). Like concanavalin, this lectin is desorbed with glucose.

Ukita and co-workers have isolated lectins by biospecific affinity chromatography. Matsumoto and Osawa (1972) have cross-linked inhibitory saccharides to starch and then purified lectins specifically attacked by the saccharides. L-Fucose–starch adsorbs lectin from *Ulex europaeus* and hemagglutinin from eel serum. The lectin can be eluted by a change of pH and ionic strength. Tomita *et al.* (1972) from the same laboratory have separated galactose-binding lectins by Sepharose chromatography.

At our institute, Sepharose has been specially prepared for more efficient adsorption of galactose-binding lectins. Ersson and Porath have hydrolyzed cross-linked Sepharose and adsorbed a lectin from Crotalaria on this adsorbent (Ersson *et al.*, 1973). Porath and Fornstedt have purified lectins isolated from different plants using lactose and fucoidin covalently bound to Sepharose with divinyl sulfone (J. Porath and N. Fornstedt, unpublished experiments).

In an early paper describing the use of the BrCN method (Kristiansen *et al.*, 1969), we showed that coupled aliphatic amines were stably bound in the pH range of 3–12. We also carried out a systematic comparison of different coupling procedures. Immunosorbents were prepared by attachment of blood group substance A to Sepharose 2B, to aminoethyl cellulose, and to lysine agarose. In the first case, Sephadex was activated as usual by BrCN, then partially deacetylated blood group substance was coupled. With this procedure up to 80 mg A substance/gm matrix could be attached. The two other coupling procedures are unusual in that the ligand was activated by BrCN and then coupled to aminoethyl cellulose or to lysine agarose. Lysine agarose is especially interesting since it illustrates how, by using a diamine (lysine), two carbohydrates can be coupled without amino groups in either the ligand or the unsubstituted matrix. The spacer provides amino groups for both the ligand and the original matrix. The lysine agarose method resulted in a yield of 23 mg A substance/gm dry gel.

Vicia cracca lectin–Sepharose has been used with great success for isolating blood group substance from ovarian cysts. This adsorbent has been especially useful for purifying and isolating active fragments from hydrolyzed blood group substance A (Kristiansen, 1974a,b).

Table VIII is a compilation of work published on lectin–carbohydrate adsorbents.

2. *Peptide Fractionation*

S-Peptide and S-protein from ribonuclease form an active complex spontaneously. Kato and Anfinsen (1969) prepared S-protein–Sepharose and could in this way purify synthetic S-peptides. The adsorption was not entirely specific but was considerable, even for closely related side products. In a similar way, synthetic peptides containing sequences present in staphylococcal nuclease were purified (Ontjes and Anfinsen, 1969). Givol *et al.* (1970) have used a similar technique for isolating DNP-peptides on Sepharose to which anti-DNP antibodies have been attached.

Wilchek (1970) has introduced a very elegant variation of biospecific affinity chromatography. Staphylococcal nuclease has affinity for thymidine 3′,5′-diphosphate. By introducing a substituent with this nucleotide by the affinity labeling technique, nuclease fragments are obtained which, after hydrolysis, show affinity to nuclease–Sepharose. All peptides which have been labeled in this way are adsorbed to a bed of nuclease–Sepharose. Desorption is effected with ammonia.

In the future it seems probable that peptides synthesized by the solid phase technique or other methods will be more often purified with Sepharose-bound antibodies directed toward the proteins (or specific hydrolytic fragments) to be synthesized. However, it is not to be ex-

TABLE VIII

Lectin–Carbohydrate Complexes

Carbohydrate or glycoprotein	Matrix	Lectin	Coupling method	Reference
Dextran	Cross-linked dextran (Sephadex G-200)	Concanavalin A	—	Agrawal and Goldstein (1965)
Blood group substance A (matrix-bound or free)	Sepharose 2B; lysine-Sepharose 2B; aminoethyl cellulose; cellulose	*Vicia cracca* anti-A lectin (free or matrix-bound)	BrCN	Kristiansen *et al.* (1969), Kristiansen (1974a,b)
Dextran	Sephadex G-100	*Vicia cracca* "nonspecific" lectin	—	Aspberg *et al.* (1968)
Various carbohydrates; blood group substance	Sepharose	Concanavalin A (matrix-bound)	BrCN	Lloyd (1970)
Glycoproteins in human plasma	Sepharose	Concanavalin A	BrCN	Aspberg and Porath (1970)
L-Fucose, tri-*N*-acetylchitotriose, thyroglobulin	Starch	*Ulex europeus* lectin, *Cytisus sessilifolius* lectin, *Phaseolus vulgaris* lectin	—	Matsumoto and Osawa (1972)
Agarose	Sepharose	*Ricinus communis* lectin, *Momordia charanta* lectin, *Abrus precatorius* lectin	—	Tomita *et al.* (1972)
Agarose (partially hydrolyzed)	Partially hydrolyzed, cross-linked Sepharose	*Crotalaria* lectin	—	Ersson *et al.* (1973)
Human plasma glycoproteins	Cross-linked Sepharose	*Phaseolus vulgaris* lectin	BrCN	J. Porath and N. Fornstedt, unpublished experiments
Blood group substance A + H	Copolymer of leucine-*N*-carboxyanhydride and blood group substance A + H	*Dolichos biflorus* lectin	—	Etzler and Kabat (1970)
N-ε-Aminocaproyl-β-D-galactopyranosylamine	Sepharose	Soybean	BrCN	Gordon *et al.* (1972)

pected that the specific methods will entirely replace conventional purification procedures.

It is possible in some cases to base the fractionation of peptides on the spontaneous tendency of similar peptides to associate to dimer or oligomer complexes. Association and dissociation occur in separate media:

$$\text{Ⓜ}-P_1+P_2+P_3+ \rightleftarrows \text{Ⓜ}-P_1\cdots P_1+P_2+P_3+$$

$$\downarrow \quad \begin{matrix}\text{(1) filtration}\\ \text{(2) change of medium}\end{matrix}$$

$$\text{Ⓜ}-P_1+P_1$$

Such a procedure resembles classical methods of purifying synthetic products by recrystallization.

3. *Miscellaneous*

The biotin–avidin system has played a role in the development of the biospecific adsorption technique. McCormick (1965) prepared biotin-cellulose by reacting cellulose with biotin chloride in pyridine. Unfortunately, simultaneous specific and nonspecific adsorption made the purification procedure ineffective. Porath and Miller-Andersson (Porath, 1967) attached biotin to Sephadex via a phenylglyceryl spacer. Using egg white as starting material a 100,000-fold purification was achieved. Since only 35% yield was obtained in the first experiment with the fresh gel and the yield dropped on subsequent use, the project was temporarily abandoned. As in the McCormick procedure, desorption was achieved by lowering the ionic strength. We ascribe the poor result to the presence of carboxyl groups in the gel. Cuatrecasas and Wilchek (1968) have improved the biospecific adsorption method by introducing agarose as the matrix for this system. They displaced avidin with 6 *M* guanidine-HCl, a severe treatment resulting in avidin denaturation, although reversible. We feel, however, that a satisfactory adsorbent for this very difficult system has not yet been devised. Cross-linked desulfated and possibly reduced agarose might be a better proposition. Purification of biotin on avidin adsorbents should be much easier. Bodanszky and Bodanszky (1970) have purified biotin-containing peptides with avidin–Sepharose.

Carrier proteins provide complex systems suitable for biospecific chromatography. Olesen and co-workers (1971) have coupled hydroxyl cobalamine albumin conjugates to bromoacetyl cellulose and obtained an adsorbent for cobalamine-binding proteins in serum and gastric juice. The adsorption was carried out batchwise and protein was desorbed

TABLE IX

Miscellaneous Biospecific Purification Procedures

Adsorbate	Adsorbent	Reference
RNA	tRNA cellulose	Bautz and Hall (1962)
RNA	DNA-nitrocellulose	Gillespie and Spiegelman (1965), Goldhaber (1965) Bautz and Reilly (1966)
tRNA	Anti-tRNA antibodies coupled to agarose	Bonavida *et al.* (1970)
tRNA	Aminoacyl-tRNA-synthetase	Denburg and DeLuca (1970)
Single-stranded DNA and RNA	Nucleic acid agarose	Poonian *et al.* (1971)
rDNA	tRNA-agarose, poly-rU agarose, and 16 S-rRNA-agarose	Robberson and Davidson (1972)
DNA-associated proteins	DNA-cellulose	Alberts *et al.* (1968)
Adenine nucleotides	Thymidine cellulose	Sander *et al.* (1966)
Deoxyribonucleotides (chemisorption)	Polynucleotide cellulose	Cozzarelli *et al.* (1967), Jovin and Kornberg (1968)
Histidyl-tRNA	Histidyl-tRNA synthetase agarose	Blasi *et al.* (1971)
Aminoacyl-tRNA synthetases	Aminoacyl-tRNA-bromo-acetylamidobutyl-Sepharose	Bartkowiak and Pawel-kiewicz (1972)
Phe-tRNA	Oxidized poly U on poly-acrylhydrazide agar	Engelhardt *et al.* (1970)
"Codases"	Aminoacyl-tRNA on poly-acrylhydrazide agar	Engelhardt *et al.* (1970)
Estrogen receptor	Polyvinyl-(*N*-phenylene-maleimide)-estradiol; Estradiol-*p*-diazobenzyl-ether of cellulose	Wonderhaar and Mueller (1969)
Testosterone-binding proteins	Androstane derivative coupled to agarose	Burstein (1969)
Estradiol-binding proteins	Estradiol derivatives coupled to agarose	Cuatrecasas and Anfinsen (1971b)
Aldosterone-binding proteins	Deoxycorticosterone bound to aminoethylagarose	Ludens *et al.* (1972)
Corticosteroid-binding globulin	Cortisol-Sepharose	Rosner and Bradlow (1971)
Retinol-binding proteins	Thyroxine-binding prealbumin coupled to agarose	Vahlquist *et al.* (1971)
Thyroxine-binding proteins	L-Thyroxine-agarose	Pensky and Marshall (1969)
B_{12}-binding proteins	Cellulose-B_{12}-derivatives	Gräsbeck *et al.* (1970)
Human complement C1′	Human IgG-agarose	Bing (1971a)
Human complement C1′s	*Meta*-aminobenzamidine-Sepharose	Bing (1971b)
Plant hormone 2,4D	2,4D-Lysine coupled to agarose	Venis (1971)

(incompletely) with hydroxycobalamine. Table IX shows other work on several transport proteins and other protein systems.

VIII. CONCLUDING REMARKS

The substances to be purified must be capable of complex formation. In nature, protein complexes are ubiquitous, and this fact immediately indicates the general character of biospecific procedures and, at the same time, the importance of initial fractionation of the components of the complex. Sometimes the biologic activity is displayed only by the complex and not by the isolated components. This may complicate the use of biospecific adsorption for isolation in some instances, but in others may facilitate the study of the interactions between separate components in multicomponent systems. Starting with one component as a matrix-bound ligand, the other components may be added, one after the other, in a deliberate sequence to form multiple complexes with different physical, chemical, and biologic properties. Frontal chromatography should be a useful technique for analyses of this kind.

It is likely that bioaffinity chromatography will be widely used in the future for the isolation of hormones and drug receptors, among others. In view of the important role of lipids in the function of membrane-bound enzymes, further development of bioaffinity chromatography may be necessary. For instance, chromatography in lipophilic or amphipatic (mixed hydrophilic–lipophilic) elution systems might be required in order to prevent irreversible conformational changes of membrane-bound proteins. Presumably the composition of the solvent system used should resemble the natural lipid environment as much as possible. As carriers of biospecific adsorbents in such systems, agarose and agar derivatives are likely to be suitable since agar gels, particularly in cross-linked form, will easily accommodate both polar and nonpolar solvents.

An interesting variant of biospecific adsorption is enzymatically directed chemisorption. In principle, an acceptor substance is anchored on a solid support and treated with a donor substrate in the presence of the corresponding enzyme, for example, a transferase. A gel-bound product is formed which in turn may act as a substrate in a new reaction. Such enzymatic solid phase synthesis has been suggested by Barker *et al.* (1972) for oligosaccharides. Great difficulties may be anticipated in this area, but this kind of enzymatic synthesis may eventually become feasible.

Biospecific affinity chromatography is likely to be useful for *in vitro* studies of the biochemical actions of xenobiotics. Drugs, carcinogens, pesticides, herbicides, and antibiotics may be tested by *in vitro* adsorption experiments designed to reveal or confirm the interactions of these substances with enzymes or hormones. For example, drugs may be specifically adsorbed in beds of immobilized enzymes or hormones which are suspected to be involved in receptor sites. Conversely, suitably devised derivatives of the xenobiotics may be covalently bound to gel matrices and be used as specific adsorbents for specifically interacting proteins and other naturally occurring substances. Biospecific fractionation procedures may become very valuable for testing drugs and chemicals used for environmental control as well as for detecting, on the biochemical level, adverse effects of synthetic organic compounds and their conversion products.

Concerning the general methodology of bioaffinity chromatography, much effort is still needed to develop matrices which, under all conceivable conditions of biologic research, have high stability and negligible leakage and retain the chemical reactivity required for derivatization. The problems of denaturation and inactivation of desorbed substances are to a large extent inherent in any separation method of high selective power. Even the most resourceful separationist will sometimes have to accept the elusive nature of his prey whenever the natural chemical environment, which he considers an impurity, is a prerequisite for the stability and biologic activity of the substance he chases.

Desorption problems tend to need individual solutions, and the imaginative reader is cordially invited to contribute to the pool of knowledge that will make life easier for present and future colleagues.

REFERENCES

Abdou, N. I., and Richter, M. (1969). *J. Exp. Med.* **130,** 141.

Agrawal, B. B. L., and Goldstein, I. J. (1965). *Biochem. J.* **96,** 23C.

Akanuma, H., Kasuga, A., Akanuma, T., and Yamasaki, M. (1971). *Biochem. Biophys. Res. Commun.* **45,** 27.

Ako, H., Ryan, C. A., and Foster, R. J. (1972). *Biochem. Biophys. Res. Commun.* **46,** 1639.

Alberts, B. M., Amodio, F. J., Jenkins, M., Gutmann, E. D., and Ferris, F. J. (1968). *Cold Spring Harbor Symp. Quant. Biol.* **33,** 289.

Araki, C. (1956). *Bull. Chem. Soc. Jap.* **26,** 543.

Arsenis, C., and McCormick, D. B. (1964). *J. Biol. Chem.* **239,** 3093.

Aspberg, K., and Porath, J. (1970). *Acta Chem. Scand.* **24,** 1839.

Aspberg, K., Holmén, H., and Porath, J. (1968). *Biochim. Biophys. Acta* **160,** 116.

Avrameas, S. (1968). *Immunochemistry* **6,** 43.

Avrameas, S. (1970a). *Int. Rev. Cytol.* **27**, 349.
Avrameas, S. (1970b). *C. R. Acad. Sci., Ser. D* **270**, 2205.
Avrameas, S., and Guilbert, B. (1971a). *C. R. Acad. Sci., Ser. D* **273**, 2705.
Avrameas, S., and Guilbert, B. (1971b). *Eur. J. Immunol.* **1**, 394.
Avrameas, S., and Ternynck, T. (1966). *C. R. Acad. Sci., Ser. C* **262**, 1175.
Avrameas, S., and Ternynck, T. (1967a). *Biochem. J.* **102**, 37C.
Avrameas, S., and Ternynck, T. (1967b). *J. Biol. Chem.* **242**, 1651.
Avrameas, S., and Ternynck, T. (1969). *Immunochemistry* **6**, 53.
Axén, R., and Ernback, S. (1971). *Eur. J. Biochem.* **18**, 351.
Axén, R., and Porath, J. (1964). *Acta Chem. Scand.* **18**, 2193.
Axén, R., and Vretblad, P. (1971a). *Protides Biol. Fluids, Proc. Colloq.* **18**, 383.
Axén, R., and Vretblad, P. (1971b). *Acta Chem. Scand.* **25**, 2711.
Axén, R., Porath, J., and Ernback, S. (1967). *Nature (London)* **214**, 1302.
Axén, R., Vretblad, P., and Porath, J. (1971). *Acta Chem. Scand.* **25**, 1129.
Baggio, B., Pinna, L., Morel, V., and Siliprandi, N. (1970). *Biochim. Biophys. Acta* **212**, 515.
Baker, B. R., and Siebeneick, H.-U. (1971). *J. Med. Chem.* **14**, 799.
Baker, B. R., and Vermeulen, N. M. J. (1970). *J. Med. Chem.* **13**, 1143.
Bar-Eli, A., and Katchalski, E. (1960). *Nature (London)* **188**, 856.
Bar-Eli, A., and Katchalski, E. (1963). *J. Biol. Chem.* **238**, 1960.
Barker, R., Olsen, K. W., Shaper, J. H., and Hill, R. L. (1972). *J. Biol. Chem.* **247**, 7135.
Barker, S. A., Somers, P. J., Epton, R., and McLaren, J. V. (1970). *Carbohyd. Res.* **14**, 287.
Bartkowiak, S., and Pawelkiewicz, J. (1972). *Biochim. Biophys. Acta* **272**, 137.
Bata, J. E., Gyenes, L., and Sehon, A. H. (1964). *Immunochemistry* **1**, 289.
Bauer, E. A., Jeffrey, J. J., and Eisein, A. Z. (1971). *Biochem. Biophys. Res. Commun.* **44**, 4.
Bautz, E. K. F., and Hall, B. D. (1962). *Proc. Nat. Acad. Sci. U. S.* **48**, 400.
Bautz, E. K. F., and Reilly, E. (1966). *Science* **151**, 328.
Beeley, J. G., and McCairns, E. (1972). *Biochim. Biophys. Acta* **271**, 204.
Benesch, R. E., and Benesch, R. (1956). *J. Amer. Chem. Soc.* **78**, 1597.
Bennich, H., and Johansson, S. G. O. (1971). *Advan. Immunol.* **13**, 1.
Bing, D. H. (1971a). *J. Immunol.* **107**, 1243.
Bing, D. H. (1971b). *Immunochemistry* **8**, 539.
Blasi, F., Goldberger, R. F., and Cuatrecasas, P. (1971). *In* "Biochemical Aspects of Reactions on Solid Supports" (G. R. Stark, ed.), Ch. 2, p. 105. Academic Press, New York.
Blumberg, S., Schechter, I., and Berger, H. (1970). *Eur. J. Biochem.* **15**, 97.
Bodanszky, A., and Bodanszky, M. (1970). *Experientia* **26**, 327.
Boegman, R. J., and Crumpton, M. J. (1970). *Biochem. J.* **120**, 373.
Böhme, H. J., Kopperschläger, G., Schultz, J., and Hofmann, E. (1972). *J. Chromatogr.* **69**, 209.
Bonavida, B., Fuchs, S., and Sela, M. (1970). *Biochem. Biophys. Res. Commun.* **41**, 1335.
Brishammar, S. (1972). Doctoral Dissertation, Uppsala, Sweden.
Brown, E., Racois, A., and Gueniffey, H. (1971). *Bull. Soc. Chim. Fr.*, p. 4341.
Burstein, L. H. (1969). *Steroids* **14**, 263.
Campbell, D. H., Luescher, E., and Lerman, L. S. (1951). *Proc. Nat. Acad. Sci. U. S.* **37**, 575.

Cardinaud, R., and Holguin, J. (1972). *Biochem. J.* **127,** 30P.
Catt, K., Niall, H. D., and Tregear, G. W. (1967). *Nature (London)* **213,** 825.
Cavalieri, L. F., and Carroll, E. (1970). *Proc. Nat. Acad. Sci. U. S.* **67,** 807.
Cebra, J. J., Givol, D., Silman, H. J., and Katchalski, E. (1961). *J. Biol. Chem.* **236,** 1720.
Centeno, E. R., and Sehon, A. H. (1971). *Immunochemistry* **8,** 887.
Chan, W. W.-C., and Mawer, H. M. (1972). *Arch. Biochem. Biophys.* **149,** 136.
Chan, W. W.-C., and Takahashi, M. (1969). *Biochem. Biophys. Res. Commun.* **44,** 347.
Chauvet, J., and Acher, R. (1972). *FEBS Lett.* **23,** 317.
Chello, P. L., Cashmore, A. R., Jacobs, S. A., and Bertino, J. R. (1972). *Biochim. Biophys. Acta* **268,** 30.
Chua, G. K., and Bushuk, W. (1969). *Biochem. Biophys. Res. Commun.* **37,** 545.
Chudzik, J., and Koj, A. (1972). *Postepy Biochem.* **18,** 73.
Claeyssens, M., Kersters-Hilderson, H., Van Wauve, J.-P., and de Bruyne, C. K. (1970). *FEBS Lett.* **11,** 336.
Clerici, E., and Schechter, I. (1970). *Int. Arch. Allergy Appl. Immunol.* **38,** 554.
Collier, R., and Kohlhaw, G. (1970). *Anal. Biochem.* **42,** 48.
Cozzarelli, N. R., Melechen, M. E., Jovin, I. M., and Kornberg, A. (1967). *Biochem. Biophys. Res. Commun.* **28,** 578.
Crumpton, M. J., and Parkhouse, R. M. E. (1972). *FEBS Lett.* **22,** 210.
Cuatrecasas, P. (1970a). *J. Biol. Chem.* **245,** 574.
Cuatrecasas, P. (1970b). *J. Biol. Chem.* **245,** 3059.
Cuatrecasas, P. (1970c). *Nature (London)* **228,** 1327.
Cuatrecasas, P. (1971a). *In* "Biochemical Aspects of Reactions on Solid Supports" (G. R. Stark, ed.), p. 79. Academic Press, New York.
Cuatrecasas, P. (1971b). *J. Agr. Food Chem.* **19,** 600.
Cuatrecasas, P., and Anfinsen, C. B. (1971a). *Annu. Rev. Biochem.* **40,** 259.
Cuatrecasas, P., and Anfinsen, C. B. (1971b). *In* "Methods in Enzymology" (W. B. Jacoby, ed.), Vol. 22, p. 345. Academic Press, New York.
Cuatrecasas, P., and Wilchek, M. (1968). *Biochem. Biophys. Res. Commun.* 33, 235.
Cuatrecasas, P., Wilchek, M., and Anfinsen, C. B. (1968). *Proc. Nat. Acad. Sci. U. S.* **61,** 636.
Dandliker, W. B., and de Saussure, V. A. (1968). *Immunochemistry* **5,** 357.
Dandliker, W. B., and de Saussure, V. A. (1971). *In* "The Chemistry of Biosurfaces" (M. L. Hair, ed.), Vol. 1, p. 1. Dekker, New York.
Denburg, J., and DeLuca, M. (1970). *Proc. Nat. Acad. Sci. U. S.* **67,** 1057.
Dennis, D. (1968). *Anal. Biochem.* **24,** 544.
Détrait, J., and Boquet, P. (1972). *C. R. Acad. Sci. Ser. D* **274,** 1765.
Deutsch, D. G., and Mertz, E. T. (1970). *Science* **170,** 1095.
Duckworth, M., and Yaphe, W. (1971). *Carbohyd. Res.* **16,** 189.
Dudai, Y., Silman, I., Kalderon, N., and Blumberg, S. (1972). *Biochim. Biophys. Acta* **268,** 138.
Edgington, T. B. (1971). *J. Immunol.* **106,** 673.
Edmonson, D., Massey, V., Palmer, G., Beacham, L. M., III, and Elion, G. B. (1972). *J. Biol. Chem.* **247,** 5.
Eldjarn, L., and Jellum, E. (1963). *Acta Chem. Scand.* **17,** 2610.
Engelhardt, W. (1924). *Biochem. Z.* **148,** 463.
Engelhardt, W., Kisselev, L. L., and Nezlin, R. S. (1970). *Monatsh. Chem.* **101,** 1510.

Engvall, E., and Perlmann, P. (1971). *Immunochemistry* **8,** 811.

Ersson, B., Aspberg, K., and Porath, J. (1973). *Biochim. Biophys. Acta* **310,** 446.

Eskenazy, M. (1970). *Nature (London)* **226,** 855.

Etzler, M. E., and Kabat, E. A. (1970). *Biochemistry* **9,** 869.

Evans, W. H., Mage, M. G., and Peterson, E. A. (1969). *J. Immunol.* **102,** 899.

Evrin, P.-E., Peterson, P. A., Wide, L., and Berggård, I. (1971). *Scand. J. Clin. Lab. Invest.* **28,** 439.

Falkbring, S. O., Göthe, P. O., Nyman, P. O., Sundberg, L., and Porath, J. (1972). *FEBS Lett.* **24,** 229.

Feinstein, G. (1970). *Biochim. Biophys. Acta* **214,** 224.

Feinstein, G. (1971a). *Biochim. Biophys. Acta* **236,** 74.

Feinstein, G. (1971b). *Naturwissenschaften* **58,** 389.

Fink, E., Jaumann, E., Fritz, H., Ingrisch, H., and Werle, E. (1971). *Hoppe-Seyler's Z. Physiol. Chem.* **352,** 1591.

Fink, E., Schiessler, H., Arnhold, M., and Fritz, H. (1972). *Hoppe-Seyler's Z. Physiol. Chem.* **353,** 1633.

Friedberg, F. (1971). *Chromatogr. Rev.* **14,** 121.

Fritz, H., Schult, H., Hutzel, M., Wiedermann, M., and Werle, E. (1966). *Hoppe-Seyler's Z. Physiol. Chem.* **348,** 308.

Fritz, H., Eckert, J., and Werle, E. (1967). *Hoppe-Seyler's Z. Physiol. Chem.* **348,** 1120.

Fritz, H., Trautschold, I., Haendle, H., and Werle, E. (1968a). *Ann. N. Y. Acad. Sci.* **146,** 400.

Fritz, H., Hochstrasser, K., Werle, E., Brey, B., and Gebhardt, M. (1968b). *Fresenius' Z. Anal. Chem.* **243,** 452.

Fritz, H., Gebhardt, M., Fink, E., Schramm, W., and Werle, E. (1969a). *Hoppe-Seyler's Z. Physiol. Chem.* **350,** 129.

Fritz, H., Brey, B., Schmal, A., and Werle, E. (1969b). *Hoppe-Seyler's Z. Physiol. Chem.* **350,** 617.

Fritz, H., Gebhardt, M., Meister, R., Illchmann, K., and Hochstrasser, K. (1970a). *Hoppe-Seyler's Z. Physiol. Chem.* **351,** 571.

Fritz, H., Gebhardt, M., Meister, R., and Schult, H. (1970b). *Hoppe-Seyler's Z. Physiol. Chem.* **351,** 1119.

Fritz, H., Fink, E., Meister, R., and Klein, G. (1970c). *Hoppe-Seyler's Z. Physiol. Chem.* **351,** 1344.

Fritz, H., Brey, B., Müller, M., and Gebhardt, M. (1971a). *Proc. Int. Res. Conf. Proteinase Inhibitors, Munich 1970,* pp. 28–37. de Gruyter, Berlin and New York.

Fritz, H., Gebhardt, M., Meister, R., and Fink, E. (1971b). *Proc. Int. Res. Conf. Proteinase Inhibitors, Munich 1970,* pp. 271–280. de Gruyter, Berlin and New York.

Fritz, H., Brey, B., and Béress, L. (1972a). *Hoppe-Seyler's Z. Physiol. Chem.* **353,** 19.

Fritz, H., Wunderer, G., and Dittmann, B. (1972b). *Hoppe-Seyler's Z. Physiol. Chem.* **353,** 893.

Garner, C. W., and Smith, L. C. (1972). *J. Biol. Chem.* **247,** 561.

Gauldie, J., and Hillcoat, B. L. (1972). *Biochim. Biophys. Acta* **268,** 35.

Giddings, J. C. (1965). "Dynamics of Chromatography. Part I: Principle and Theory." Dekker, New York.

Gillespie, D., and Spiegelman, S. (1965). *J. Mol. Biol.* **12,** 829.

Givol, D., Weinstein, Y., Gorecki, M., and Wilchek, M. (1970). *Biochem. Biophys. Res. Commun.* **38**, 825.

Goldhaber, P. (1965). *Science* **147**, 407.

Goldman, R., Goldstein, L., and Katchalski, E. (1971). *In* "Biochemical Aspects of Reactions on Solid Supports" (G. R. Stark, ed.), pp. 1–78. Academic Press, New York.

Goldstein, G. B., Underdown, B. J., Heiner, D. C., Rose, B., and Goodfriend, L. (1969). *Int. Arch. Allergy Appl. Immunol.* **36**, 573.

Goldstein, L., Pecht, M., Blumberg, S., Atlas, P., and Levin, Y. (1970). *Biochemistry* **9**, 2322.

Gordon, J. A., Blumberg, S., Lis, H., and Sharon, N. (1972). *FEBS Lett.* **24**, 193.

Gräsbeck, R., Puutula, L., and Aro, H. (1970). *Scand. J. Clin. Lab. Invest.* **25**, Suppl. 113, 64.

Gribnau, A. A. M., Schoenmakers, J. G. G., van Kraaikamp, M., and Bloemendal, H. (1970). *Biochem. Biophys. Res. Commun.* **38**, 1064.

Grubhofer, H., and Schleith, L. (1954). *Hoppe-Seyler's Z. Physiol. Chem.* **297**, 108.

Guilford, H., Larsson, P.-O., and Mosbach, K. (1972). *Chem. Scripta* **2**, 1.

Gurvich, A. E. (1957). *Biokhimiya* **22**, 977.

Gurvich, A. E., and Drizlikh, G. J. (1964). *Nature (London)* **203**, 648.

Gurvich, A. E., Kapner, R. B., and Nezlin, R. S. (1959). *Biokhimiya* **26**, 935.

Guyda, H. J., and Friesen, H. G. (1971). *Biochem. Biophys. Res. Commun.* **42**, 1068.

Hamaguchi, K., and Geiduschek, E. P. (1962). *J. Amer. Chem. Soc.* **84**, 1329.

Hanstein, W. G., Davis, K. A., and Hatefi, Y. (1971). *Arch. Biochem. Biophys.* **147**, 534.

Hatefi, Y., and Hanstein, W. G. (1969). *Proc. Nat. Acad. Sci. U. S.* **62**, 1129.

Haynes, R., and Walsh, K. A. (1969). *Biochem. Biophys. Res. Commun.* **36**, 235.

Hedin, S. G. (1907a). *Biochem. J.* **II**.

Hedin, S. G. (1907b). *Hoppe-Seyler's Z. Physiol. Chem.* **50**, 497.

Hjertén, S. (1961). *Biochim. Biophys. Acta* **53**, 514.

Hjertén, S. (1964). *Biochim. Biophys. Acta* **79**, 393.

Hjertén, S. (1971). *J. Chromatogr.* **61**, 73.

Hoffman, A. S., Schmer, G., Harris, G., and Kraft, W. G. (1972). *Trans. Amer. Soc. Artif. Intern. Organs* **18**, 10.

Inman, J. K., and Dintzis, H. M. (1969). *Biochemistry* **8**, 4074.

Isliker, H. C. (1953). *Ann. N. Y. Acad. Sci.* **57**, 225.

Isliker, H. C. (1957). *Advan. Protein Chem.* **12**, 387.

Ito, K., Wicher, K., and Arbesman, C. E. (1969). *J. Immunol.* **103**, 622.

Izumi, K. (1971). *Carbohyd. Res.* **17**, 227.

Jagendorf, A. J., Patchornik, A., and Sela, M. (1963). *Biochim. Biophys. Acta* **78**, 516.

Jameson, G. W., and Elmore, D. T. (1971). *Biochem. J.* **124**, 66P.

Jerina, D. M., and Cuatrecasas, P. (1970). *Proc. Int. Congr. Pharmacol. 4th, Basel, Switzerland, 1969,* Vol. 1, p. 236. Schwabe, Basel and Stuttgart.

Jervis, L. (1972). *Biochem. J.* **127**, 29P.

Jovin, I. M., and Kornberg, A. (1968). *J. Biol. Chem.* **243**, 250.

Kågedal, L., and Åkerström, S. (1970). *Acta Chem. Scand.* **24**, 1601.

Kalderon, N., Silman, I., and Blumberg, S. (1970). *Biochim. Biophys. Acta* **207**, 560.

Kaplan, M. E., and Kabat, E. A. (1966). *J. Exp. Med.* **123**, 1061.

Kasai, K., and Ischi, S. (1972). *J. Biochem. (Tokyo)* **71**, 363.

Kasche, V. (1971). Doctoral Dissertation, Uppsala, Sweden.
Kassel, B., and Harciniszyn, M. B. (1971). *Proc. Int. Res. Conf. Proteinase Inhibitors, Munich 1970,* p. 43.
Katchalski, E., Silman, I., and Goldman, R. (1971). *Advan. Enzymol. Relat. Areas Mol. Biol.* **34,** 445.
Kato, I. (1969). *Tampakushitsu, Kakusan, Koso* **14,** 1145.
Kato, I., and Anfinsen, C. B. (1969). *J. Biol. Chem.* **244,** 5849.
Kaufman, B. T., and Pierce, J. V. (1971). *Biochem. Biophys. Res. Commun.* **44,** 608.
Kay, G., and Crook, E. M. (1967). *Nature (London)* **216,** 514.
Kay, G., and Lilly, M. D. (1970). *Biochim. Biophys. Acta* **198,** 276.
Kent, L. H., and Slade, J. H. R. (1960). *Biochem. J.* **77,** 12.
Kocemba-Sliwowska, U. (1972). *Postepy Biochem.* **18,** 59.
Koj, A., Chudzik, J., Pajdak, W., and Dubin, A. (1972). *Biochim. Biophys. Acta* **268,** 199.
Kostner, G., and Holasek, A. (1970). *Lipids* **5,** 501.
Kristiansen, T., Sundberg, L., and Porath, J. (1969). *Biochim. Biophys. Acta* **184,** 93.
Kristiansen, T., Einarsson, M., Sundberg, L., and Porath, J. (1970). *FEBS Lett.* **7,** 294.
Kristiansen, T. (1974a). *Biochim. Biophys. Acta* **338,** 246.
Kristiansen, T. (1974b). *Biochim. Biophys. Acta* **338,** 374.
Kursanow, D. N., and Solodkow, P. A. (1943). *Zh. Prikl. Khim. (Leningrad)* **16,** 351.
Lååls, T. (1972). *J. Chromatogr.* **66,** 347.
Lang, H. (1972). *Nachr. Chem. Tech.* **20,** 71.
Laurent, T. C., and Killander, J. (1964). *J. Chromatogr.* **14,** 317.
Lehmann, H. E., Sanwald, R., and Rapp, W. (1971). *Klin. Wochenschr.* **49,** 943.
Lerman, L. S. (1953). *Proc. Nat. Acad. Sci. U. S.* **39,** 232.
Levin, Y., and Katchalski, E. (1968). *Biochem. Prep.* **12,** 21.
Levin, Y., Pecht, M., Goldstein, L., and Katchalski, E. (1964). *Biochemistry* **3,** 1913.
Levison, S. A., Kierszenbaum, F., and Dandliker, W. B. (1970). *Biochemistry* **9,** 322.
Lindberg, V., and Eriksson, E. (1971). *Eur. J. Biochem.* **18,** 474.
Liu, T. H., and Mertz, E. T. (1971). *Can. J. Biochem.* **49,** 1055.
Lloyd, K. O. (1970). *Arch. Biochem. Biophys.* **137,** 460.
Lowe, C. R., and Dean, P. D. G. (1971). *FEBS Lett.* **14,** 313.
Lowe, C. R., Mosbach, K., and Dean, P. D. G. (1972). *Biochem. Biophys. Res. Commun.* **48,** 1004.
Ludens, J. H., de Vries, J. R., and Fanestil, D. D. (1972). *J. Steroid Biochem.* **3,** 193.
McCormick, D. B. (1965). *Anal. Biochem.* **13,** 194.
Mage, M. G., Evans, W. H., and Peterson, E. A. (1969). *J. Immunol.* **102,** 908.
Manecke, G. (1962). *Pure Appl. Chem.* **4,** 507.
Manecke, G., and Günzel, G. (1967). *Naturwissenschaften* **54,** 531.
Mannik, M., and Stage, D. E. (1971). *J. Immunol.* **106,** 1670.
Mapes, C. A., and Sweeley, C. C. (1972). *FEBS Lett.* **25,** 279.
Marcus, S. L., and Balbinder, E. (1972). *Biochem. Biophys. Res. Commun.* **47,** 438.
Martin, A. J. P., and Porter, R. R. (1951). *Biochem. J.* **49,** 215.
Matsumoto, I., and Osawa, T. (1972). *Biochem. Biophys. Res. Commun.* **46,** 1810.
Mawal, R., Morrison, J. F., and Ebner, K. E. (1971). *J. Biol. Chem.* **246,** 7106.
Melrose, G. J. H. (1971). *Rev. Pure Appl. Chem.* **21,** 83.
Micheel, F., and Evers, J. (1949). *J. Makromol. Chem.* **3,** 200.
Miller, J. V., Cuatrecasas, P., and Thomson, E. B. (1971). *Proc. Nat. Acad. Sci. U. S.* **64,** 1014.

Mitz, M. A., and Summaria, L. J. (1961). *Nature (London)* **189**, 576.
Mosbach, K., Guilford, H., Larsson, P. O., Ohlsson, R., and Scott, M. (1971). *Biochem. J.* **125**, 20P.
Mosbach, K., Guilford, H., Ohlsson, R., and Scott, M. (1972). *Biochem. J.* **127**, 625.
Mosolov, V. V., and Lushnikova, E. V. (1970). *Biokhimiya* **35**, 440.
Murphy, R. F., Elmore, D. T., and Buchanan, K. D. (1971). *Biochem. J.* **125**, 61P.
Nevaldine, B., and Kassell, B. (1971). *Biochim. Biophys. Acta* **250**, 207.
Newbold, P. C. H., and Harding, N. G. L. (1971). *Biochem. J.* **124**, 1.
Nezlin, R. S. (1960). *Dokl. Akad. Nauk SSSR* **131**, 676.
Nezlin, R. S. (1970). "Biochemistry of Antibodies," p. 42. Plenum, New York.
Nicolas, J. C., Pons, M., Descomp, B., and Crastes de Paulet, A. (1972). *FEBS Lett.* **23**, 175.
Nishi, S., and Hirai, H. (1972). *Biochim. Biophys. Acta* **278**, 293.
Nossal, G. I. V., and Ada, G. L. (1971). "Antigens, Lymphoid Cells, and the Immune Response." Academic Press, New York.
O'Carra, P., and Barry, S. (1972). *FEBS Lett.* **21**, 281.
Ohlsson, R., Brodelius, P., and Mosbach, K. (1972). *FEBS Lett.* **25**, 234.
Olesen, H., Hippe, E., and Haber, E. (1971). *Biochim. Biophys. Acta* **243**, 66.
Olivecrona, T., and Egelrud, T. (1971). *Biochem. Biophys. Res. Commun.* **43**, 524.
Onkelinx, E., Meuldermans, W., Jonian, M., and Lontie, R. (1969). *Immunology* **16**, 35.
Ontjes, D. A., and Anfinsen, C. B. (1969). *Proc. Nat. Acad. Sci. U. S.* **64**, 428.
Orth, H. O., and Brümmer, W. (1972). *Angew. Chem., Int. Ed. Engl.* **11**, 249.
Pensky, J., and Marshall, J. S. M. (1969). *Arch. Biochem. Biophys.* **135**, 304.
Pihar, O. (1971). *Chem. Listy* **65**, 713.
Poe, M., Greenfield, N. J., Hirschfield, J. M., Williams, M. N., and Hoogsteen, K. (1972). *Biochemistry* **11**, 1023.
Pogell, B. M. (1966). *In* "Methods in Enzymology" (W. A. Wood, ed.), Vol. 9, p. 9. Academic Press, New York.
Poillon, W. N. (1971). *Biochem. Biophys. Res. Commun.* **44**, 64.
Poonian, M. S., Schlabach, A. J., and Weissbach, A. (1971). *Biochemistry* **10**, 424.
Porath, J. (1962). *Nature (London)* **196**, 47.
Porath, J. (1967). *Gamma Globulins, Proc. Nobel Symp., 3rd, 1967*, p. 287.
Porath, J. (1968). *Nature (London)* **218**, 834.
Porath, J., and Flodin, P. (1959). *Nature (London)* **183**, 1657.
Porath, J., and Fryklund, L. (1970). *Nature (London)* **226**, 1169.
Porath, J., and Sundberg, L. (1970). *Protides Biol. Fluids, Proc. Colloq.* **18**, 401.
Porath, J., and Sundberg, L. (1972a). *In* "The Chemistry of Biosurfaces" (M. L. Hair, ed.), Vol. 2, p. 653. Dekker, New York.
Porath, J., and Sundberg, L. (1972b). *Nature (London)* **238**, 261.
Porath, J., Axén, R., and Ernback, S. (1967). *Nature (London)* **215**, 1491.
Porath, J., Janson, J.-C., and Låås, T. (1971). *J. Chromatogr.* **60**, 167.
Poretz, R. D., and Goldstein, I. J. (1970). *Biochemistry* **9**, 2890.
Pressman, D., and Grossberg, A. L. (1968). "The Structural Basis of Antibody Specificity." Benjamin, New York.
Rademecker, M., and Goodfriend, M. (1969). *Immunochemistry* **6**, 484.
Rapp, W. J., Lehmann, H. E., and Sanwald, R. (1972). *Amer. J. Dis. Child.* **123**, 324.
Reeck, G. R., Walsh, K. A., and Neurath, H. (1971). *Biochemistry* **16**, 4690.
Rees, D. A. (1961). *J. Chem. Soc., London* p. 5168.

Rees, D. A. (1963). *J. Chem. Soc., London* p. 1821.
Rees, D. A. (1969). *Advan. Carbohyd. Chem.* **25**, 321.
Reiner, R. H., and Walch, A. (1971). *Chromatographia* **4**, 578.
Rexová-Benková, L., and Tivensky, V. (1971). *Biochim. Biophys. Acta* **268**, 187.
Rickwood, D. (1972). *Biochim. Biophys. Acta* **269**, 47.
Rivat, C., Bourgignon, J., Rivat, L., and Ropartz, C. (1971). *Proc. Congr. Int. Soc. Blood Transfusion, 12th, 1969* Bibl. Haematol. No. 38, Part I.
Robberson, D. L., and Davidson, N. (1972). *Biochemistry* **11**, 533.
Robbins, J. B., Haimovich, J., and Sela, M. (1967). *Immunochemistry* **4**, 11.
Robinson, N. C., Tye, R. W., Neurath, H., and Walsh, K. A. (1971). *Biochemistry* **10**, 2743.
Robinson, P. J., Dunnil, P., and Lilly, M. D. (1971). *Biochim. Biophys. Acta* **242**, 659.
Rohde, W., and Lezius, A. G. (1971). *Hoppe-Seyler's Z. Physiol. Chem.* **353**, 1507.
Röschlau, P., and Hess, B. (1972). *Hoppe-Seyler's Z. Physiol. Chem.* **353**, 441.
Rosenberry, T. L., Chang, H. W., and Chen, Y. T. (1971). *J. Biol. Chem.* **247**, 1555.
Rosner, W., and Bradlow, H. L. (1971). *J. Clin. Endocrinol. Metab.* **33**, 193.
Rovery, M., and Bianchetta, J. (1972). *Biochim. Biophys. Acta* **268**, 212.
Rutishauser, U., Millette, C. F., and Edelman, G. M. (1972). *Proc. Nat. Acad. Sci. U. S.* **69**, 1596.
Ryan, E., and Fottrell, P. F. (1972). *FEBS Lett.* **23**, 73.
Salter, D. N., Ford, J. E., Scott, K. J., and Andrews, P. (1972). *FEBS Lett.* **20**, 302.
Sander, E. G., McCormick, D. B., and Wright, L. D. (1966). *J. Chromatogr.* **21**, 419.
Sanderson, C. J., and Wilson, D. V. (1971). *Immunology* **20**, 1061.
Sato, N., and Cargille, C. M. (1972). *Endocrinology* **90**, 302.
Schaller, H., Nüsslein, C., Bonhoeffer, F. L., Kurz, C., and Nietzschmann, I. (1972). *Eur. J. Biochem.* **26**, 474.
Schell, H., and Ghetie, V. (1968). *Rev. Roum. Biochim.* **5**, 295.
Schenkein, I., Bystryn, J.-C., and Uhr, J. W. (1971). *J. Clin. Invest.* **50**, 1864.
Schmer, G. (1972). *Hoppe-Seyler's Z. Physiol. Chem.* **353**, 810.
Sehon, A. H. (1963). *Brit. Med. Bull.* **19**, 183.
Seki, T., Yang, H. Y. T., Levin, Y., Jenssen, T. A., and Erdós, E. G. (1970). *Advan. Exp. Med. Biol.* **8**, 23.
Semenza, G., and Kolínská, J. (1967). *Protides Biol. Fluids, Proc. Colloq.* **15**, 581.
Silman, I. H., and Katchalski, E. (1966). *Annu. Rev. Biochem.* **35**, 873.
Skogseid, A. (1948). Doctoral Dissertation, University of Oslo, Oslo, Norway.
Skogseid, A. (1952). U. S. Patent 2,592,350.
Sokolovsky, M., and Zisapel, N. (1971). *Biochim. Biophys. Acta* **250**, 203.
Sordi, S., and Piazzi, S. E. (1970). *Ann. Sclavo* **12**, 16.
Sprossler, B., and Lingens, F. (1970). *FEBS Lett.* **6**, 232.
Stanislawski, M., and Coeur-Joly, G. (1972). *Biochimie* **54**, 203.
Starkenstein, E. (1910). *Biochem. Z.* **24**, 210.
Steers, E., Jr., Cuatrecasas, P., and Pollard, H. B. (1970). *J. Biol. Chem.* **246**, 196.
Stevenson, K. J., and Landman, A. (1970). *Can. J. Biochem.* **49**, 119.
Stewart, K. K., and Doherty, R. F. (1971). *FEBS Lett.* **16**, 226.
Sundberg, L., Porath, J., and Aspberg, K. (1970). *Biochim. Biophys. Acta* **221**, 394.
Svensson, S., Hammarström, S. G., and Kabat, E. A. (1970). *Immunochemistry* **7**, 413.
Swanljung, P. (1971). *Anal. Biochem.* **43**, 382.

Tallberg, T., Eskola, L., and Borgström, G. H. (1971). *Protides Biol. Fluids, Proc. Colloq.* **18**, 415.
Tannenbaum, M., and Goodfriend, L. (1971). *Int. Arch. Allergy Appl. Immunol.* **41**, 778.
Ternynck, T., and Avrameas, S. (1972). *FEBS Lett.* **23**, 24.
Tesser, G. I., Fisch, H.-U., and Schwyzer, R. (1972). *FEBS Lett.* **23**, 56.
Thompson, A. R., and Davie, E. W. (1971). *Biochim. Biophys. Acta* **250**, 210.
Tiselius, A. (1948). *Ark. Kemi, Mineral. Geol.* **26B**, No. 1.
Tobiška, J. (1964). "Die Phythämagglutinine." Akademie-Verlag, Berlin.
Tomino, A., and Paigen, K. (1970). "The Lac Operon." Cold Spring Harbor Lab., Cold Spring Harbor, New York.
Tomita, M., Kurokawa, T., Onozaki, K., Ichiki, N., Osawa, T., and Ukita, T. (1972). *Experientia* **28**, 84.
Tripatzis, I., Warecka, K., and Man-Chung, W. (1971). *Nature (London), New Biol.* **230**, 250.
Truffa-Bachi, P., and Wofsy, L. (1970). *Proc. Nat. Acad. Sci. U. S.* **66**, 685.
Uren, J. R. (1970). *Biochim. Biophys. Acta* **236**, 67.
Vahlquist, A., Nilsson, S. F., and Peterson, P. A. (1971). *Eur. J. Biochem.* **20**, 160.
Varró, R., and Bátory, G. (1972). *Z. Immunitaetsforsch., Exp. Klin. Immunol.* **142**, 418.
Venis, M. A. (1971). *Proc. Nat. Acad. Sci. U. S.* **68**, 1824.
Vretblad, P., and Axén, R. (1971). *FEBS Lett.* **18**, 254.
Warecka, K., Möller, H. J., Vogel, H.-M., and Tripatzis, I. (1972). *J. Neurochem.* **19**, 719.
Weetall, H. H. (1967). *J. Bacteriol.* **93**, 1876.
Weetall, H. H. (1970). *Biochem. J.* **117**, 257.
Weetall, H. H. (1972). *In* "The Chemistry of Biosurfaces" (M. L. Hair, ed.), Vol. 2, p. 597. Dekker, New York.
Weetall, H. H., and Weliky, N. (1964). *Nature (London)* **204**, 896.
Weibel, M. K., Weetall, H. H., and Bright, H. J. (1971). *Biochem. Biophys. Res. Commun.* **44**, 347.
Weigel, H. (1963). *Advan. Carbohyd. Chem.* **16**, 61.
Weliky, N., Brown, F. S., and Dale, E. C. (1969). *Arch. Biochem. Biophys.* **131**, 1.
Weston, P. D., and Avrameas, S. (1971). *Biochem. Biophys. Res. Commun.* **45**, 1574.
Wichterle, O., and Lim, D. (1960). *Nature (London)* **185**, 117.
Wide, L. (1969a). *Lancet* **2**, 863.
Wide, L. (1969b). *Acta Endocrinol. (Copenhagen)* **63**, Suppl. 142, 207.
Wide, L. (1972). *Scand. J. Clin. Lab. Invest.* **29**, Suppl. 126, 14.4a.
Wide, L., and Killander, J. (1971). *Scand. J. Clin. Lab. Invest.* **27**, 151.
Wide, L., and Porath, J. (1966). *Biochim. Biophys. Acta* **130**, 257.
Wide, L., Axén, R., and Porath, J. (1967a). *Immunochemistry* **4**, 381.
Wide, L., Bennich, H., and Johansson, S. G. O. (1967b). *Lancet* **2**, 1105.
Wigzell, H. (1970). *Transplant. Rev.* **5**, 76.
Wigzell, H. (1971). *In* "Progress in Immunology" (B. Amos, ed.), p. 1105. Academic Press, New York.
Wigzell, H., and Andersson, B. (1971). *Annu. Rev. Microbiol.* **25**, 291.
Wilchek, M. (1970). *FEBS Lett.* **7**, 161.
Wilchek, M., Salomon, Y., Lowe, M., and Selinger, Z. (1971). *Biochem. Biophys. Res. Commun.* **45**, 1177.

Wofsy, L., Kimura, I., and Truffa-Bachi, P. (1971). *J. Immunol.* **107**, 725.
Wonderhaar, B., and Mueller, G. C. (1969). *Biochim. Biophys. Acta* **179**, 626.
Yon, R. J. (1972). *Biochem. J.* **126**, 765.
Zolton, R. P., and Mertz, E. T. (1971). *Can. J. Biochem.* **50**, 529.
Zolton, R. P., Mertz, E. T., and Russel, H. T. (1972). *Clin. Chem.* **18**, 654.

3

Proteins and Sodium Dodecyl Sulfate: Molecular Weight Determination on Polyacrylamide Gels and Related Procedures

KLAUS WEBER AND MARY OSBORN

I. INTRODUCTION

The detergent sodium dodecyl sulfate (SDS)[1] dissociates proteins into their constituent polypeptide chains. Polyacrylamide gel electrophoresis in the presence of SDS separates the polypeptide chains according to their molecular weights. Thus the molecular weights of the polypeptide chains of a given protein can be determined by comparing their electrophoretic mobilities on SDS gels to the mobilities of marker proteins with polypeptide chains of known molecular weights. SDS polyacrylamide gel electrophoresis is easily and rapidly performed, requires only inexpensive equipment, and can be used with microgram amounts of protein. The method is reliable and reproducible, and results obtained in the molecular weight range of 15,000 to 200,000 are generally within 10% of those obtained by other techniques (Shapiro *et al.*, 1967; Weber and Osborn, 1969).

SDS gel electrophoresis is particularly useful for the analysis of multicomponent systems such as complex enzymes, viruses, and membranes which are often difficult to dissociate by other methods but are readily dissociable at 100° in the presence of SDS and 2-mercaptoethanol. Gel electrophoresis provides not only the molecular weights of the polypeptide chains but also an approximate stoichiometry of the multicomponent system.

Proteins separated on SDS gels can be subjected to further characterization if the separated polypeptide chains are eluted from the gels and the SDS is removed. Elution of only a few gels will generally yield nanomole quantities of the individual components. Using the techniques described later in this chapter, these amounts of proteins can be sufficient for amino acid analysis, peptide mapping, amino-terminal sequence study, carboxyl-terminal analysis, or for characterization of cyanogen bromide fragments. Furthermore, individual protein components can

[1] Abbreviations: SDS, sodium dodecyl sulfate; MBA, *N,N'*-methylenebisacrylamide; TEMED, *N,N,N',N'*-tetramethylethylenediamine; PMSF, phenylmethanesulfonyl fluoride.

sometimes be renatured and the enzymatic and antigenic properties studied. Other uses of SDS gel electrophoresis are discussed by Weber and Osborn (1969), Maizel (1971), and Weber *et al.* (1972).

II. INTERACTION OF PROTEINS AND SDS

A. Empirical Findings

That the mobility of a polypeptide chain during SDS gel electrophoresis depends primarily upon its molecular weight was first indicated by

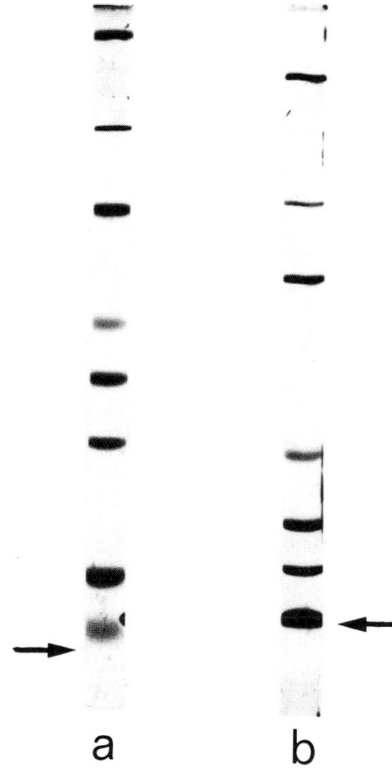

Fig. 1 Separation of proteins on SDS polyacrylamide gels. (a) Standard SDS–phosphate system (Table II). (b) SDS–Tris system with stacking gel (Table III). Both gels contain 7.5% acrylamide and 0.2% methylenebisacrylamide. The dye position is marked by the arrow (the dye moved approximately 9 cm). Protein standards from top to bottom: myosin (molecular weight 200,000), phosphorylase *a* (100,000), bovine serum albumin (68,000), ovalbumin (43,000), glyceraldehyde-3-phosphate dehydrogenase (36,000), carbonic anhydrase (29,000), myoglobin (17,200), and cytochrome *c* (11,700). In gel b, myoglobin and cytochrome *c* moved with the dye and were not resolved.

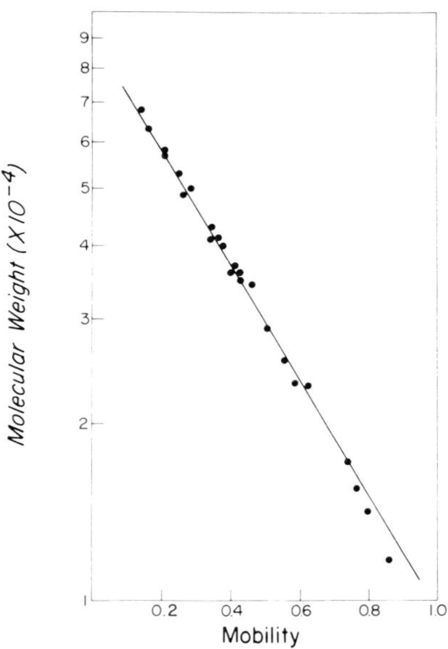

Fig. 2 Comparison of the molecular weights of the standard marker proteins listed in Table I with their electrophoretic mobilities on standard 10% gels in the SDS–phosphate system (Table II). This plot covers the molecular weight range from 14,000 to 70,000.

the results of Shapiro *et al.* (1967). This observation was confirmed and extended by Weber and Osborn (1969) in a study of some forty proteins with accurately known polypeptide chain molecular weights. This and subsequent work in many laboratories has extended the list of proteins studied by SDS gel electrophoresis to more than one hundred. A typical separation of eight proteins is shown in Fig. 1. Figure 2 shows the typical plot of the logarithms of the molecular weights against the electrophoretic mobilities, using the proteins given in Table I as standard markers. These proteins were prepared according to Method 1 (see Section III,D) and in most cases the results were verified by use of one of the other methods for sample preparation. If sufficient amounts of SDS and 2-mercaptoethanol are added and the solution is heated to 100°, dissociation into polypeptide chains occurs quantitatively and oligomeric forms or "aggregates" are not observed.

Different molecular weight ranges may be studied by varying the amount of acrylamide or of methylenebisacrylamide or of both in the gel (Weber and Osborn, 1969; Maizel, 1971; Weber *et al.*, 1972). The plot obtained for log molecular weight vs. mobility is usually a straight line, but with some gel systems it may be slightly concave. Figure 3

TABLE I

Molecular Weights of the Polypeptide Chains of Standard Marker Proteins[a]

Protein	MW of polypeptide chain
Myosin (heavy chain)[b]	200,000
RNA polymerase (*E. coli*)[c]	150,000 and 160,000
γ-Globulin (unreduced)[d]	150,000
Serum albumin (dimer)[d]	136,000
β-Galactosidase (*E. coli*)*	130,000
Phosphorylase *a* (muscle)*	100,000[e]
Serum albumin*	68,000
L-Amino acid oxidase (snake venom)	63,000
Catalase (liver)*	58,000
Pyruvate kinase (muscle)	57,000
Glutamate dehydrogenase (liver)*	53,000
γ-Globulin (H chain)	50,000
Fumarase (muscle)*	49,000
Ovalbumin*	43,000
Alcohol dehydrogenase (liver)*	41,000
Enolase (muscle)	41,000
Aldolase (muscle)*	40,000
D-Amino acid oxidase (kidney)	37,000
Glyceraldehyde-3-phosphate dehydrogenase (muscle)*	36,000
Lactate dehydrogenase (muscle)*	36,000
Pepsin[f]	35,000
Carboxypeptidase A[f]	34,600
Carbonic anhydrase*	29,000
Chymotrypsinogen*	25,700
γ-Globulin (L chain)	23,500
Trypsin*[f]	23,300
Myoglobin*	17,200
Hemoglobin	15,500
Lysozyme (egg white)*	14,300
Cytochrome *c* (muscle)*	11,700

[a] The table lists reliable standard proteins, most of which are commercially available. Those which have been used routinely are marked by an asterisk. References for the molecular weight values are given by Weber and Osborn (1969).

[b] Myosin can be obtained from one of the laboratories working on muscle proteins.

[c] Miles Laboratories produces a partially purified enzyme preparation in which the two larger polypeptide chains (MW approximately 150,000 and 160,000; see Burgess, 1969) can easily be distinguished.

[d] For the reliability of such a marker, see the text.

[e] P. Cohen, T. Duewer, and E. Fischer, personal communication.

[f] Care must be taken to avoid proteolysis of these proteins during the preparation of samples.

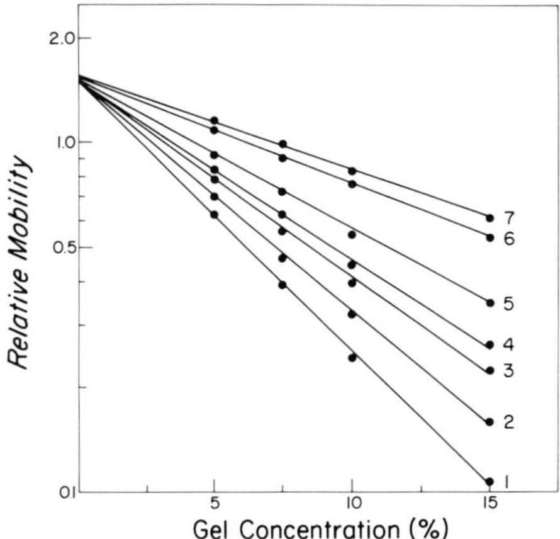

Fig. 3 Electrophoretic mobilities of seven standard proteins on SDS polyacrylamide gels plotted on a logarithmic scale against the gel concentration. Extrapolation of the different lines to zero gel concentration indicates that the different proteins have approximately the same value of "free mobility" (see Section II). The weight ratio of acrylamide to methylenebisacrylamide is 37:1 and the SDS–phosphate system described in Table II is used for all experiments. The acrylamide concentrations illustrated are 15%, 10%, 7.5%, and 5%. The numbers refer to the following proteins: 1, catalase; 2, fumarase; 3, aldolase; 4, glyceraldehyde-3-phosphate dehydrogenase; 5, carbonic anhydrase; 6, myoglobin; 7, bacteriophage R17 coat protein. The molecular weights of the first six proteins are given in Table I. Bacteriophage R17 coat protein has a molecular weight of 14,000.

shows the molecular weight plots for several proteins on gels of varying acrylamide concentration in which the ratio of acrylamide to methylenebisacrylamide has been kept constant. Instructions for preparing such gels are given in Tables II and III. Good separation is obtained on 10% gels in the molecular weight range of 10,000 to 70,000 and on 5% gels in the range of 25,000 to 200,000. The very loose 3.3% gels allow the separation of very high molecular weight components and may be used for the study of chemically cross-linked polypeptide chains, since components with molecular weights as high as 1,000,000 can enter these gels. Gels containing 15% acrylamide may be useful in the molecular weight range below 50,000. Plots of log molecular weight vs. mobility for gels with varying methylenebisacrylamide concentrations and a constant content of acrylamide have been given elsewhere (Weber and Osborn, 1969). On any given type of gel it may be difficult to differentiate the various components that migrate together very slowly or very rapidly. In such cases the porosity of the gel should be changed so that the molecules of interest are in the center of the gel.

Resolution is also affected by the choice of the buffer system. Instructions for preparing gels using SDS–Tris–glycine buffer are given in Table III (Laemmli, 1970; Maizel, 1971). These gels include a stacking gel and often provide better resolution than the standard SDS–phosphate system of Table II. Another system which was recently introduced makes use of a SDS–Tris–borate buffer (Neville, 1971). However, for most purposes the standard SDS–phosphate system has quite adequate resolving power and has been used extensively in the last few years (Shapiro *et al.*, 1967; Weber and Osborn, 1969; Dunker and Rueckert, 1969). For very low molecular weight components (<15,000) resolution is significantly improved by using highly cross-linked gels and the SDS–urea system of Swank and Munkres (1971). Instructions for preparing these gels are given in Table IV.

B. Theoretical Background

Previous work with native proteins on starch gels (Ferguson, 1964) as well as on polyacrylamide gels (Hedrick and Smith, 1968) in buffer solutions has shown that molecular weights of proteins can be estimated provided relative mobilities (m_R) are available at several gel concentrations. The dependence of the relative mobility m_R upon the molecular weight is given by the Ferguson equation

$$\log m_R = -K_R C + \log m_0$$

where m_R is the relative mobility at gel concentration C, m_0 is the relative mobility at zero gel concentration ("apparent free relative mobility"), and K_R is the "retardation coefficient." K_R depends upon the cross-linking of the gel system, the shape of the molecules, and their molecular weight.

Hedrick and Smith (1968) have shown that several globular proteins can be treated by such an approach and have proposed a simple molecular weight determination of globular proteins on polyacrylamide gels. Plotting the log m_R values for standard proteins against the known gel concentrations, they obtained straight lines with slopes of $-K_R$ ("Ferguson" plot; see Ferguson, 1964). Replotting the K_R values against the molecular weights provided a linear relationship from which molecular weights of unknown proteins could be obtained.

In this approach m_0, the relative mobility at zero gel concentration ("apparent free mobility"), is a function of the protein and depends on charge, shape, and size. Following this interpretation, Fig. 3 shows a Ferguson plot of several proteins on SDS gels using the standard SDS–phosphate system. Neville (1971) has recently published a similar plot using an SDS–borate system. The straight lines of the Ferguson plots intersect at approximately the same value for zero gel concentration for all typical proteins studied on a given type of SDS gel, provided the percentage of cross-linking is held constant.

That m_0 is a nearly constant value in the SDS system reflects the empirical finding (Shapiro *et al.*, 1967; Weber and Osborn, 1969) that a linear relationship exists between log m_R and K_R at any given gel concentration. If the cross-linking and the shape factor are constants, log m_R is a direct function of the molecular weight.

The small variation in the value of m_0 in the Ferguson plots is most likely larger than that due to experimental error (Neville, 1971). The small deviations therefore reflect contributions of the individual protein species. Atypically behaving proteins (see Section II,C and the discussion of glycoproteins below) can therefore readily be recognized using a Ferguson plot.

The "constant value" of m_0 for SDS gel electrophoresis indicates that protein–SDS complexes behave such that their free mobility is independent of charge and size. Independence of charge demands a very high degree of SDS binding. Independence of size demands that the ratio of effective charge to frictional coefficient must be independent of the molecular weight. Thus the protein–SDS complex must have a structure in which the free mobility (m_0) of a large molecule is identical with the free mobility of a segment or a smaller molecule under conditions of free electrophoresis. This situation is very similar to the behavior of nucleic acids on gel electrophoresis, and indeed several laboratories have shown that nucleic acid molecules are separated on polyacrylamide gels according to size (see, for instance, Peacock and Dingman, 1968). Thus the separation of both nucleic acids and SDS–proteins can be described as the separation of a set of homopolymers. The intrinsic negative charge of the nucleic acid molecules "parallels" the charge of the protein–SDS complexes.

The empirically established relationship between electrophoretic mobility and polypeptide chain molecular weight (Shapiro *et al.*, 1967; Weber and Osborn, 1969; Dunker and Rueckert, 1969) indicates that for protein–SDS complexes (1) the charge per unit mass is approximately constant, and (2) the ratio of effective charge to frictional coefficient is approximately constant and independent of the molecular weight. Experimental work on protein–SDS complexes supports these assumptions. Binding studies of a variety of different proteins indicate that above an SDS monomer concentration of 8×10^{-4} *M*, 1.4 gm of SDS are bound per gram of protein (Pitt-Rivers and Impiombato, 1968; Reynolds and

Tanford, 1970a,b). Thus the number of SDS molecules bound to a polypeptide chain is approximately half the number of amino acid residues present in the polypeptide chain. This high level of binding and the constant binding ratio will generally "swamp out" the intrinsic charge contribution of most proteins, and an approximately constant negative charge per unit mass will be obtained. SDS binding measurements indicate that the percentage of SDS as free monomer or as micelles varies with the ionic strength, and the amount of monomer increases at low ionic strength. Binding of SDS to proteins depends upon the equilibrium concentration of the monomer and not upon total SDS.

The hydrodynamic and optical properties of the resulting protein–SDS complexes have been studied in some detail (Fish *et al.*, 1970; Reynolds and Tanford, 1970b). It has been proposed that the complexes behave like prolate ellipsoids or rods. The short axis of such complexes is constant and of the order of 18 Å, while the long axis varies in proportion to the molecular weight or to the number of amino acid residues per polypeptide chain. Therefore the protein–SDS complexes should have a conformation in which the particle length varies uniquely with the molecular weight.

The uniformity of SDS binding, the hydrodynamic properties of the protein–SDS complexes, and the uniformity of "mobility in free solution" suggest that the basis for our understanding of SDS gel electrophoresis is sound. A more detailed theoretical discussion of gel electrophoresis has been provided by several workers and the reader is referred to these studies (Rodbard and Chrambach, 1970; Chrambach and Rodbard, 1971; Neville, 1971).

C. Complications in Molecular Weight Determinations

1. Abnormal SDS Binding

Failure to reduce disulfide bonds restricts the conformational freedom of the polypeptide and diminishes the extensive SDS binding. Bovine serum albumin and pancreatic ribonuclease are found to bind only 0.9 gm of SDS per gram of protein instead of the usual value of 1.4, though the latter value may be obtained after extensive reduction (Pitt-Rivers and Impiombato, 1968; Reynolds and Tanford, 1970a). Similarly, the introduction of artificial chemical cross-links will restrict the binding of SDS and lower the electrophoretic mobilities (Davies and Stark, 1970). Glycoproteins show impaired SDS binding (Pitt-Rivers and Impiombato, 1968), since only the protein component is expected to bind SDS. The chemical nature of the carbohydrate component may be important for the electrophoretic mobility (Schubert, 1970; Bretscher, 1971a). Thus glycoproteins *cannot* be expected to behave normally on SDS gels.

It has also been reported that the primary structure of a protein influences SDS binding. Highly-charged polylysylglutamic acid binds only 0.4 gm of SDS per gram of polypeptide (Pitt-Rivers and Impiombato, 1968). Succinylation of bovine serum albumin leads to decreased SDS binding (Pitt-Rivers and Impiombato, 1968). Maleylation reduces the electrophoretic mobility of several proteins, presumably due to decreased SDS binding (Arndt and Berg, 1970; Tung and Knight, 1971). In a careful study Tung and Knight (1971) showed that maleylated proteins bind fewer SDS molecules than the unmaleylated counterparts. The decrease in SDS binding correlates fairly well, in general, with the decrease in electrophoretic mobility on SDS gels. With the exception of pepsin and several low molecular weight proteins, the decrease in electrophoretic mobility corresponds to an increase of 10% in the apparent molecular weight. A large decrease in SDS binding is expected to have at least two effects on the electrophoretic mobility. (1) The reduced net charge will lower the electrophoretic mobility. (2) Reduced binding will, however, decrease the frictional drag because of fewer SDS molecules and thereby increase the mobility. For most of the proteins studied, the first effect is more pronounced. Such a simplified explanation does not take into consideration the conformation of a maleylated protein in SDS solution that may not be the same as that of normal proteins in the same solvent. Tung and Knight (1971) have presented preliminary experiments indicating alterations in the conformation of maleylated proteins. These results clearly indicate that decreased SDS binding induces complicated changes in the electrophoretic mobility which most likely will yield an increased apparent molecular weight upon SDS gel electrophoresis.

2. *Contribution of the Intrinsic Net Charge of the Protein*

Because of the high level of binding of SDS, the net charge contribution of most proteins will rarely be more than 10% of the charge introduced by the binding of the detergent. However, protein molecules with exceptionally high net charges are expected to show deviations from the typical relationship of molecular weight vs. mobility. Thus histones, with their very high positive net charge, show the usual SDS binding (Reynolds and Tanford, 1970a) but atypically low electrophoretic mobilities (Panyim and Chalkley, 1971). Histone F_1 (molecular weight 21,000) has an apparent molecular weight of 35,000 on SDS–phosphate gels (K. Weber, unpublished observations). Some of the effects discussed under "abnormal SDS binding" may also reflect contributions of the net charge of the protein. The experimental evidence obtained with maleylated proteins (Tung and Knight, 1971) and with histones (Panyim and Chalkley, 1971) reinforces the idea that small differences in molecular weight between different polypeptide chains should be interpreted with great caution. In general, the electrophoretic mobility of most proteins is given by the length of the polypeptide chain alone. However, a complex interplay of net charge, abnormal SDS binding, conformational changes, and molecular length may govern the electrophoretic mobility of some proteins. Taking the current results as a basis for a generalization, one may expect that an apparent molecular weight obtained by SDS gel electrophoresis is generally within 10% of the true molecular weight. Proteins of exceptional amino acid composition (e.g., histones) will show much greater deviation. In such a case the amino acid composition of a particular protein will give an indication of abnormality. In a multicomponent system, an analysis of the amino acid composition of the individual polypeptide chains (see Section IV,B) should be performed in order to detect a possible abnormal composition.

3. *Low Molecular Weight Proteins*

Gel filtration experiments in SDS solution indicate that small polypeptide chains (molecular weight $< 15{,}000$) behave abnormally. Rodlike particles begin to approximate spheres when their lengths approach the magnitude of their diameters (Fish *et al.*, 1970). Hence it is not surprising that typical plots of molecular weight vs. mobility change slope in this molecular weight range, as first reported by Dunker and Rueckert (1969). Insufficient standardization of the plot can therefore give rise to incorrect results (Swank and Munkres, 1971; Dunker and Rueckert, 1969).

In addition, some abnormal behavior can be expected from small polypeptide chains because the factors responsible for the atypical behavior of higher molecular weight proteins have an even greater influence on lower molecular weight proteins. For instance, small intrinsic charge differences among various mutant coat proteins of bacteriophage Qβ (molecular weight 14,000) produce an extremely small but reproducible separation on SDS gels (Strauss and Kaesberg, 1970). In this case, the difference in the intrinsic charges of the proteins at pH 7 is probably

50%–100%, but the overall charge differences among the protein–SDS complexes are only about 3%. Similarly, the separation of some (but not all) hemoglobin α and β chains is very likely due to the contributions of charge and conformation as well as to the differences in molecular weight (M. Osborn and M. Mathews, unpublished observations).

These results indicate that the intrinsic charge and the conformation of small polypeptide chains may be more important in determining their electrophoretic mobilities on SDS gels than is the case with large proteins. Increased concentrations of methylenebisacrylamide and addition of 8 *M* urea seem to improve the results. Swank and Munkres (1971) have demonstrated the usefulness of such modifications for low molecular weight proteins. However, even under these conditions, deviations of 20% (Swank and Munkres, 1971) or greater (K. Weber, unpublished observations) can easily be found, indicating that other procedures must be used to corroborate the apparent molecular weights obtained by SDS gel electrophoresis.

4. *Presence of Proteolytic Enzymes*

Precautions must be taken to avoid proteolytic degradation during the preparation of samples for electrophoresis (Weber *et al.*, 1972; Pringle, 1970a,b). Several proteolytic enzymes fail to give bands on SDS gels if incubated at room temperature (rather than at 100°) with SDS and 2-mercaptoethanol prior to electrophoresis (Pringle, 1970b). In addition, many protein preparations which supposedly consist of a single highly purified component have been shown to contain small amounts of contaminating proteases, and extensive degradation has been observed when such preparations are incubated with SDS and 2-mercaptoethanol at room temperature. These findings, which are only new illustrations of the classic observation that the rate of proteolysis may be increased by several orders of magnitude by denaturation of the protein substrate, make appropriate controls necessary, particularly when a new system is studied. Illustrations of the types of patterns obtained with and without proper precautions to eliminate proteolysis during sample preparation are given by Pringle for the enzyme hexokinase (Pringle, 1970b). A recent study by Nelson (1971) shows that some proteins lose enzymatic activity very slowly at room temperature when exposed to SDS. In our experience, a 3-min incubation in 1% SDS and 1% 2-mercaptoethanol at 100° almost always denatures proteases so rapidly that very little degradation can occur. Alkylation of a protein in guanidine–HCl is probably even more effective in minimizing proteolysis. To date, extensive proteolysis during the preparation of samples by this procedure has been observed only when examining the protease subtilisin which has exceptional resistance to denaturation. In this case, only performic acid oxidation, or treatment with phenylmethanesulfonyl fluoride prior to denaturation by SDS or guanidine–HCl at 100°, was effective in preventing extensive self-digestion (Weber *et al.*, 1972; Pringle, 1970b).

5. *Conclusions*

Complications such as proteolytic degradation and incomplete reduction of disulfide bonds can be readily recognized and controlled by use of the various methods of sample preparation given in Section III,D. Caution is necessary in the lower molecular weight range, and independent molecular weight determinations by other procedures are required. Atypically behaving proteins (e.g., glycoproteins) can sometimes be recognized, since their apparent molecular weights will vary when gels of different porosities are used. This behavior is in direct contrast to that of standard proteins, such as those listed in Table I, which have apparent molecular weights that are not affected by the porosity of the gel (Fig. 3). In certain cases, abnormalities in SDS binding and protein conformation or large differences in intrinsic protein charge may lead to slightly increased or decreased electrophoretic mobilities. Therefore, caution is advisable in interpreting small differences in electrophoretic mobilities as being due solely to differences in molecular weights.

The regular behavior of most globular proteins and enzymes within the molecular weight range of 15,000 to 100,000 has been well documented in the last few years. In their native states these proteins display considerable differences in size, in net charge, and in secondary structure, yet in virtually all cases the molecular weights inferred from the electrophoretic mobilities are in excellent agreement (5%–10%) with the values obtained by more laborious physicochemical methods (Shapiro *et al.*, 1967; Weber and Osborn, 1969; Dunker and Rueckert, 1969). The use of other procedures for the determination of molecular weights is always advisable, especially if sufficient material is available. In the determination of the subunit structure of an enzyme, SDS gel electrophoresis is extremely helpful; this determination can then be readily verified by gel filtration in guanidine–HCl or equilibrium centrifugation in the same solvent. Although these latter procedures clearly lack the resolution of SDS gel electrophoresis for multicomponent systems, they rest on a different theoretical basis and provide a highly desirable independent estimate.

III. MOLECULAR WEIGHT DETERMINATIONS ON SDS GELS

A. Gel Systems

1. Apparatus

The initial choice is between a slab gel or a disc gel apparatus. Slab gels allow the comparison of up to 25 samples on the same gel, require less material per sample, and produce dye fronts which move more uniformly. They are especially advantageous for analytical work when large numbers of samples are processed routinely. A simple design, which may be modified for proteins, has been described by Adams *et al.* (1969); other designs are discussed by Studier (1972) and by Maizel (1971). Disc gel apparatuses are advantageous because the number of

TABLE II

Instructions for Preparing SDS–Phosphate Gels of Various Porosities[a,b]

	Final concentration of acrylamide (%)					
Solution (ml)	20	15	10	7.5	5	3.3
Acrylamide A	—	—	13.5	10.1	6.75	4.5
Acrylamide B	13.5	10.1	—	—	—	—
Water	0	3.4	0	3.4	6.75	9.0
Gel buffer	15.0	15.0	15.0	15.0	15.0	15.0
Ammonium persulfate	1.5	1.5	1.5	1.5	1.5	1.5
TEMED	0.045	0.045	0.045	0.045	0.045	0.045

[a] Each column lists the amounts (in ml) of the various solutions necessary to make at least 12 gels, 7 cm long (6 mm diameter) having the indicated final acrylamide concentrations. In this series, the ratio of acrylamide to methylenebisacrylamide (MBA) is kept constant (37:1).

[b] Acrylamide A: 22.2 gm acrylamide, 0.6 gm MBA, water to 100 ml.

Acrylamide B: 44.4 gm acrylamide, 1.2 gm MBA, water to 100 ml.

Gel buffer (0.2 *M*, pH 7.2): 2.0 gm SDS, 7.8 gm $NaH_2PO_4 \cdot H_2O$, 38.6 gm $Na_2HPO_4 \cdot 7H_2O$, water to 1 liter.

Ammonium persulfate: 15 mg/ml in water.

Reservoir buffer: 1 part gel buffer, 1 part water.

Sample buffer: 0.01 *M* sodium phosphate (pH 7.2), 1% SDS, and 1% 2-mercaptoethanol.

Running temperature: Room temperature.

Current: 8 mA/gel (6 mm diameter). If the samples contain large amounts of protein, use 4 mA/gel until samples have entered the gel.

Time: 3 hr for 5% gel; 5 hr for 10% gel; 8 hr for 15% gel.

samples run at a given time can be varied and larger amounts of proteins can be processed. This is particularly important if recovery of the material for further chemical characterization is desired. A variety of simple designs for disc gel apparatus, adapted from that described by Davis (1964), have been suggested. It is desirable to choose an apparatus in which the length of the gel can be varied. In disc gel

TABLE III

Instructions for Preparing SDS–Tris Gels of Various Porosities[a,b]

	Final concentration of acrylamide (%)				
Solution (ml)	20	15	10	7.5	5
Lower gel					
Acrylamide A	—	—	13.50	10.10	6.75
Acrylamide B	13.50	10.10	—	—	—
Water	8.40	11.80	8.40	11.80	15.15
Lower gel buffer (4×)	7.50	7.50	7.50	7.50	7.50
Ammonium persulfate	0.30	0.30	0.30	0.30	0.30
TEMED	0.03	0.03	0.03	0.03	0.03
10% SDS	0.30	0.30	0.30	0.30	0.30
Upper gel					
Water	6.00				
Upper gel buffer (4×)	2.50				
Acrylamide A	1.35				
Ammonium persulfate	0.10				
10% SDS	0.10				
TEMED	0.01				

[a] Each column lists the amount (in ml) of the various solutions necessary to make at least 12 gels of the indicated acrylamide concentrations. In this series, the ratio of acrylamide to methylenebisacrylamide (MBA) is kept constant (37:1). The volume of the stacking gel is usually 0.4 ml.

[b] Acrylamide A: 22.2 gm acrylamide, 0.6 gm MBA, water to 100 ml.
Acrylamide B: 44.4 gm acrylamide, 1.2 gm MBA, water to 100 ml.
Ammonium persulfate: 100 mg/ml in water.
SDS (10%): 1 gm SDS/10 ml water.
Lower gel buffer (4× concentrated): 1.5 *M* Tris–HCl buffer (18.15 gm Tris base adjusted to pH 8.8 with 6 *N* HCl; final volume 100 ml).
Upper gel buffer (4× concentrated): 0.5 *M* Tris–HCl buffer (6 gm Tris base adjusted to pH 6.8 with 6 *N* HCl; final volume 100 ml).
Reservoir buffer: 6 gm Tris base, 28.8 gm glycine, 1 gm SDS, water to 1000 ml (pH 8.3).
Sample buffer: 0.01 *M* Tris–HCl (pH 6.8), 0.1% SDS, 0.1% 2-mercaptoethanol.
Running temperature: Room temperature or 4°.
Current: During stacking 1 mA/gel, then increase to 2–3 mA/gel.
Time: 2.5–4 hr.

TABLE IV

Instructions for Preparing Highly Cross-Linked Gels for Low Molecular Weight Proteins

Mix:	Acrylamide C	16.6 ml
	Buffer B	6.25 ml
	Urea	24.20 gm

Add 35 mg ammonium persulfate and enough Tris base to bring the pH to 6.8; adjust the volume to 50 ml.

Acrylamide C: 37.6 gm acrylamide, 3.76 gm MBA, water to 100 ml.[a]
Buffer B: 0.6 ml TEMED, 0.8 gm SDS, 0.8 M H_3PO_4 adjusted to pH 5 with Tris; final volume 100 ml (store at 4°).
Reservoir buffer: 0.1 M H_3PO_4 adjusted to pH 6.8 with Tris base; 1 gm SDS/liter.
Sample buffer: 0.01 M H_3PO_4 adjusted to pH 6.8 with Tris, 8 M urea, 1% SDS, and 1% 2-mercaptoethanol.
Current: 2 mA/gel.
Time: ~15 hr on long gels.

[a] The acrylamide concentration is 12.5% and the ratio of acrylamide to methylene-bisacrylamide is 10:1. For results with different amounts of both reagents, see Swank and Munkres (1971).

apparatus this is done by supporting the upper vessel on a ring stand. In addition, since the pH of some running buffers changes appreciably during prolonged electrophoresis, large volumes of buffer should be employed and hence the apparatus should have large buffer trays; alternatively, a pump allowing continuous exchange of buffers between the upper and lower reservoirs should be used.

Separate power supplies for electrophoresis and destaining are useful. For electrophoresis, a model that produces 80–100 mA and at least 200 V is sufficient. For destaining electrophoretically a power supply with a high current output (0.25–1.0 A) is useful.

Instructions for preparing the gels differ according to which buffer system is chosen. Three gel systems are described in Tables II, III, and IV. The standard SDS–phosphate gel system (Table II), which has been extensively used over the last few years (Shapiro *et al.*, 1967; Weber and Osborn, 1969; Dunker and Rueckert, 1969), does not make use of a stacking gel; however, it generally gives an adequate resolution of complex mixtures of polypeptide chains (Fig. 1). The SDS–Tris–glycine system (Table III) uses a stacking gel and often provides a better resolution than the standard SDS–phosphate gel (Maizel, 1971; Laemmli, 1970). The SDS–urea system (Table IV) uses highly cross-linked gels and thus provides a much better resolution for low molecular weight polypeptide chains (Swank and Munkres, 1971).

2. *Chemicals*

Acrylamide, *N,N'*-methylenebisacrylamide (MBA), and *N,N,N',N'*-tetramethylethylenediamine (TEMED) can generally be used without further purification. Solutions of acrylamide and MBA should be colorless; if not, they should be recrystallized from ethyl acetate and acetone, respectively. Acrylamide solutions should be dissolved by stirring and if necessary filtered through Whatman No. 1 filter paper. The solutions are stable for several months if stored in a dark bottle at 4°. (*Caution:* Unpolymerized acrylamide is a skin irritant and a poison. Care should be taken to avoid contact, i.e., wear gloves and do not pipette by mouth.) SDS (95%) should be recrystallized from ethanol. Ammonium persulfate solutions are unstable and should be made up just before use. TEMED is stable if kept in a dark bottle at 4°.

B. Preparation of Gels

1. Use glass tubes 3 cm longer than the required gel length. Soak them overnight in chromic acid cleaning solution, rinse well with distilled water, and oven-dry. Mark the required gel length on each tube.
2. Close one end of each tube with a serum cap (Scientific Glass Co., Bloomfield, N. J.) or with Parafilm, and place the tubes in a rack in vertical position.
3. From Tables II, III, or IV select the appropriate gel system. Following the instructions, mix all ingredients except ammonium persulfate and TEMED in a vacuum flask at room temperature. Deaerate for 1 min with an aspirator or vacuum pump. Add ammonium persulfate and TEMED and mix.
4. Using a propipette and a 10-ml pipette, fill each tube up to the mark. Tap the tubes to eliminate air bubbles.
5. Carefully layer distilled water on top of the gel solution in each tube. Add the water with a syringe that has had the bevelled tip of a 2-inch No. 25 gauge needle bent at 90°, or with a Pasteur pipette bent in similar fashion. This procedure prevents the formation of an air/gel meniscus and ensures a flat gel surface. Isobutanol can be used instead of water. It can be squirted onto the gel solution and assures an even gel surface.
6. After 10–20 min an interface will be seen, indicating that the gel has polymerized. The gels can be used immediately or may be kept for days at room temperature, capped with Parafilm or a serum cap to prevent evaporation of water.

7. For the SDS–Tris system, a stacking gel is required. This should be poured just before use. Mix the ingredients for the upper gel (Table III) and deaerate before adding the ammonium persulfate and TEMED. Remove the water from the top of the gel by shaking the tubes. Apply the upper gel, layer with water, and allow to polymerize.

8. If the tubes are dirty, if the solutions are not at room temperature before being poured, or if polymerization occurs too rapidly, the gel may pull away from the surface or contain internal bubbles. In the latter case, reduce the amount of TEMED and/or ammonium persulfate. The gel surface may be uneven as a result of excessively rapid polymerization or of improper layering of the water. Siliconized gel tubes may also be used to avoid wall effects.

C. Molecular Weight Standards

In order to obtain reliable molecular weights by the SDS gel technique, it is essential that the proteins used as standards have polypeptide chains of accurately known molecular weight. The selection of standards is usually dictated by (1) their availability, either commercially or from other workers using these proteins, and (2) the particular molecular weight range to be studied.

A list of satisfactory protein standards, most of which are commercially available, is given in Table I. A more extensive list of molecular weight markers has been published previously (Weber and Osborn, 1969). These proteins cover the molecular weight range from 12,000 to 70,000 very well (Weber and Osborn, 1969). Unfortunately, in the molecular weight range above 70,000, only a few proteins are available commercially. As an alternative in this molecular weight range, a series of oligomers can be prepared by chemically cross-linking proteins with diethylpyrocarbonate (Wolf *et al.*, 1970) or dimethylsuberimidate (Davies and Stark, 1970). The reliability of such markers is discussed in Section II. Although it is difficult to obtain accurate molecular weights below 12,000 (see Section II,C), a sufficient number of marker proteins is available for this molecular weight range. Cyanogen bromide fragments of proteins with known amino acid sequence can also be used (Swank and Munkres, 1971).

D. Protein Sample Preparation

The protein samples must be prepared in a way that allows complete denaturation and reduction of disulfide bonds. Precautions must be taken to prevent proteolysis due to impurities in the sample or due to inherent enzymatic activity in the protein itself (Pringle, 1970a,b). Several methods of sample preparation have been reviewed recently (Weber *et al.*, 1972) and will be described only briefly. In nearly all cases, Method 1 will give satisfactory results, and it is recommended for routine use. Method 2, the most important control, should be used at least once with each new system studied. If Methods 1 and 2 do not give identical results, the other methods described may be useful in determining the cause of the discrepancy.

Method 1 (Standard Procedure): SDS at 100°

Add 0.01 *M* sodium phosphate (pH 7.0) to a lyophilized, precipitated, or particulate protein sample. Normally proteins already dissolved in other buffers may be used directly. Place nine parts of 0.01 *M* sodium phosphate (pH 7.0) containing 1% SDS and 1% 2-mercaptoethanol in a tube in a 100° bath. Add one part of the protein solution and cap the tube. The final protein concentration can range from 0.05 to 1.0 mg/ml, and volumes as small as 50 μl may be used. It is essential that the weight ratio of SDS to protein be at least 3:1 (see Section II). After 2 min of incubation, cool the sample to room temperature. The sample may then be used directly or dialyzed against sample buffer (see above). Dialysis is particularly useful if the ionic strength of the sample is high or if the sample volumes are large.[2]

Method 1 (Variations)

Two useful variations of Method 1 are (a) omission of the 100° treatment, and (b) extension of the 100° treatment to 5 min. If low molecular weight components are observed in the unheated samples but are not present in the heated samples, enzymatic proteolysis is a problem. The 100° treatment will eliminate proteolytic artifacts in most cases, but it is desirable to ensure this by employing other procedures designed to minimize proteolysis (see Methods 2–4 below). Occasionally some nonenzymatic hydrolysis of peptide bonds may occur during the preparation of a sample by Method 1, generating spurious low molecular weight bands. In such cases the spurious bands will be more prominent if the 100° treatment is extended; the most reliable result will be ob-

[2] We have used buffers of various ionic strengths, at pH values ranging from 4 to 9, without apparent difficulty. Ammonium sulfate up to 10% saturation had no ill effects. If appreciable concentrations of potassium, guanidinium, or other SDS-precipitating cations are present, dialysis against 0.01 *M* sodium phosphate (pH 7.0) is necessary. Buffers of high ionic strength can decrease SDS binding.

tained with an unheated or very briefly heated sample (J. R. Pringle, personal communication).

Method 2 (Standard Control): Denaturation by Guanidine Hydrochloride Followed by Alkylation

This procedure nearly always guarantees complete denaturation of the protein and rapid inactivation of any proteases present (Weber *et al.*, 1972). To approximately 1 mg of protein (a lyophilisate, a precipitate, or a solution at 10 mg/ml) add 1 ml of hot (100°) 0.1 *M* Tris–HCl (pH 8.5) containing 7–8 *M* guanidine–HCl. Immediately transfer the sample to a boiling water bath, add 15 μl of 2-mercaptoethanol, and cap the tube. After 3–5 min, lower the temperature to 37° and alkylate the protein.[3] Dissolve 260 mg of iodoacetic acid in 1 ml of 1.0 *M* NaOH, and add 0.25 ml to 1 ml of protein solution. Raise the pH to 8–9 (using pH paper) by dropwise addition of 2 *M* NaOH. When the nitroprusside test[4] becomes negative (about 2 min), add several more drops of the iodoacetate solution, raise the pH to 10.5, and incubate for 10 min. Then add excess 2-mercaptoethanol (30–35 μl) and readjust the pH to about 7. Prepare the sample for electrophoresis by prolonged dialysis (guanidinium dodecyl sulfate is insoluble), first against 8 *M* urea in 0.01 *M* Tris–HCl (pH 8), and then against 0.01 *M* sodium phosphate (pH 7) containing 0.1% SDS.[5]

Method 3 (Alternative Control): Performic Acid Oxidation

Mix 1 part of 30% H_2O_2 with 9 parts of 88% formic acid and let the mixture stand for 2 hr at 25° in a capped tube. Dissolve 0.5–2.0 mg of protein in 1 ml of reagent and leave for at least 1 hr at 0°. Then dilute the solution fifteenfold with distilled water, freeze immediately, and lyophilize. Redissolve the powder in 0.2 *M* sodium phosphate

[3] The procedure described above is designed to alkylate various functional groups of proteins in order to enhance irreversible inactivation. Small changes in electrophoretic mobility may occur after such extensive alkylation. The purpose of the procedure is to avoid major artifacts rather than to obtain highly accurate mobility values.

[4] Grind 1 part of sodium nitroprusside and 2 parts of sodium carbonate in a mortar until a "homogenous mixture" is obtained. For assay, add a drop of solution to a small amount of the solid powder. The presence of 2-mercaptoethanol is indicated by a dark purple color.

[5] If necessary, protein solutions containing SDS, guanidine–HCl, or urea can be concentrated by vacuum dialysis, by covering a dialysis bag with Sephadex G-200, or by dialysis against distilled water followed by lyophilization.

(pH 8) containing 1% SDS, incubate at 37° for several hours, and dialyze against 0.01 *M* sodium phosphate (pH 7) containing 0.1% SDS.

Method 4 (Special Controls): Inhibition of Proteases prior to Denaturation

The following measures may be taken to inhibit proteases if there is any reason to suspect that proteolysis has occurred (Weber *et al.*, 1972; J. R. Pringle, personal communication). These procedures are performed before denaturation with SDS. After the treatment, proceed as in Method 1.

(a) Most proteases of the "serine type" can be effectively inactivated by treatment with phenylmethanesulfonyl fluoride (PMSF). Dissolve the reagent in 95% ethanol at a concentration of 6 mg/ml and add, with continuous mixing, 1 part of this solution to 20 parts of the protein solution. Let stand for 10–30 min.

(b) Divalent cation-dependent proteases can be inhibited by treatment with 1,10-phenanthroline. Dissolve the reagent in 95% ethanol at a concentration of 20 mg/ml and add, with continuous mixing, 1 part of this solution to 20 parts of protein solution. Incubate for 30 min.

(c) Proteases of the "cysteine type" can be inactivated by treatment with iodoacetate or with *N*-ethylmaleimide.

(d) In addition, the activities of various types of proteolytic enzymes can be decreased by adjusting the pH of the solution. For example, if degradation occurs at pH 7, pH 4 should be tried.

E. Electrophoresis

For each gel, mix 5 μl of tracking dye solution (0.05% bromphenol blue in 0.01 *M* phosphate, pH 7.0), a drop of glycerol (or several crystals of sucrose), and 5 μl of 2-mercaptoethanol in a small tube or on a small square of Parafilm. Add the protein sample which should contain 1–20 μg of each of the polypeptides to be visualized on the gel and should yield a total sample volume of 25–150 μl. If the volume is less than this, sample buffer may be added. Larger protein loads or larger sample volumes lead to some broadening of the bands. Relatively small volumes should be used with samples of high ionic strength.

Remove serum caps or Parafilm and shake the water layers from the gels. Place the tubes in the electrophoresis apparatus, add reservoir buffer to the lower tray, and tap the gels to remove air bubbles. Add the samples to the tops of the gels using a Pasteur pipette or micropipette. Carefully layer reservoir buffer over each sample using a

Pasteur pipette with a bent tip or a syringe with a bent tip, and add reservoir buffer to the upper tray. Alternatively, reservoir buffer may be added to the upper and lower trays before loading the sample. The sample is then layered on the gel with a Pasteur pipette or a micropipette through the reservoir buffer, and because of its increased density, forms a compact layer on top of the gel.

Connect the power supply to the electrodes with the anode in the lower tray and begin electrophoresis (see Tables II, III, and IV). Discontinue electrophoresis when the tracking dye approaches the bottom of the gel. The current may be adjusted for convenience without significantly affecting the resolution.

Gels can be removed from glass tubes by squirting water between the gel and the tube wall using a syringe with a long, fine needle. The gel should slip out easily but, if necessary, a pipette bulb can be attached to one end of the tube to exert pressure. For gels with acrylamide concentrations greater than 10%, break the tube with a hammer (Maizel, 1971). The dye band will be 4–5 mm wide in the SDS–phosphate system and narrower in the SDS–Tris system. Since it will disappear during the staining and destaining procedure, mark the center of the dye band either with a needle dipped in waterproof India ink or with fine copper wire. Alternatively, measure the length of the gel and the distance migrated by the center of the dye band in each gel.

F. Staining and Destaining

The following procedures are recommended for the staining and destaining of gels.

1. Staining

The staining solution contains:

Coomassie Brilliant Blue R250	1.25 gm
Methanol	227 ml
Glacial acetic acid	46 ml
Water	to 500 ml

Dissolve the dye in methanol, add acetic acid and water, and filter through Whatman No. 1 filter paper. The staining solution can be stored for several months at room temperature.

Coomassie Brilliant Blue is several times more sensitive than other dyes such as Amido Black (Fazekas de St. Groth *et al.*, 1963). To stain, place the gels in test tubes filled with staining solution and leave for 2–12 hr at room temperature. Then rinse the gels with distilled water and destain. In this procedure, fixation and staining occur simultaneously.[6] The use of separate staining and destaining steps offers maximum sensitivity. A more rapid but less sensitive staining procedure has been described by Chrambach *et al.* (1967).[7]

2. *Destaining*

The destaining solution contains:

Methanol	50 ml
Glacial acetic acid	75 ml
Water	to 1 liter

Visualization of the protein bands requires removal of excess dye from the gels. Several procedures are available (Weber *et al.*, 1972):

a. **Diffusion.** For the best results, gels should be stained for only 2–3 hr. Add the gel to a culture tube with 10–20 ml of destaining solution and 2–3 gm of the ion exchange resin AG 501-X8 (BioRad Laboratories). Cap the tubes and put them on a bacterial roller at 37°. The resin absorbs the dye as it diffuses from the gel, and optimal destaining is obtained overnight. Gels treated in this manner show no dye background (Weber *et al.*, 1972).

b. **Electrophoretic Methods.** Either transverse or longitudinal electrophoresis may be used for destaining (Weber and Osborn, 1969; Ward, 1970; Maizel, 1971). Both methods are fast (1 hr), but the backgrounds produced are usually not as good as those obtained by the diffusion method (Weber *et al.*, 1972). There is also the danger that small polypeptides may be removed (Swank and Munkres, 1971).

Stained bands remain visible almost indefinitely if they are not exposed to direct light and if the gels are stored in stoppered tubes containing 7.5% acetic acid. Gels can be photographed effectively using transmitted light and a Polaroid camera.

[6] Some workers (e.g., Shapiro *et al.*, 1967) fix the gels for 12–24 hr in 20% sulfosalicylic acid or in 10% trichloroacetic acid before staining. We have omitted this step without apparent problems. However, it may be advisable to include it in the case of proteins that are slightly soluble in the staining solution (e.g., very small molecules).

[7] In this procedure the gel is immersed in 0.05% Coomassie Brilliant Blue dissolved in 10% trichloroacetic acid. The dye is relatively insoluble in this solution and therefore partitions into protein–dye complexes. If enough protein is present in each band, the pattern will appear immediately after a short staining procedure, and no separate destaining step is necessary.

G. Calculation of Electrophoretic Mobilities

After destaining, measure the distance from the top of the gel to the protein bands. If India ink or copper wire was used to mark the dye position before staining, measure the distance from the top of the gel to this position. Mobilities may then be calculated directly. If the dye position was measured before staining but not marked, the change in length of the gels during the staining and destaining procedure must be taken into account (Weber and Osborn, 1969; Weber *et al.*, 1972). If mobilities are calculated relative to the tracking dye rather than to a standard protein added to each sample, results from different laboratories can be compared directly.

H. Alternative Methods for Detection of Protein Bands

1. Reaction of reduced and alkylated proteins with 1-dimethylamino-naphthalene-5-sulfonyl (Dansyl) chloride introduces a fluorescent label into the proteins. Dansylated proteins (in quantities as small as 0.1 μg) can be detected by their fluorescence under ultraviolet light (Shelton, 1971). Except for the low molecular weight region, the introduction of the Dansyl group has little effect on the mobility of the protein.

2. If sufficient protein has been used, the gels can be scanned directly at 280 nm. The relative positions of the different components can be determined from a densitometer tracing. This procedure requires that preformed gels be soaked extensively in Reservoir Buffer to remove ultraviolet-absorbing material from the gel matrix. The gels are then "slipped" back into gel tubes and loaded. Such preeluted gels do not show the same mobilities as normal gels.

3. Glycoproteins can be detected by the use of a carbohydrate-specific stain (Zacharius *et al.*, 1969). Care must be taken to eliminate SDS from the gel and to keep the gel at acid pH during staining. A suitable procedure has been reported recently by Glossman and Neville (1971).

4. Proteins which have been radioactively labeled can be detected by liquid scintillation counting or by autoradiography. The most common procedures for introducing radioactive labels into isolated proteins are alkylation with [^{3}H]- or [^{14}C]iodoacetate or iodoacetamide, reductive alkylation with [^{3}H]- or [^{14}C]formaldehyde (Rice and Means, 1971), or iodination with ^{131}I (Greenwood *et al.*, 1963; Marchalonis, 1969).

a. Liquid Scintillation Counting. After electrophoresis, the gel is put on Parafilm and frozen at $-20°$. A block of razor blades is used to slice the gel into 1-mm sections. The slices are transferred to counting vials and covered with 0.5 ml of 0.1% SDS. After the vials have been shaken for 6–12 hr at 37°, 3 ml of "AQUASOL" (New England Nuclear) are added and the samples are ready for liquid scintillation counting. Elution from gels that are highly cross-linked is sometimes difficult. Very high molecular weight polypeptides may show decreased elution under normal conditions. Very good recovery of radioactive material seems possible if gel slices are placed in vials and 4.5 ml of 3% "PROTOSOL" (New England Nuclear) in toluene-"Omnifluor" (4 gm Omnifluor per liter) are added. The capped vials are incubated overnight at 37° before counting (R. Horvitz, personal communication). Alternatively, the slices are soaked in 0.07 ml of 2 *M* piperidine for 30 min, then 0.5 ml of NCS–reagent is added, and the slices are allowed to swell for 4 hr at 65°. Ten ml of 0.5% Omnifluor (New England Nuclear) in toluene are added before counting (Rice and Means, 1971). An automatic gel divider for the whole procedure has been described by Maizel (1971) and is available from Savant Co. For a discussion of other counting procedures, see Maizel (1971) and Helleiner and Wunner (1971).

b. Autoradiography. Autoradiography can be used for proteins labeled with ^{14}C or ^{35}S. The procedure can be performed either on a longitudinal slice of the gel (see Fairbanks *et al.*, 1965) or on a slab gel. In order to prepare the gel for autoradiography, the gel material must be immobilized and dried *in vacuo* on a sheet of filter paper or on a dialysis membrane. The proper performance of the drying step is crucial in order to prevent cracked and curled gels. Simple drying assemblies which can be built in the laboratory have been described by Fairbanks *et al.* (1965), Maizel (1971), and Studier (1972). The basic principle is to arrange gel slices or slab gels on a piece of Mylar or Saran Wrap placed on a silicone rubber sheet, and to cover the gel material with a sheet of Whatman 3MM paper. The paper is covered with a porous polypropylene sheet and then with another silicone rubber sheet containing a hole into which a polypropylene tubing connector has been tightly inserted. Air is evacuated from the assembly by means of a vacuum pump. The drying process is enhanced by placing the entire assembly in a boiling water bath, and is usually completed in 30 min. The dried gels are autoradiographed on X-ray films (Kodak Blue Brand) which are developed in the usual manner. A band of 50,000 dpm usually requires 20 hr of exposure. The simple design of slab gels and the ease of autoradiography make slab gels the obvious choice if multiple samples are routinely processed. Autoradiography may be performed directly after electrophoresis or after staining and destaining.

5. Localization of protein bands in preparative gels has been recently described by Weiner *et al.* (1972).

6. A very convenient staining procedure makes use of the fluorometric detection of proteins that have been treated with "fluorescamine" {4-phenylspiro[furan-2(3*H*),1′-phthalan]-3,3′-dione}. The usefulness of fluorescamine as a reagent for determining amino acids, peptides, and proteins has been discussed by Udenfriend *et al.* (1972). Recently Vandekerckhove and Van Montagu (1973) have described the use of the reagent in SDS gel electrophoresis. The protein sample is dissolved in 0.02 *M* sodium phosphate (pH 8.0) containing 1% SDS and 1% 2-mercaptoethanol. A stock solution of fluorescamine (1 mg/ml in dry acetone) is added to the protein solution (10 nmoles) in a volume ratio of 1:10 or 1:20. The addition is made under continuous shaking on a Vortex mixer. After 10 sec the reaction is complete. Glycerol and bromphenol blue are then added and the sample is ready for electrophoresis. The protein bands can be visualized by illuminating the gels with ultraviolet light. The best results are obtained when electrophoresis is performed in the dark.

I. Reproducibility, Sensitivity, and Resolving Power

1. Reproducibility

The electrophoretic mobility displayed by a particular polypeptide during SDS gel electrophoresis is highly reproducible when several samples are run in parallel. For example, the two protein bands of aspartate transcarbamylase run simultaneously on twelve separate gels gave mobilities of 0.445 ± 3% and 0.745 ± 3%, respectively (Weber and Osborn, 1969). Since individual gels differ in composition, age, or current load, the actual distances migrated may vary considerably more than 3%, and hence it is essential to include tracking dye in every sample. Mobility values determined for a particular polypeptide on different occasions usually vary by 5%, and sometimes variations as high as 10% are encountered. These variations seem to occur when different batches of acrylamide are used. They have no effect on the determination of molecular weight, provided that standard proteins are always run at the same time as the unknown whose molecular weight is to be determined. Under no circumstances should the mobilities shown in Fig. 2 or Fig. 3 be used as absolute standards.

2. Sensitivity and Resolving Power

The sensitivity and resolving power of the technique are dependent upon the sharpness of the bands obtained. Band sharpness is determined by the volume of the sample, the amount of protein, the effectiveness of "stacking" at the top of the gel, the type of gel used, and the duration of electrophoresis. In addition, the band width increases with the mobility, especially in the SDS–phosphate system. Very poor stacking is experienced only if high concentrations of electrolytes have not been removed by dialysis prior to electrophoresis. Better stacking, and hence sharper bands, will result if a stacking gel system is used (Table III; see also Section II,A).

The amount of protein necessary to give a visible band after staining depends upon the band sharpness, the stain used, the amount of background color which remains after destaining, and the diameter of the gel. Under the conditions we have described, 0.5 μg of protein should give a detectable band and 0.1 μg may do so in the more favorable cases. Sensitivity may be further increased by decreasing the diameter of the gels. Micro-SDS gels have been run successfully.

The amount by which two polypeptide chains must differ in molecular weight to be resolvable by SDS gel electrophoresis is influenced by all the factors that affect band sharpness. Resolution can be drastically improved by using longer gels (15 cm), running small amounts of protein, and dialyzing the samples against sample buffer prior to electrophoresis (see Fig. 1). If the tracking dye is allowed to run to the bottom of a 10% gel that is 15 cm long, and if components with molecular weights near 40,000 are examined, a difference in mobility corresponding to molecular weight differences of 1500–2000 is sufficient to permit visual resolution of the components. However, such very small differences in mobility may not necessarily reflect *only* a molecular weight difference (see Section II).

A further possibility is the use of acrylamide concentration gradients. These can be produced as step gradients or by means of gradient generators. Such systems may be extremely useful for special separation problems.

J. Determination of an Unknown Molecular Weight

1. Select as standards several proteins with polypeptide molecular weights ranging from 12,000 to 100,000 (Table I). Prepare each protein sample by Method 1. If the protein preparations have not been used before, it is advisable to run them on separate gels to ensure that the techniques are working. Once this precaution has been taken, several standards can be run on the same gel column.[8]

2. Prepare the unknown protein by Method 1 and perform electro-

[8] Alternatively, the "split gel" technique allows two sets of proteins to be run separately on the same gel (Traut *et al.*, 1967).

phoresis on 7.5% gels. Plot the mobilities vs. the known molecular weights of the standards (Table I) on semilogarithmic paper (Fig. 2). Interpolate the molecular weight of the unknown from its mobility. The optimal gel concentration and set of standards can now be selected for further experiments (see Fig. 3). A mobility value between 0.2 and 0.8 is desirable, and several standards should cover this region.

3. Prepare a sample of the unknown protein by Method 2. This is an absolutely necessary control to avoid possible complications due to the presence of a proteolytic enzyme (see Section II). For most proteins, samples prepared by Methods 1 and 2 will give identical results. If different values are obtained, Methods 3 and 4 may be of use (see Section III,D).

4. For very accurate results, the protein standards and the unknown should be treated identically. Standards and the unknown can be run in parallel on different gels or on the same gel. Ideally, all samples should be dialyzed against sample buffer and the amount of standards and unknown used should be the same. The most precise results will be obtained by using longer gels (15 cm), low protein concentrations (1–5 μg), and standards with polypeptide chain molecular weights very close to that of the unknown. It is also advisable to determine the apparent molecular weight of the unknown on gels of several different porosities to aid in recognizing a protein that behaves atypically (see Section II,C).

IV. METHODS FOR FURTHER CHARACTERIZATION OF PROTEINS SEPARATED ON SDS GELS

As previously mentioned, it is possible to recover from SDS gels nanomole quantities of the separated polypeptide chains and to subject them to further characterization by amino acid composition, peptide mapping, amino-terminal sequence analysis, or carboxyl-terminal analysis. It is sometimes also possible to renature the individual protein components and to study their enzymatic or antigenic properties. The following section describes these procedures as applied to proteins separated by SDS gel electrophoresis.

A. Stoichiometry and Subunit Structure of Multicomponent Systems

1. Enzymes and Viruses

Complex enzymes and viruses can be dissociated at 100° in SDS solution in the presence of 2-mercaptoethanol. Individual components can then be separated according to their molecular weights using SDS gel electrophoresis. The amount of protein in each band can be determined by scanning a destained gel. Densitometers suitable for this purpose are available commercially or can be built as attachments to various spectrophotometers. The relative amounts of the different proteins derived from the scan and the molecular weights of the individual components provide the stoichiometric relationship of components in the multicomponent system. The stoichiometries of several multichain enzymes, e.g., RNA polymerase (Burgess, 1969), Qβ replicase (Kamen, 1970), and aspartate transcarbamylase (Rosenbusch and Weber, 1971), have been successfully determined in this way.

This procedure relies on two assumptions: (a) that different polypeptide chains have very similar color yields (stain intensity per unit mass of protein), and (b) that the intensity of stain for a given component is a linear function of the amount of protein present. Such a linear relationship has been verified for Coomassie Brilliant Blue at 550 nm in the range of 0.5–15 μg of protein (Fazekas de St. Groth *et al.*, 1963) and for Fast Green in the range of 10–150 μg of protein (Gorovsky *et al.*, 1970). In a recent study by Bickle and Traut (1971), differences of up to 20% in color yield were found for eight different proteins stained with Coomassie Brilliant Blue. Although a smaller deviation would be highly desirable, it seems that the procedure can yield an approximation of the stoichiometry which is sufficiently accurate for most purposes. Some difficulty may be expected if the individual components are present in drastically different weight ratios. Maizel (1971) has discussed the potential use of Procian Brilliant Blue. This dye seems to give more uniform staining with different proteins but is unfortunately less sensitive than Coomassie Brilliant Blue.

An independent procedure for the determination of subunit structure of oligomeric enzymes has been employed by Davies and Stark (1970). Using the bifunctional reagent dimethyl suberimidate, they have shown that cross-linking occurs predominantly within oligomers. SDS gel electrophoresis of the modified proteins revealed a set of species with molecular weights equal to integral multiples of the molecular weight of the polypeptide chain. For example, the tetramer glyceraldehyde-3-phosphate dehydrogenase yielded four protein species, whereas the hexamer L-arabinose isomerase showed six protein species (see Fig. 4). Application of this procedure to multichain enzymes composed of dissimilar polypeptide chains should yield information about the spatial arrangement of the individual chains. Another useful technique involves cross-linking reagents, which can be cleaved reversibly after SDS gel electrophoresis has been performed. This approach should allow a detailed identification of the components in a cross-linked species.

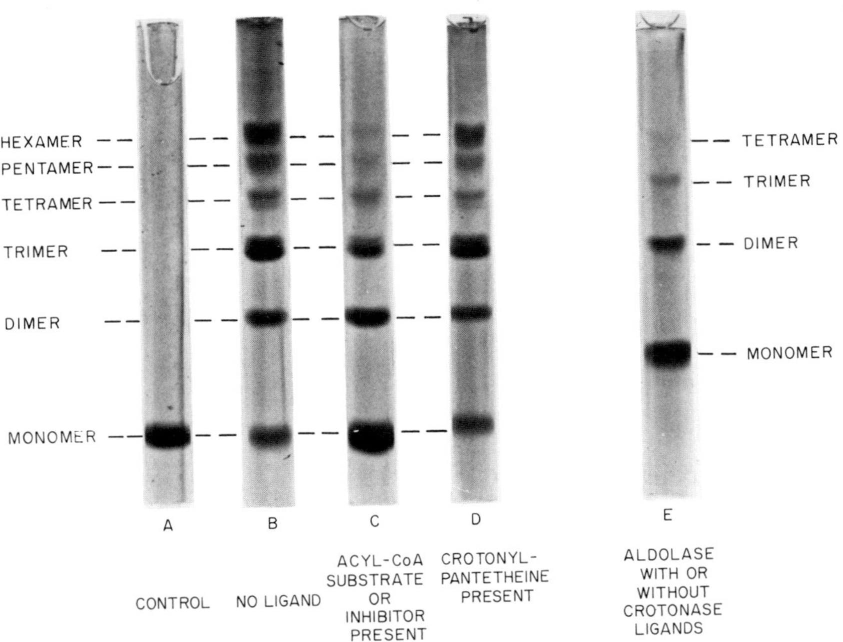

Fig. 4 SDS gel patterns of oligomeric enzymes cross-linked with dimethyl/suberimidate. The position of the cross-linked peptide chains is indicated. The first four gels show the patterns obtained with the hexameric enzyme crotonase. Gel A, without cross-linking. Gel B, cross-linked in the absence of ligand. Gels C and D show patterns of crotonase cross-linked in the presence of ligands. Gel E is a control and shows cross-linking of the tetrameric enzyme aldolase. The picture was kindly provided by R. M. Waterson.

SDS gel electrophoresis is a standard technique for protein analysis in studies of phages and viruses. A major advantage of the procedure is that noncovalent interactions between the different proteins and the nucleic acid as well as protein–protein interactions are disrupted. A few examples are given in Fig. 5. Note that Qβ bacteriophage and the animal EMC virus show relatively simple patterns, whereas bacteriophage T4 shows a rather complex pattern. Since the resolution is based on the molecular weights of the individual components, situations can arise in which a mixture of several proteins of nearly identical size will yield a single protein band. Comparison of the bands of urea buffer gels and SDS gels is therefore necessary to characterize a complex structure in more detail (see Section IV,H). An example of such an analysis is given by Traut and co-workers for *E. coli* ribosomes (Traut *et al.*, 1969). The gel pattern shown in Fig. 6 clearly demonstrates the existence of a large number of individual proteins in a ribosomal structure; however, gel

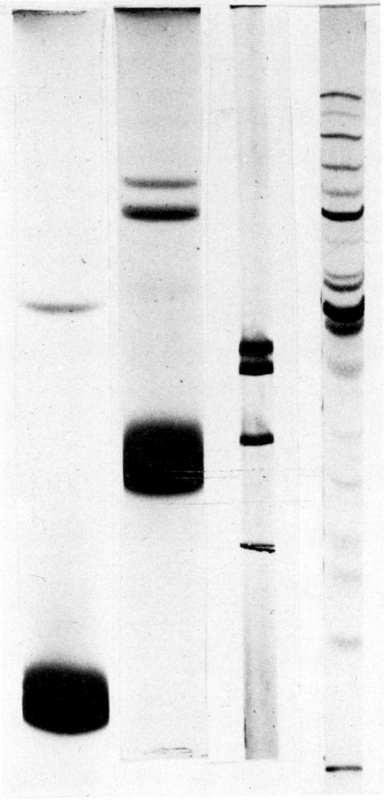

Fig. 5 SDS gel electrophoresis of different virus preparations. (a) Virion of *E. coli* bacteriophage R17. One minor component (maturation protein, MW 38,000) and the major capsid protein (MW 14,000) are found. For details, see Osborn *et al.* (1970). (b) Virion of *E. coli* bacteriophage Qβ. The major capsid protein has a molecular weight of 14,000. The minor component with a molecular weight of 41,000 is the maturation protein; it is present in one copy per virion. The second minor component (MW 36,000) is due to a leaky stop signal at the end of the coat protein gene. For details, see Weiner and Weber (1971). (c) Virion of encephalomyocarditis (EMC) virus. The three major capsid proteins, with molecular weights of 34,000, 30,000 and 23,000, are present in approximately equal molar amounts. Two minor components with molecular weights of 40,000 and 9,000 are not visible on this gel. For details, see Butterworth *et al.* (1971). (d) Virion of *E. coli* bacteriophage T4. The virion probably contains more than 20 unique proteins which range in molecular weight from 10,000 to 150,000. For details, see Laemmli (1970) and Dickson *et al.* (1970). The picture was kindly provided by U. K. Laemmli.

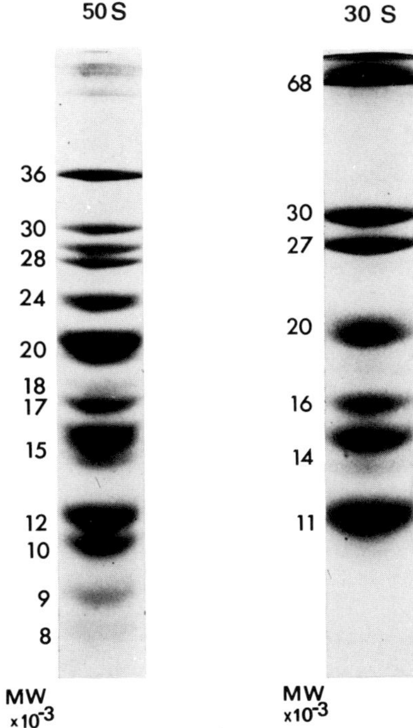

Fig. 6 SDS gel pattern of the ribosomal proteins of the 50 S and 30 S subunits of *E. coli* ribosomes. The numbers on the left indicate the approximate molecular weights. Electrophoresis proceeded from top to bottom on 15% polyacrylamide gels using the SDS–phosphate system (for further details, see Traut *et al.*, 1969). The picture was kindly provided by R. R. Traut.

electrophoresis in urea buffer gives much better resolution since separation occurs on the basis of the size and charge of the individual proteins.

An alternative method of obtaining the stoichiometric ratios is by determining the amino acid compositions of the individual separated bands of a multicomponent system (see below).

2. *Cells and Membranes*

Whole cells can be extracted directly with hot SDS solution, and usually all the proteins are solubilized, since all noncovalent interactions between different macromolecules are eliminated. At the present time no other denaturing agent seems to allow such an instantaneous solubilization of the cellular proteins. The cell extract can be directly processed on SDS gels and usually more than 90% of the protein material will clearly enter the gel. Obviously such a procedure is only useful if the number of proteins to be studied is small or if a particular protein species can be specifically labeled. Such situations often arise in virus- or phage-infected cells.

When *E. coli* is infected with T-even phages, host protein biosynthesis stops and phage-specific proteins can then be specifically labeled with radioactive amino acids. Using this approach, Laemmli (1970) studied the maturation phenomenon of the major head protein of bacteriophage T4. This protein is synthesized as a precursor polypeptide chain which is proteolytically cleaved during assembly (Laemmli, 1970). A more difficult situation is encountered after *E. coli* is infected with phage λ, since host protein biosynthesis is not stopped. Strong irradiation of the cells with ultraviolet light prior to phage infection suppresses host protein biosynthesis. Infection of such ultraviolet-treated cells with phage λ in the presence of radioactive amino acids allows the study of the phage-specific proteins (see, for instance, Hendrix, 1971). In the case of *E. coli* infected with RNA bacteriophage, host (but not phage) protein biosynthesis can be abolished by specifically inhibiting host messenger RNA synthesis with the drugs actinomycin or rifamycin (see Viñuela *et al.*, 1967; Weiner and Weber, 1971).

Similar approaches have been used in studies involving animal cells infected with different viruses. The RNA-containing poliovirus shuts off host protein biosynthesis in infected HeLa cells. In a series of elegant experiments using pulse-chase techniques and SDS gel electrophoresis, Baltimore and his collaborators showed that the virus RNA is translated into one long polypeptide chain. This precursor is then enzymatically cleaved to yield various phage-specific polypeptide chains (for a review, see Baltimore, 1971). Experiments with the DNA virus SV40 are more difficult since this virus does not turn off host protein biosynthesis. However, even in this case Anderson and Gesteland (1972) were able to follow the time course of production of at least one major virus capsid protein.

Another field of potential application is opened if it is possible to specifically label a particular enzyme or protein component *in vivo* with an "active site" label.

SDS gel electrophoresis has become an accepted method of characterizing the major protein components of a membrane. Since membranes usually have a high lipid content, it is important to add enough SDS to completely disrupt the membrane and dissociate all the proteins. Figure 7 shows a complex gel pattern characteristic of the red blood cell membrane. Combining SDS gel electrophoresis with some of the chemical techniques discussed later in this chapter enabled Bretscher (1971b)

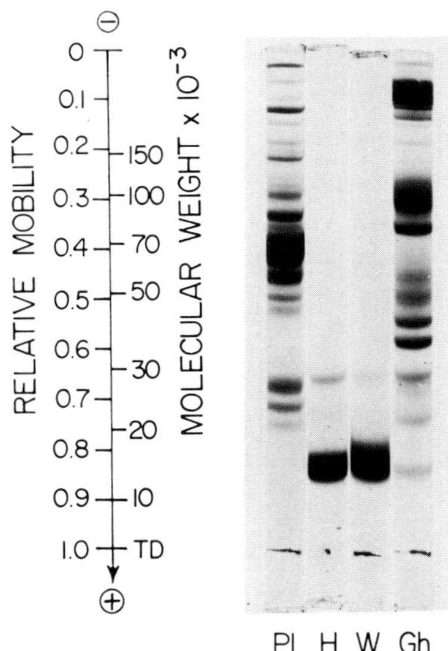

Fig. 7 Complexity of the red blood cell membrane as revealed by SDS gel electrophoresis. The gel patterns shown are for proteins from plasma (Pl), hemolysate supernatant (H), wash of ghosts (W) and ghosts (red cell membranes) (Gh). For more details, see Fairbanks *et al.* (1971) and Bretscher (1971a,b). The picture was kindly provided by G. Fairbanks.

to show by an elegant labeling technique that at least two of the major components of the red blood cell membrane span the membrane.

B. Amino Acid Analysis

The amino acid composition of the stained protein bands may be determined directly, following the procedures described by Kyte (1971). A routine amino acid analysis usually requires less than 5 nmoles. Generally this amount can be obtained from several gels. Alternatively, gels with larger cross-sectional areas can be used. Gels with a diameter of 1 cm can accommodate several milligrams of protein, provided the band pattern allows such overloading.

The stained gel slices are added to a hydrolysis tube containing 3 ml of 6 *N* HCl and 1% mercaptoacetic acid. Norleucine is added in a known quantity to monitor the recovery. After acid hydrolysis at 110° for 24 hr, the tube is refrigerated to precipitate the large amount of polyacrylic acid which is the hydrolysis product of the gel matrix. The supernatant is dried and dissolved directly in the starting buffer for amino acid analysis. Entry of the ninhydrin into the amino acid analyzer coil should be postponed for 20 min when using the long column to avoid the reaction of mercaptoacetic acid with ninhydrin. The large amounts of ammonia derived from the hydrolysis of polyacrylamide generally prevent the detection of arginine and may cause precipitation in the coil. If an arginine value is important, it seems advisable to remove the ammonia with NaOH after removal of the hydrochloric acid. Amino acid compositions obtained for standard proteins by this procedure usually agree well with those obtained by conventional methods. Serine and glycine values are sometimes too high, probably due to byproducts from the gel. Methionine and tyrosine may be partially destroyed even in the presence of the protective mercaptoacetic acid. The use of preeluted gels (Weiner *et al.*, 1972) seems to allow good recovery of tyrosine. In order to avoid the presence of stain and gel matrix during hydrolysis, the protein can be eluted from the gel and then processed for acid hydrolysis (see below).

Recently a method of amino acid analysis has been described which employs the conventional chromatographic separation procedure but uses "fluorescamine" to determine amino acids fluorometrically (Stein *et al.*, 1973). This method is reported to require only 1 μg of protein for a complete amino acid analysis.

C. Recovery of Proteins from SDS Gels

Several procedures are available to localize individual protein bands on a gel without subjecting the entire gel to staining and destaining procedures (for a detailed discussion, see Weiner *et al.*, 1972). The most useful approach is to cut a longitudinal guide strip from the edge of the preparative gel with the slicer described by Fairbanks *et al.* (1965). Localization of the proteins in about 30 min is possible by staining and destaining the guide strip while the bulk of the gel is wrapped in Saran Wrap and held at 4°. The corresponding regions of the gel are then cut out with a razor blade and collected. Alternative methods involve the use of an analytical gel run in parallel or the addition of small quantities of a fluorescent derivative (see Section III,H). In the case of a multicomponent system in which the individual components are well-separated and present in similar amounts, one can visualize the individual bands by simply chilling the gel to 4° for several hours. The protein bands form white precipitation lines if the amount of protein is in the range of 20–40 μg.

The gel sections cointaining the protein are cut into 1-mm pieces or macerated by passing them through a syringe. The protein is then eluted by shaking at 37° into 10 volumes of 0.05 *M* NH_4HCO_3 containing 0.05% SDS. After 6 hr the polyacrylamide fragments are removed by centrifugation or by passage through a small syringe stuffed with glass wool, then washed once with a small volume of elution buffer. The combined supernatants can be dialyzed against SDS in order to remove the buffer originally present in the gel, though usually this is not necessary. Lyophilization will remove the NH_4HCO_3, leaving a protein–SDS mixture.[9] Using different radioactively-labeled proteins, Weiner *et al.* (1972) showed that the recovery of proteins eluted by this procedure from 5% and 10% acrylamide gels was nearly complete. At this stage, the isolated protein can be directly subjected to amino-terminal analysis (see below) or amino acid analysis. However, if peptide mapping or renaturation is planned, the detergent has to be removed.

D. Removal of SDS from Proteins

Quantitative removal of SDS from protein–SDS solution is possible using a simple ion-exchange step (Weber and Kuter, 1971). The lyophilized protein–SDS mixture is dissolved in 6 *M* urea, 0.1 *M* NH_4HCO_3 to yield a 0.1% to 1% SDS solution. Dowex 1X2 (200–400 mesh from Bio-Rad) is equilibrated with 6 *M* urea, 0.05 *M* NH_4HCO_3. Under these conditions 1 ml of settled resin will easily bind 100 mg of SDS. The protein solution is passed through a microcolumn of Dowex 1X2 developed with the urea buffer. Alternatively, the resin is added as a thick slurry to the solution, and after occasional shaking the resin is removed by centrifugation or passage through a syringe plugged with glass wool. The resulting protein solution in 6 *M* urea is quantitatively free of detergent. Protein freed from SDS can then be subjected to peptide mapping or renaturation. The procedure is reviewed in detail elsewhere (Weber and Kuter, 1971).

An alternative procedure provides protein material sufficiently free of SDS to be used for enzymatic digestion, but does not remove the SDS quantitatively. The protein is precipitated with trichloroacetic acid (10%, or higher, depending upon the SDS concentration), and after removal of the SDS-containing supernatant, the precipitate is redissolved in 0.2 *M* NaOH and the procedure repeated. Residual trichloroacetic acid can be removed by ethanol or ether.

[9] A simple procedure for collecting protein from an SDS solution is to adjust the solution to 0.4 *M* NH_4HCO_3 and then to add 9 volumes of cold acetone (Weiner *et al.*, 1972). Precipitation of the salt produces a flocculent white precipitate of inorganic carrier and protein. After centrifugation and removal of the supernatant, the acetone is blown off with nitrogen and the NH_4HCO_3 is removed in a heated desiccator under vacuum. The inorganic carrier allows nearly quantitative recovery of the protein, even from very dilute protein solutions (original concentration, 50 μg/ml), provided the SDS concentration is 0.1%.

In both procedures, carrier protein should be added if a small amount of radioactive protein is processed.

E. Peptide Mapping

Peptide mapping generally requires approximately 5 nmoles of protein. If thin-layer carriers are used rather than paper, the spots are more compact, allowing a greater sensitivity. The use of "fluorescamine" instead of ninhydrin should routinely allow the detection of 0.1 nmoles of peptide material (Udenfriend *et al.*, 1972; Vandekerckhove and Van Montagu, 1973). If SDS is removed by the Dowex 1 × 2 procedure described above (Section IV,D), the protein may be digested with trypsin after diluting the urea solution threefold. After digestion and lyophilization to remove NH_4HCO_3, the sample is applied to a small Dowex 50 column in the H^+ form. The urea is removed by washing with water, and the peptides are eluted with 5 *M* ammonia. The eluate is lyophilized immediately. These procedures can be used with radioactively labeled proteins; however, if the amount of labeled protein is small, carrier protein has to be added after elution from the gel.

An extremely useful procedure for comparative studies of very small amounts of protein is peptide mapping of iodinated proteins. Radioactive iodine can be introduced into the original protein sample by a chemical method using chloramine T (Greenwood *et al.*, 1963) or by an enzymatic method using lactoperoxidase (Marchalonis, 1969). Alternatively, the protein can be eluted from the gel and iodinated by a chemical procedure. In the latter case, protein from stained or unstained gels can be used. The iodination procedures introduce radioactive iodine into tyrosine residues of the protein and allow detection of minimal amounts of protein (for a discussion of sensitivity, see Greenwood *et al.*, 1963; Marchalonis, 1969). Extensive manipulation of the material may require the addition of carrier protein after the iodination step.

Proteins can also be characterized by their cyanogen bromide fragments. The SDS–urea system (Table IV) of Swank and Munkres (1971) gives excellent resolution in the lower molecular weight range and can be used for such experiments.

F. Amino-Terminal Sequence Analysis

A simple, rapid, manual technique for the determination of amino-terminal sequences of proteins on a nanomole scale is described by Weiner *et al.* (1972). In this modification of the Dansyl–Edman degradation, inorganic carriers (e.g., $NaHCO_3$) permit convenient manipulation of small amounts of protein, and the use of SDS throughout the procedure maintains protein solubility. Nanomole quantities of pure protein for this analysis can be readily isolated from multicomponent systems by analytical polyacrylamide gel electrophoresis in SDS. Proteins are generally recovered in good yield from the gel by elution. The method is suitable for the chemical characterization of individual polypeptide chains derived from multichain enzymes and viruses. Usually 5–10 residues can be determined from nanomole amounts. The details of the procedure are described elsewhere (Weiner *et al.*, 1972).

The procedure has been used as a micro-method on several proteins eluted from SDS gels using multicomponent systems as starting material, and some results are summarized in Table V. Proteins of both known and unknown amino-terminal sequence have been subjected to the SDS–Dansyl–Edman degradation. For proteins of known sequence, the results agree with those found by conventional techniques. For proteins of unknown sequence (e.g., minor capsid protein from the bacteriophages

TABLE V

Amino-Terminal Sequences of Proteins Eluted from SDS Polyacrylamide Gels[a]

Protein	nmoles	μg	MW	Amino-terminal sequence
lac Repressor[b,c]	1.5	60	38,000	Met-Lys-Pro-Val-Thr-Leu-Tyr-Asx-
lac Repressor mutant i[100 c]	2.6	90	34,000	Ala-Glx-Leu-Asx-Tyr-Ile-Pro-
Bacteriophage Qβ[b,d]				
Major coat protein III	5.0	70	14,000	Ala-Lys-Leu-Glx-Thr-Val-Thr-Leu-
Maturation protein IIa	1.0	40	41,000	Pro-Lys-Leu-Pro-
Read-through protein IIb	2.0	70	36,000	Ala-Lys-Leu-Glx-Thr-Val-Thr-Leu-
Bacteriophage R17[b]				
Major coat protein	5.0	70	14,000	Ala-Ser-Asx-Phe-Thr-Glx-Phe-Val-Leu-Val-
Maturation protein	3.0	120	38,000	Met-Arg-Ala-Phe-Ser-Ala-Leu-Asx-
Qβ Replicase subunit II[b,d]	0.7	45	64,000	Ser-Lys-Thr-Ala-
Qβ Replicase subunit IV (T_S factor)[e]	1.1	40	35,000	Ala-Glx-Ile-Thr-Ala-Ser-Leu-

[a] Amino-terminal sequences of these proteins were determined by the SDS–Dansyl–Edman microtechnique described by Weiner *et al.* (1972).

[b] Weiner *et al.* (1972).

[c] Platt *et al.* (1972).

[d] Weiner and Weber (1971).

[e] Blumenthal *et al.* (1972).

$Q\beta$ and R17), the amino-terminal sequence data are compatible with known RNA sequences from the phage genome.

Several interesting results have been obtained with this technique. The *E. coli* bacteriophage $Q\beta$ has three capsid proteins with molecular weights of 14,000 (III), 36,000 (IIb), and 41,000 (IIa) (weight ratios 95:4:1). The virus is known to have only two genes for structural proteins. Characterization of the two minor proteins eluted from SDS gels by the procedure described above showed that proteins III and IIb have identical amino-terminal sequences for at least eight residues (Weiner and Weber, 1971). A similar conclusion was reached independently using conventional sequencing techniques with 100 times as much starting material (Moore *et al.*, 1971). In further studies, Weiner and Weber (1971) showed that the IIb protein is due to a leaky stop signal of the coat gene of the phage, and they identified the translational stop signal as the codon UGA.

G. Digestion with Carboxypeptidases

Carboxypeptidases A and B are remarkably resistant to SDS at room temperature in the absence of 2-mercaptoethanol (Guidotti, 1960). This property should allow proteins eluted from gels to be characterized by their carboxyl-terminal residue in the presence of SDS.

H. Combination of SDS Gel Electrophoresis and Urea Gel Electrophoresis

Protein–SDS complexes are readily dissociated by electrophoresis on urea-containing acrylamide gels at alkaline pH, since the onset of the electrophoretic migration separates the slowly moving protein molecules from the faster moving SDS ions. It is therefore possible to further characterize proteins eluted from SDS gels using urea gel electrophoresis. The design of suitable two-dimensional gel systems can be expected. Such systems should allow the characterization of individual components not only by molecular weight but also by net charge.

I. Renaturation of Biologic Activity

Successful renaturation of several enzymes after denaturation with SDS has been described by Weber and Kuter (1971). The SDS is removed as described above (see Section IV,D) in a urea–Tris buffer

TABLE VI

Renaturation of Enzymes after Dissociation in SDS Solution

Protein	Recovery from SDS solution (%)	Recovery after SDS gel electrophoresis (%)
Aspartate transcarbamylase[a]	60	—
Catalytic subunit of aspartate transcarbamylase[a,b]	40	25
lac Repressor (inducer-binding)[a]	50	15
Aldolase[a]	30	20
β-Galactosidase[a]	60	15
Qβ Replicase subunit IV[c]	—	25
σ Factor of RNA polymerase[a]	—	45

[a] Proteins were dissociated in SDS solution and renatured either directly or after SDS gel electrophoresis. The procedure is given in detail by Weber and Kuter (1971).
[b] Rosenbusch and Weber (1971)
[c] Blumenthal *et al.* (1972).

using Dowex 1 × 2. The resulting protein–urea solution is free of SDS. After dilution of the urea, enzymatic activity can be recovered in good yield from several oligomeric enzymes (Weber and Kuter, 1971; Rosenbusch and Weber, 1971). Table VI summarizes the recovery of enzymatic activity from several proteins which were denatured with SDS and then renatured by the above procedure. When the same procedure is applied to proteins eluted from SDS gels, much lower yields are generally found. However, better results can be obtained if the gels used for the separation are preeluted as described elsewhere (Weiner *et al.*, 1972; Blumenthal *et al.*, 1972).

Thus it seems that the renaturation procedure can be used to identify an individual component in a multicomponent system by enzymatic activity and polypeptide chain molecular weight. An example of this approach has been recently reported by Blumenthal *et al.* (1972). The smallest polypeptide chain of Qβ replicase, an RNA-synthesizing enzyme, was separated from the other three polypeptide chains on SDS gels. After elution from the gel, the protein was renatured essentially as described above, and the enzymatic activity, typical for the protein biosynthesis factor T_S, was recovered with an overall yield of approximately 20%.

After SDS treatment certain proteins may be detected by their biologic activity, if sufficient SDS is removed by the addition of a carrier protein (e.g., serum albumin) which competes with the original protein for the detergent. Gill and Pappenheimer (1971) reported that diptheria toxin can be assayed after elution from SDS gels in the presence of serum albumin. Although this procedure may not be generally applicable, it provides a possible tool for the study of individual components in a complex mixture.

J. Antigen–Antibody Complexes and Immunologic Studies

Antigen–antibody complexes are efficiently dissociated by treatment with SDS and 2-mercaptoethanol at 100°, as originally shown by Horwitz and Scharff (1969). If the dissociated material is then separated on SDS gels, the antigen can be recovered in chemically pure form.

Antibodies are especially useful when a protein antigen must be rapidly isolated for chemical characterization. Such a need can often arise in a comparative study of mutant proteins. If antibody against the pure antigen is available and if the mutant antigen is present in a cell-free extract at a level of 0.1% of the total protein and also shows immunologic cross-reactivity, a direct isolation procedure seems possible. The antigen–antibody complex is allowed to form and then the precipitate is spun out. The complex is dissociated in SDS and subjected to SDS gel electrophoresis. The antigen eluted from the gel is then sufficiently pure for chemical analysis by the procedures outlined above. This technique is especially useful if the mutant protein is so different from the wild-type antigen that a new purification procedure would have had to be designed to obtain the new protein in pure form.

Platt *et al.* (1972) have used this approach to show that a mutant of the *lac* repressor in *E. coli* synthesizes a polypeptide chain which, in comparison to the wild-type molecule, is missing the 42 amino-terminal residues. They showed that this mutant protein arises from a nonsense mutation in position 26 of the polypeptide chain, followed by chain termination and reinitiation of protein biosynthesis at the first internal methionine residue after the nonsense block. Although amino-terminal sequence studies were used in this work, the amount of mutant protein isolated by the procedure would have allowed characterization by amino acid composition or peptide mapping. Figure 8 shows SDS gel patterns of *lac* repressor antibody complexes. The antigen was purified from the cell-free extract by ammonium sulfate fractionation and then subjected to antibody precipitation.

Antibody precipitation followed by SDS gel electrophoresis is a useful procedure for studying the *in vivo* synthesis of an antigen, the possible covalent modification of the antigen, and *in vivo* stability. This approach has been applied to studies of γ-globulin biosynthesis (Horwitz and

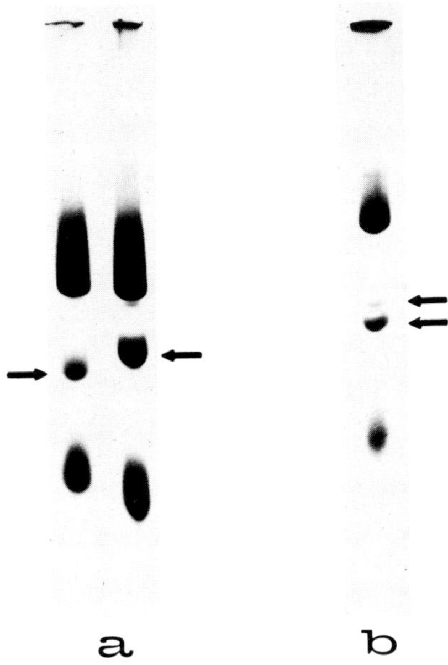

Fig. 8 SDS gel electrophoresis of antigen–antibody complexes. (a) Guide strips from preparative gels, which are heavily overloaded. The middle band is the i^{100} mutant *lac* repressor (left-hand gel) or the wild-type repressor (right-hand gel). The heavy bands at the top and bottom correspond to the heavy and light chains of γ-globulin, respectively. (b) An analytical gel used for the molecular weight determination. The four bands (from top to bottom) are the polypeptide chains of heavy IgG (molecular weight 50,000), *lac* repressor (molecular weight 38,000), i^{100} mutant repressor (molecular weight 34,000), and light IgG (molecular weight 23,000). For this gel 10 μg of pure *lac* repressor were added to the i^{100} protein–antibody complex. The two repressor species are clearly separated. The figure is taken from Platt *et al.* (1972) and illustrates how antibody precipitation and SDS gel electrophoresis can be combined to isolate a mutant protein in pure form.

Scharff, 1969; Schubert and Cohn, 1970), *in vivo* modification of RNA polymerase of *E. coli* after infection with phage T4 (Goff and Weber, 1970), and in a study of a *lac* repressor mutant which is degraded *in vivo* (Platt *et al.*, 1970).

In cases where an antigen is in the molecular weight range of the heavy chain of the γ-globulin molecule (MW 50,000), the use of papain-treated antibody allows isolation of the antigen. If the antigen is in the molecular weight range of the antibody light chain (MW 23,000), dissociation of the antigen–antibody complex is performed in SDS, avoiding the use of 2-mercaptoethanol. The blocking of cysteine groups can be achieved by iodoacetate or a mercurial (J. P. Rosenbusch and K. Weber, unpublished observations).

Two other promising immunologic techniques should be mentioned. If antibodies prepared against certain antigens are available, proteins eluted from SDS gels may be tested for their antigenic activity. Certain proteins can be applied to an immunodiffusion plate in 0.1% SDS solution and still precipitate with the antibody (J. P. Rosenbusch and K. Weber, unpublished observations; Roberts, 1969). Alternatively, the detergent can be removed prior to immunodiffusion. These techniques can obviously be exploited to identify a particular antigen in a multicomponent system. A recent example of this approach is the demonstration that subunit IV of the RNA-synthesizing enzyme Qβ replicase is identical to the protein biosynthesis factor T_S (Blumenthal *et al.,* 1972).

Proteins eluted from SDS gels as well as the protein–polyacrylamide mixture can be used as antigens for immunization even without prior removal of detergent (J. P. Rosenbusch and K. Weber, unpublished observations).[10] Quite often antibodies are obtained which precipitate the native antigen as well as the denatured protein.

REFERENCES

Adams, J. M., Jeppesen, P. G. N., Sanger, F., and Barrell, B. G. (1969). *Nature (London)* **223**, 1009.

Anderson, C. W., and Gesteland, R. F. (1972). *J. Virol.* **9**, 758.

Arndt, D. J., and Berg, P. (1970). *J. Biol. Chem.* **245**, 665.

Baltimore, D. (1971). *Bacteriol. Rev.* **35**, 235.

Bickle, T. A., and Traut, R. R. (1971). *J. Biol. Chem.* **246**, 6828.

Blumenthal, T., Landers, T. A., and Weber, K. (1972). *Proc. Nat. Acad. Sci. U. S.* **69**, 1313.

Bretscher, M. S. (1971a). *Nature (London), New Biol.* **231**, 229.

Bretscher, M. S. (1971b). *J. Mol. Biol.* **59**, 351.

Burgess, R. R. (1969). *J. Biol. Chem.* **244**, 6168.

Butterworth, B. E., Hall, L., Stoltzfus, C. M., and Rueckert, R. R. (1971). *Proc. Nat. Acad. Sci. U. S.* **68**, 3083.

Chrambach, A., and Rodbard, D. (1971). *Science* **172**, 440.

Chrambach, A., Reisfeld, R. A., Wyckoff, M., and Zaccari, J. (1967). *Anal. Biochem.* **20**, 150.

Davies, G. E., and Stark, G. R. (1970). *Proc. Nat. Acad. Sci. U. S.* **66**, 651.

Davis, B. J. (1964). *Ann. N. Y. Acad. Sci.* **121**, 404.

Dickson, R. C., Barnes, S. L., and Eiserling, F. A. (1970). *J. Mol. Biol.* **53**, 461.

Dunker, A. K., and Rueckert, R. R. (1969). *J. Biol. Chem.* **244**, 5074.

Fairbanks, G., Levinthal, C., and Reeder, R. H. (1965). *Biochem. Biophys. Res. Commun.* **20**, 393.

Fairbanks, G., Steck, T. L., and Wallach, D. F. H. (1971). *Biochemistry* **10**, 2606.

[10] If substantial amounts of polyacrylamide byproducts are injected with the protein, the γ-globulin will also contain antibodies against material of the gel matrix. These antibodies have to be preabsorbed before further use.

Fazekas de St. Groth, S., Webster, R. G., and Datyner, A. (1963). *Biochim. Biophys. Acta* **71**, 377.
Ferguson, K. A. (1964). *Metab., Clin. Exp.* **13**, 985.
Fish, W. W., Reynolds, J. A., and Tanford, C. (1970). *J. Biol. Chem.* **245**, 5166.
Gill, D. M., and Pappenheimer, A. M. (1971). *J. Biol. Chem.* **246**, 1492.
Glossman, H., and Neville, D. M., Jr. (1971). *J. Biol. Chem.* **246**, 6339.
Goff, C. G., and Weber, K. (1970). *Cold Spring Harbor Symp. Quant. Biol.* **35**, 101.
Gorovsky, M. A., Carlson, K., and Rosenbaum, J. L. (1970). *Anal. Biochem.* **35**, 359.
Greenwood, F. C., Hunter, W. M., and Glover, J. S. (1963). *Biochem. J.* **89**, 114.
Guidotti, G. (1960). *Biochim. Biophys. Acta* **42**, 177.
Hedrick, J. L., and Smith, A. J. (1968). *Arch. Biochem. Biophys.* **126**, 155.
Helleiner, C. W., and Wunner, W. H. (1971). *Anal. Biochem.* **39**, 333.
Hendrix, R. W. (1971). *In* "The Bacteriophage Lambda" (A. D. Hershey, ed.), p. 355. Cold Spring Harbor Lab., Cold Spring Harbor, New York.
Horwitz, M. S., and Scharff, M. D. (1969). *In* "Fundamental Techniques in Virology" (K. Habel and N. P. Salzman, eds.), Vol. 1, p. 253ff. and p. 297ff. Academic Press, New York.
Kamen, R. (1970). *Nature (London)* **228**, 527.
Kyte, J. (1971). *J. Biol. Chem.* **246**, 4157.
Laemmli, U. K. (1970). *Nature (London)* **227**, 680.
Maizel, J. V., Jr. (1971). *In* "Methods in Virology" (K. Maramorosch and H. Koprowski, eds.), Vol. 5, p. 179. Academic Press, New York.
Marchalonis, J. J. (1969). *Biochem. J.* **113**, 299.
Moore, C. H., Farron, F., Bohnert, D., and Weissmann, C. (1971). *Nature (London), New Biol.* **234**, 204.
Nelson, C. A. (1971). *J. Biol. Chem.* **246**, 3895.
Neville, D. M. (1971). *J. Biol. Chem.* **246**, 6328.
Osborn, M., Weiner, A. M., and Weber, K. (1970). *Eur. J. Biochem.* **17**, 63.
Panyim, S., and Chalkley, R. (1971). *J. Biol. Chem.* **246**, 7557.
Peacock, A. C., and Dingman, C. W. (1968). *Biochemistry* **7**, 668.
Pitt-Rivers, R., and Impiombato, F. S. A. (1968). *Biochem. J.* **109**, 825.
Platt, T., Miller, J. H., and Weber, K. (1970). *Nature (London)* **228**, 1154.
Platt, T., Weber, K., Ganem, D., and Miller, J. H. (1972). *Proc. Nat. Acad. Sci. U. S.* **69**, 897.
Pringle, J. R. (1970a). *Biochem. Biophys. Res. Commun.* **39**, 46.
Pringle, J. R. (1970b). Ph.D. Thesis, Harvard University, Cambridge, Massachusetts.
Reynolds, J. A., and Tanford, C. (1970a). *Proc. Nat. Acad. Sci. U. S.* **66**, 1002.
Reynolds, J. A., and Tanford, C. (1970b). *J. Biol. Chem.* **245**, 5161.
Rice, R. H., and Means, G. E. (1971). *J. Biol. Chem.* **246**, 831.
Roberts, D. B. (1969). *J. Mol. Biol.* **45**, 221.
Rodbard, D., and Chrambach, A. (1970). *Proc. Nat. Acad. Sci. U. S.* **65**, 970.
Rosenbusch, J. P., and Weber, K. (1971). *J. Biol. Chem.* **246**, 1644.
Schubert, D. (1970). *J. Mol. Biol.* **51**, 287.
Schubert, D., and Cohn, M. (1970). *J. Mol. Biol.* **53**, 305.
Shapiro, A. L., Viñuela, E., and Maizel, J. V., Jr. (1967). *Biochem. Biophys. Res. Commun.* **28**, 815.
Shelton, K. R. (1971). *Biochem. Biophys. Res. Commun.* **43**, 367.
Stein, S., Böhlen, P., Stone, J., Dairman, W., and Udenfriend, S. (1973). *Arch. Biochem. Biophys.* **155**, 202.
Strauss, E. G., and Kaesberg, P. (1970). *Virology* **42**, 437.

Studier, F. W. (1972). *Science* **176**, 367.
Swank, R. T., and Munkres, K. D. (1971). *Anal. Biochem.* **39**, 462.
Traut, R. R., Moore, P. B., Delius, H., Noller, H., and Tissières, A. (1967). *Proc. Nat. Acad. Sci. U. S.* **57**, 1294.
Traut, R. R., Delius, H., Ahmad-Zadeh, C., Bickle, T. A., Pearson, P., and Tissières, A. (1969). *Cold Spring Harbor Symp. Quant. Biol.* **34**, 25.
Tung, J.-S., and Knight, C. A. (1971). *Biochem. Biophys. Res. Commun.* **42**, 1117.
Udenfriend, S., Stein, S., Böhlen, P., Dairman, W., Leingruber, W., and Weigele, M. (1972). *Science* **178**, 871.
Vanderkerckhove, J., and Van Montagu, M. (1973). In press.
Viñuela, E., Algranati, I. D., and Ochoa, S. (1967). *Eur. J. Biochem.* **1**, 3.
Ward, S. (1970). *Anal. Biochem.* **33**, 259.
Weber, K., and Kuter, D. J. (1971). *J. Biol. Chem.* **246**, 4505.
Weber, K., and Osborn, M. (1969). *J. Biol. Chem.* **244**, 4406.
Weber, K., Pringle, J. R., and Osborn, M. (1972). *In* "Methods in Enzymology" (C. H. W. Hirs and S. N. Timasheff, eds.), Vol. 26, p. 3. Academic Press, New York.
Weiner, A. M., and Weber, K. (1971). *Nature (London), New Biol.* **234**, 206.
Weiner, A. M., Platt, T., and Weber, K. (1972). *J. Biol. Chem.* **247**, 3242.
Wolf, B., Lausarot, P. M., Lesnaw, J. A., and Reichmann, M. E. (1970). *Biochim. Biophys. Acta* **200**, 180.
Zacharius, R. M., Zell, T. E., Morrison, J. H., and Woodlock, J. J. (1969). *Anal. Biochem.* **30**, 148.

4

Sedimentation Analysis of Proteins

K. E. VAN HOLDE

I. INTRODUCTION: THE ROLE OF ANALYTICAL SEDIMENTATION METHODS IN PROTEIN CHEMISTRY

A. Scope of the Chapter

In writing a chapter on sedimentation studies of proteins, one is conscious that over fifty years have passed since Svedberg began his pioneering work. Much has changed in protein chemistry, and the changes have been especially startling in the past twenty years. So it will not do to repeat the standard treatment that has appeared in so many places. As a result, many of the conventional topics are given short shrift; some are deleted entirely. In their place, I want to discuss the problems contemporary protein chemists face and the sedimentation techniques they are actually using.

Another problem arises from the enormous diversity of present applications of the ultracentrifuge. To keep the chapter to a reasonable length, I have restricted it entirely to *analytical* applications, omitting the vast field of preparative ultracentrifugation. To many, for whom the ultracentrifuge is primarily a preparative tool, this will seem an unpardonable restriction.

The chapter does not pretend to be comprehensive, even in those topics covered. There is neither space to derive equations nor to survey all the literature of recent years. My apologies to the many whose important contributions I slight.

As a prelude to the rest of the chapter, it seems worthwhile to outline in a few paragraphs the history of the ultracentrifugation of proteins.

B. The Past

It is, perhaps, only a slight exaggeration to claim that the physical study of proteins really began with Svedberg's development of the ultracentrifuge. In the period of a few years, this remarkable scientist and his collaborators developed the principal techniques still used today and established two critical facts about proteins—that they were indeed macromolecules and that they were homogeneous.

The first had already been suspected by many, but the demonstration that these molecules could be sedimented at appreciable velocities in moderate centrifugal fields made the fact vivid, and the application of absolute methods for the determination of particle weights provided a scale for comparison with smaller molecules. The demonstration of homogeneity (although crude at first) was more surprising and contained implications concerning the biosynthesis of proteins that were not to be fully appreciated for many years.

Even our current interest in subunit structure was anticipated by Svedberg, albeit in a somewhat oversimplified way. For some time he held to the hypothesis that a subunit having a molecular weight of approximately 17,000 might be common to all proteins. The enormous tables of sedimentation coefficients and molecular weights given in "The Ultracentrifuge" (Svedberg and Pedersen, 1940) attest, in their organization, to this interest in subunit structure.

TABLE I

Early Developments in Ultracentrifugation[a]

Development	Reference
Centrifuge with optical system—first analytical ultracentrifuge	Svedberg and Nichols (1923)
Theory of the transient state in approach to sedimentation equilibrium	Mason and Weaver (1924)
Ultracentrifuge with sector cells, improved optical system, T control	Svedberg and Rinde (1924)
Analysis of sedimentation, including radial dilution effect	
Theory of molecular weight measurement from s and D, or sedimentation equilibrium	Svedberg (1925)
Thermodynamic theory of sedimentation equilibrium, including discussion of nonideality, associating systems, and charge effects	Tiselius (1926)
First measurements of protein molecular weights (hemoglobin and ovalbumin) by sedimentation equilibrium	Svedberg and Fahraeus (1926)
Definition of time to attain sedimentation equilibrium under certain conditions	Weaver (1926)
Development of oil turbine ultracentrifuge; application to measurement of sedimentation coefficients of proteins	Svedberg (1927)
Molecular weight of hemoglobin from s and D	Svedberg and Nichols (1927)
First sedimentation equilibrium and velocity studies on very large proteins—demonstration of homogeneity of *Helix pomatia* hemocyanin	Svedberg and Chinoaga (1928)
Development and first solutions of differential equation for sedimentation transport	Faxén (1929), Lamm (1929)

[a] This is not an exhaustive list for the period covered; rather it is intended to list only those developments which, in this author's opinion, were most significant for later work.

The progress in technique and theory made between 1924 and 1930, largely by members of the Svedberg group, was truly remarkable. Instruments evolved from the first crude, gear-driven machines to the massive oil turbine instruments capable of extremely high fields. The rapid development of ultracentrifugation is chronicled in Table I, which shows that most of the foundations were laid by 1929.

Much has been accomplished since 1930, but the primary emphasis has been in refinements of apparatus and technique. To the classic methods of sedimentation velocity and sedimentation equilibrium, modern workers have added only two really new ideas: the Archibald approach-to-equilibrium method, and the density gradient technique in its many variations. The former is actually little used today, whereas density gradient methods have become exceedingly popular and surely represent the most significant single advance since 1930.

C. The Present

Since our knowledge of proteins has advanced so dramatically since Svedberg's time and so many other powerful techniques have become available, we might wonder why a protein chemist today should turn to sedimentation methods.

Classically, the problem was rather like this: After isolation of a protein, the first requirement was the demonstration of its homogeneity and the second was the determination of its molecular weight as precisely as possible by a combination of sedimentation and diffusion measurements or by sedimentation equilibrium techniques. Along with this information, it was customary to provide a rough measure of "axial ratio" or some other indicator of particle shape.

At the present time the emphasis has shifted. In most cases, separation of proteins heterogeneous in weight is quite easy, and such methods as gel electrophoresis or isoelectric focusing provide highly sensitive tests for heterogeneity in composition. The *physical* determination of molecular weight is only a first stage in analysis today. It tells us the approximate size of the unit which has been isolated. The modern protein chemist is then most likely to use sodium dodecyl sulfate gel electrophoresis (see Ch. 3) to determine the number and rough size of the polypeptide chains. *The exact molecular weight determination of these chains will ultimately be based on the determination of their amino acid sequences.* Subsequent use of sedimentation methods will be directed toward studies of the various states of aggregation into which these chains may reversibly or irreversibly enter, the thermodynamic analysis of these reactions, or the detection of conformational changes in the molecules. I belabor this point, for sedimentation methods have classically been thought of primarily as means of determining molecular weight. I do not believe that this will much longer be their primary function in protein chemistry except in determining the stoichiometry of aggregated states. Again, the kinds of information about particle shape which hydrodynamic methods can provide seem trivial in comparison to the results of X-ray diffraction studies. It should be emphasized, however, that in the studies of macromolecules *other* than proteins, the old uses and values still pertain, for with many such systems (polysaccharides and synthetic polymers, for example) there do not exist such homogeneous, structurally specific subunits as the polypeptide chains of proteins. But this review deals with the use of sedimentation analysis in protein studies.

In understanding better the basic simplicity of protein structure we have become more sensitive to the crucial biochemical role played by interactions of polypeptide chains with one another and with other molecules. It would seem likely that this trend will continue, and that we will recognize more and more complexity in the multienzyme complexes of cells. At a certain level of size, electron microscopy becomes more revealing of morphology (see Ch. 6), but in studying the stoichiometry, thermodynamics, and kinetics of assembly of such structures, the ultracentrifuge can play an invaluable role.

II. MOVING-BOUNDARY SEDIMENTATION TRANSPORT

In this section, I shall discuss the "classic" sedimentation velocity experiment in which the motion of a solvent–solution boundary is followed. The newer "band" or "zonal" techniques will be treated in Section III.

A. Homogeneous Solutes

1. The Measurement and Uses of Sedimentation Coefficients

In the conventional moving-boundary sedimentation velocity experiment, an initially uniform solution in a sector-shaped ultracentrifuge cell is subjected to a centrifugal field so intense that a boundary is formed between solution and pure solvent. The velocity of motion of this boundary is measured and yields the sedimentation coefficient

$$s = (dr_b/dt)/\omega^2 r_b \tag{1}$$

Here, as in all subsequent discussion, r is a distance from the center of rotation (r_b is the position of the boundary, defined below), and ω is the radial velocity of the rotor, in radians/sec. The sedimentation coefficient is measured in units of 10^{-13} sec, or *Svedbergs* (S). The meaningfulness of sedimentation coefficient determinations stems from the well-known relation

$$s^0 = M(1 - \bar{v}^0\rho)/Nf^0 \quad (2)$$

where M is the molecular weight and $\bar{v}$ the partial specific volume of the sedimenting solute, ρ is the density of the solution, f is the translational frictional coefficient, and N is Avogadro's number. The superscript zeros are to remind us that Eq. (2) strictly applies only to a solute at infinite dilution. Concentration effects will be discussed in Section II,*A*,*2*.

Thus the sedimentation coefficient of a solute depends upon its molecular mass, its specific volume, and, through the frictional coefficient, upon its size, shape, and hydration. It is *not* a unique function of M, a fact that is forgotten (sometimes with disastrous results) by those who translate sedimentation velocities determined in sucrose gradient experiments directly into "molecular weights."

To see how s should be measured in such an experiment, we must consider the behavior of a solution subjected to a high centrifugal field. The shape of the concentration gradient in the boundary will depend upon the sedimentation coefficient, the diffusion coefficient (D), and the rotor speed. A complete mathematical analysis involves the solution of differential equations for the transport process. This has been accom-

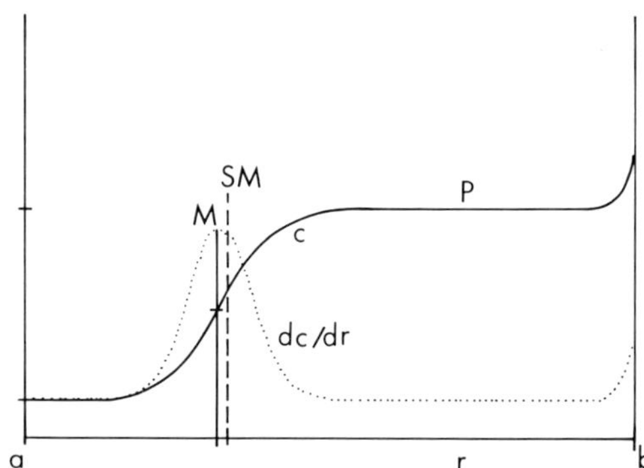

Fig. 1 A schematic diagram of the concentration profile (c) and the concentration gradient (dc/dr) in a moving-boundary sedimentation experiment. Direction of sedimentation is from left to right. The following symbols are used: a, meniscus; b, cell bottom; M, peak maximum; SM, (approximate) second moment position; P, plateau region.

plished, for certain idealized cases, by a number of workers (see Fujita, 1962; Dishon *et al.*, 1967). The kind of result obtained is diagramed in Fig. 1, which also resembles very nearly what we observe when we watch the sedimentation of a protein with the ultracentrifuge optical system. The top portion of Fig. 2 shows the experimentally observed refractive index gradient curve for a protein, as seen by schlieren optics, and the bottom portion, the absorbance curves for the same experiment, as shown by the scanner absorption system. In general, the refractive index gradient is proportional to the concentration gradient (dc/dr); the absorbance is proportional to c.

If D is small and $\omega^2 s$ large, the boundary will remain relatively sharp as it traverses the cell. But if D is not zero, the boundary will spread, and we must decide which point in the boundary should be taken as the boundary position, r_b, to be used in Eq. (1). The exact boundary point was first determined by Goldberg (1953), who showed that it should be the second moment of the gradient curve

$$r_b^2 = \int_a^{r_p} r^2\, dc/dr\, dr \Big/ \int_a^{r_p} dc/dr\, dr \tag{3}$$

Here a is the value of r corresponding to the meniscus and r_p is a point somewhere in the plateau region (see Fig. 1). Equation (3) is valid so long as the boundary is completely free of the meniscus. Even if the

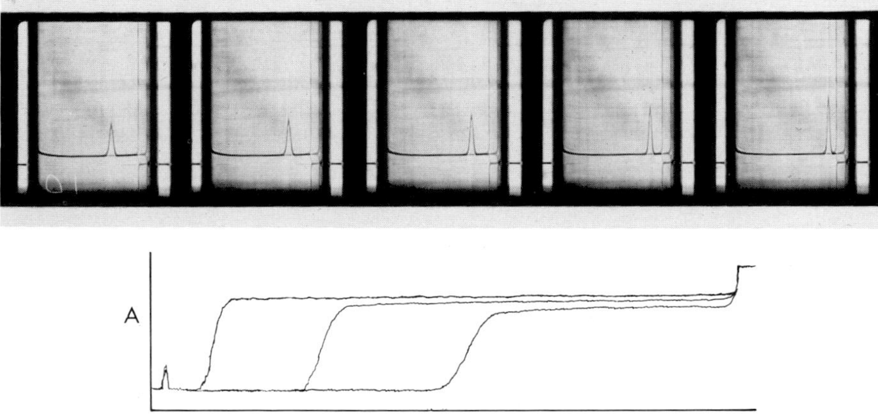

Fig. 2 Sedimentation of crab hemocyanin at 40,345 rpm, 20.7°, as depicted by schlieren photographs and scanner traces. Top: Successive schlieren photographs (right to left) taken at 5, 9, 13, 17, and 21 min after reaching speed. Sedimentation is to the *left* in these pictures. Bottom: Scans are shown for 5, 21, and 41 min after reaching speed. These are drawn with the direction of sedimentation from *left to right*. Note the radial dilution. Scans were made at a wavelength of 296 nm. The coordinates denote absorption (A) and distance from the center of rotation (r). Data of D. Carpenter.

meniscus concentration (c_a) is not zero, the equivalent boundary position can be obtained; the quantity $c_a a^2$ need only be added to the numerator of Eq. (3). However, Eq. (3) is awkward to use (though it may become less so with computer-linked ultracentrifuges), and it is fortunate that a much easier measurement is possible. Fujita (1962) has shown that the *rate of motion of the maximum in the gradient curve approximates, in most cases of interest to the protein chemist, very closely the rate of motion of the second moment.* Dishon *et al.* (1967) provide graphs which quantitate the error incurred. However r_b is measured, calculation of s normally proceeds from the integrated form of Eq. (1):

$$\ln r_b = \omega^2 st + C \tag{4}$$

where C is a constant of integration. Most current analyses proceed via a least-squares fitting of $\ln r_b$ vs. time. While not strictly justified if errors in r_b rather than in $\ln r_b$ are randomly distributed, it is adequate in most cases and better than "eyeballing."

One other prediction of the theory of sedimentation transport is of importance. As is indicated by the curves in the bottom portion of Fig. 2, the concentration (c_p) in the plateau region decreases steadily with time. Specifically, theory predicts and experiments confirm (see Richards and Schachman, 1959) that

$$c_p/c_0 = (a/r_b)^2 = e^{-2s\omega^2 t} \tag{5}$$

where a is the meniscus position and c_0 the original concentration. This "radial dilution" is due to both the sector shape of the cell and the inhomogeneity of the centrifugal field. The effect will turn out to be of importance in the analysis of paucidisperse systems (Section II,B).

What does the protein chemist get from a sedimentation coefficient measurement? Not a direct measure of M, but a number which gives an indication of M with about the same precision as does gel chromatography. If the protein molecule were an unhydrated, spherical particle, one could apply Stokes' law in the calculation of f^0:

$$f_{\text{sp}}{}^0 = 6\pi\eta R = 6\pi\eta(3M\bar{v}/4\pi N)^{1/3} \tag{6}$$

where η is the solvent viscosity. Then, from Eq. (2),

$$s_{\text{sp}}{}^0 = \frac{M^{2/3}(1 - \bar{v}\rho)}{6\pi\eta N^{2/3}(3\bar{v}/4\pi)^{1/3}} \tag{7}$$

$$\frac{s_{\text{sp}}{}^0\bar{v}^{1/3}}{(1 - \bar{v}\rho)} = \frac{M^{2/3}}{6\pi\eta N^{2/3}(3/4\pi)^{1/3}} \tag{8}$$

$$= (12.0 \times 10^{-3})M^{2/3} \text{ in water at } 20^\circ \tag{9}$$

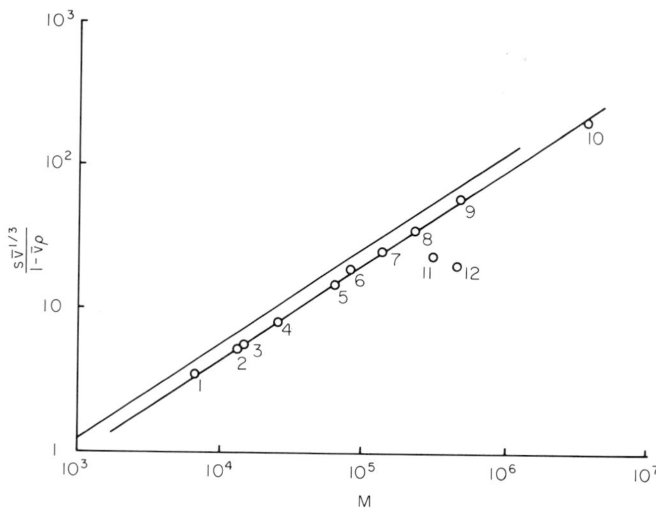

Fig. 3 A graph, according to Eq. (8), of $s^0\bar{v}^{1/3}/(1 - \bar{v}\rho)$ vs. M for a number of proteins. The upper line is the theoretical graph for spherical, unhydrated molecules [Eq. (9)]. The lower line is the empirical graph for "globular" proteins and follows the equation $s^0\bar{v}^{1/3}/(1 - \bar{v}\rho) = (9.2 \times 10^{-3})M^{2/3}$ when s^0 is in Svedbergs. Key to proteins: 1, lipase; 2, ribonuclease; 3, lysozyme; 4, chymotrypsinogen A; 5, bovine serum albumin; 6, enolase; 7, lactic dehydrogenase; 8, catalase; 9, urease, crab hemocyanin; 10, squid hemocyanin; 11, fibrinogen; 12, myosin.

when s_{sp}^0 is expressed in Svedbergs and all other quantities are in cgs units. In actuality, most globular proteins roughly follow this relationship. Figure 3 shows collected data on a number of proteins of known M, together with the line given by Eq. (9). The indication is that an empirical correlation can be made which fails only in the cases of radically nonspherical proteins. It is *not* recommended that M be "determined" in this way, but it is about as good (and subject to about the same pitfalls) as the use of gel filtration for molecular weight "determination." The regularity of the correlation shown in Fig. 3 should not be overemphasized; even for the globular proteins, an uncertainty of about 10% is to be expected. Such regularity as there is simply reflects the fact that there is a large class of proteins with axial ratios not too far from unity and with about the same level of hydration. The danger in using such an empirical correlation with a newly isolated protein is twofold: the protein might be like myosin or fibrinogen, or it might be undergoing reversible association. If so, a single sedimentation coefficient measurement could give meaningless results (see Section II,C).

Of course, the molecular weight of a homogeneous protein can be unambiguously determined from a combination of s^0 and the diffusion coefficient, D^0, using the classical Svedberg equation:

$$s^0/D^0 = M(1 - \bar{v}\rho)/RT \tag{10}$$

This is little used at the present time. Diffusion coefficient measurements are awkward, and there are better ways of obtaining M (sedimentation equilibrium, for example). It is possible that the method will come back into use if the rapid, precise measurement of D^0 by laser scattering methods gains wide acceptance (for a review of these methods, see Dubin, 1972).

If M is known from independent measurements, Eq. (2) may be used to obtain the frictional coefficient, f^0. The problems of interpreting f^0, which depends upon size, shape, and hydration, are detailed in many texts and reviews (for example, Schachman, 1959; Tanford, 1961; Van Holde, 1971). In most cases only the most qualitative information can be obtained, for the number of parameters upon which f^0 can depend is too great.

There exist two special kinds of problems in protein chemistry for which sedimentation velocity measurements of the frictional coefficient can give very useful information. The first utilizes the great sensitivity of the *difference* sedimentation technique to detect small conformational changes (see Section II,A,3). The second involves the comparison of the sedimentation coefficient of an oligomeric protein with that of its subunit to provide a provisional value of the number of subunits, or, if this is known, to distinguish between different oligomeric structures. The latter use depends upon Kirkwood's (1954) expression for the frictional coefficient of a multiunit structure in terms of the subunit frictional coefficient and oligomer geometry (Bloomfield *et al.*, 1967). This relationship, which is only approximately correct (see Zwanzig, 1966) states that the frictional coefficient of an n-mer will be given by

$$f_n{}^0 = \frac{n\xi}{1 + (\xi/6\pi\eta n)\Sigma_i\Sigma_j\langle 1/R_{ij}\rangle} \tag{11}$$

Here ξ is the frictional coefficient of one subunit, and the $\langle 1/R_{ij}\rangle$ values are the averaged reciprocals of intersubunit distances. The summation is carried over all pairs of subunits. If we assume further that the subunits are all spherical, of radius R, and in fixed geometry, and let $\alpha_{ij} = R_{ij}/R$, Eq. (2) yields the simple form

$$s_n{}^0/s_1{}^0 = 1 + \frac{1}{n}\sum_i\sum_j (1/\alpha_{ij}) \tag{12}$$

TABLE II

Predicted Sedimentation Coefficients of Oligomers

n	Geometry	s_n^0/s_1^0
2	Linear	1.500
3	Linear	1.833
	Triangular	2.000
4	Linear	2.208
	Square planar	2.353
	Tetrahedral	2.500

where s_1^0 is the subunit sedimentation coefficient. Table II gives the results for some simple oligomers; it is evident that different tetramer models, for example, lead to quite different results for s_4^0/s_1^0. It must be remembered that the equations are approximate, and that they neglect any volume changes in subunit conformation on association. Nonetheless, the results for the hemoglobin system, as an example, indicate that the method is useful. If we take the sedimentation coefficient of individual hemoglobin subunits (α or β) to be 1.77 ± 0.05 S (Antonini *et al.*, 1966), we calculate 3.91 ± 0.11 S for a linear tetramer, 4.18 ± 0.12 S for a square planar tetramer, and 4.43 ± 0.13 S for a tetrahedral tetramer. The latter value is close to experimental results for hemoglobin, which range around 4.4 S.

2. *Concentration Dependence, Solvent, and Charge Effects*

Most proteins exhibit some concentration dependence in their hydrodynamic properties; this becomes a special problem when one is dealing with fibrous or denatured proteins. In general, the sedimentation coefficient appears to vary with concentration as

$$s = s^0/1 + K_s c \tag{13a}$$

or

$$s \cong s^0(1 - K_s c) \tag{13b}$$

In these equations, the appropriate concentration is that in the plateau region. Since c_p is not exactly constant [Eq. (5)], s for such a system will vary slightly during the run; this is usually neglected and the initial concentration is used in Eq. (13a,b). Equation (13a) appears to be more satisfactory if K_s is large; it reduces to Eq. (13b) if K_s is small. No really adequate theory exists to predict the magnitude of K_s for particles of arbitrary shape. However, Creeth and Knight (1965) have reviewed a large amount of data and have found that for most globular particles $K_s \cong 1.6[\eta]$ where $[\eta]$ is the intrinsic viscosity. This is the empirical result first pointed out by Wales and Van Holde (1954) for random coil polymers. Since the intrinsic viscosities for most globular proteins are small, the concentration dependence is correspondingly small. For a globular protein, K_s will be of the order of magnitude of 5 ml/gm. This means that the difference between s and s^0 will be about 3% at $c = 6$ mg/ml (in the concentration range normally used with schlieren optics) and 0.5% at 1 mg/ml (a convenient concentration for the ultraviolet scanner optical system). In the latter case, the error is no greater than the usual uncertainty in determination of s values and may be ignored. Nonetheless, in investigating a protein whose behavior is unknown, a series of concentrations should be studied, for one is never sure, *a priori*, that the concentration dependence is negligible, and such experiments can also reveal reversible association or dissociation (see Section II,C).

One other consequence of the concentration dependence of s is of great importance. If this dependence is extreme, molecules on the high concentration side of the boundary will sediment appreciably more slowly than those on the low concentration side. The latter will tend to catch up, thus artificially sharpening the boundary. Equations for boundary shape in such circumstances are given by Fujita (1962) and by Dishon *et al.* (1967). Artificial boundary sharpening complicates the calculation of D from boundary spreading during sedimentation. Although equations have been derived to correct D for the concentration-dependent sharpening (Fujita, 1959; Van Holde, 1960), the procedure is not generally recommended.

Finally, it should be mentioned that in some cases an *increase* in s with increasing c is observed. This is presumptive evidence for reversible association (see Section II,C).

In considering the effects of solvent environment on the sedimentation velocity, we should consider two domains. First, there are those situations, usually encountered with dilute buffers, in which the only perceptible effect on the sedimentation coefficient arises from the effect of the buffer in modifying the density and/or viscosity of the solvent. It often can be assumed that selective interaction of solvent components with the protein is negligible and that the conformation of the protein is not sensitive to the presence of these components. Then a sedimentation coefficient measured at temperature T in buffer can be "corrected" to some standard conditions (such as water at 20°) by use of the classical relation

$$s_{20,w} = s_{T,b} \frac{(1 - \bar{v}\rho)_{20,w}}{(1 - \bar{v}\rho)_{T,b}} \frac{\eta_{T,b}}{\eta_{20,w}} \tag{14}$$

It is difficult to know *a priori* when Eq. (14) is sufficient, but wide experience seems to indicate that it is generally valid for solutions containing buffers or other salts up to a few tenths ionic strength.

When one attempts sedimentation velocity studies in concentrated salt solutions or in the presence of denaturing agents such as urea or guanidine hydrochloride, the situation becomes much more complex. In the first place, the effects of preferential interaction of either the second solute or water with the protein can become severe, and the molecular weight and specific volume in Eq. (2), which refer to anhydrous components (Williams *et al.*, 1958), are no longer appropriate to the situation. A detailed analysis has been given by Eisenberg (1962).

In the presence of concentrated denaturing solutes, the sedimentation coefficient will be altered not only by preferential interaction but even more so by the profound conformational change which is induced. Tanford *et al.* (1967) have investigated such systems in considerable detail. They find that most protein molecules are dissociated into single polypeptide chains *and* unfolded in 6 M guanidine hydrochloride (Gu·HCl) containing 0.1 M mercaptoethanol (EtSH). As a consequence, the sedimentation coefficient varies with the degree of polymerization (n) of the polypeptide chain in the manner expected for random coil polymers in good solvents. Tanford *et al.* (1967) find the empirical relation

$$\frac{s^0}{1 - \phi'\rho} = 0.286n^{0.473} \tag{15}$$

in 6 M Gu·HCl, 0.1 M EtSH, at 25°. Here ϕ' is the effective partial specific volume in the mixed solvent. This value is usually found to be about 0.01 ml/gm lower in 6 M Gu·HCl than the $\bar{v}$ observed in buffer (Kirby-Hade and Tanford, 1967). While Eq. (15) can be used to estimate chain weights, it is probably both less precise and less convenient than the corresponding equation obtained by Tanford *et al.* (1967) for intrinsic viscosity.

Finally, it is important to point out that the sedimentation coefficients of proteins may depend upon the charge if the protein is dissolved in a buffer of very low ionic strength. The situation is rather similar to the "Donnan effects" which are of importance in osmotic pressures and sedimentation equilibrium studies (see Section IV,B). As in the cases of sedimentation equilibrium and osmotic pressure, the effect is to lower the apparent molecular weight, but the error can be reduced by the presence of low molecular weight electrolyte. Since most proteins, if not too far from their isoelectric points, have a relatively small charge/mass ratio, the effects of charge on the sedimentation coefficient can usually be made negligible by carrying out experiments in buffers of about 0.1 ionic strength.

In summary, it would seem advisable, if possible, to carry out sedimentation transport studies in buffer solutions of the order of 0.1 ionic strength. If the buffer is much more dilute, charge effects could be significant; if it is much more concentrated, preferential interaction may have to be reckoned with.

3. *Difference Sedimentation*

In some kinds of studies, it would be desirable to be able to measure very small changes in the sedimentation coefficient. For example, addition of an effector to an allosteric enzyme may well be expected to produce a slight change in the tertiary or quaternary structure of the protein. Such changes will usually be too minute to be detectable within the ordinary precision of sedimentation velocity studies (ca. 1%). In response to this need, techniques of *difference sedimentation* measurement have been developed.

A first criterion for such measurements is that the two solutions being compared be placed in the same rotor and run simultaneously. This immediately eliminates or minimizes two of the principal sources of error in the determination of s, the rotor speed and temperature. Second, it is desirable to have a method in which the difference in boundary velocity is measured directly.

The simplest experimental technique is that which was utilized by Gerhart and Schachman (1968) and by Schumaker and Adams (1968). The two solutions are placed in separate cells, one or both of which have "wedge windows" which deflect the images so that the schlieren patterns for the two boundaries are displaced and distinguishable on the photographic plate. The limiting source of error is claimed to be the accuracy with which the position of the gradient maxima can be determined (Schumaker and Adams, 1968). Great caution must be taken, however, for the use of wedge windows may lead to artifacts if the windows are distorted or misaligned. Any "difference" should be checked by repeating the experiment with each cell refilled with the opposite solution. A precision of 0.3%–0.5% in s can be obtained by this method (Schumaker and Adams, 1968; Gerhart and Schachman, 1968; Kirschner and Schachman, 1971a).

A much more sophisticated technique has been developed by Schachman and co-workers (Richards and Schachman, 1957, 1959; Kirschner and Schachman 1971a,b). This makes use of the fact that the Rayleigh interferometric system measures the concentration *difference* between conjugate levels in the two channels of a double-sector cell. Thus, if a sedimentation transport experiment is performed with two identical solutions in sectors that are equally filled, no deviation in the straight interference fringes would be seen, even though steep boundaries were traversing each sector. If one boundary moves even a little faster than the other, a "peak" will be observed in the fringe pattern in the region of concentration mismatch. Needless to say, the technique requires, for success, exceedingly careful alignment of the optical system. In particular, the slits in the Rayleigh mask must be exactly parallel and symmetrically placed, otherwise nonconjugate levels in the two sectors will be compared and spurious results obtained. Successful application of the technique in Schachman's laboratory required the development of a whole new Rayleigh mask system. It is not, at the present time, a method for the uninitiated!

The precision which can be obtained by the Richards–Kirschner–Schachman technique is astonishing. It has been possible to routinely measure differences of as little as 0.01 S, with a precision of ± 0.0005 S. With this method, the effects of substrates and effectors on the allosteric enzyme aspartate transcarbamylase have been investigated in great detail (Kirschner and Schachman, 1971b). If the method can be made generally available, it will open a whole new field of application of the sedimentation transport experiment.

B. Paucidisperse, Nonassociating Proteins

In certain instances, the investigator may wish to study mixtures containing two or more proteins. In some cases, these will be species in chemical equilibrium; such mixtures are considered in Section III,C. Here we treat *nonreacting* systems. The need for this kind of analysis has diminished with the development of modern separation methods. For example, gel exclusion chromatography can, in most instances, perform such an analysis and yield a preparative separation as well. However, a few comments are appropriate about the few cases for which such sedimentation studies are still advantageous.

If two materials differ sufficiently in sedimentation coefficient and if their diffusion coefficients are not too large, resolution of boundaries by refractometric optical systems (schlieren or Rayleigh) is possible. The criteria for minimal resolution (nonoverlap at the two maxima in the gradient curve) were worked out many years ago by Pedersen

(Svedberg and Pedersen, 1940). Pedersen's analysis indicates that resolution depends upon a quantity which may be written as $(b^2 - a^2)\omega^2$. This shows, as might be expected, that a high rotor speed is a prerequisite. This is one of the few reasons which could be advanced (see also Sections III,C and IV,D) for developing ultracentrifuges capable of higher radial velocity than those currently available. But considering the decreasing importance of this kind of study, it is certainly not a compelling reason.

How the *amounts* of the two or more components will be determined will depend upon the optical system employed. If schlieren optics are used, the areas under the individual peaks must be measured, usually by numerical integration of comparator-measured peaks or planimetry of tracings. With Rayleigh optics, one can simply count fringes across each boundary (including fringe fraction). In either case, reliable analysis requires clear resolution of the boundaries; much better resolution than the minimal criteria given by Svedberg and Pedersen (1940) is needed. The assumption is usually made with refractometric optics that every component has the same specific refractive index increment; this is usually a good approximation for most proteins, but likely to fail with lipoproteins, for example.

With scanner optics, more flexibility is possible. If the absorption spectra of the individual components are known and differ, it may be possible to use wavelengths that specifically select for one component or the other. In any event, an extinction coefficient must be known for each component at the wavelength used.

Whatever optical system is used, one must correct for the radial dilution effect, according to Eq. (5). It is a good idea to measure a series of photographs or scans to assure that radial dilution is proceeding as the equation predicts. Marked deviations from such behavior can be indicative of certain types of reacting systems (see Section II,C).

A further complication lies in the Johnston–Ogston effect (Johnston and Ogston, 1946). If there is appreciable concentration dependence in the sedimentation coefficients, the apparent proportions of the components will err from the true values. The most common observation is that s decreases with increasing c; in this event, the slow components will tend to "pile up" behind the fast boundaries, and the analysis will tend to indicate too much of the slow components, too little of the fast. Since the effect will be most pronounced with those macromolecules which exhibit a strong dependence of s upon c, it is usually, though not always, of less consequence with proteins than with highly concentration-dependent macromolecules such as nucleic acids. A more detailed discussion and many experimental examples are given by Schachman

(1959). The reader who contemplates such analyses should study Chapter IV of Schachman's book in detail.

C. Reacting Systems

Recent interest in reversible association and dissociation reactions of proteins has stimulated corresponding interest in the behavior of such systems during the sedimentation transport experiment. Experimentally, studies of this kind have not to date proved as fruitful as the corresponding sedimentation equilibrium experiments (see Section IV,C). However, it is of importance for the protein chemist to be familiar with the behavior of reversibly interacting systems during sedimentation; certain aspects of this behavior can be diagnostic for or informative of reversible association, and certain deceptive situations can arise. First, I shall consider reversible isomerization and association (or dissociation) reactions, then the interaction of small molecules or ions with proteins, and finally, combine these two and consider those association reactions which are *mediated* by the binding of small molecules or ions. Unless otherwise noted, it will be generally assumed that the reactions involved proceed at rates so large that chemical equilibrium can be maintained at every point in the ultracentrifuge cell during the transport process.

1. Reversible Isomerization or Association Reactions

Much of our understanding of the behavior of such systems in transport processes is due to the work of Gilbert and collaborators. In two pioneering papers (Gilbert, 1955, 1959) the behavior of rapid monomer–n-mer reactions

$$nP_1 \rightleftarrows P_n; \qquad K = c_n/c_1^n \tag{16}$$

was analyzed by solving the differential equations for mass transport coupled with the chemical reaction. These initial studies involved severely restrictive assumptions: diffusion-free sedimentation of all components was assumed to occur in a rectangular cell with a uniform centrifugal (or gravitational) field. Nevertheless, the essential features of the behavior are displayed. The following points are to be particularly noted:

(a) A monomer–dimer system will exhibit a concentration gradient curve which has but a single maximum, although the curve may be asymmetric. The position of this maximum will correspond neither to the sedimentation coefficient of the dimer nor that of the monomer, but will lie between them, approaching one or the other as the equilibrium is shifted to either side.

(b) A monomer–n-mer system ($n > 2$) may exhibit two maxima in its concentration gradient. One of these will travel with the monomer velocity, the other with a velocity intermediate between that of monomer and n-mer. Fujita (1962) has extended the Gilbert theory to consider a sector-shaped cell in a centrifugal field. He finds for the fast boundary

$$s_f = s_1 \frac{1 + nK(s_n/s_1)(c_1^0)^{n-1}}{1 + nK(c_1^0)^{n-1}} \tag{17}$$

where s_1 and s_n are the sedimentation coefficients of pure monomer and n-mer, and c_1^0 is the concentration of monomer in the original mixture. When $n = 2$, Eq. (17) gives the apparent s value for the single peak of a monomer–dimer system. Figure 4 shows experimental results for a presumed monomer–tetramer system and the kind of pattern predicted by the Gilbert theory. Note that these data were obtained with scanner optics and represent concentration rather than its gradient. While the two boundaries are spread by diffusion, they correspond roughly in position to what is expected from Eq. (17). The deviations in concentration in the intermediate regions suggest that other species (dimers and/or trimers) may also be present.

Going beyond the Gilbert approximation (i.e., to include diffusion,

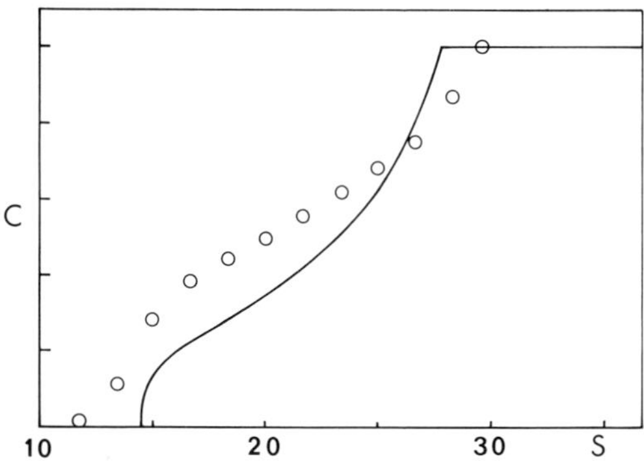

Fig. 4 The theoretical curve for the concentration profile of a rapidly equilibrating monomer–tetramer system in sedimentation transport (solid line), compared to experimental data for ghost shrimp hemocyanin (circles). The abscissa denotes s values (Svedbergs) and the ordinate, concentration (C). The sedimentation coefficient of pure monomer is 14 S, that of pure tetramer, 37 S. The equilibrium constant, obtained from the weight-average sedimentation coefficient, was used to calculate the theoretical curve according to the Gilbert theory. The differences are partly due to diffusion, but may also reflect the existence of intermediate states of aggregation. Data of R. Roxby.

concentration dependence of s, or to allow finite reaction rate constants) leads one into serious mathematical difficulties. In most cases, it has not been possible to solve the differential equations analytically. For this reason, many theorists have turned to computer simulation of the behavior. This is advantageous because the behavior of almost any system can be investigated (if enough ingenuity, time, and money are available), but disadvantageous because we see results only for particular values of the parameters chosen. Nonetheless, considerable progress has been made, particularly in the study of how diffusion, finite rate constants, and pressure affect isomerization and association reactions. For details of a number of such studies, the reader is referred to the book by Cann (1970).

Reacting systems hold traps for the unwary. We have already seen that a two-species mixture (monomer–dimer) may exhibit fewer peaks in the gradient curve than there are species present. It is not widely recognized that the converse can also be true. For example, a monomer–dimer mixture or even an isomerizing macromolecule can, under some circumstances, give as many as *three* maxima in the gradient curve. This kind of behavior is to be expected when the reactions are so slow that chemical equilibrium cannot keep pace with transport but so fast that appreciable changes take place during the course of the centrifugation (Fig. 5). A general note of caution: *If chemical reactions are suspected, the number of peaks in the schlieren pattern should not be taken as necessarily equal to the number of species present.*

How does one know if association–dissociation reactions are involved? A most useful indicator is the concentration dependence of s. Since increasing the total concentration of an associating mixture will displace the equilibrium toward the associated species, one will generally observe

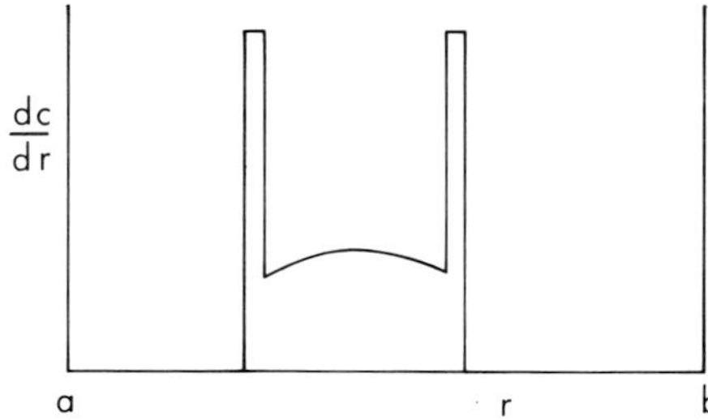

Fig. 5 Hypothetical concentration gradient curve for the sedimentation transport of a macromolecule undergoing a relatively slow isomerization reaction. Diffusion has been neglected. The two sharp boundaries correspond to the two states of the molecule. The gradient between these boundaries results from the tendency to reequilibrate. Adapted from Van Holde (1962).

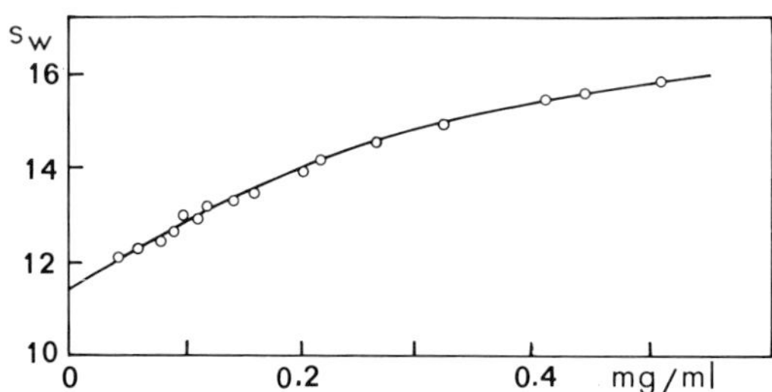

Fig. 6 The weight-average sedimentation coefficient (s_w) of glutamic dehydrogenase (GDH) as a function of concentration. At 20° GDH is known to undergo unlimited association (see Section IV). Adapted from Reisler *et al.* (1970).

an *increase* in the apparent sedimentation coefficient of one or more boundaries with increasing c. Even in the case of an equilibrating system, the sedimentation coefficient calculated from the second moment of the entire gradient region will correspond to the weight-average sedimentation coefficient of the mixture:

$$s_w = \Sigma s_i c_i / \Sigma c_i \tag{18}$$

An example of this behavior is shown in Fig. 6. Since such behavior is the opposite of the usual decrease of s with c (Section II,A,2), it is easy to recognize.

To date, relatively little use has been made of sedimentation transport experiments for determining equilibrium constants or other thermodynamic quantities for such equilibria. This is principally because general analytic solutions for realistic situations do not exist. For the sedimentation of diffusible, reactive species in sector-shaped cells in the ultracentrifuge, we have only computer-simulated solutions for specific cases. These are informative, but difficult as yet to use to calculate the parameters characteristic of the reaction.

2. *Sedimentation Studies of the Binding of Small Molecules to Proteins*

Recent development of absorption optics, particularly the scanning system, has made possible another kind of application of sedimentation transport experiments. Suppose a protein, P, binds a small molecule, A

$$\mathrm{P} + \mathrm{A} \rightleftarrows \mathrm{PA} \tag{19}$$

If A is essentially nonsedimentable, then sedimentation transport of P will leave behind free A which will be very nearly at the concentration [A] present in the original mixture, that given by the equilibrium relationship

$$K = [\mathrm{PA}]/[\mathrm{A}][\mathrm{P}] \tag{20}$$

Note that radial dilution of P and PA will not influence or disturb the equilibrium as long as the sedimentation coefficient of PA is practically identical to that of P and A does not sediment. If these assumptions hold, the ratio [PA]/[P] will remain constant in the plateau region. If [A] is measured in the region above the protein boundary and the total initial concentrations of both A and P are known, one can calculate the number of moles of A bound per mole of P, and the binding constant. While for simplicity we have considered the binding of a single ligand, the analysis is equally valid if any number of ligands are bound. For a much more detailed discussion including situations in which A sediments appreciably or in which the equilibrium is modified by pressure, see Cann (1970).

The original application of the scanner-equipped ultracentrifuge to this kind of problem was by Steinberg and Schachman (1966), who studied the binding of methyl orange to bovine serum albumin. An example of a more recent experiment is shown in Fig. 7. Note that the wavelength used corresponds to an absorption band of the dye.

The method is rapid and potentially accurate. Steinberg and Schachman (1966) found results almost identical with those from the more laborious technique of equilibrium dialysis. With multiplexer operation of the scanner, up to five solutions can be studied simultaneously. For best results, careful calibration of the scanner with solutions of known optical density is required.

3. *Ligand-Mediated Association*

Many of the association or dissociation reactions of proteins are mediated by the binding of ions or small molecules. For example, a number of cases are known in which addition of divalent ions promotes the

Fig. 7 The binding of NADH to beef liver glutamic dehydrogenase as measured by sedimentation velocity. NADH concentration was recorded by the scanning optical system at 360 nm. The vertical distance (a) corresponds to the concentration of free NADH and the vertical distance (b) to the approximate concentration of bound NADH. The latter is not exact since the extinction coefficient changes as a result of binding. Data of D. Malencik and S. Anderson.

association of subunits. It has recently been recognized, largely through the efforts of Cann and his collaborators, that such reactions can behave in a quite peculiar fashion in sedimentation transport. While such reactions may be quite complex, the principles can be illustrated by a simple prototype:

$$nP + mX \rightleftarrows P_nX_m \tag{21}$$

where X is a small molecule or ion. This reaction has been analyzed in considerable detail (see Cann and Goad, 1968; Goad and Cann, 1969; Cann, 1970). The behavior of such systems during sedimentation depends upon whether or not the transport of protein components can induce appreciable gradients in the concentration of X. If such gradients are established, they will, of course, modify the moving-boundary pattern, for the equilibrium is sensitive to the concentration of X. The establishment of appreciable X gradients will be facilitated by (a) a high molar ratio of P to X; (b) a large value of m; (c) a low diffusion coefficient for X; and (d) high centrifugal fields, with correspondingly rapid protein transport.

An extreme example of the kind of behavior which can be expected for a monomer–dimer reaction is shown in Fig. 8. The reaction is assumed to be rapidly reversible, and thus, according to the Gilbert theory, should exhibit a gradient with a single maximum. Yet under the conditions chosen here, two well-defined maxima with s values close to those of the monomer and dimer species are observed. In simplest terms, the explanation is this: Under the condition chosen and with the large

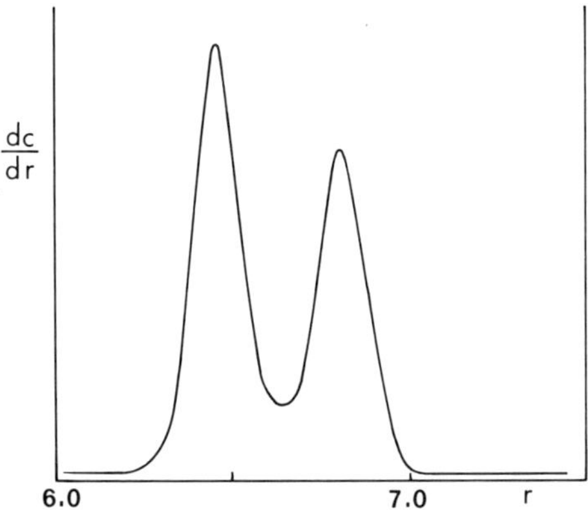

Fig. 8 The theoretical sedimentation pattern of a system in which rapid dimerization is mediated by a small ligand X: $2P_1 + 30X \rightleftarrows P_2X_{30}$. The positions of the peaks correspond approximately to the s values of monomer and dimer. Redrawn from Cann (1970).

value of m assumed, the dimer species carries much of the ligand X with it. This leaves a very low concentration of free X in the region behind the dimer boundary. Monomer molecules which find themselves in this region are essentially unable to dimerize; there is just not enough X left. If the initial concentration of X were made larger, the presence of dimer would have little effect on the relative concentration of X. Then the two peaks would merge, leading in the limit of high X concentration to Gilbert behavior.

Such systems present a hazard to the unwary experimenter, for not only may the existence of rapid equilibration be obscured, but the "sedimentation coefficients" and "species concentrations" deduced from the partially resolved boundaries will in some cases bear little relationship to the true values. Such systems may be more common than realized, for it is precisely the case where X is present in trace amounts that the most deceptive effects will occur. For many more examples of this kind and a thorough discussion of the theory, the reader is referred to Chapter IV of the book by Cann (1970).

4. *Pressure Effects*

In everything we have considered so far, it has been assumed, explicitly or implicitly, that the chemical reactions occur without volume change. In general, one would not expect this to be the case. Since the gradient of hydrostatic pressure in the ultracentrifuge cell will range from 1 atmosphere at the meniscus to several hundred atmospheres at the bottom in an experiment at high rotor speed, the sensitivity of the equilibrium to these pressure changes may be far from negligible. By Le Chatelier's principle we would expect, for example, that an association reaction involving an increase in volume would be shifted toward the side of dissociation at the high pressure existing at the cell bottom. The pressure gradient in the ultracentrifuge cell is $dP/dr = \rho w^2 r$, which leads to the following expression for the dependence of the equilibrium constant upon r:

$$K(r) = K(a) \exp\left[-\frac{\omega^2(r^2 - a^2)\rho\, \Delta v}{2RT}\right] \tag{22}$$

where $K(a)$ is the equilibrium constant at the meniscus ($P = 1$ atm) and Δv is the change in volume per mole of product. As has been pointed out by a number of investigators (see, for example, Kegeles *et al.*, 1967; Ten Eyck and Kauzmann, 1967), this can lead to enormous changes in K if Δv and ω^2 are both large.

The effects of such pressure-dependent shifts in the equilibrium can be dramatic. An example is found in the experiments of Josephs and Harrington (1966, 1967) on the polymerization of myosin. At low rotor speeds this system yields two well-resolved boundaries, a monomer boundary moving at about 6.5 S, and a polymer boundary at about 150 S. Such a pattern is expected by the Gilbert theory for a monomer–n-mer equilibrium when n is large. The fast peak moves with approximately the n-mer velocity [as can be seen from Eq. (17) if n and K are very large], and is hypersharp because of concentration dependence. If the pressure in the cell is increased by increasing the rotor speed or by layering oil onto the solution, a positive concentration gradient appears in the region between the two peaks, and the slower boundary increases in relative area. This can be explained as a result of dissociation of the polymer caused by the increasing pressure in lower levels of the solution column.

As has been demonstrated theoretically by Kegeles *et al.* (1967), pressure-dependent reactions can frequently lead to density instabilities, and unless a compensating positive density gradient is somehow provided, convection must result in at least part of the cell. This need not always be obvious to the experimenter, for the instabilities may occur only on the edges of the boundaries, eroding these and modifying their shape. Therefore, it behooves the careful investigator to vary the rotor speed in experiments of this kind.

III. ZONAL SEDIMENTATION TRANSPORT IN DENSITY GRADIENTS

A. General Principles

Zonal sedimentation transport differs from moving-boundary sedimentation in much the same way as zonal electrophoresis differs from the classic Tiselius method. In these zonal methods, a band of the macromolecular substance migrates through the medium. Whereas in zonal electrophoresis, stability against convection is usually gained by the use of a supporting medium (gel, paste, etc.), in zonal centrifugation, density gradients are almost always employed. The general conditions for stability of a migrating zone are diagramed in Fig. 9. What is essential is that at all times the *total* density gradient be positive throughout the system. Negative gradients will lead to convection which will continue until they are abolished. All of the techniques discussed in this section

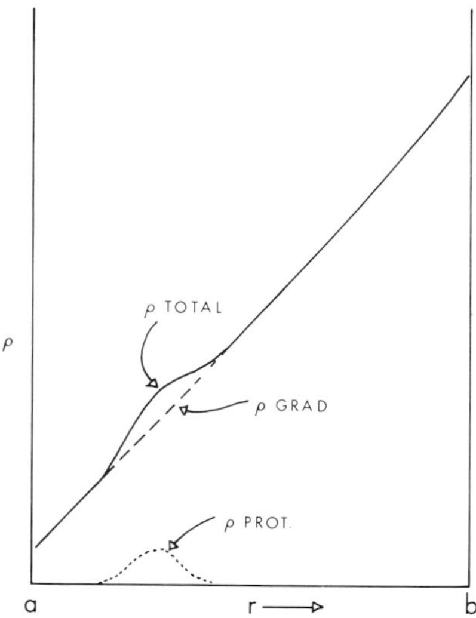

Fig. 9 A schematic drawing to show the conditions for stability in a zonal sedimentation experiment. The symbol ρ denotes density. For further details, see the text.

depend upon this principle, and all require the establishment, in some way, of a stabilizing gradient. Since the danger of convection always exists, whether from overloading or inadvertent distortion of the stabilizing gradient, these techniques are potentially subject to artifacts and should be employed with more caution than they are usually given. Svensson *et al.* (1957), Vinograd and Bruner (1966), and Schumaker (1967) have given theoretical treatments of the requirements for stability.

Putting aside until Sections III,B and III,C questions as to how the density gradient is formed and how the migration of the zones is to be followed, we may first ask how results are to be interpreted. In particular, what point in a zone of finite width should be followed to measure the sedimentation coefficient? The problem has been considered in some detail by Vinograd *et al.* (1963) and somewhat more generally by Schumaker and Rosenbloom (1965). The latter authors have obtained results for a number of experimental conditions. For example, they show that for concentration-independent sedimentation in a sector-shaped cell in a centrifugal field one has

$$d\,\langle \ln r \rangle / dt = \omega^2 s \tag{23}$$

While this closely resembles Eq. (1) for the classic sedimentation transport, two points should be emphasized:

1. The quantity $\langle \ln r \rangle$ in Eq. (23) is the *mass average* of ln r over the zone. That is, if one defines dm as the increment in solute mass across a radial distance dr,

$$\langle \ln r \rangle = \int \ln r \, dm / \int dm \tag{24}$$

where integration is across the solute zone. In practice, for nearly symmetrical zones, $\langle \ln r \rangle$ will be very close to $\ln r_m$, where r_m is the maximum concentration point.

2. Not only must s in Eq. (23) refer to the values of density and viscosity which exist at the point $\langle \ln r \rangle$, but in fact Eq. (23) is not exact if a large gradient exists in density and/or viscosity. Schumaker and Rosenbloom (1965) show that in a sector cell, the effect of variation of ρ and η may be approximately taken into account by writing

$$\frac{d \langle \ln r \rangle}{dt} \cong s_{20,w}\omega^2 \frac{(1 - \bar{v}\rho_c)}{(1 - \bar{v}\rho_{20,w})} \frac{\eta_{20,w}}{\eta_c} - D_{20,w} \frac{1}{r_c} \frac{\eta_{20,w}}{\eta_c^2} \left(\frac{\partial \eta}{\partial r} \right)_{t,c} \tag{25}$$

Here the quantities subscripted with c are those which obtain at the "zone center" defined by $\langle \ln r \rangle$. The second term on the right is usually very small.

If, on the other hand, a parallel-walled cell is used (such as a tube in a swinging-bucket rotor), Eq. (23) is to be replaced by

$$d \ln \langle r \rangle / dt = \omega^2 s \tag{26}$$

where $\langle r \rangle$ is the mass-average of r (e.g., the center of mass of the zone) and s again refers to prevailing values of viscosity and density. The question of how one should obtain an $s_{20,w}$ value in such experiments will be deferred to the next section, since these conditions pertain mainly to sucrose-gradient experiments.

In practice, it appears doubtful whether experimental precision in most cases justifies distinctions between various ways of measuring s [Eqs. (1), (23), and (26)]. For narrow zones, the concentration maximum will probably do quite well for either cell geometry. However, if one encounters broad, rapidly spreading zones, such distinctions should be kept in mind.

B. Gradient Centrifugation of Proteins in Swinging-Bucket Tubes

While the use of density gradients for stabilization of sedimenting systems goes back many years, wide application of the zonal technique to analytical studies of *proteins* dates from the classic paper of Martin and Ames (1961). Such experiments are almost always performed using preformed gradients in swinging-bucket rotors. We shall be concerned here with the analytical, rather than the preparative aspects of such work. In Martin and Ames' original work, and in most studies since, sucrose was used as the gradient-forming material. Hence the term "sucrose gradient centrifugation" is often used for this technique. However, a wide variety of substances can be and have been used. (For lists of materials used and their properties, see Moore, 1969.)

While countless minor variants of the technique have been devised, the methods still used by most workers approximate those of Martin and Ames (1961). A linear or nearly linear sucrose gradient is prepared in a swinging-bucket tube by use of a simple mixing device (Moore, 1969). The steepness and range of the gradient needed will depend to some extent upon the conditions of the experiment, but 5%–20% gradients of sucrose are most common. The sample is then carefully pipetted onto the top of the gradient. Its density must be appreciably less than that of 5% sucrose. After centrifuging for an appropriate time, the tubes are removed and the position of the sedimented zones determined. This may be done in a variety of ways: by dripping the contents from the bottom, by careful siphoning, by scanning of the tube in a spectrophotometer, etc. Needless to say, there are also an almost infinite number of ways of analyzing the various fractions obtained by dripping or siphoning; this is the unique advantage of the method. Since such specific tests as enzymatic activity assay can be used, it is possible to locate the zone corresponding to a particular enzyme, even though it be grossly contaminated. Radioactivity can be used to detect labeled compounds.

Converting the distance the zone has traveled to a sedimentation coefficient is a somewhat imprecise operation. For a direct computation, one must take into account not only the acceleration and deceleration periods but also the fact that the viscosity and density of the medium have varied continuously through the distance the zone has traveled. Martin and Ames (1961) point out that with use of 5%–20% sucrose gradients between 3° and 15°, the velocity of the zone is almost constant. They then provide methods for calculation of s. A more general approach is given by Schumaker and Rosenbloom (1965). Neglecting a small term corresponding to the diffusion coefficient term in Eq. (25) they obtain

$$\int_{\langle r \rangle_0}^{\langle r \rangle_f} \left[\frac{(1 - \bar{v}\rho_{20,w})}{(1 - \bar{v}\rho_c)} \frac{\eta_c}{\eta_{20,w}} \right] d \ln \langle r \rangle = s_{20,w}\omega^2 t_{\text{eff}} \tag{27}$$

Here $\langle r \rangle$ is the center of mass position used in Eq. (26), and the sedimentation is presumed to have occurred between an initial position $\langle r \rangle_0$ and a final position $\langle r \rangle_f$ in the time t_{eff}. Of course, it is necessary to know the sucrose concentration at each point in order to carry out the integration.

The awkwardness of such procedures has led many to the much simpler method of using "marker" substances of known sedimentation coefficient. Advantage is then taken of the nearly linear relation between s and distance traveled to interpolate for the unknown s. However, as Martin and Ames (1961) were careful to point out, the linear interpolation is valid *only* if the substance being studied and the markers have the same value of $\bar{v}$. Methods of correction for differences in $\bar{v}$ are given by Martin and Ames (1961), but many workers have failed to take such corrections into consideration.

C. Band Sedimentation and the Cohen Technique

The first successful technique for zonal centrifugation in the analytical ultracentrifuge was developed by Vinograd and co-workers in 1963 (Vinograd *et al.*, 1963; Vinograd and Bruner, 1966). The method is very simple in concept. Using a special type of synthetic boundary cell, a thin lamella of solution containing the macromolecule is layered onto a denser buffer solution (henceforth called the lower solution) which nearly fills the sector (see Fig. 10). The density difference can be established by adding NaCl (1 M) to the lower solution or by using D_2O as part of the solvent in this solution and using H_2O in that containing the macromolecule. The sedimenting zone is stabilized against convection by two effects: (1) diffusion between the denser and less dense solution creates a density gradient near the meniscus, and (2) sedimentation of the salt (if used) creates a gentle density gradient which increases with time (see Fig. 10b). The conditions for stability have been investigated in considerable detail by Vinograd and Bruner (1966). The stabilizing gradient required (and hence the salt concentration needed in the lower solution) will depend upon the concentration of the macromolecule used. The lower limit of practical concentration depends upon the limit of detectability of the macromolecule. Since proteins have very much lower extinction coefficients (at least near 260–280 nm) than nucleic acids, experiments with proteins require a large density difference if one wishes to be sure of stability. Since the stabilizing gradients are changing with time, the situation is complex, for a rapidly sedimenting protein can "outrun" the gradient near the meniscus. If this happens, the band may "sink" as diagramed in Fig. 10c. Hence one must be wary of artifacts. Further complications include the fact that band-spreading can be serious, especially if s is strongly c-dependent (unlikely with proteins) since the leading edge of the

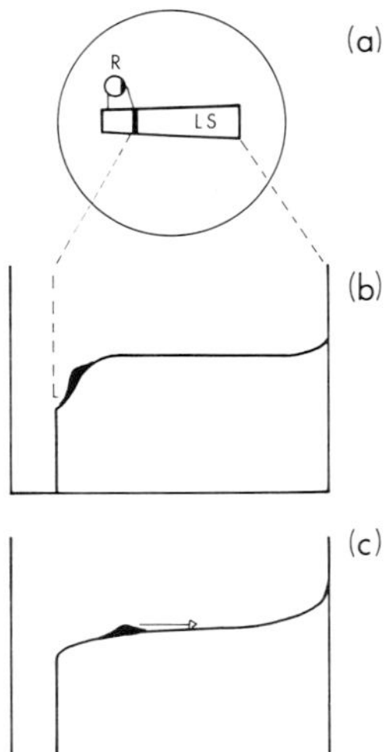

Fig. 10 Band sedimentation by the technique of Vinograd *et al.* (1963). (a) The kind of centerpiece used. As the ultracentrifuge is accelerated, the reservoir, R, empties and layers the band on top of the lower solution, LS. (b) The situation shortly after banding. This is a stable situation. The macromolecular contribution to the density is shown by the dark area. (c) An unstable situation in which the band has "outrun" the gradient. Convection would actually have occurred before this point.

boundary will tend to "run away." Probably for these reasons and because it offers few advantages over the conventional sedimentation transport experiment with proteins, the Vinograd technique has not been used as much by protein chemists as by those studying nucleic acids.

On the other hand, a different gradient technique has been developed which holds great promise for protein studies. This method, called "active enzyme sedimentation" or the "Cohen Method," was first described by Cohen in 1963. In a series of subsequent papers Cohen and his collaborators describe in considerable detail the theory and practice of the method (Cohen and Hahn, 1965; Cohen *et al.*, 1967; Cohen and Mire, 1971a,b). In essence, the method is similar to the Vinograd band technique, with the following important difference: the lower solution contains a substrate of the enzyme, which undergoes a color change when acted upon by the sedimenting enzyme. Thus, as the zone of enzyme traverses the cell, the product of the reaction is left behind the zone. Figure 11 diagrams what will be observed if the monochromator of the ultracentrifuge optical system is set for a wavelength at which the product absorbs and the substrate does not.

The elegance of the technique is at once evident. It allows us to ob-

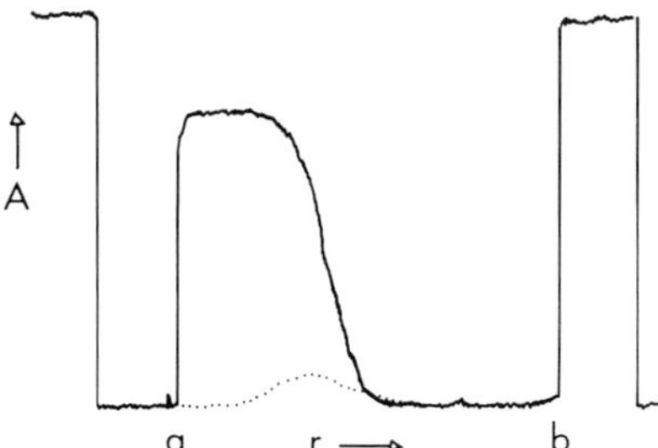

Fig. 11 A schematic view of what one would observe in an experiment of the Cohen type. The product absorbs at the wavelength used; the substrate does not. The dotted curve represents the approximate position of the enzyme band (which would not itself be observable). *A* is absorbance.

serve sedimentation of an enzyme at just those concentrations commonly used for enzyme kinetics studies. Thus it can tell us what the molecular state of the enzyme is at those concentrations at which we have studied it *as an enzyme.* Furthermore, the very low protein concentrations required (often between 1 and 10 μg/ml) make stability of the sedimenting zone easy to ensure. Finally, studies can be made with impure preparations, for only the sedimentation of the specific enzyme whose substrate is used will be observed.

With all of these advantages, it seems surprising that the technique has not been more widely used to date. The reason is probably two-fold: (1) The full theoretical treatment and the method of analysis presented by Cohen *et al.* (1967) are formidably mathematical and may have frightened away many potential users. (2) The method can be prone to artifacts, unless the user fully understands its limitations. Happily, the most recent articles by Cohen and Mire (1971a,b) have clearly described a simple and reasonably accurate computational method and have outlined in considerable detail the potential pitfalls and how to avoid them. One may expect, especially with the growing availability of scanning optical systems, that the method will rapidly gain in popularity.

Figure 12 illustrates the essence of the Cohen–Mire method. The product concentration $[P_t(r)]$ as a function of radius in the cell at different times is the given data. If the initial lamella were infinitely thin and there were no diffusion of enzyme product, the product concentration at any time would be a step function. But the layer of enzyme solution is of finite thickness and broadens further due to diffusion. Diffusion of low molecular weight product can also be quite rapid. If every enzyme molecule in the layer can consume substrate at the same rate (which essentially requires that the substrate concentration be saturating), a

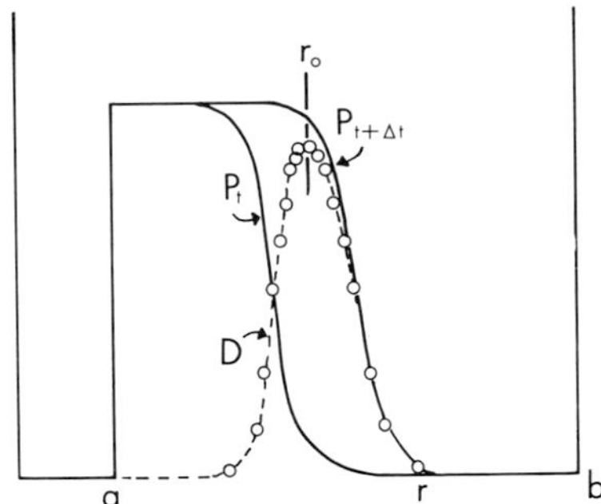

Fig. 12 The Cohen–Mire (1971a) difference method. For details, see the text.

reasonably simple analysis is possible. For a rigorous analysis, it is still necessary to correct for sedimentation and especially diffusion of the product. The computations in this case are complicated and best carried out by digital computer. In principle, it is possible to obtain both s and D for the enzyme from this calculation. A much simpler and reasonably accurate method for calculating s proceeds as follows: The difference is taken between the product distribution at time t and at $t + \Delta t$, where Δt is some fixed small increment:

$$D_t^{t+\Delta t}(r) = P_{t+\Delta t}(r) - P_t(r) \tag{28}$$

This difference will give a bell-shaped curve as shown in Fig. 12. It can be shown that

$$\ln r_0 \cong s\omega^2 t + \text{constant} \tag{29}$$

where r_0 is the position of the maximum in the difference curve.

It is critical that the substrate concentration be sufficiently high so that every molecule of enzyme in the sedimenting zone can react with substrate with equal velocity. Consider what can happen when the ratio of enzyme to substrate is too high. Then molecules in the leading edge of the band will react with great velocity, and the substrate will be largely consumed by the leading edge of the band. Since the band spreads, the leading edge travels with a greater apparent sedimentation rate than the band center. The product concentration distribution will reflect mainly the behavior of the leading edge. Too high an s will be obtained, and it will get higher at greater enzyme concentration. The apparent s vs. c curve will resemble that of an associating system! Other artifacts can appear if the enzyme denatures during sedimentation or if a coupled-enzyme assay system is used. Thus considerable care is needed in using the Cohen technique.

Results of a number of experiments in various laboratories (Hathaway and Criddle, 1966; Cohen *et al.*, 1967; Cohen and Mire, 1971b; Hoagland and Teller, 1969) indicate that the method is capable of good precision if used with care. Cohen and Mire (1971a) claim a precision of 1% for s and 10% for D using the "rigorous" method of analysis, and a precision of 3%–5% for s using the approximate method described here.

As an example of the application of this technique to an enzymological problem, recent studies of glyceraldehyde-3-phosphate dehydrogenase by Hoagland and Teller (1969) may be cited. This tetrameric enzyme was studied at concentrations as low as 4×10^{-3} mg/ml, using the absorbance change accompanying NAD reduction as an indicator. While data at high concentrations tended to give somewhat high values of s, all experiments at enzyme concentrations below about 0.1 mg/ml gave $s_{20,w}$ values in good agreement with those commonly found for the tetramer. This was taken as confirmation that the tetramer is indeed an active form of the enzyme.

IV. SEDIMENTATION EQUILIBRIUM

In most of this section, I shall be considering the conventional sedimentation equilibrium experiment. The density gradient sedimentation equilibrium experiment will be considered briefly in Section IV,D.

A. Ideal, Nonreacting Solutes

1. Fundamental Equations and Methods of Analysis

The sedimentation equilibrium method has become the preferred technique for the physical determination of molecular weight and weight homogeneity of proteins and for quantitative study of their association reactions. This is because it is based on firm theoretical foundations and allows great precision. The basic concept must be familiar to all readers. If a solution is subjected to a constant centrifugal field for a sufficient time, the concentration gradients produced by the sedimentation transport will generate a counter diffusion; when this diffusion transport balances the sedimentation transport, a new state of equilibrium is attained. In this state, the concentration gradient of a homogeneous solute species in an ideal two-component system will be given by

$$\frac{d \ln c}{dr^2} = \frac{M(1 - \bar{v}\rho)\omega^2}{2RT} \tag{30}$$

where all symbols have been defined in the preceding sections, except for the gas constant, $R = 8.314 \times 10^7$ erg/degree·mole. For the derivation of Eq. (30), the reader is referred to any of a number of standard texts (i.e., Svedberg and Pedersen, 1940; Tanford, 1961; Van Holde, 1971). Equation (30) is also frequently given in integrated form:

$$c(r) = c(r_0) \exp \left[\frac{M(1 - \bar{v}\rho)\omega^2}{2RT} (r^2 - r_0^2) \right] \tag{31}$$

or

$$\ln c(r) = \ln c(r_0) + \frac{M(1 - \bar{v}\rho)\omega^2}{2RT} (r^2 - r_0^2) \tag{32}$$

where r_0 is any reference position (most often the meniscus or cell bottom).

It is evident from Eq. (32) that the molecular weight may be obtained from the slope of a graph of $\ln c$ vs. $(r^2 - r_0^2)$. (We shall henceforth use $r_0 = a$, the meniscus position, as reference point.) The problem of determining the molecular weight of a homogeneous protein then reduces to precise determination of c as a function of r, plus knowledge of the auxiliary quantities $\bar{v}$, ρ, ω, and T. The latter three are usually available with high precision; the partial specific volume $\bar{v}$ is a somewhat more difficult problem and will be discussed in Section IV,A,*4*. To determine c, one may use any of the conventional optical systems. With the scanner, the evaluation is simple and direct (see Fig. 13). In fact, if Beer's law holds, the absorbance itself can be used in Eq. (32), since the quantity of interest is the rate of change of $\ln c$ with distance. Unfortunately, to date, scanner data are not as precise as those obtained from other optical systems (see Section V). The schlieren system can be used, if Eq. (30)

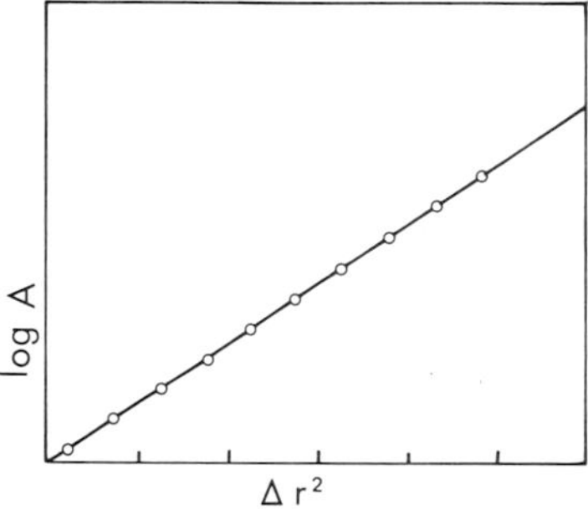

Fig. 13 Sedimentation equilibrium of ribonuclease A as observed by the scanner method. Conditions: 20,000 rpm, 20°, 0.8 mg/ml. The slope of the line drawn through the points corresponds to the known molecular weight (13,683). The data look very good, but see Fig. 14.

is rewritten in terms of dc/dr (see Van Holde and Baldwin, 1958). However, most current studies employ the Rayleigh interferometric system. This has the advantage of high precision, but the disadvantage that the fringe difference between any two points in the cell yields only the *difference* in concentration between these points. This means that one must either know the concentration at some one point, or invoke the conservation of mass to provide the extra data. In the "high speed" method of Yphantis (1964), the ultracentrifuge is run at so high a speed that the concentration becomes essentially zero at the meniscus; this is then the reference point. In the "low speed" method in which the concentration does not go to zero at the meniscus, a number of techniques have been developed to obtain a reference concentration. These methods are described in detail in a number of papers (see Van Holde and Baldwin, 1958; Schachman, 1963; LaBar, 1965; Charlwood, 1967) and shall not be dealt with in detail here.

One technique that has been widely used and is capable of high precision involves using a synthetic boundary cell to obtain the initial concentration, c_0, in terms of interference fringes or schlieren plate area. Taking into account the conservation of mass, integration of Eq. (30) across the cell yields

$$\frac{c(b) - c(a)}{c_0} = \frac{\Delta c}{c_0} = \frac{M(1 - \bar{v}\rho)\omega^2(b^2 - a^2)}{2RT} \tag{33}$$

TABLE III

Some Protein Molecular Weights Obtained by Sedimentation Equilibrium (Sed. eq.)

Protein	Molecular weights: Sequence	Molecular weights: Sed. eq.	% Error	Method[a]	Reference
Ribonuclease A	13,683	13,740	+0.4	LS, Schl	Van Holde and Baldwin (1958)
Lysozyme	14,305	14,500	+1.3	LS, Schl	Sophianopoulos *et al.* (1962)
Metmyoglobin	17,816	17,450	−2.0	HS,[b] Sc	Kellett (1971)
Chymotrypsinogen A	25,767	25,960	+0.7	LS, Ray	LaBar (1965)
β-Lactoglobulin (dimer)	36,844	36,500	−0.9	HS, Ray	Yphantis (1964)
Hemoglobin	64,500	61,300–63,800	−5.0 to 1.0	HS,[b] Sc	Kellett (1971)

[a] Key: LS, low speed; HS, high speed; Schl, schlieren optics; Sc, scanner optics; Ray, Rayleigh optics.

[b] These experiments were done at speeds intermediate between the traditional HS and LS techniques.

Table III presents some experimental results obtained with proteins for which exact molecular weights are available from amino acid sequencing. It would appear that the low speed method is generally capable of somewhat higher accuracy, but it is more consuming of an experimenter's time and material. Details of experimental procedures for sedimentation equilibrium studies are given by Van Holde (1967) and by Teller (1973).

2. *Heterogeneous Solutes*

If a mixture of solute components, each of which behaves ideally, is present in the ultracentrifuge cell, Eqs. (30)–(32) apply to each component individually. In this case, one can obtain average molecular weights for the mixture present at each point in the cell:

$$\text{weight-average:} \quad M_{\text{wr}} = \frac{\Sigma c_i(r) M_i}{\Sigma c_i(r)} = \frac{2RT}{(1 - \bar{v}\rho)\omega^2} \frac{d \ln c}{dr^2} \tag{34}$$

$$\text{number-average:} \quad M_{\text{nr}} = \frac{\Sigma c_i(r)}{\Sigma c_i(r)/M_i} = \frac{c(r)}{\displaystyle\int_0^c \frac{dc(r)}{M_{\text{wr}}}} \tag{35}$$

$$\text{z-average:} \quad M_{\text{zr}} = \frac{\Sigma c_i(r) M_i^2}{\Sigma c_i(r) M_i} = \frac{2RT}{(1 - \bar{v}\rho)\omega^2} \frac{d \ln (cM_{\text{wr}})}{dr^2} \tag{36}$$

Equation (33), on the other hand, will yield the weight-average molecular weight of the original mixture placed in the ultracentrifuge cell. Point averages calculated from Eqs. (34)–(36) can be used to test homogeneity, since with a homogeneous ideal solute each value should be constant over the solution column and all should identically yield the molecular weight, M. Two practical points should be noted: (1) The calculation of M_{zr} requires, in effect, taking the *second* derivative of $\ln c$ vs. r^2, with accompanying loss of precision: (2) To obtain M_{nr} one must integrate from zero concentration. This is possible only with the high speed method, in which $c \to 0$ at the meniscus.

One point in interpretation should be strongly emphasized. Frequently, investigators present a curve of $\ln c$ vs. r^2 which approximates a straight line and state that the material appears to be homogeneous. This can be deceptive, for the $\ln c$ vs. r^2 graph is notoriously insensitive to heterogeneity, and nonideality can produce a downward curvature in the graph which compensates for the upward curvature produced by heterogeneity (see Section IV,B). Figure 14 provides an example using data for a mixture of ovalbumin and serum albumin. The heterogeneity becomes apparent only when one uses Eqs. (34)–(36) to calculate an apparent average molecular weight at each point in the cell. Such calculations are routinely done in our laboratories (as in many) by a com-

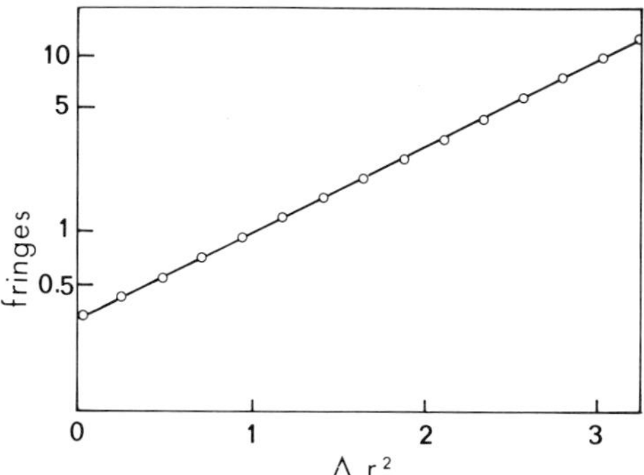

Fig. 14 This is meant to serve as a warning. It illustrates sedimentation equilibrium data (using Rayleigh optics) for a *mixture* of 0.5 mg/ml bovine serum albumin and 0.7 mg/ml ovalbumin at 4.9°, 18,054 rpm. A "best straight line" has been drawn through the points. Actually, the data are better than the scale of the figure can indicate, and calculation of point average molecular weights shows that M_w, for example, varies from 51,000 to 61,000 across the cell. These data have been analyzed by the method of Dyson and Isenberg (1971) and results are given in Table IV. Taken from data of Dyson and Isenberg (1971).

puter program in which the data are fitted to a polynomial function over a limited region about each point. The high speed method is particularly suited to reveal small amounts of a low molecular weight contaminant (Yphantis, 1964). On the other hand, unless the experiment is conducted at so low a concentration that fringes are visible for the whole solution column, a high molecular weight contaminant may be missed entirely. In the low speed method, resolution of components is generally less complete, but very high molecular weight impurities are readily detected.

For systems containing a small number of protein components, it is in principle possible to determine the amount and molecular weight of each from sedimentation equilibrium. The special case of equilibrating mixtures will be dealt with in detail in Section IV,C, but the general principles are the same for paucidisperse, nonassociating systems. Basically, one must resolve a multiexponential curve:

$$c = \sum_i c_i(a) \exp [M_i(1 - \bar{v}_i\rho)\omega^2(r^2 - a^2)/2RT] \tag{37}$$

If the individual values of $\bar{v}_i$ are known or identical, one can in principle obtain a unique set of values for M_i and $c_i(a)$. Once the individual concentrations of species of known M_i are known at one point, the proportion in the original mixture can be deduced. The analysis of such multiexponential functions is a common problem in physics and chemistry and

TABLE IV

Analysis of a Two-Component Mixture[a]

	Ovalbumin		Serum albumin		
	c_1^0	M_1	c_2^0	M_2	RMS[b]
Separately determined[c]	5.52	44,380	14.1	67,640	—
Analysis of combined data	5.66	43,230	14.0	68,590	0.022
Analysis of actual mixture[d]	(4.23)	41,420	(9.90)	68,261	0.011

[a] Concentrations are given as fringe displacements, where 1 mg/ml is approximately four fringes.

[b] Root-mean-square deviation, in fringes.

[c] From individual data sets, by a linear least-squares procedure.

[d] Note that these amplitudes (c_j^0) cannot be compared to those above since exact starting concentrations were not known. Data of Dyson and Isenberg (1971).

presents severe difficulties. The basic problem is that in many cases, great accuracy is required to obtain correct results. For even a two-species solute there are four parameters [two for M_i and two for $c_i(a)$], and many combinations of values will give approximately the same c vs. r^2 curve. Many methods have been tried for this kind of analysis with varying degrees of success for various situations. A general technique for such analyses has been recently developed (Isenberg and Dyson, 1969; Dyson and Isenberg, 1971). Application of this "method of moments" to the data shown in Fig. 14 yields the results given in Table IV. Considering the fact that the molecular weights of the components differ by a factor of only 1.5, the results are encouraging. They point out, however, the need for *exceedingly* precise data in the study of multicomponent solutes. Other critical studies of the problem of analysis of such paucidisperse systems have been made by Jeffrey and Pont (1969) and by Goldberg and Edelstein (1969).

3. *The Approach to Equilibrium*

The state of sedimentation equilibrium is approached asymptotically with time. The differential equations describing this process were first solved many years ago by Mason and Weaver (1924). They considered a very simple case: The homogeneous, ideal solute was assumed to be initially distributed uniformly in a rectangular cell. It was then subjected to a steady field which was independent of r. The solutions may be expressed in the general form

$$c(r, t) = c(r, \infty) + \sum_{m=1}^{\infty} Q_m(r)e^{-P_m t} \tag{38}$$

Here $c(r, \infty)$ is the equilibrium distribution, and the second term on the right represents the transient, which vanishes as $t \to \infty$. The terms $Q_m(r)$ represent complicated functions of r, and P_m represent constants whose values depend upon the parameters of the experiment (s, D, ω, etc.). The Mason–Weaver equation was used by Van Holde and Baldwin (1958) to show that for the low speed experiment, the time "to equilibrium" is given by

$$t_\epsilon = F(\alpha, \epsilon)\frac{(b-a)^2}{D} \tag{39}$$

where $b - a$ is the column height and $F(\alpha, \epsilon)$ is a function of the parameter $\alpha = 2RT/M(1 - \bar{v}\rho)\omega^2(b^2 - r^2)$. Actually, since equilibrium is never attained exactly, t_ϵ represents the time required to approach equilibrium so closely that the fractional error in measured molecular weight will have the value ϵ, defined from Eq. (33)

$$\epsilon = (\Delta c_{\text{eq}} - \Delta c)/\Delta c_{\text{eq}} \tag{40}$$

where Δc is the concentration difference between the ends of the column. The importance of Eq. (39) is that it shows that for the low speed experiment, t_ϵ depends upon $(b - a)^2$; hence it pays to keep the column short. For the high speed experiment the situation is more complex (see Yphantis, 1964); equilibrium is usually approached faster but is not so sensitive to column length, being roughly proportional to $(b - a)$.

A number of more sophisticated analyses of the approach to equilibrium have followed upon the Mason–Weaver solution (see Archibald, 1938, 1942, 1947; Pasternak *et al.*, 1957; Hexner *et al.*, 1961). The requirements for constant field and rectangular cell have been dropped, different initial distributions have been considered, and experiments have been proposed in which the rotor speed is varied. While the solutions are of the same general form as Eq. (38), the functions $Q_m(r)$ and the constants P_m are quite sensitive to the initial distribution of solute and to the way in which the rotor speed is varied during the experiment. A general rule appears to be this: Any variation in the experiment which produces at early times a concentration distribution that more nearly approximates the equilibrium distribution than does a uniform distribution will shorten the time required for sedimentation. The experiment can be varied either by using a synthetic boundary cell to produce one or more "steps" in the initial distribution (see, for example, Pasternak *et al.*, 1957; Charlwood, 1967; Griffith, 1967; Chervenka, 1970) or by running the centrifuge initially at a higher speed than the expected equilibrium speed ("overspeeding"). Careful analysis of the behavior of the transient state terms in equations such as Eq. (38) has led to criteria by which the time may be reduced severalfold over that required for a uniform initial distribution at constant field. Such techniques have been described by Hexner *et al.* (1961), Richards *et al.* (1968), and Teller *et al.* (1969).

4. The Partial Specific Volume

One of the thorniest problems in doing really precise molecular weight measurements of proteins by the sedimentation equilibrium technique lies in obtaining accurate values of $\bar{v}$, the partial specific volume. Since the equations for sedimentation equilibrium involve the factor $1 - \bar{v}\rho$, and $\bar{v}$ is usually 0.70–0.75, a 1% error in $\bar{v}$ leads to an error of about 3% in M.

In the strictest sense, the quantity $\bar{v}$ should not be used at all in most protein studies. Proteins are almost invariably analyzed in the presence of buffers and/or neutral salts, and thus the system always contains at least three components. Casassa and Eisenberg (1961, 1964) have demonstrated that in such circumstances the $(1 - \bar{v}\rho)$ term in the sedimentation equilibrium equation should be replaced by $(\partial\rho/\partial c)_\mu$, which represents the rate of change of solution density with protein concentration as determined in experiments in which the low molecular weight components in the solution are in dialysis equilibrium with buffer. One can continue to use equations of the traditional form by defining an "effective" specific volume, ϕ', by $(1 - \phi'\rho) = (\partial\rho/\partial c)_\mu$. However, for dilute buffers the difference may generally be neglected, and one may call the term $(1 - \bar{v}\rho)$. In this case $\bar{v}$ is usually determined from a series of precise density measurements on solutions in which different weight fractions w of protein have been dissolved in buffer (see Charlwood, 1957):

$$\bar{v} \cong \frac{1}{\rho_0}\left(1 - \frac{\rho - \rho_0}{w\rho}\right) \tag{41}$$

Here ρ is solution density and ρ_0 solvent density. The equation is only approximate, since the right side of Eq. (41) actually measures the *apparent* specific volume. However, if this quantity is independent of protein concentration, as it usually is for dilute protein solutions, it will be identical to $\bar{v}$ (Dayhoff *et al.*, 1952).

Extremely precise density measurements are needed to obtain accurate values of $\bar{v}$. For example, to obtain 1% accuracy in $\bar{v}$ (required for 3% accuracy in M), the density values must be more precise than 1 part in 10^5 at protein concentrations of 1 mg/ml. To obtain 0.1% accuracy in $\bar{v}$ (required for 0.3% accuracy in M), densities must be more precise than 1 part in 10^6 at this concentration (Charlwood, 1957). Of course, the

TABLE V

Some Precise Methods for Density Measurements

Method	Claimed accuracy (1 part in parts)	Volume of solution (ml)	Reference or manufacturer
Pycnometry	10^5	~30	Bauer and Lewin (1959)
Magnetic float devices	10^5–10^6	0.3	Ulrich *et al.* (1964)
	10^5–10^6	0.5	Goodrich *et al.* (1969)
Electrobalance with plummet	10^5	1.0	Cahn Instrument Co., Paramount, California
Vibrating tube densitometer	10^5–10^6	0.7	Anton Paar A. G., Graz, Austria

requirements are less stringent at higher protein concentrations. It is possible but difficult to obtain the necessary precision with most of the standard techniques described in Table V.

In any event, measurements of $\bar{v}$ require either painstaking experiments with large quantities of protein or special apparatus. Thus it is understandable that attempts have been made to eliminate this measurement, especially for experiments in which the greatest accuracy in M is not required. Two alternative techniques have come into use.

***a.* Calculation of $\bar{v}$ from Amino Acid Composition.** Cohn and Edsall (1943) pointed out that one can apparently obtain quite satisfactory values of $\bar{v}$ for a number of proteins by assigning specific volumes, $\bar{v}_i$, to amino acid residues. Then if ω_i is the weight fraction of the ith kind of residue

$$\bar{v} \cong \Sigma \omega_i \bar{v}_i \tag{42}$$

Such a calculation neglects volume changes due to transfer of residues from aqueous solution to protein, volume changes due to specific interactions, and the effects of electrostriction. It is remarkable that it works at all, and, in general, it works quite well. McMeekin and Marshall (1952) and Charlwood (1957) have carefully compared calculated values of $\bar{v}$ with those measured experimentally. If the proteins are studied near neutral pH, the agreement is in most cases within 1%. One finds, however, appreciable deviations at high and low pH. There are also known to be changes in $\bar{v}$ upon association or dissociation of some multisubunit proteins or when proteins are unfolded in denaturing solvents. Such differences are hard to predict and can lead to errors if the "calculation" method is used indiscriminately.

Kirby-Hade and Tanford (1967) have found that the value of ϕ' for a number of proteins in a solution of 6 M Gu·HCl was about 0.01–0.02 cm^3/gm *less* than $\bar{v}$ measured in buffer. However, there were exceptions to this rule, and some recent studies (see, for example, Thomas and Edelstein, 1971) indicate that ϕ' will vary with the Gu·HCl concentration. Caution is required!

***b.* Determination of $\bar{v}$ from Sedimentation Equilibrium in Mixed Solvents.** Edelstein and Schachman (1967) have pointed out that if sedimentation equilibrium experiments are performed with the same protein in solvents of different density one should be able to obtain both M and $\bar{v}$. This method, at least in its simplest form, depends upon the assumption that the mixture of components used to establish the different densities does not exhibit preferential interaction with the protein. In most cases the D_2O–H_2O system has been used. Good results have been obtained, both with proteins in dilute buffer (Edelstein and Schachman, 1967) and with proteins in denaturing solvents (Thomas and Edelstein, 1971). While not capable of the same accuracy as can be obtained from separate measurements of M and $\bar{v}$, the method has the advantage of requiring exceedingly small amounts of protein—as little as a few micrograms.

As sequencing of proteins continues, the interest in highly precise sedimentation equilibrium measurements of the molecular weights of purified proteins may diminish. In this event, interest in precise values of $\bar{v}$ may also fall off. But it should be remembered that really careful studies of associating systems will require either assurance that $\bar{v}$ is independent of the degree of association or data on how it changes. So the basic problem remains.

B. Nonideal Protein Solutions; Effects of Charge and Mixed Solvents

Most globular proteins are very close to thermodynamic ideality, especially at the low concentrations accessible to the interferometric and absorption optical systems. However, there are situations in which the protein chemist is apt to encounter appreciable nonideality. These include (1) highly asymmetric molecules (such as myosin) in which the excluded volume contributes appreciably to the nonideality; (2) highly charged proteins in water or in very dilute buffer; and (3) proteins dissolved in denaturing solvents such as guanidine hydrochloride.

The sedimentation equilibrium theory for a heterogeneous, nonideal mixture is quite complex (see Williams *et al.*, 1958). Fortunately, such systems are rarely encountered in protein chemistry, except with some associating systems (see Section IV,C,2,*b*). For a *homogeneous*, nonideal solute, the apparent average molecular weight will depend upon solute concentration, and hence will vary with position in the cell. Furthermore, the average molecular weights calculated by Eqs. (34)–(36) will vary with c and hence with r. One may usually express the variation of these averages in terms of a virial expansion:

$$\frac{1}{M_{\mathrm{nr}}^{\mathrm{app}}} = \frac{1}{M} + \frac{B}{2}c(r) + \cdots \tag{43}$$

$$\frac{1}{M_{\mathrm{wr}}^{\mathrm{app}}} = \frac{1}{M} + Bc(r) + \cdots \tag{44}$$

$$\frac{1}{M_{\mathrm{zr}}^{\mathrm{app}}} = \frac{1}{M} + 2Bc(r) + \cdots \tag{45}$$

Here B is the so-called second virial coefficient. Usually the first two terms of this expansion are sufficient in the concentration range studied with Rayleigh or scanner optics, but at higher concentrations, terms of $c^2(r)$, etc. may be needed. The behavior described by Eqs. (43)–(45)

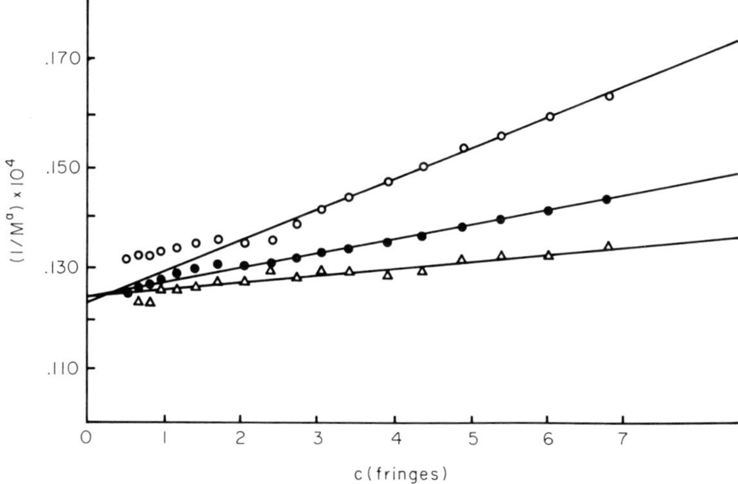

Fig. 15 Sedimentation equilibrium of a nearly homogeneous substance, crab hemocyanin, dissolved in 6 M Gu·HCl at 34,508 rpm, 22.0°. Number- (△), weight- (●), and z- (○) average molecular weights have been calculated at different points in the cell and their reciprocal values plotted against concentration. The three curves give the same value of B, the second virial coefficient, to within 10%. Data of D. Carpenter.

is illustrated in Fig. 15, which shows data for subunits of crab hemocyanin dissolved in 6 M Gu·HCl. Here the reciprocal apparent average molecular weights from a single experiment are plotted against the concentration at various points in the cell. Since the experiment was performed by the high speed method, the concentration approaches zero near the meniscus. The fact that all three averages approach nearly the same intercept and that the slopes are approximately in agreement with Eqs. (43)–(45) is good evidence for homogeneity. However, the inherent limitations of sedimentation equilibrium studies of nonideal substances are shown by the fact that SDS gel electrophoresis reveals the existence of *two* components, differing about 10% in molecular weight (Loehr and Mason, 1973; Carpenter and Van Holde, 1973).

It has been pointed out by Roark and Yphantis (1969) that combinations of Eqs. (43)–(45) can be used to eliminate nonideality effects if these effects can be adequately described by a single virial coefficient. If, for example, Eq. (43) is multiplied by 2 and Eq. (44) is subtracted from it, one obtains

$$\frac{2}{M_{\text{nr}}^{\text{app}}} - \frac{1}{M_{\text{wr}}^{\text{app}}} = \frac{1}{M} \tag{46}$$

if terms in c^2 and higher powers of c are negligible. Thus the quantity on the left should be independent of concentration. A number of such equations can be derived (see Roark and Yphantis, 1969).

In working with concentrated denaturing solvents, it is important to remember that the $(1 - \bar{v}\rho)$ term must be replaced by $(1 - \phi'\rho) = (\partial\rho/\partial c)_\mu$ (see Section IV,A,*4*).

If a protein molecule carries a high net charge, it will behave nonideally. In fact, two effects are involved. The *primary charge effect* contributes to the concentration dependence of the apparent molecular weight, giving an effective virial coefficient proportional to the square of the net charge and inversely proportional to the concentration of low molecular weight electrolyte present in the solution. Thus the primary charge effect can be repressed by increasing the ionic strength of the medium. Furthermore, it vanishes by extrapolation to zero protein concentration. There is also, however, a *secondary* charge effect which does not vanish when the protein concentration goes to zero. Fortunately, it will be small in most cases of interest to the protein chemist. For most protein studies, using a buffer with an ionic strength of 0.1 or greater and protein concentrations below a few milligrams per milliliter will assure that charge effects are negligible. For further details, the reader is referred to Williams *et al.* (1958).

C. Associating Systems

1. General Principles and the Nature of the Problem

In recent years, protein chemists have become increasingly aware of and interested in the association–dissociation reactions of proteins. Evidently, most high molecular weight proteins consist of subunits (see Ch. 5 and Klotz *et al.*, 1970) and many proteins that are commonly found in monomeric form can associate under appropriate conditions. The physical study of such systems usually involves answering the following questions:

(a) What is the *stoichiometry* of the association? Are specific aggregates of the monomer formed, and if so, what are these?

(b) How strong is the association? Specifically, what are the values of the *association equilibrium constants* or the corresponding free energies?

(c) What are the *thermodynamic parameters* of the association? Specifically, what are the signs and magnitudes of enthalpy and entropy changes? Is the association process driven by enthalpy or entropy changes? (Such information can be revealing of the molecular mechanism involved.)

(d) What are the *kinetics* of the processes involved?

Sedimentation equilibrium can be used to answer the first three questions. In terms of accuracy, range of application, and information obtained, it is probably the most powerful technique available today. For this reason, much attention has been devoted in recent years to the necessary theory, and application has been made to a wide variety of proteins. The literature on this subject is vast and cannot be reviewed completely here; for a more comprehensive survey, the reader is referred to several recent reviews (Teller *et al.*, 1969; Roark and Yphantis, 1969; Van Holde *et al.*, 1969; Teller, 1973).

A point of frequent misunderstanding is the idea that the centrifugal field in an ultracentrifuge cell perturbs the chemical equilibrium. This is true only when the equilibrium constant is pressure-dependent because of a volume change in the reaction (see Section II,C,*4*). In any case, as Tiselius (1926) showed many years ago, both chemical equilibrium and sedimentation equilibrium can be maintained at every point in the ultracentrifuge cell. Consider the simple example of a monomer–dimer reaction without volume change. Denoting monomer by subscript 1, dimer by subscript 2, we can write Eq. (31) for each, whether or not they are in equilibrium:

$$c_1(r) = c_1(a) \exp M_1(1 - \bar{v}\rho)\omega^2(r^2 - a^2)/2RT \tag{47}$$

$$c_2(r) = c_2(a) \exp M_2(1 - \bar{v}\rho)\omega^2(r^2 - a^2)/2RT \tag{48}$$

We have used the same $\bar{v}$ for both, since there is assumed to be no volume change. Now suppose that the system is an equilibrium mixture, characterized by an equilibrium constant, K. The concentrations at the meniscus must obey the relation

$$\frac{c_2(a)}{[c_1(a)]^2} = K \tag{49}$$

Since $M_2 = 2M_1$, Eqs. (47)–(48) yield

$$\frac{c_2(r)}{[c_1(r)]^2} = \frac{c_2(a)}{[c_1(a)]^2} \frac{\exp 2M_1(1 - \bar{v}\rho)\omega^2(r^2 - a^2)/2RT}{[\exp M_1(1 - \bar{v}\rho)\omega^2(r^2 - a^2)/2RT]^2} \tag{50}$$

The exponential terms on the right are identical; hence

$$\frac{c_2(r)}{[c_1(r)]^2} = \frac{c_2(a)}{[c_1(a)]^2} = K \tag{51}$$

Thus the same concentration relationship is satisfied at every level, r, in the cell. The individual species concentrations will vary with r and their *ratio* will vary as r varies, but they will always be present at concentrations which satisfy the equilibrium relation. This points up one of the great advantages of the sedimentation equilibrium technique: in a single experiment, the investigator studies the reaction over a wide range of total concentrations. Thus a single sedimentation equilibrium experiment is equivalent to a whole series of individual light-scattering experiments.

Another important point emerges which allows us to use sedimentation equilibrium to distinguish between a reversibly reacting system and a heterogeneous, nonreacting system. In any equilibrating system, the composition of the mixture (the relative concentrations of the various solute species) will be determined entirely by the total concentration, the temperature, and the pressure. If we restrict our attention for the moment to pressure-independent reactions and keep T fixed, only the total concentration at a particular point in the cell determines the composition, and hence *any* average molecular weight at that point. Conversely, in a nonequilibrating heterogeneous mixture, the amount of any species at any point in the cell is determined only by its molecular weight and its concentration in the original mixture. Now suppose we investigate a system at two different initial concentrations. As Yphantis (1964) has pointed out, we need only determine whether M_{wr} (or M_{nr}, or M_{zr}, etc.) is the same at points of equal *radius* or at points of equal *concentration*. Figure 16 diagrams the situation. If M_{wr} (for example) in these different experiments is identical at equal concentration points, the system must be in chemical equilibrium. An extension of this idea is that we

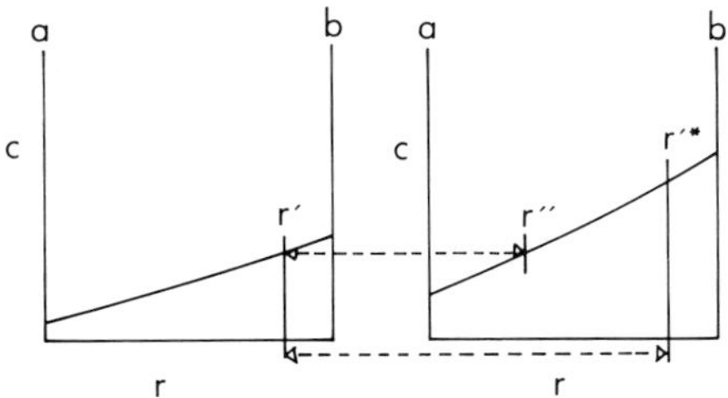

Fig. 16 A schematic drawing of the sedimentation equilibrium of a heterogeneous substance at two different initial concentrations. If the mixture is in equilibrium, points r' and r'', which are at the same c, will yield the same values of average molecular weights. On the other hand, if the mixture is not in equilibrium, points r' and r'^{*}, which are taken to be at equivalent positions in the cells, will yield the same average molecular weight values.

should be able to superimpose results from a number of experiments, at different initial concentrations and rotor speeds, on a single graph, if we plot an average molecular weight against the local concentration. A good example is shown in Fig. 17, which illustrates data on the dimerization of β-lactoglobulin A at low pH.

Figure 17 illustrates some other points as well. First, note that the molecular weight approaches the monomer weight at low concentration; this is expected behavior for any reversible association. On the other hand, at higher concentrations M_{wr} does not come even close to the dimer weight (36,844). This is because the solute species are quite non-

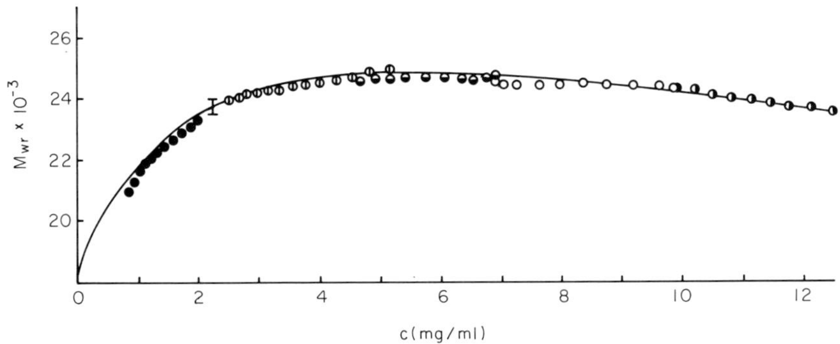

Fig. 17 Sedimentation equilibrium of a reversibly dimerizing system (β-lactoglobulin A at pH 2.46). Different symbols denote data from experiments at different initial concentrations. The solid line corresponds to the values obtained for K and B. Drawn from data of Tang (1971).

ideal. At high c the effects of nonideality and association compete, making the apparent molecular weight first level off and then decrease. (See Section IV,C,2,*b* for a discussion of the effects of nonideality.) This example points up the hazard in attempting to judge the stoichiometry of association reactions by inspection. Finally, if only experiments in the range of 5–10 mg/ml had been performed with β-lactoglobulin at this pH and T, one would conclude that β-lactoglobulin A was a homogeneous protein having a molecular weight of about 24,500! Nothing could better emphasize the necessity of carrying out experiments over a wide concentration range and under diverse conditions.

I now return to the specific questions raised at the beginning of this section.

What is the stoichiometry of the association? First, one must consider what stoichiometries are likely to be found. These can be roughly grouped into three classes:

(i) Monomer–n-mer associations:

$$nP_1 \rightleftarrows P_n \tag{52}$$

Such a scheme assumes that no aggregates larger than n-mer will form and that there is no appreciable equilibrium concentration of intermediates if $n > 2$. At first sight, such processes might appear unlikely, but in fact there are good reasons to think them to be fairly common. The geometry of subunit interactions (see Ch. 5) will often force the subunits to come together in finite, closed structures (rings, tetrahedra, etc.). Furthermore, as Van Holde (1966) has pointed out, the extra free energy released in ring closure may make the completion of such structures highly favored, thus decreasing the concentration of intermediates. Alternatively, Klotz *et al.* (1970) have shown how even moderate cooperativity in association can yield what is close to a monomer–n-mer system. For example, all of the evidence found by Langerman and Klotz (1969) indicates that hemerythrin is a cyclic octamer, which can be in equilibrium with monomers alone.

(ii) Limited associations: These include all situations in which a finite number of species is involved, such as

$$\begin{aligned} 2P_1 &\rightleftarrows P_2 \qquad K_2 \\ 2P_2 &\rightleftarrows P_4 \qquad K_4 \end{aligned} \tag{53}$$

An example would be hemoglobin, in which formation of the $\alpha\beta$ dimer is highly favored (Schachman and Edelstein, 1966) and precedes association to the tetramer.

(iii) Unlimited associations:

$$\begin{array}{ll} 2P_1 \rightleftarrows P_2 & K_1 \\ P_2 + P_1 \rightleftarrows P_3 & K_2 \\ P_3 + P_1 \rightleftarrows P_4 & K_3 \\ \vdots & \vdots \\ P_n + P_1 \rightleftarrows P_{n+1} & K_n \end{array} \tag{54}$$

Such association processes are much more common with proteins than might be expected; they seem to result when the subunit carries two association sites on opposite sides. If the subunits are identical, one might expect that the free energy increment would be about the same for the addition of each unit to the chain. In this case (*isodesmic* association; Van Holde *et al.*, 1969), the individual equilibrium constants in Eq. (54) become identical, greatly simplifying the analysis. A number of protein systems appear to behave in just this way; perhaps the best documented is the association of glutamate dehydrogenase, which has been investigated both by light-scattering (Eisenberg and Tomkins, 1968) and sedimentation equilibrium (Reisler *et al.*, 1970).

2. *Analysis of the Sedimentation Equilibrium of Associating Systems*

Over the past few years there has been an explosive increase in the literature on this subject. To attempt to review adequately the theoretical and experimental papers on this topic would require far more space than this entire chapter, so I shall only try to outline what seem to me to be the main trends in analysis, with a few experimental examples to illustrate the problems. For simplicity, I shall discuss ideal systems first, but the reader who wishes to perform such an analysis should be forewarned that almost all real systems turn out to be nonideal. Further details can be found in the review papers listed at the beginning of Section IV,C,*1*.

a. **Ideal Associating Systems without Volume Change.** There are two broad approaches that can be followed in order to obtain information about stoichiometry and association constants.

i. *Analysis of the* $c(r)$ *vs.* r *curve.* Whatever the stoichiometry of an associating system, one can write the total concentration at any point r as a sum of species concentrations:

$$c(r) = c_1(r) + c_2(r) + c_3(r) + \cdots \tag{55}$$

where the subscripts 1, 2, 3, etc., refer to monomer, dimer, trimer, etc. Since the system is assumed to be in equilibrium, this may be rewritten as

$$c(r) = c_1(r) + K_1 c_1^2(r) + K_2 c_1^3(r) + \cdots \tag{56}$$

where the K_i terms are equilibrium constants. Of course, some of the K_i terms may be zero, corresponding to species present in negligible amounts. Since

$$\begin{aligned} c_1(r) &= c_1(a) \exp M_1[(1 - \bar{v}\rho)\omega^2(r^2 - a^2)/2RT] \\ &= c_1(a)\Gamma \end{aligned} \tag{57}$$

the concentration may be written as a power series in Γ.

$$c(r) = c_1(a)\Gamma + K_1 c_{1a}^2 \Gamma^2 + K_2 c_{1a}^3 \Gamma^3 + \cdots \tag{58}$$

If Γ is known (which depends upon M_1 being known), the analysis reduces to finding the set of coefficients $K_i c_{1a}^{i+1}$ which best fits the data. This will automatically give both the stoichiometry and all of the equilibrium constants.

Unfortunately, it is much more difficult than it sounds to obtain a unique set of parameters and show that it is unique. A form of this technique was first used by Reinhardt and Squire in 1965, and since that time a number of variations have been developed including, in some cases, extension to nonideal systems (see, for example, Teller *et al.*, 1969; Van Holde *et al.*, 1969; Hashmeyer and Bowers, 1970; Chun and Kim, 1970a; Rosenthal, 1971; Holladay and Sophianopoulos, 1972). The "method of moments" used by Dyson and Isenberg (1971) is a method for solving this same kind of problem. Unfortunately, few authors to date have published tests of their methods with real data; computer-simulated examples (usually with random noise added) have been employed by most. Most studies of real protein systems have been carried out with the kinds of analysis described below.

ii. *Methods involving average molecular weights.* Since the sedimentation equilibrium experiment can yield several average molecular weights (i.e., M_n, M_w, M_z), it would seem possible to make use of the way in which these averages vary with concentration to deduce the stoichiometry and the required association constants. Figure 18 shows how weight-average molecular weight, for example, will vary with total solute concentration for simple monomer–*n*-mer associations. Indeed, for such associations a rather simple equation describes the variation of M_w with c:

$$\frac{Q}{[1 - (Q/n - 1)]^n} = K(n - 1)c^{n-1} \tag{59}$$

where $Q = (M_w/M_1) - 1$. Graphs of the left side of Eq. (59) vs. c^{n-1} can be tried for various values of n. The right choice (if the association

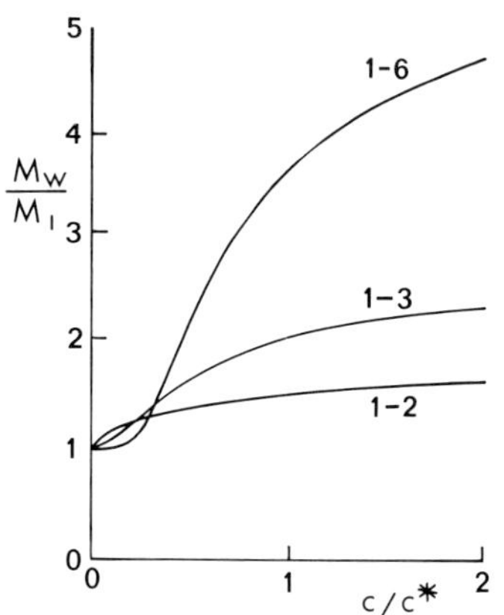

Fig. 18 Theoretical curves of M_w/M_1 for some ideal monomer–n-mer systems; c^* is the concentration at which monomer and n-mer are present in equal quantity. The symbols 1–2, 1–3, and 1–6 signify monomer–dimer, monomer–trimer, and monomer–hexamer, respectively. Redrawn from Howlett (1972).

is of this type and the system is ideal) should give a straight line having a slope of $K(n - 1)$. Similarly, Van Holde and Rossetti (1967) and Elias and Bareiss (1967) have shown that the isodesmic association can be described by an equation of the form

$$(M_w/M_1)^2 - 1 = 4Kc \tag{60}$$

again allowing an easy test.

A rather different approach, which makes use of rather general interrelationships between average molecular weights, was first suggested for monomer–dimer reactions by Sophianopoulos and Van Holde (1964) and then rediscovered and greatly generalized by Yphantis and coworkers. It is possible to show (Roark and Yphantis, 1969) that the following relationship must hold for any ideal two-component system, associating or not.

$$M_k = -(M_1M_2/M_{k-1}) + M_1 + M_2 \tag{61}$$

where M_1 and M_2 are the molecular weights of the two species, and $M_k = M_n$, M_w, M_z, etc., as $k = 0, 1, 2$, etc. The practical consequence of this relationship is shown in Fig. 19. Data from any ideal two-species system will lie on a straight line on such a "two-species plot." The intercepts of the line with the hyperbola $M_k = 1/M_{k-1}$ will yield the values M_1 and M_2. For well-behaved systems, such graphs allow determination of the stoichiometry and the molecular weights. Evaluation of the equilibrium constants is then facilitated. Roark and Yphantis (1969) have

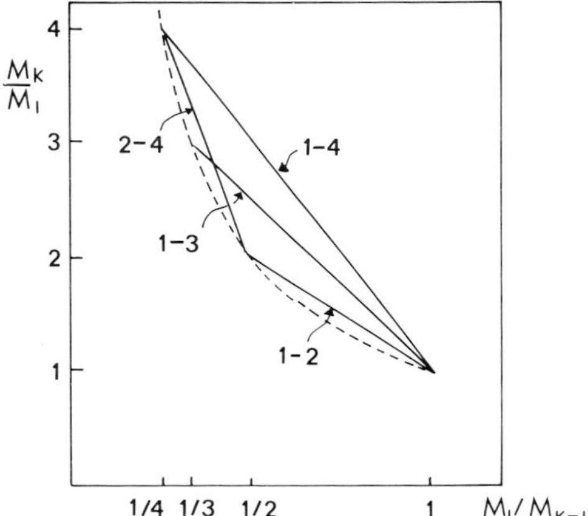

Fig. 19 Schematic "two-species plots" for monomer–n-mer associations. (1–2, monomer–dimer; 2–4, dimer–tetramer, etc.). For details, see the text.

further generalized the technique to devise "three-species plots" and have developed methods for dealing with nonideal solutions.

There exists one other class of methods, which is based fundamentally on a rather elegant equation derived by Steiner (1952, 1954). For any associating system, it is possible to calculate the concentration of monomer in a mixture of total concentration c if the weight-average molecular weight is known over the entire concentration range from zero to c:

$$\ln c_1/c = \int_0^c [(M_1/M_w) - 1] d \ln c \tag{62}$$

Of course the precise value of M_1 must be known. Steiner's equation provides immediate access to a primary datum, for once both c_1 and M_w are known at the same concentration, numerous possibilities for analysis emerge. As an example, consider an ideal monomer–n-mer reaction. We may write

$$c = c_1 + c_n = c_1 + Kc_1^n \tag{63}$$

With this equation we may determine the mechanism and evaluate K in a straightforward manner. This method has been used, for example, by Langerman and Klotz (1969).

Adams and co-workers (Adams and Williams, 1964; Adams and Filmer, 1966; Adams, 1967a) have made use of the Steiner equation in developing a number of techniques for the analysis of associating systems. A review of these and other methods has been published (Adams, 1967b). This group (see Adams, 1969) has also considered in some detail the methods for investigation of reactions of the type $A + B \rightleftarrows AB$. Here the analysis becomes considerably more complex, especially if the partial specific volumes of A and B are unequal, as they will usually be. However, this difficulty is not severe in studies of *hybrid* molecules involving similar subunits. A very nice example of this kind of experiment is given by Behnke *et al.* (1970) in studies of the interaction of subunits of procarboxypeptidase A. Both sedimentation equilibrium and sedimentation velocity techniques were employed in this work.

***b.* Nonideality.** The sedimentation equilibrium of nonideal heterogeneous systems, whether associating or merely polydisperse, presents a formidable problem. If we describe the nonideality in terms of virial coefficients (B_i) for the individual molecular species, we can write an expression for the experimentally available "apparent" weight-average molecular weight:

$$\frac{2RT}{\omega^2(1-\bar{v}\rho)}\frac{d\ln c}{dr^2} = M_w^{\text{app}} = \frac{M_w}{1+\Sigma_i B_i c_i} \tag{64}$$

Adams and Fujita (1963) pointed out that this equation could be made tractable by the simple assumption that $B_i = M_i B$; that is, that the virial coefficients are proportional to the molecular weight. One then obtains

$$M_w^{\text{app}} = \frac{M_w}{1+BcM_w} \tag{65}$$

or

$$\frac{1}{M_w^{\text{app}}} = \frac{1}{M_w} + Bc \tag{66}$$

which is of the same form as Eq. (44), which gave the apparent weight-average for the homogeneous, nonideal protein. Corresponding equations for M_n^{app} and M_z^{app} are

$$\frac{1}{M_n^{\text{app}}} = \frac{1}{M_n} + \frac{B}{2}c \tag{67}$$

and

$$\frac{1}{M_z^{\text{app}}} = \frac{1}{M_z}(1+BM_w c)^2 \tag{68}$$

Note that whereas Eq. (67) is identical to Eq. (43), Eq. (68) differs from Eq. (45). The apparent molecular weights are all calculated from Eqs. (34)–(36).

The complications of dealing with nonideal associating systems now become apparent. The results obtained from the ultracentrifuge via Eqs. (34)–(36) will be the apparent average molecular weights given in

Eqs. (66)–(68). In general, even if the Adams–Fujita assumption holds, the value of B is unknown and introduces one more unknown parameter into the analysis. All of the equations of the kind given above (Section IV,C,*2,a*) must be rewritten in terms of apparent average molecular weights.

There is not space here to detail the methods which have been developed to analyze such systems. Suffice it to mention that many of the methods described in Section IV,C,*2,a* have been generalized to include nonideal systems. For example, Dyson's method, given by Van Holde *et al.* (1969), is applicable to such situations. Equations (59) and (60) may be rewritten in terms of apparent average molecular weights and a virial coefficient. Roark and Yphantis (1969) have described "two-component plots" in terms of combinations of apparent average molecular weights which (approximately) eliminate nonideality. Adams and Williams (1964) have extended the Steiner equation (Eq. (62)) to nonideal solutions. Equation (62) will yield an apparent weight fraction monomer. Adams and Williams (1964) have shown that

$$\ln c_1/c = \ln (c_1/c)^{\text{app}} + BM_1c \tag{69}$$

A rather nice example of the analysis of apparent weight-average molecular weight data is given in a recent publication by Chun *et al.* (1972). These authors (see also Chun and Kim, 1970b) have developed graphical methods somewhat similar to those of Roark and Yphantis (1969). Taking data of Eisenberg and Tomkins (1968) on beef liver glutamate dehydrogenase, they first use a graph of $\eta[=(M_1/M_w^{\text{app}}) - \ln(c_1/c)^{\text{app}}]$ vs. $\xi[=(2M_1/M_n^{\text{app}}) - (M_1/M_w^{\text{app}})]$ to show that the association is indeed an isodesmic one (Fig. 20); they then use other graphs to obtain the association constants and B (Fig. 21). Although these data are from light-scattering experiments, the same techniques can be used with sedimentation equilibrium.

Finally, it should be reemphasized that nonideal behavior may be exhibited by highly charged macromolecules. This presents an especially difficult problem with associating systems. It has been recently solved, at least to a good approximation, by Roark and Yphantis (1971).

c. **Pressure Effects.** As indicated by Eq. (22) (Section II,C,*4*), a volume change in the reaction will lead to a dependence of the equilibrium constant on r. Although this was recognized very early by Tiselius (1926), it has received little attention in sedimentation equilibrium studies until recently. In general, it has been assumed that pressure effects in sedimentation equilibrium would be much less serious than in sedimentation velocity studies, since the centrifugal fields employed

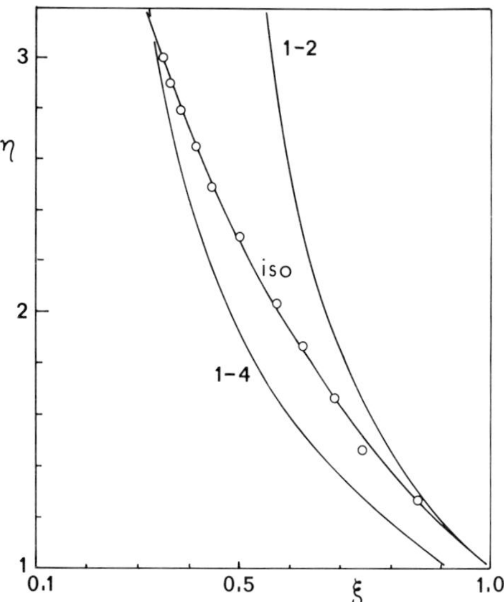

Fig. 20 A graph to demonstrate that glutamic dehydrogenase associates by isodesmic association. The solid lines are functions defined by Chun *et al.* (1972) and the graph is drawn from data given in their paper. The circles, which are experimental points taken from Eisenberg and Tomkins (1968), fall almost exactly on the curve for an isodesmic association. Association curves for monomer–dimer and monomer–tetramer are given for comparison.

are much smaller and the solution column is commonly quite short. Nonetheless, such effects should be taken into account, especially when the complexity of the reaction requires that the data be analyzed in a complicated way. A small systematic drift in the equilibrium constants might completely mislead or confuse such an analysis.

The situation has been analyzed in greatest detail by Jeffrey and Howlett and their co-workers (Howlett *et al.*, 1970; Jeffrey *et al.*, 1971). They point out that if a volume change occurs in an association reaction, one would expect the "superposition principle" to fail; that is, data obtained at different rotor speeds or at different initial concentrations will not fall on the same M_w vs. c curve. Nonsuperposition has been observed for a number of proteins, including adrenocorticotropin (Squire and Li, 1961), insulin (Jeffrey and Coates, 1966), lysozyme (Adams and Filmer, 1966; Deonier and Williams, 1970), and several others. While other explanations can be and have been given (e.g., the presence of components which do not participate in the reaction), it is possible that volume changes are occurring in some of these reactions.

In fact, very recent studies of lysozyme have led Howlett *et al.* (1972)

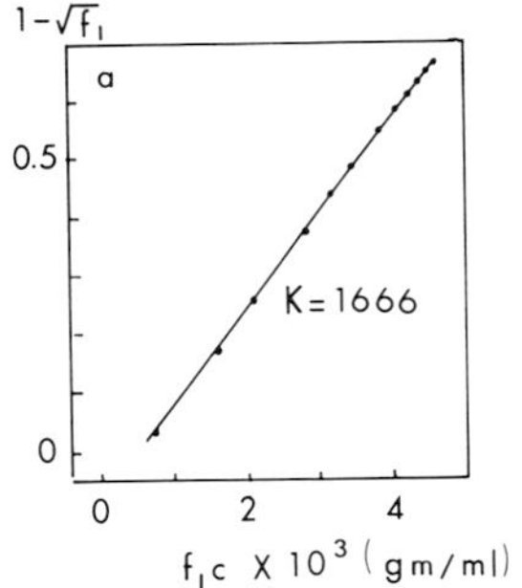

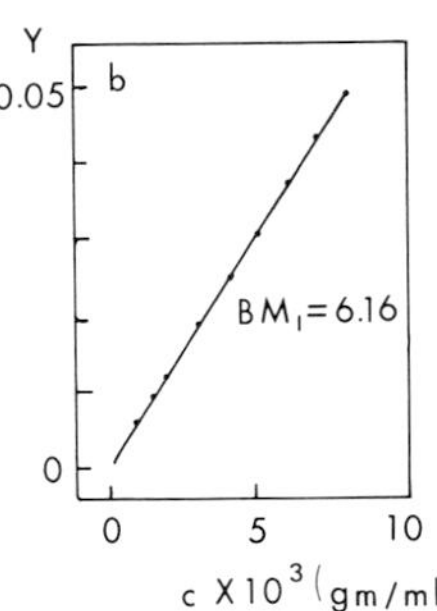

Fig. 21 Graphs showing the determination of (a) the association constant, *K*, and (b) the virial coefficient, *B*, for the data shown in Fig. 20. The quantity *Y* is defined as $(M_1/M_w^{\text{app}}) - \sqrt{f_1}/(2 - \sqrt{f_1})$, where f_1 is the weight fraction monomer. Both *K* and BM_1 can be measured accurately from the slopes. Taken from Chun *et al.* (1972).

to conclude that there is actually a decrease in volume accompanying the dimerization of lysozyme at pH 8.0. It remains to be seen in how many other cases such effects will be found. If significant pressure dependence is a common phenomenon, it may turn out that the low speed method will be more suitable than the high speed technique for sedimentation equilibrium studies of reacting systems.

D. Sedimentation Equilibrium of Proteins in Density Gradients

The density gradient equilibrium experiment, which has played so important a role in nucleic acid research, has been but little employed by protein chemists. Perhaps this is because as an analytical property, density is not as useful for proteins as it is for polynucleotides. However, there are situations (with lipoproteins and glycoproteins, for example) where analysis by density can be most revealing, and one may expect further work with this technique. Applications of the method to proteins have been reviewed by Ifft (1969).

The basic idea underlying the experiment is simple. If a macromolecular substance is present in a solution in which a low molecular weight substance (like a heavy-metal salt) sediments to equilibrium to yield a density gradient, the macromolecular component will, under proper conditions, collect in a concentrated band at some point in the cell (Fig. 22). This will occur if there is some point in the solution column at which the density of the solution equals the "density" of the macromolecular component. Concentration in the band will be of Gaussian form (Meselson *et al.*, 1957):

$$c = c_0 \exp (r - r_0)^2/2\sigma^2 \tag{70}$$

where c_0 is the concentration at the band center, r_0. This expression is exact only for narrow bands over which the density change is very small. The standard deviation, σ, given by Hearst and Vinograd (1961), is

$$\sigma^2 = \frac{RT}{M_s \bar{v}_s (d\rho/dr)_{\text{eff}} \omega^2 r_0} \tag{71}$$

Here $(d\rho/dr)_{\text{eff}}$ is the "effective" density gradient, defined below. The molecular weight and partial specific volume in Eq. (71) are not the values for the anhydrous macromolecule [as in Eqs. (2) and (30), for example] but are corrected for the interaction of the macromolecule with the gradient-forming ions and water. This complication arises because the required salt concentrations necessitate analysis of the mixture as a three-component system. It can be shown (Williams *et al.*, 1958; Hearst and Vinograd, 1961) that

$$\bar{v}_s = \frac{\bar{v} + \Gamma' \bar{v}_w}{1 + \Gamma'}; \qquad M_s = M(1 + \Gamma') \tag{72}$$

Here Γ' is a preferential solvation parameter and $\bar{v}_w$ is the partial specific volume of water. Essentially, Γ' can measure the excess binding of

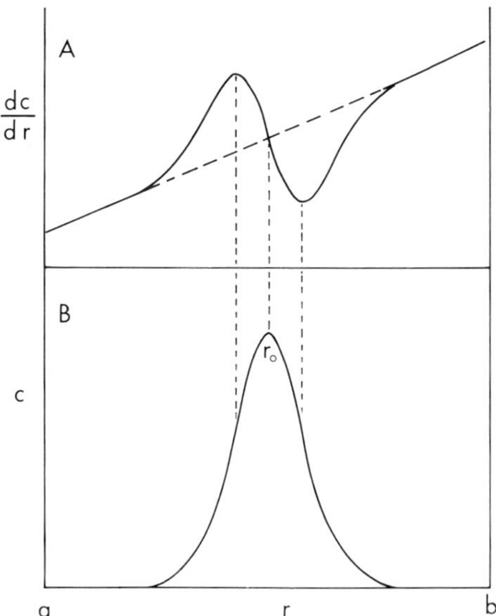

Fig. 22 A schematic representation of (A) the refractive index gradient curve, and (B) the absorbance curve for a single component in density gradient sedimentation equilibrium. It is assumed that the gradient-forming salt contributes to the refractive index gradient in (A) but not to the absorbance in (B).

water (gram water per gram protein) over salt ions. Ifft and Vinograd (1966) have pointed out that if the salt binding is known from independent measurements, then the $\bar{v}_s$ can be used to measure the net hydration of the macromolecule–salt complex:

$$\bar{v}_s = \frac{\bar{v} + z_x\bar{v}_x + \Gamma_*'\bar{v}_w}{1 + z_x + \Gamma_*'} \tag{73}$$

Here z_x is the number of grams of neutral salt bound per gram of protein. If the ions are bound unequally, a more complicated expression is required.

The solvated specific volume, $\bar{v}_s$, can be determined from the band position. The band will be centered at the point in the gradient, r_0, where $\rho = \rho_0$, and

$$\rho_0 = 1/\bar{v}_s \tag{74}$$

Location of this "isopycnic density" allows, in principle, calculation of the net hydration of a protein. In order to know the densities at various points in the solution column, somewhat involved calculations are required; these are fully explained by Ifft (1969) and will not be detailed here.

Equation (71) indicates that the molecular weight can be determined from the breadth (standard deviation) of the band. For this calculation, the "effective density gradient" is required. This is the physical density gradient (that which would be calculated from the known molecular weight and nonideality of the salt), corrected for differences in compressibility of the solution and macromolecule. Again, the calculations are somewhat complex. The value of molecular weight obtained is not M, but $M(1 + \Gamma')$.

An example of the kind of information that can be obtained from such experiments is found in the excellent study of isoelectric bovine mercaptalbumin by Ifft and Vinograd (1966). Using seven different gradient-forming salts (CsCl, CsBr, CsI, KBr, RbBr, CrBr, and Cs_2SO_4), they obtained values of Γ_*' M_s, and M in each. The hydration parameter Γ_*' (which refers to the protein–salt complex) ranged from 0.37 to 0.71 gm/gm, with concomitant variation in M_s. But the values of the anhydrous molecular weight, deduced from these data, varied only between 63,000 and 72,000, in good agreement with literature values of approximately 67,000.

Studies have also been made of the isopycnic banding of denatured proteins in salt gradients. Since the "molecular weight" of the aggregated protein is very large, the bands which are formed are very narrow (Cox and Schumaker, 1961).

Density gradient equilibrium experiments seem to hold unusual promise for the study of conjugated proteins such as the lipoproteins and glycoproteins. Since the nonprotein moiety (especially lipid) has a different density than the protein, it is possible to separate lipoproteins which differ in protein content. Pioneering experiments with serum lipoproteins were performed by Adams and Schumaker (1964, 1969, 1970a,b,c) using sucrose–NaBr gradients. The schlieren optical system provides excellent resolution for mixtures of lipoproteins, and computer methods have been developed for their analysis. Figure 23 shows the observed protein gradient and the computer-calculated pattern for a mixture of two low-density lipoproteins. Similar studies of glycoproteins have been reported by Creeth and Denborough (1970), who showed that such proteins can be cleanly resolved by this technique.

The application of the density gradient equilibrium technique to analytical ultracentrifugation has been handicapped by the rather long times required for establishment of equilibrium (often of the order of 60 hr) and the fact that dissolved proteins form rather broad bands, even at the highest rotor speeds. While the use of preformed gradients has been helpful in reducing the time in *preparative* banding experiments, no really adequate way of accomplishing this in the *analytical* cell has been devised. For theoretical discussions of the approach to equilibrium in preformed and nonpreformed gradients, the reader is referred to studies by Meselson and Nazarian (1963), Baldwin and Shooter (1963), and Dishon *et al.* (1971).

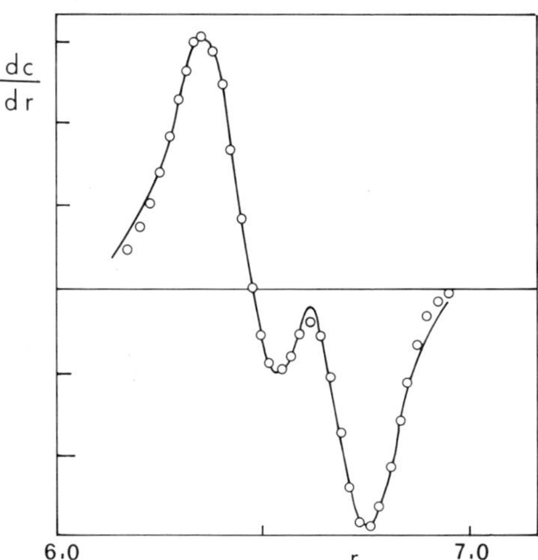

Fig. 23 The observed refractive index gradient (solid line) and the computer-fitted data (circles) for the equilibrium banding of a mixture of two low-density lipoproteins. Taken from Adams and Schumaker (1969).

When the protein gradient extends all the way to the limits of the cell, quantitative analysis becomes difficult (Adams and Schumaker, 1964, 1970b) and steeper gradients are required. This is one of the few situations where rotor speeds higher than those currently available might be useful.

In summary, while the equilibrium density gradient technique holds little attraction as a way of measuring protein molecular weights, it does appear to be potentially useful in two areas—the study of protein hydration and the analysis of mixtures of conjugated proteins.

V. ADVANCES AND TRENDS IN ULTRACENTRIFUGE TECHNOLOGY

Recurring themes in the preceding sections have been the need for more precision and flexibility in concentration distribution measurements, and the increasing complexity of theories and their attendant calculations. Such needs have in the past spurred the developments of the Rayleigh and scanner optical systems and the introduction of desk-top computers and large computers for analysis of data.

Currently a new wave of development is sweeping through the field, and I will attempt in this section to assess its likely effects on sedimentation studies of proteins. At the moment the major advances appear to be centered in two areas, the development of improved optical systems and the design of computer-linked ultracentrifuges.

A. Advances in Optical System Design

The Rayleigh interferometric system has long been employed for the most accurate sedimentation equilibrium measurements. Good as it is, it has deficiencies in the sharpness of fringes and in the resolution of fringes in high gradient regions. Several investigators have turned to the use of lasers as light sources in order to improve the quality of fringe patterns.

For example, a helium–neon CW laser has been employed by Williams (1972). This represents a simple modification of the optical system which improves the resolution of fringes as well as the light intensity at the photographic plate. The latter consideration is by no means trivial, since the low light levels produced by the conventional mercury lamp require long exposures and fast emulsions. Fast emulsions are coarse-grained, and the grain itself places a limitation on resolution.

A more dramatic advance was the development of a pulsed-laser interferometric system by Paul and Yphantis (1972). The use of high intensity light pulses, synchronized with the rotor motion, has some special advantages. Since the cell is illuminated *only* when the slits are in front of both the solution and reference column, the "background" produced by single-slit illumination is eliminated, and contrast is improved. Multiple cell operation is easy; the pulses are simply timed to illuminate whichever cell is to be photographed. The superb quality of the interference pattern can be seen in Fig. 24. This results in part from the better contrast, and in part from the fine-grained film. It seems likely that this system, or something like it, will be the interferometric method of choice in the future.

There would be many advantages in developing scanner systems that have precision comparable to the Rayleigh method. These include (i) measurement of absolute concentration rather than concentration differences; (ii) the possibility of working at *very* low concentrations; (iii) discrimination between different solutes with different absorption bands; and (iv) ease of direct interfacing with computers (see Section V,B).

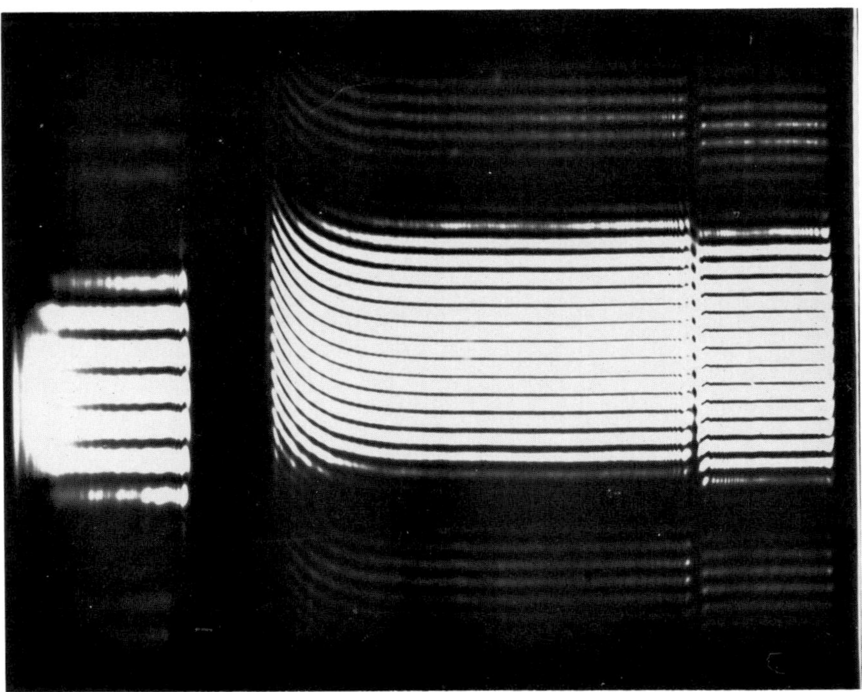

Fig. 24 A Rayleigh fringe pattern for the sedimentation equilibrium of insulin, as recorded by the pulsed-laser apparatus of Paul and Yphantis (1972). The meniscus is to the right. For details, see the text.

While photoelectric scanning units are commercially available from a number of ultracentrifuge manufacturers, several investigators have constructed their own. Foremost in this area have been Schachman and his collaborators. In a series of papers (Schachman, 1960; Hanlon *et al.*, 1962; Lamers *et al.*, 1963; Schachman and Edelstein, 1966) they have described increasing refinements which have enabled them to carry out both sedimentation equilibrium and sedimentation velocity measurements at protein concentrations as low as a few micrograms per milliliter. A number of other workers have also described scanner assemblies (see, for example, Van Rapenbusch and Deschepper, 1966; Spragg *et al.*, 1965). A simple device that may improve scanner performance in equilibrium studies is the installation of a "stepping" motor (Wampler, 1971), which allows the scanner to record for any desired period at any point. Despite the considerable effort in this area, it does not appear that the scanner method can yet equal the Rayleigh method in precision, at least in the concentration range in which the latter is optimal.

B. Computers and the Ultracentrifuge

Some individuals have been using digital computers for many years. (For a review of the applications as of 1966, see Trautman, 1966.) Initially, this took two forms, the use of desk-top computers for simple programs, and the "batch" analysis of data for more complex problems (e.g., the analysis of sedimentation equilibrium data). The limitations of both methods are obvious. Both are time-consuming if more than the simplest analysis is desired, and batch processing is inherently inflexible; if a different kind of analysis is needed, the whole data set must be run over again. Until a few years ago, all data were laboriously transcribed by hand from comparator measurement of plates and then punched onto cards for analysis. It has been obvious to many that the ideal arrangement would be one in which the raw data were fed automatically from the ultracentrifuge into a computer large enough to handle any desired computations, with instantaneous operator control of the analysis. As we shall see, such systems are being developed. First, however, I would like to describe the current scheme of operation in our laboratories at Oregon State University. The methods we use, brought into our laboratory through the efforts of Dr. R. Dyson, represent a kind of "half-way" step in the approach to complete automation. They are easily accessible to most laboratories and have proved efficient and flexible.

First, we have equipped the comparator with a digitizer (Ansevin,

1967), which either punches the coordinates on paper tape or feeds them directly into the University computer. The cross-hair alignment on points in the pattern is set by hand. We have considered installing a simple device similar to the type recently developed by Richards (1971) for photoelectrically judging fringe center positions. This would eliminate an element of subjectivity in the analysis. Automatically punching the coordinates on tape is fast, eliminates transcription error, and provides data in a handy form for computer analysis.

Second, we have installed keyboard terminals (with tape readers) in the laboratory. These terminals connect with the University's central computer, using a time-sharing system. All of the programs we use, which have many variations and options, are available in the computer files and can be obtained on demand. Results of the analysis are printed on the terminal console and/or plotted on a linked x-y plotter. The results can be seen almost immediately, and the analysis can be modified as desired by the experimenter at the terminal. Furthermore, new programs can be entered and tested at the terminal. We find that this system can easily keep up with two analytical ultracentrifuges which are operating full time. The main disadvantages at present are that plate reading is still a somewhat tedious chore and the reading of scanner data is still done in a rather primitive way. This could be improved quite easily by feeding the scanner output into an analog–digital converter and then into a tape punch (Beckwith *et al.*, 1971).

A number of laboratories are using arrangements more or less similar to ours. In a few places, attempts are being made to connect computers even more directly to the ultracentrifuge. Since no one has yet devised a suitable way to directly digitize the optical patterns produced by schlieren or Rayleigh systems, most of these efforts are dedicated to direct analysis of scanner data. In such instances a small "dedicated" computer is installed in the laboratory. At this point, there are two modes of operation which may be chosen. The simplest is to put the output of the scanner unit (i.e., the signal which normally goes to the recorder) into an analog–digital converter and then into the scanner (Pekar *et al.*, 1971). Alternatively, one may digitize the photomultiplier signal itself and make all conversions to absorbance, base line corrections, etc., in the computer (Spragg and Goodman, 1969; Rehmar and Edelstein, 1971). In either case, there is a potential for storing and averaging data so as to improve the signal/noise ratio.

It seems to me that this kind of direct linkage to a small computer, coupled with an improved scanning system, may well represent the most promising line of development in the next decade or so. If, in this way, the scanner can be made to provide precision equal to or better than the Rayleigh system, the inherently greater flexibility of absorption optics and the ease with which multiple samples can be handled would offer distinct advantages. The pulsed-laser Rayleigh system of Paul and Yphantis (1972) (Section V,A) can also allow multiple cell operation. Very recently, an automated system for analysis of Rayleigh patterns was developed by DeRosier *et al.* (1972). This technique utilizes Fourier transform analysis of data obtained from a computer-linked densitometer.

The ultracentrifuge is being called upon to perform increasingly sophisticated analyses. At the present time, theory has outstripped the capabilities of the traditional measuring and data-handling systems. The increased interest in new systems is a reflection of this. With the availability of more accurate and faster methods of data collection, we may expect that problems that currently can be handled only with difficulty (complicated association reactions, for example) will become routine. It is my opinion that the ultracentrifuge will become an instrument used primarily in the study of *interactions* between macromolecules and between macromolecules and small molecules. At the moment it would appear that sedimentation equilibrium will be the most important technique, but the potentialities of transport methods may not yet be fully realized.

ACKNOWLEDGMENTS

I wish to thank all of those who have allowed me to make use of unpublished data in the preparation of this chapter. These include Drs. D. Carpenter, R. Roxby, D. Malencik, S. Anderson, R. Dyson, and D. Yphantis. The opportunity to read unpublished manuscripts from Drs. D. Teller and D. Yphantis was of great help. Finally, I wish to thank Barbara Hanson, who rapidly and precisely typed the manuscript under difficult circumstances.

REFERENCES

Adams, E. T., Jr. (1967a). *Biochemistry* **6,** 1864.

Adams, E. T., Jr. (1967b). "Fractions," No. 3. Spinco Division, Beckman Instruments, Inc., Palo Alto, California.

Adams, E. T., Jr. (1969). *Ann. N. Y. Acad. Sci.* **164,** 226.

Adams, E. T., Jr., and Filmer, D. L. (1966). *Biochemistry* **5,** 2971.

Adams, E. T., Jr., and Fujita, H. (1963). *In* "Ultracentrifugal Analysis in Theory and Experiment" (J. W. Williams, ed.), p. 119. Academic Press, New York.

Adams, E. T., Jr., and Williams, J. W. (1964). *J. Amer. Chem. Soc.* **86,** 3454.

Adams, G. H., and Schumaker, V. N. (1964). *Nature (London)* **202,** 490.

Adams, G. H., and Schumaker, V. N. (1969). *Ann. N. Y. Acad. Sci.* **164,** 130.

Adams, G. H., and Schumaker, V. N. (1970a). *Biochim. Biophys. Acta* **202,** 305.

Adams, G. H., and Schumaker, V. N. (1970b). *Biochim. Biophys. Acta* **202,** 315.
Adams, G. H., and Schumaker, V. N. (1970c). *Biochim. Biophys. Acta* **202,** 462.
Ansevin, A. T. (1967). *Anal. Biochem.* **19,** 498.
Antonini, E., Bucci, E., Fronticelli, C., Chiancone, E., Wyman, J., and Rossi-Fanelli, A. (1966). *J. Mol. Biol.* **17,** 29.
Archibald, W. J. (1938). *Phys. Rev.* **54,** 371.
Archibald, W. J. (1942). *Ann. N. Y. Acad. Sci.* **43,** 211.
Archibald, W. J. (1947). *J. Appl. Phys.* **18,** 362.
Baldwin, R. L., and Shooter, E. M. (1963). *In* "Ultracentrifugal Analysis in Theory and Experiment" (J. W. Williams, ed.), p. 143. Academic Press, New York.
Bauer, N., and Lewin, S. Z. (1959). *In* "Physical Methods of Organic Chemistry" (A. Weissberger, ed.), 3rd ed., p. 131. Wiley (Interscience), New York.
Beckwith, A. C., Nielsen, H. C., and Butterfield, R. O. (1971). *Anal. Chem.* **43,** 1471.
Behnke, W. D., Teller, D. C., Wade, R. D., and Neurath, H. (1970). *Biochemistry* **9,** 4189.
Bloomfield, V., Van Holde, K. E., and Dalton, W. O. (1967). *Biopolymers* **5,** 149.
Cann, J. R. (1970). "Interacting Macromolecules." Academic Press, New York.
Cann, J. R., and Goad, W. B. (1968). *Advan. Enzymol. Relat. Areas Mol. Biol.* **30,** 139.
Carpenter, D., and Van Holde, K. E. (1973). *Biochemistry* **12,** 2231.
Casassa, E. F., and Eisenberg, H. (1961). *J. Phys. Chem.* **65,** 427.
Casassa, E. F., and Eisenberg, H. (1964). *Advan. Protein Chem.* **19,** 287.
Charlwood, P. A. (1957). *J. Amer. Chem. Soc.* **79,** 776.
Charlwood, P. A. (1967). *Biopolymers* **5,** 663.
Chervenka, C. H. (1970). *Anal. Biochem.* **34,** 24.
Chun, P. W., and Kim, S. J. (1970a). *J. Phys. Chem.* **74,** 899.
Chun, P. W., and Kim, S. J. (1970b). *Biochemistry* **9,** 1957.
Chun, P. W., Kim, S. J., Williams, J. D., Cope, W. T., Tang, L. H., and Adams, E. T., Jr. (1972). *Biopolymers* **11,** 197.
Cohen, R. (1963). *C. R. Acad. Sci.* **256,** 3513.
Cohen, R., and Hahn, C. (1965). *C. R. Acad. Sci.* **260,** 2077.
Cohen, R., and Mire, M. (1971a). *Eur. J. Biochem.* **23,** 267.
Cohen, R., and Mire, M. (1971b). *Eur. J. Biochem.* **23,** 276.
Cohen, R., Giraud, B., and Messiah, A. (1967). *Biopolymers* **5,** 203.
Cohn, E. J., and Edsall, J. T. (1943). "Proteins, Amino Acids, and Peptides." Van Nostrand-Reinhold, Princeton, New Jersey.
Cox, D. J., and Schumaker, V. N. (1961). *J. Amer. Chem. Soc.* **83,** 2439.
Creeth, J. M., and Denborough, M. A. (1970). *Biochem. J.* **117,** 89.
Creeth, J. M., and Knight, C. G. (1965). *Biochim. Biophys. Acta* **102,** 549.
Dayhoff, M. O., Perlmann, G. E., and MacInnes, D. A. (1952). *J. Amer. Chem. Soc.* **74,** 2515.
Deonier, R. C., and Williams, J. W. (1970). *Biochemistry* **9,** 4260.
DeRosier, D. J., Munk, P., and Cox, D. J. (1972). *Anal. Biochem.* **50,** 139.
Dishon, M., Weiss, G., and Yphantis, D. A. (1967). *Biopolymers* **5,** 697.
Dishon, M., Weiss, G. H., and Yphantis, D. A. (1971). *Biopolymers* **10,** 2095.
Dubin, S. (1972). *In* "Methods in Enzymology" (C. H. W. Hirs and S. N. Timasheff, eds.), Vol. 26, Part C, p. 119. Academic Press, New York.
Dyson, R. D., and Isenberg, I. (1971). *Biochemistry* **10,** 3233.
Edelstein, S. J., and Schachman, H. K. (1967). *J. Biol. Chem.* **242,** 306.

Eisenberg, H. (1962). *J. Chem. Phys.* **36,** 1837.
Eisenberg, H., and Tomkins, G. (1968). *J. Mol. Biol.* **31,** 37.
Elias, H. G., and Bareiss, R. (1967). *Chimia* **21,** 53.
Faxén, O. H. (1929). *Ark. Mat., Astron. Fys.* **21B,** 1.
Fujita, H. (1959). *J. Phys. Chem.* **63,** 1092.
Fujita, H. (1962). "The Mathematical Theory of Sedimentation Analysis." Academic Press, New York.
Gerhart, J. C., and Schachman, H. K. (1968). *Biochemistry* **7,** 538.
Gilbert, G. A. (1955). *Discuss. Faraday Soc.* **20,** 68.
Gilbert, G. A. (1959). *Proc. Roy. Soc., Ser. A* **250,** 377.
Goad, W. B., and Cann, J. R. (1969). *Ann. N. Y. Acad. Sci.* **164,** 172.
Goldberg, M. E., and Edelstein, S. J. (1969). *J. Mol. Biol.* **46,** 431.
Goldberg, R. J. (1953). *J. Phys. Chem.* **57,** 194.
Goodrich, R., Swinehart, D. F., Kelly, M. J., and Reithel, F. J. (1969). *Anal. Biochem.* **28,** 25.
Griffith, O. M. (1967). *Anal. Biochem.* **19,** 243.
Hanlon, S., Lamers, K., Lauterbach, G., Johnson, R., and Schachman, H. K. (1962). *Arch. Biochem. Biophys.* **99,** 157.
Hashmeyer, R. H., and Bowers, W. F. (1970). *Biochemistry* **9,** 435.
Hathaway, G., and Criddle, R. S. (1966). *Proc. Nat. Acad. Sci. U. S.* **56,** 680.
Hearst, J. E., and Vinograd, J. (1961). *Proc. Nat. Acad. Sci. U. S.* **47,** 999.
Hexner, P. E., Radford, L. E., and Beams, J. W. (1961). *Proc. Nat. Acad. Sci. U. S.* **47,** 1848.
Hoagland, V. D., Jr., and Teller, D. (1969). *Biochemistry* **8,** 594.
Holladay, L. A., and Sophianopoulos, A. J. (1972). *J. Biol. Chem.* **247,** 427.
Howlett, G. J. (1972). Ph.D. Thesis, University of Melbourne, Melbourne, Australia.
Howlett, G. J., Jeffrey, P. D., and Nichol, L. W. (1970). *J. Phys. Chem.* **74,** 3607.
Howlett, G. J., Jeffrey, P. D., and Nichol, L. W. (1972). *J. Phys. Chem.* **76,** 3429.
Ifft, J. B. (1969). *In* "A Laboratory Manual of Analytical Methods of Protein Chemistry" (P. Alexander and H. P. Lundgren, eds.), Vol. 5, Chapter 4, p. 151. Pergamon, Oxford.
Ifft, J. B., and Vinograd, J. (1966). *J. Phys. Chem.* **70,** 2814.
Isenberg, I., and Dyson, R. D. (1969). *Biophys. J.* **9,** 1337.
Jeffrey, P. D., and Coates, J. H. (1966). *Biochemistry* **5,** 489.
Jeffrey, P. D., and Pont, M. J. (1969). *Biochemistry* **8,** 4597.
Jeffrey, P. D., Howlett, G. J., and Nichol, L. W. (1971). *Proc. Aust. Biochem. Soc.* **4,** 25.
Johnston, J. P., and Ogston, A. G. (1946). *Trans. Faraday Soc.* **42,** 789.
Josephs, R., and Harrington, W. F. (1966). *Biochemistry* **5,** 3474.
Josephs, R., and Harrington, W. F. (1967). *Proc. Nat. Acad. Sci. U. S.* **58,** 1587.
Kegeles, G., Rhodes, L., and Bethune, J. L. (1967). *Proc. Nat. Acad. Sci. U. S.* **58,** 45.
Kellett, G. L. (1971). *J. Mol. Biol.* **59,** 401.
Kirby-Hade, E. P., and Tanford, C. (1967). *J. Amer. Chem. Soc.* **89,** 5034.
Kirkwood, J. G. (1954). *J. Polym. Sci.* **12,** 1.
Kirschner, M. W., and Schachman, H. K. (1971a). *Biochemistry* **10,** 1900.
Kirschner, M. W., and Schachman, H. K. (1971b). *Biochemistry* **10,** 1919.
Klotz, I. M., Langerman, N. R., and Darnall, D. W. (1970). *Annu. Rev. Biochem.* **39,** 25.
LaBar, F. E. (1965). *Proc. Nat. Acad. Sci. U. S.* **54,** 31.

Lamers, K., Putney, F., Steinberg, I. Z., and Schachman, H. K. (1963). *Arch. Biochem. Biophys.* **103**, 379.
Lamm, O. (1929). *Ark. Mat., Astron. Fys.* **21B**, 2.
Langerman, N. R., and Klotz, I. M. (1969). *Biochemistry* **8**, 4746.
Loehr, J. S., and Mason, H. S. (1973). *Biochem. Biophys. Res. Comm.* **54**, 741.
McMeekin, T. L., and Marshall, K. (1952). *Science* **116**, 142.
Martin, R. G., and Ames, B. N. (1961). *J. Biol. Chem.* **236**, 1372.
Mason, M., and Weaver, W. (1924). *Phys. Rev.* **23**, 412.
Meselson, M., and Nazarian, G. M. (1963). *In* "Ultracentrifugal Analysis in Theory and Experiment" (J. W. Williams, ed.), p. 131. Academic Press, New York.
Meselson, M., Stahl, F. W., and Vinograd, J. (1957). *Proc. Nat. Acad. Sci. U. S.* **43**, 581.
Moore, P. H. (1969). *Phys. Tech. Biol.* **2(Part B)**, 285.
Pasternak, R. A., Nazarian, G. M., and Vinograd, J. R. (1957). *Nature (London)* **179**, 92.
Paul, C. H., and Yphantis, D. A. (1972). *Anal. Biochem.* **48**, 588 and 605.
Pekar, A. H., Weller, R. E., Byers, R. A., and Frank, B. H. (1971). *Anal. Biochem.* **42**, 516.
Rehmar, M. J., and Edelstein, S. J. (1971). *Fed. Proc., Fed. Amer. Soc. Exp. Biol.* **30**, 1181 (Abstr.).
Reinhardt, W. P., and Squire, P. G. (1965). *Biochim. Biophys. Acta* **94**, 566.
Reisler, E., Pouyet, J., and Eisenberg, H. (1970). *Biochemistry* **9**, 3095.
Richards, E. G. (1971). *Fed. Proc., Fed. Amer. Soc. Exp. Biol.* **30**, 1181 (Abstr.).
Richards, E. G., and Schachman, H. K. (1957). *J. Amer. Chem. Soc.* **79**, 5324.
Richards, E. G., and Schachman, H. K. (1959). *J. Phys. Chem.* **63**, 1578.
Richards, E. G., Teller, D. C., and Schachman, H. K. (1968). *Biochemistry* **7**, 1054.
Roark, D. E., and Yphantis, D. A. (1969). *Ann. N. Y. Acad. Sci.* **164**, 245.
Roark, D. E., and Yphantis, D. A. (1971). *Biochemistry* **10**, 3241.
Rosenthal, A. (1971). *Macromolecules* **4**, 35.
Schachman, H. K. (1959). "Ultracentrifugation in Biochemistry." Academic Press, New York.
Schachman, H. K. (1960). *Brookhaven Symp. Biol.* **13**, 49.
Schachman, H. K. (1963). *Biochemistry* **2**, 887.
Schachman, H. K., and Edelstein, S. J. (1966). *Biochemistry* **5**, 2681.
Schumaker, V. N. (1967). *Advan. Biol. Med. Phys.* **11**, 245.
Schumaker, V. N., and Adams, P. (1968). *Biochemistry* **7**, 3422.
Schumaker, V. N., and Rosenbloom, J. (1965). *Biochemistry* **4**, 1005.
Sophianopoulos, A. J., and Van Holde, K. E. (1964). *J. Biol. Chem.* **243**, 1804.
Sophianopoulos, A. J., Rhodes, C. K., Holcomb, D. N., and Van Holde, K. E. (1962). *J. Biol. Chem.* **237**, 1107.
Spragg, S. P., and Goodman, R. F. (1969). *Ann. N. Y. Acad. Sci.* **164**, 294.
Spragg, S. P., Travers, S., and Saxton, T. (1965). *Anal. Biochem.* **12**, 259.
Squire, P. G., and Li, C. H. (1961). *J. Amer. Chem. Soc.* **83**, 3521.
Steinberg, I. Z., and Schachman, H. K. (1966). *Biochemistry* **5**, 3728.
Steiner, R. F. (1952). *Arch. Biochem. Biophys.* **39**, 333.
Steiner, R. F. (1954). *Arch. Biochem. Biophys.* **49**, 400.
Svedberg, T. (1925). *Kolloid-Z. (Zsigmondy-Festschr.)* **36**, 53.
Svedberg, T. (1927). *Z. Phys. Chem.* **127**, 51.
Svedberg, T., and Chinoaga, E. (1928). *J. Amer. Chem. Soc.* **50**, 1399.
Svedberg, T., and Fahraeus, A. (1926). *J. Amer. Chem. Soc.* **48**, 430.

Svedberg, T., and Nichols, J. B. (1923). *J. Amer. Chem. Soc.* **45,** 2910.
Svedberg, T., and Nichols, J. B. (1927). *J. Amer. Chem. Soc.* **49,** 2920.
Svedberg, T., and Pedersen, K. O. (1940). "The Ultracentrifuge." Oxford Univ. Press, London and New York.
Svedberg, T., and Rinde, H. (1924). *J. Amer. Chem. Soc.* **46,** 2677.
Svensson, H., Hagdahl, L., and Lerner, K. D. (1957). *Sci. Tools* **4,** 1.
Tanford, C. (1961). "Physical Chemistry of Macromolecules." Wiley, New York.
Tanford, C., Kawahara, K., and Lapanje, S. (1967). *J. Amer. Chem. Soc.* **89,** 729.
Tang, L. H. (1971). Ph.D. Thesis, Illinois Institute of Technology, Chicago.
Teller, D. C. (1973). *In* "Methods in Enzymology" (C. H. W. Hirs and S. N. Timasheff, eds.), Vol. 27, p. 346. Academic Press, New York.
Teller, D. C., Horbett, T. A., Richards, E. G., and Schachman, H. K. (1969). *Ann. N. Y. Acad. Sci.* **164,** 66.
Ten Eyck, L. F., and Kauzmann, W. (1967). *Proc. Nat. Acad. Sci. U. S.* **58,** 888.
Thomas, J. O., and Edelstein, S. J. (1971). *Biochemistry* **10,** 477.
Tiselius, A. (1926). *Z. Phys. Chem.* **124,** 449.
Trautman, R. (1966). "Fractions," No. 2. Spinco Division, Beckman Instruments, Inc., Palo Alto, California.
Ulrich, D. V., Kupke, D. W., and Beams, J. W. (1964). *Proc. Nat. Acad. Sci. U. S.* **52,** 349.
Van Holde, K. E. (1960). *J. Phys. Chem.* **64,** 1582.
Van Holde, K. E. (1962). *J. Chem. Phys.* **37,** 1922.
Van Holde, K. E. (1966). *In* "Molecular Architecture in Cell Physiology" (T. Hayashi and A. Szent-Györgyi, eds.), p. 81. Prentice-Hall, Englewood Cliffs, New Jersey.
Van Holde, K. E. (1967). "Fractions," No. 1. Spinco Division, Beckman Instruments, Inc., Palo Alto, California.
Van Holde, K. E. (1971). "Physical Biochemistry." Prentice-Hall, Englewood Cliffs, New Jersey.
Van Holde, K. E., and Baldwin, R. L. (1958). *J. Phys. Chem.* **62,** 734.
Van Holde, K. E., and Rossetti, G. P. (1967). *Biochemistry* **6,** 2189.
Van Holde, K. E., Rossetti, G. P., and Dyson, R. D. (1969). *Ann. N. Y. Acad. Sci.* **164,** 279.
Van Rapenbusch, R., and Deschepper, J. C. (1966). *C. R. Acad. Sci.* **262,** 1365.
Vinograd, J., and Bruner, R. (1966). *Biopolymers* **4,** 131 and 157.
Vinograd, J., Bruner, R., Kent, R., and Weigle, J. (1963). *Proc. Nat. Acad. Sci. U. S.* **49,** 902.
Wales, M., and Van Holde, K. E. (1954). *J. Polym. Sci.* **14,** 81.
Wampler, D. E. (1971). *Anal. Biochem.* **44,** 528.
Weaver, W. (1926). *Phys. Rev.* **23,** 499.
Williams, J. W., Van Holde, K. E., Baldwin, R. L., and Fujita, H. (1958). *Chem. Rev.* **58,** 715.
Williams, R. C. (1972). *Anal. Biochem.* **48,** 164.
Yphantis, D. A. (1964). *Biochemistry* **3,** 297.
Zwanzig, R. (1966). *J. Phys. Chem.* **45,** 1858.

5

Quaternary Structure of Proteins

IRVING M. KLOTZ, DENNIS W. DARNALL, AND NEAL R. LANGERMAN

I. INTRODUCTION

Protein interactions, in their widest manifestations, provide the molecular mechanism of operation and control of cellular reactions as well as the basis of many aspects of the macroscopic functions of tissues and organs. These interactions may be viewed as an interlocked set of discrete types of macromolecular dynamic rearrangements such as are illustrated schematically in Fig. 1. Two major classes of *intrinsic protein interactions* are represented at the top of Fig. 1, those within an individual subunit, and those between the subunits in an ensemble that constitutes the quaternary structure. There are also many types of *extrinsic protein interactions,* of which Fig. 1 presents only one—formation of complexes between proteins and small molecules. Small-molecule binding has been singled out because it usually provides the pivotal coupling device for interlinking the macromolecular rearrangements within and between the two classes of intrinsic interactions.

This chapter will focus primarily on those interactions involving subunits within a quaternary ensemble, but the others will be considered in passing, particularly when they express themselves in the functional aspects of quaternary structure.

That proteins generally might be constituted of subunits seems to have been first clearly recognized by Svedberg (Svedberg and Fahraeus,

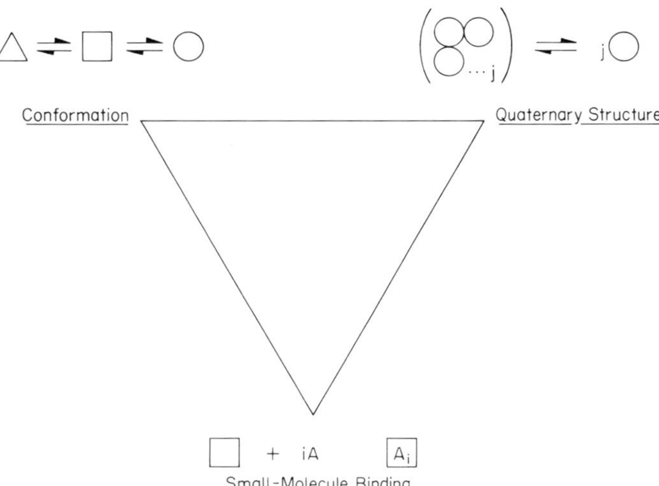

Fig. 1 Some types of protein interactions and their interlinkages.

1926; Svedberg, 1929; Svedberg and Heyroth, 1929; Svedberg and Hedenius, 1934; Svedberg and Pedersen, 1940). This became apparent when the ultracentrifuge revealed that the molecular weights of many proteins, particularly oxygen carriers of marine origin, varied (in integral relationships) with changes in aqueous solvent, changes that did not cleave covalent bonds. As a simplifying principle of protein architecture —that all proteins must be constituted of subunits from a limited number of molecular weight classes—the Svedberg proposal has not survived in its original form. As the precision of molecular weight measurements increased and as the details of primary and tertiary structure were unraveled, it became clear that the polypeptide unit building blocks did not fall into a few classes but rather had a wide spectrum of molecular weights. Nevertheless, in its less restrictive form—that large proteins are built up from smaller entities noncovalently linked—the Svedberg perception has proven valid.

To our knowledge the term *quaternary* structure was first used by Bernal (1958) as a supplement to the terms *primary, secondary,* and *tertiary* structure (Linderstrøm-Lang, 1952; see also Wetlaufer, 1961). As used and illustrated by Bernal, quaternary structure refers to a macromolecular system built up from *noncovalently* linked subordinate entities which we shall call *subunits.* According to this definition,[1] a given protein may have several different kinds of subunit. The smallest subunit that can be released from a quaternary structure without cleavage

[1] Some investigators also call constituent polypeptide chains which are covalently linked to each other by S—S bonds "subunits" of the complete protein and consider the complete macromolecule a "quaternary structure." For example, antibody proteins, IgG, are constituted of two heavy chains and two light chains covalently cross-linked (Smith *et al.,* 1971). These individual chains, having molecular weights near 50,000 and 25,000, respectively, would be called subunits of the quaternary structure (MW about 160,000). The A and B chains of insulin could also be considered subunits on this basis. From this viewpoint, consistency would require that the three chains of chymotrypsin, which are linked by S—S bonds, also be considered subunits of the quaternary structure (MW 26,000). We believe that in the interest of clarity we should adhere to Bernal's definition of a quaternary structure as one constituted of subordinate entities that interact *noncovalently* with each other. These noncovalently linked subordinate entities we shall call subunits. In contrast, polypeptide chains that are covalently linked by S—S bonds or by the ϵ-(γ-Glu)—Lys bonds that have become increasingly apparent (Williams-Ashman *et al.,* 1972) might be called *constituent* units of a tertiary structure. A quaternary structure (in contrast to a tertiary structure) can be dissociated into its subordinate entities by denaturants and other reagents that do not disrupt covalent bonds and, in principle if not always in practice, by continuous dilution. A quaternary structure (in contrast to a tertiary structure) can be hybridized *in vitro,* in principle if not always in practice, by exposure to conditions that do not disrupt covalent bonds.

of covalent bonds is called the *monomer*. Monomers of different proteins may differ quite markedly. For example, insulin (MW 11,500) contains two identical monomers with a molecular weight of 5700 each. Aldolase is constituted of four identical monomers with a molecular weight of 40,000 each. A protein may also be constituted of two different types of monomer. For example, fully dissociated aspartate transcarbamylase releases six identical catalytic (C) monomers (MW 33,000 each) and six identical regulatory (R) monomers (MW 17,000 each): each type of monomer has a different amino acid sequence. A monomer of a given protein may be a single polypeptide chain or may contain two or more chains linked by covalent bonds [e.g., S—S or ϵ-(γ-Glu)—Lys]. Thus the monomer of aldolase is a single chain with no interchain covalent bonds. In contrast, each monomer of insulin contains two polypeptide chains (designated A and B) that are connected by disulfide bonds.

It is also sometimes convenient to speak of a *protomer* (Monod *et al.*, 1965) which is the minimum-size subunit that, on association with an integral number of identical subunits, will generate the quaternary structure. For aldolase, the protomer is the same as the monomer. On the other hand, for aspartate transcarbamylase, the protomer is the combination CR, containing one catalytic and one regulatory subunit; six protomers generate the quaternary structure (Gerhart and Schachman, 1968; Wiley and Lipscomb, 1968; Nelbach *et al.*, 1972; Wiley *et al.*, 1972). For a protein such as NADP-sulfite reductase (Siegel and Kamin, 1971), which can be designated by the monomer stoichiometry $\alpha_4\beta_8$, the protomer is $\alpha\beta_2$.

In some circumstances, subunits other than the monomer or protomer may also exist as definite entities. For example, among the subordinate entities observed for aspartate transcarbamylase are a catalytic subunit with a molecular weight of 100,000 (C_3) and a regulatory subunit with a molecular weight of 34,000 (R_2). In general, any associated state containing 2, 3, . . . n monomers or protomers will be referred to as an *aggregate*, an *ensemble*,[2] or an *oligomer*.[2] When the associated state contains all n monomers, it will be designated as the *quaternary structure, quaternary oligomer, quaternary aggregate*, or *quaternary ensemble*.

[2] According to Webster's *Third New International Dictionary, Unabridged* (1961), an *ensemble* is "a system of items that constitute an organic unity; a congruous whole; aggregate." When n becomes large, in principle the term oligomer should be replaced by polymer, i.e., the prefix "few" should be replaced by "many." However, we will not use the term polymer for an association of monomers, since polymer has also been used to designate the covalently linked amino acid residues in a polypeptide chain.

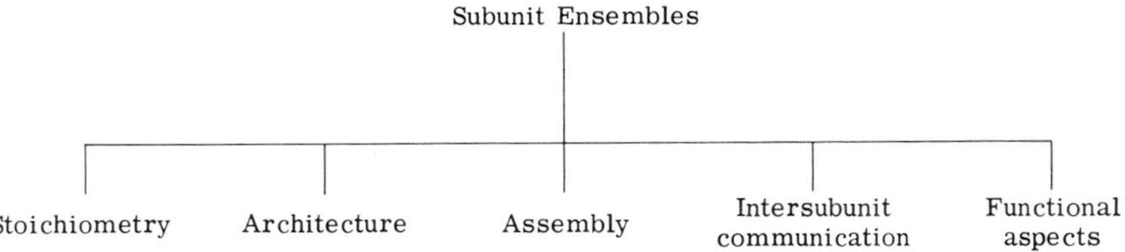

Fig. 2 Aspects of quaternary structure and function.

The specific aspects of quaternary structure that will be examined in this chapter are outlined in the chart in Fig. 2. We shall first review the methods used to determine the number of subunits in a quaternary structure and summarize the data on stoichiometry available in the literature. The geometric arrangements of these subunits, i.e., the architecture of the ensemble, will then be described, with particular emphasis on the types of symmetry that may be found. The assembly of subunits will then be analyzed from the point of view of the energetics of subunit interactions. This will be followed by a description of the molecular nature of interface contacts between subunits insofar as they have been definitively established. Finally, all of these features will provide a background for an examination of the functional aspects of quaternary structure in terms of subunit interactions.

II. STOICHIOMETRIC CONSTITUTION

The first piece of structural information needed to establish the quaternary structure of a protein is the number of subunits in the ensemble. Of the many procedures used for this purpose, some are highly specific for individual or limited types of proteins and some are very general, at least in principle.

Ideally, several procedures should be utilized to determine the number of subunits in an oligomer. Each individual method has its pitfalls, and only by combining results from several methods can the stoichiometry be established unequivocally. In addition, the material itself may be a source of deception. For instance, traces of proteolytic enzymes in supposedly pure protein preparations can cause limited proteolysis of particularly susceptible peptide bonds, leading to low apparent molecular weights for the subunits. Trace amounts of proteolytic enzymes have resulted in erroneous conclusions regarding the subunit structure of hexokinase (Pringle, 1970) and cytochrome b_2 (Labeyrie and Baudras, 1972; Jacq and Lederer, 1972).

TABLE I

Ligand-Promoted Association or Dissociation of Proteins

Protein	Ligand	Ligand promotes	Reference
Pyruvate decarboxylase	Thiamine pyrophosphate, Mg^{2+}	Association	Gounaris *et al.* (1971)
Cytidine triphosphate synthetase	ATP or UTP	Association	Levitzki and Koshland (1972)
DDT-dehydrochlorinase	DDT	Association	Dinamarca *et al.* (1971)
Fumarase	Malate, phosphate, citrate or ATP	Association	Teipel and Hill (1971)
N^{10}-Formyltetrahydrofolate synthetase	Monovalent cations or tetrahydropteroyl triglutamate	Association	MacKenzie and Rabinowitz (1971), Curthoys *et al.* (1972)
Glucose-6-phosphate dehydrogenase	NADPH NADP	Dissociation Association	Bonsignore *et al.* (1971), Scott (1971)
Chorismate mutase-prephenate dehydratase	Phenylalanine	Association	Schmit and Zalkin (1971)
Fatty acid synthetase complex	NADPH	Association	Kumar and Porter (1971)
α-Isopropylmalate synthase	Leucine Acetyl-CoA and bromopyruvate	Dissociation Association	Leary and Kohlhaw (1972)
Tryptophan synthetase (B component)	Pyridoxal phosphate	Association	Hathaway (1972)
Tryptophanase	Pyridoxal phosphate Monovalent cations	Association Dissociation	Hoch and DeMoss (1972), Morino and Snell (1967a)
Phosphoenolpyruvate carboxytransphosphorylase	Phosphoenol pyruvate, phosphate and carbon dioxide	Dissociation	Haberland *et al.* (1972)
Hexokinase	Glucose	Dissociation	Gazith *et al.* (1968)
Glyceraldehyde-3-phosphate dehydrogenase	ATP or cyclic AMP	Dissociation	Stancel and Deal (1968), Yang and Deal (1969), Stancel (1969)
Isocitrate dehydrogenase	Citrate or isocitrate	Dissociation	LeJohn *et al.* (1969)

TABLE I (*Continued*)

Protein	Ligand	Ligand promotes	Reference
Arginine decarboxylase	Monovalent cations	Dissociation	Boeker and Snell (1968)
Hemerythrin	Monovalent anions	Dissociation	Klapper and Klotz (1968)
δ-Aminolevulinate dehydratase	Potassium ions	Association	van Heyningen and Shemin (1971)
Homoserine dehydrogenase	Threonine, isoleucine, methionine, divalent cations or monovalent anions	Association	Mankovitz and Segal (1969)
Threonine dehydrase	AMP	Association	Whanger *et al.* (1968)
Alkaline phosphatase	Zinc ions	Association	Reynolds and Schlesinger (1969)
α-Amylase	Zinc ions	Association	Isemura and Kakiuchi (1962), Stein and Fischer (1960)
L-Aspartate β-decarboxylase	Pyridoxal phosphate	Association	Bowers *et al.* (1970), Tate and Meister (1970)
Carbamyl phosphate synthetase	IMP or ornithine UDP	Association Dissociation	Anderson and Marvin (1970)
Phosphorylase	AMP	Association or dissociation	Wang *et al.* (1970), Kastenschmidt *et al.* (1968), Helmreich *et al.* (1967)
Deoxycytidylate deaminase	dCTP dTTP	Association Dissociation	Maley and Maley (1968)
Acetyl-CoA carboxylase	Citrate, isocitrate, malonate, phosphate or acetyl CoA	Association	Gregolin *et al.* (1968)
Acetyl-CoA carboxylase (carboxylated form)	Citrate or isocitrate	Dissociation	Gregolin *et al.* (1968)
D-Amino acid oxidase	FAD	Association	Antonini *et al.* (1966)
Aspartokinase	Threonine	Association	Wampler and Westhead (1968)
Glutaminase	Phosphate	Association	Olsen *et al.* (1970b)
Glutamate dehydrogenase	NADPH, GTP ADP or ATP	Dissociation Association	Frieden (1963)

(*Continued*)

TABLE I (*Continued*)

Protein	Ligand	Ligand promotes	Reference
Deoxythymidine kinase	dTTP, dCDP, dCTP, or dADP	Association	Iwatsuki and Okazaki (1967)
α-Acetylgalactos-aminidase	*N*-Acetylgalactosaminide	Association	Wang and Weissmann (1971)
Phycocyanin	Monovalent anions	Dissociation	MacColl *et al.* (1971a)
Arginase	Manganese ions	Association	Carvajal *et al.* (1971)
Tyrosinase	Calcium ions	Association	Jolley *et al.* (1969)
Glutaminase	Phosphate	Association	Kvamme *et al.* (1970)
	Glutamate, citrate, or malonate	Dissociation	
Neurophysin	Calcium ions	Association	Burford *et al.* (1971)
Hemocyanin	Calcium ions	Association	Morimoto and Kegeles (1971)
Tubulin	Magnesium or calcium ions; vinblastine	Association	Weisenberg and Timasheff (1970)

A. Dissociation by Specific Substances

1. Metabolites, Cofactors or Inorganic Ions

In many instances, binding of a metabolite is followed by dissociation of an enzyme. Classic examples include the dissociation of yeast hexokinase by glucose (Gazith *et al.*, 1968), of glyceraldehyde-phosphate dehydrogenase by adenosine triphosphate or by 3′,5′-cyclic adenosine monophosphate (Stancel and Deal, 1968; Yang and Deal, 1969; Stancel, 1969), of isocitrate dehydrogenase by citrate or isocitrate (LeJohn *et al.*, 1969), and of deoxycytidylate deaminase by deoxythymidine triphosphate (Maley and Maley, 1968). A comprehensive list of enzymes dissociated by metabolites and cofactors is given in Table I.

In all of these circumstances, binding perturbs an association-dissociation equilibrium between the subunit ensemble, P_n, and its stepwise-dissociated species:

$$P_n \leftrightarrows P_{n-1} + P \leftrightarrows P_{n-2} + 2P \leftrightarrows \cdots \leftrightarrows {}_nP \tag{1}$$

If the addition of a metabolite shifts the equilibrium in Eq. (1) toward the formation of monomers, it follows that the metabolite combines more strongly with the monomer (P) than with the oligomer (P_n). Conversely, if the metabolite shifts the equilibrium toward the oligomer, then one would conclude that the metabolite is bound more strongly by the oligomer than by the monomer. Representative examples of the latter are the binding of tetrahydropteroyl triglutamate by formyltetrahydrofolate synthetase (Curthoys *et al.*, 1972) and the binding of adenosine triphosphate by cytidine triphosphate synthetase (Levitzki and Koshland, 1972). Other examples of the ligand-promoted association of subunits are listed in Table I.

The binding of small inorganic ions may also lead to the dissociation of specific oligomers. Monovalent cations facilitate the dissociation of tryptophanase (Morino and Snell, 1967a) and of arginine decarboxylase (Boeker and Snell, 1968); monovalent anions promote the dissociation of hemerythrin (Klapper and Klotz, 1968) and of phycocyanin (MacColl *et al.*, 1971a). The opposite effect is also possible (see Table I). Thus potassium ions favor the formation of oligomers of δ-aminolevulinate dehydratase (van Heyningen and Shemin, 1971), zinc ions favor the formation of oligomers of alkaline phosphatase (Reynolds and Schlesinger, 1969) and of α-amylase (Stein and Fischer, 1960; Isemura and Kakiuchi, 1962), and manganese ions have a similar effect on arginase (Carvajal *et al.*, 1971).

2. *Sulfhydryl Reagents*

A wide selection of proteins can be dissociated by reagents that block sulfhydryl groups. Examples are phosphorylase (Madsen and Cori, 1956; Chignell *et al.*, 1968), hemerythrin (Keresztes-Nagy and Klotz, 1963), pyruvate carboxylase (Palacian and Neet, 1969), fetal and adult hemoglobin (Ioppolo *et al.*, 1969), aspartate transcarbamylase (Gerhart and Schachman, 1965, 1968), hexokinase (Lazarus *et al.*, 1968), glyceraldehyde-phosphate dehydrogenase (Smith and Schachman, 1971), formyltetrahydrofolate synthetase (Nowak and Himes, 1971), and RNA polymerase (Ishihama, 1972). The most commonly used sulfhydryl reagents are *p*-mercuribenzoate, salyrganic acid, *N*-ethylmaleimide, and iodoacetate.

B. Addition of Denaturants

Whatever forces may be involved in subunit interactions, if they are perturbed by urea, guanidine hydrochloride, or sodium dodecyl sulfate, the dissociation is ascribed to denaturation. Urea causes dissociation of pyruvate kinase (Cottam *et al.*, 1969), carbamyl phosphate synthetase (Trotta *et al.*, 1971), glutamine synthetase (Wilk *et al.*, 1969), malate-lactate transhydrogenase (Allen and Patil, 1972), and *Bacillus alvei* tryptophanase (Hoch and DeMoss, 1972). Guanidine hydrochloride causes disaggregation of formyltetrahydrofolate synthetase (Welch *et al.*,

1971), α-L-fucosidase (Carlsen and Pierce, 1972), pyruvate kinase (Kuczenski and Suelter, 1970), ornithine transcarbamylase (Marshall and Cohen, 1972), and α-isopropylmalate synthase. Sodium dodecyl sulfate dissociates a wide selection of proteins into subunits including tryptophanase (Morino and Snell, 1967b), fructose diphosphatase (Sia *et al.*, 1969), and glutamate dehydrogenase (Page and Godin, 1969). Particularly of interest is the recent widespread use of sodium dodecyl sulfate in combination with a mercaptan for the reduction of covalent disulfide linkages between polypeptide chains (Shapiro *et al.*, 1967; Weber and Osborn, 1969). This combination of detergent and reducing agent seems to be a reliable method for insuring the release of single polypeptide chains, whose molecular weights can then be estimated by gel electrophoresis in the presence of detergent and mercaptan (see Ch. 4). Sodium dodecyl sulfate gel electrophoresis has recently been used to determine the subunit molecular weights of isocitrate dehydrogenase (Barnes *et al.*, 1971), α-isopropylmalate synthetase (Leary and Kohlhaw, 1972), α-L-fucosidase (Carlsen and Pierce, 1972), *N*-formimino-L-glutamate iminohydrolase (Wickner and Tabor, 1972), 6-phosphogluconate dehydrogenase (Proscal and Holten, 1972), aspartokinase 1-homoserine dehydrogenase (Clark and Ogilvie, 1972), phosphotransacetylase (Whiteley and Pelroy, 1972), ornithine transcarbamylase (Marshall and Cohen, 1972), and glutamine transaminase (Cooper and Meister, 1972).

C. Acylation with Dicarboxylic Anhydrides

Various dicarboxylic acid anhydrides have been used to dissociate proteins. Dicarboxylic anhydride reacts with protein amino groups and places two negative charges per acylated amino group on the modified protein. In proteins composed of subunits, the increased electrostatic repulsion may cause the aggregate to dissociate into its constituent chains. Succinic anhydride has been used to dissociate hemerythrin (Klotz and Keresztes-Nagy, 1962), glyceraldehyde-3-phosphate dehydrogenase (Jaenicke *et al.*, 1968), hemocyanin (Konings *et al.*, 1969), malate–lactate transhydrogenase (Allen and Patil, 1972), phosphofructokinase (Uyeda, 1969), carbamyl phosphate synthetase (Trotta *et al.*, 1971), and luciferase (Meighen *et al.*, 1971). Maleic anhydride has been successfully used to dissociate aldolase, transaldolase, and fructose diphosphatase (Sia and Horecker, 1968), glutamine synthetase (Wilk *et al.*, 1969), pyruvate kinase (Kuczenski and Suelter, 1970), carbamyl phosphate synthetase (Trotta *et al.*, 1971), and phosphofructokinase

(Uyeda, 1969). Dissociation with maleic anhydride has an added advantage over succinic anhydride in that the reaction is reversible under certain conditions. The maleyl half-amide can be released at pH 3.5 to regenerate free amino groups (Butler *et al.*, 1969), but the reaction is often inconveniently slow. A related anhydride, citraconic anhydride (Dixon and Perham, 1969), forms an adduct which is more acid labile. The half-amide derived from the reaction of the protein with tetrafluorosuccinic anhydride may be hydrolyzed readily at pH 9.5 (Braunitzer *et al.*, 1968).

D. Cross-Linking with Dimethyl Suberimidate

The number of polypeptide chains in a native protein can frequently be determined by means of a reaction with the cross-linking reagent dimethyl suberimidate (Davies and Stark, 1970). Imido esters are highly specific reagents that modify amino groups in proteins. They react easily under mild conditions and produce a species of unaltered charge. Bifunctional dimethyl suberimidate may modify amino groups in the following manner:

$$\begin{matrix} \overset{\oplus}{NH_2} \\ \| \\ C-OCH_3 \\ | \\ (CH_2)_6 \\ | \\ C-OCH_3 \\ \| \\ \underset{\oplus}{NH_2} \end{matrix} \quad + \quad P\begin{matrix} \diagup NH_2 \\ \diagdown NH_2 \end{matrix} \longrightarrow P\begin{matrix} \diagup HN-C \\ \\ \diagdown HN-C \end{matrix}\!\!\!\begin{matrix} \overset{\oplus}{NH_2} \\ \| \\ \cdot \\ | \\ (CH_2)_6 \\ | \\ \cdot \\ \| \\ \underset{\oplus}{NH_2} \end{matrix} \quad + \quad 2\,CH_3OH \qquad (2)$$

With an oligomeric protein this reaction may yield the following derivatives: (1) cross-links between oligomers (under appropriate conditions this may be reduced to insignificance); (2) cross-links within each chain of the protomer; (3) an adduct wherein one imido ester group remains intact; or, (4) cross-links between protomers of the oligomeric protein. It is the last derivative that is useful in determining the number of polypeptide chains in a native protein. Under denaturing conditions, the species with cross-links between protomers will have molecular weights different from the monomer, and if the protein contains identical subunits, these molecular weights should all be multiples of the monomeric molecular weight. Thus polyacrylamide gel electrophoresis of these cross-linked proteins in sodium dodecyl sulfate (Weber and Osborn, 1969) should give a set of bands corresponding to molecular weights equal to integral multiples of the protomer molecular weight.

The number of bands seen should also correspond to the number of subunits in the native protein. A complication in subunit counting may arise when oligomers are cross-linked. However, this possibility can be checked by cross-linking at a lower protein concentration and applying an equivalent amount of protein in the electrophoretic analysis. There should be a reduction in the amount of higher molecular weight species without a corresponding decrease in the amount of species having molecular weights equal to or lower than that of the oligomer. Cross-linking between oligomers becomes less probable at lower protein concentrations, whereas cross-linking within an oligomer is unaffected.

Davies and Stark (1970) have shown that the cross-linking method yields the expected number of subunits for aldolase, glyceraldehyde-3-phosphate dehydrogenase, arabinose isomerase, and tryptophan synthetase B protein. Cross-linking of aspartate transcarbamylase, a protein composed of nonidentical subunits, produced most of the eighteen bands expected from a protein with six catalytic subunits (MW 33,000) and six regulatory subunits (MW 17,000).

Since Davies and Stark's (1970) original report of the cross-linking method, several other subunit proteins have been examined using dimethyl suberimidate. Experiments with α-isopropylmalate synthetase (Kohlhaw and Boatman, 1971; Leary and Kohlhaw, 1972) revealed the expected four protein species. The subunit structures of phosphotransacetylase (Whiteley and Pelroy, 1972), 17β-hydroxysteroid dehydrogenase (Jarabak and Street, 1971), human serum high density lipoprotein (Scanu *et al.*, 1972), and carbamyl phosphate synthetase (Trotta *et al.*, 1971) have also been established using cross-linking experiments.

E. Peptide Mapping

Very often the number of peptides released by tryptic hydrolysis can yield information concerning the number of subunits in a protein. From a single polypeptide chain, the number of peptides released by trypsin should be given by Eq. (3) unless lysine or arginine are the carboxyl-terminal amino acid residues:

$$\text{Number of peptides} = L + A + 1 \tag{3}$$

Here L and A are the number of lysine and arginine residues, respectively, per mole of oligomer. If, on the other hand, a protein is composed of N identical subunits per oligomer, the number of tryptic peptides should be given by Eq. (4):

$$\text{Number of peptides} = \frac{L + A + N}{N} \tag{4}$$

In turn, if the oligomer contains nonidentical subunits, the number of tryptic peptides should be given by Eq. (5):

$$\text{Number of peptides} = L + A + N \tag{5}$$

Other situations, e.g., two pairs of identical subunits (an $\alpha_2\beta_2$ tetramer), would yield values between the extremes of Eqs. (4) and (5), depending upon the number of lysine and arginine residues in each subunit.

Peptide counting has given insight into the subunit constitution of a variety of proteins. The number of tryptic peptides observed is consistent with a decameric structure for arginine decarboxylase (Boeker *et al.*, 1969), with a hexameric structure for glutamate decarboxylase (Strausbauch and Fischer, 1970), with a tetrameric structure for DDT-dehydrochlorinase (Dinamarca *et al.*, 1971), prealbumin (Gonzales and Offord, 1971), formyltetrahydrofolate synthetase (MacKenzie and Rabinowitz, 1971) and β-galactosidase (Fowler and Zabin, 1970; Steers *et al.*, 1965), and with a dimeric structure for ATP-creatine transphosphorylase (Yue *et al.*, 1967) and enolase (Brewer *et al.*, 1970). Mann and Vestling (1970) have presented evidence from tryptic digests that mitochondrial malate dehydrogenase is a dimer composed of nonidentical subunits.

Occasionally the number of tryptic peptides observed may be smaller than the number expected. On the basis of its amino acid composition, *Bacillus alvei* tryptophanase, a tetramer having a molecular weight of 208,000 and four identical subunits, would be expected to yield fifty-one tryptic peptides; in fact, only twenty-two ninhydrin-positive spots have been observed on the peptide map (Hoch and DeMoss, 1972). Several possible sources of error could account for the reduced number of peptides: (1) there are trypsin-resistant bonds (e.g., Lys-Pro or Arg-Pro) in the protein, (2) several peptides remain at the origin on the peptide map or there are several overlapping peptides comprising one spot, or (3) the theoretical number of peptides does not include the possibility of identical peptide fragments from different segments of the monomer. In fact this last possibility seems to occur in tryptophanase, for Morino and Snell (1967b) have shown that the *E. coli* tryptophanase monomer is composed of two nearly identical polypeptide chains covalently bound by disulfide bonds.

More than the expected number of tryptic peptides may also be observed. For the turnip yellow mosaic virus subunit, sixteen peptides were found (Harris and Hindley, 1965) instead of the eleven expected. Two of the five additional peptides were due to incomplete tryptic hydrolysis, producing two nonunique peptides; the other three peptides arose from hydrolytic cleavage by contaminating amounts of chymotrypsin in the sample of trypsin.

Thus tryptic peptide maps must be interpreted with a certain amount of caution. Nevertheless, the essential point remains: a protein composed of identical subunits will yield a peptide map with less than the number of peptides predicted on the basis of the oligomer molecular weight and amino acid composition.

The peptide mapping technique is not restricted to enzymatic hydrolysis. Cyanogen bromide, which cleaves the peptide chain at methionine residues, has also been used with some success to establish the subunit constitution of enolase (Brewer *et al.*, 1970) and β-galactosidase (Fowler and Zabin, 1970; Steers *et al.*, 1965).

F. Dilution

The elementary chemical principle of Le Chatelier, applied to an association–dissociation equilibrium such as that represented by Eq. (1), compels the equilibrium to go toward the monomer upon dilution of the protein. This is a requirement for every subunit ensemble. In practice, however, the dilution needed to achieve appreciable equilibrium dissociation, even if the rate is fast, may be so great as to place the concentration outside of the experimentally feasible range. Proteins that dissociate into subunits in dilute solutions are tryptophanase (Hoch and DeMoss, 1972), hemerythrin (Langerman and Klotz, 1969; Klapper *et al.*, 1966), enolase (Keresztes-Nagy and Orman, 1971), arginine decarboxylase (Boeker and Snell, 1968), glyceraldehyde-3-phosphate dehydrogenase (Hoagland and Teller, 1969), glutamate decarboxylase (Strausbauch and Fischer, 1970), and pig liver esterase (Barker and Jencks, 1969).

G. Hybridization

Upon mixing variants of the same protein, one may observe the formation of one or more new quaternary species. This phenomenon, called *hybridization,* occurs with proteins composed of two or more noncovalently linked polypeptide chains. It can be detected when the hybrid species differ from one another in some property that can serve as a basis of separation. The most convenient technique for separation is electrophoresis on a solid support, which is simple, rapid in operation, and requires only a small amount of protein. Thus a difference in net charge between variant proteins is the property most often exploited in hybridization experiments.

For successful hybridization experiments, the variant proteins must have appreciably different electrophoretic mobilities. The variants often used are isozymes isolated from natural sources. There are proteins, however, for which such variants are not readily available. In these cases, a charge alteration may be artificially introduced by chemical means. Protein modification with succinic anhydride can provide the necessary chemically modified species (Keresztes-Nagy *et al.*, 1965).

Hybridization procedures have been used to determine whether a protein oligomer exists in equilibrium with its subunits and to establish the number of subunits in the aggregate.

1. *Association–Dissociation Equilibrium*

Hybridization experiments with naturally occurring variant hemoglobins have demonstrated that the tetramer, $\alpha_2\beta_2$, dissociates into asymmetrical dimers, $2\alpha\beta$, rather than into two symmetrical pairs, α_2 and β_2 (Singer and Itano, 1959). Similar experiments provided the first evidence that the mutation in hemoglobin S is in the β chain (Vinograd *et al.*, 1959) and that hemoglobin H is a tetramer, $\beta_4{}^{\mathrm{A}}$, of normal adult β subunits. More recently, asymmetric hybrids of human hemoglobins such as $\alpha^{\mathrm{A}}\alpha^{\mathrm{I}}\beta^{\mathrm{A}}\beta$ Richmond have also been found (Efremov *et al.*, 1969). In some circumstances it may be possible to estimate the equilibrium constant for a hybridization reaction. Thus for

$$2\alpha_2\gamma_2 + \beta_4 = 2\alpha_2\beta_2 + \gamma_4 \tag{6}$$

the equilibrium constant is of the order of 10 to 20 (Jones and Schroeder, 1963).

Association–dissociation equilibria may also be examined in hybridization experiments with chemically modified proteins. In such studies there is no need for a homogeneous population of chemically modified species. Succinylated hemerythrin, prepared without any attempt to obtain a homogeneous protein, was used to determine whether near neutrality the octamer is in a mobile equilibrium with its subunits (Keresztes-Nagy *et al.*, 1965). Electrophoresis on starch gels of a mixture of succinylated and native protein produced a diffuse band with a range of mobilities extending from the low value of the native protein to the very high value of the succinylated octamer.

In general, in any hybridization experiment the formation of new species or an alteration in electrophoretic mobilities following the mixing indicates that the rate of equilibrium is comparable to, or faster than, the experimental separation time. On the other hand, the absence of changes in electrophoretic mobilities is not conclusive evidence of a slow equilibrium.

In some cases it may be possible to measure the rate of dissociation of an oligomer into its subunits by means of hybridization techniques. If two similar but distinguishable variants of an oligomeric protein are available and if their subunits are reversibly interchangeable, then the rate at which hybrids are formed when the two species are mixed under dissociating conditions may provide a measure of the rate of dissociation of the protein. Meighen and Schachman (1970b) have shown that an electrophoretic variant of glyceraldehyde-3-phosphate dehydrogenase (GPD) can be obtained by succinylating native GPD with succinic anhydride. This modified species forms electrophoretically distinct hybrids with native GPD in the presence of urea, which provides a dissociating environment. Smith and Schachman (1971) showed, furthermore, that *p*-mercuribenzoate causes dissociation of GPD and that its effect is reversed by dithiothreitol. These authors therefore attempted to measure the rate of this dissociation by adding the mercurial to a mixture of native and succinylated enzyme molecules and then by introducing dithiothreitol at subsequent intervals of time in order to effect reconstitution and production of hybrid sets. Subsequent electrophoretic separation of the hybrids from the native and succinylated components and measurement of the amount of hybrid formed as a function of time between addition of mercurial and dithiothreitol gave a measure of the rate of dissociation of GPD by the mercurial. Preliminary results of these experiments showed that the dissociation of tetrameric GPD to dimers was inappreciable in five seconds, but that considerable amounts of the dimer were formed in one minute. As judged from the electrophoretic pattern, no monomers were formed during this time.

2. *Determination of Number of Subunits*

***a.* Statistics of Hybridization.** When each of the variants contains identical subunits, it is easy to show that the number of possible hybrids, N, is one more than the number of subunits s, in the oligomer. For example, for a dimeric protein ($s = 2$), hybridization of AA with the variant A′A′ will yield three different species: AA, A′A, and A′A′. (Note that A′A and AA′ would have the same physical properties and would be indistinguishable.) Similarly for any s-mer, since the physical properties are the same no matter where an A′ takes the place of an A, the hybrids will consist of the species A_s, $A_{s-1}A'$, $A_{s-2}A_2'$, . . . , A_s', that is, there will be $s + 1$ of them. Hence we may write (Shaw, 1964)

$$N = s + 1 \tag{7}$$

If the subunits are not identical even though they are structurally and functionally equivalent, then the number of hybrids that can be obtained by mixing two forms is larger than that specified by Eq. (7), and is given by

$$N = \frac{(m + s - 1)!}{s!(m - 1)!} \tag{8}$$

where m is the total number of different types of subunits (Shaw, 1964; Markert and Whitt, 1968).

The situation becomes even more complicated when the protein consists of two (or more) ensembles of subunits, each ensemble being structurally and functionally distinct from the others. An example of such a system is the enzyme aspartate transcarbamylase (Gerhart and Schachman, 1965), which contains catalytic and regulatory subunit ensembles. In essence, for each ensemble, Eq. (8) is applicable. Hence in the general case, for the entire system of j ensembles

$$N = \prod_j \frac{(m_j + s_j - 1)!}{s_j!(m_j - 1)!} \tag{9}$$

***b.* Limitations of Methodology.** Some caution is necessary in interpreting hybridization patterns. Too few protein bands may be counted if the resolution of the protein species is inadequate, or if the concentration of one or more species is very low. This problem may be alleviated somewhat by using different initial ratios of the protein variants since the amount of each randomly formed hybrid depends upon the initial concentration. Under certain conditions, Eqs. (7)–(9) will not apply, and erroneous interpretations may result. For example, hybridization of hemoglobin variants does not yield the expected number of new species (Itano and Singer, 1958). Such a result could arise from a nonrandom substitution of the chains in each subunit class. It is also essential that the association–dissociation equilibrium of the aggregate be slow relative to the time in which separation of hybrids occurs. When the association–dissociation rates are rapid enough, the number of bands formed after hybridization will not be described by Eqs. (7)–(9), but will instead be dictated by the equilibrium dynamics and the transport properties of the system (Gilbert and Jenkins, 1959).

Furthermore, for Eqs. (7)–(9) to be applicable, there must be complete equilibration between chain variants. Thus the mixture of the variants generally must be reversibly denatured. Hence successful hybridization depends also upon finding conditions of reversible denaturation. Among the techniques that have been used for reversible denaturation are lowering the pH (Itano and Singer, 1958), freezing and thawing in concentrated salt (Markert, 1963), and adding denaturants such as urea (Chilson *et al.*, 1964, 1965; Meighen and Schachman, 1970a,b).

When a chemically modified variant is to be used, additional problems may be encountered. Criteria for a satisfactory chemical modification have been enumerated by Meighen and Schachman (1970a). The modification should yield a homogeneous population of protein molecules, and, therefore, should be fairly specific. The modified protein should have an electrophoretic mobility substantially different from that of the native protein. The quaternary structure of the modified and native protein must be identical. Finally, it must be possible to reconstitute the altered enzyme after dissociation.

These criteria are met by protein modification with succinic anhydride (Keresztes-Nagy *et al.*, 1965). Substitution at amino groups occurs preferentially, and a charge difference of close to -2 is introduced per succinyl residue. Although extensive succinylation may lead to protein dissociation, conditions for partial succinylation with no disaggregation have been found for at least five proteins: hemerythrin (Keresztes-Nagy *et al.*, 1965), aldolase (Meighen and Schachman, 1970a), glyceraldehyde-3-phosphate dehydrogenase (Meighen and Schachman, 1970b), aspartyl transcarbamylase (Meighen *et al.*, 1970), and bacterial luciferase (Meighen *et al.*, 1971).

Partial succinylation may, however, yield a nonhomogeneous preparation. If a protein contains i different classes of reactive amino groups, each of which reacts with a characteristic rate, and if the reaction of groups within each class is random, then the distribution of protein molecules with varying amounts of attached succinyl residues is characterized by the standard deviation

$$\sigma = \sum_i (n_i f_i (1 - f_i))^{1/2} \tag{10}$$

where n_i is the total number of amino groups in each class, and f_i the fraction of reacted groups in each class (Meighen and Schachman, 1970a). Inspection of Eq. (10) reveals that the greatest homogeneity will be found at high and low values of f, i.e., when very few or almost all of the residues are modified. The greatest inhomogeneity will occur at intermediate values of f. If extensive succinylation results in dissociation, preparations with high values of f are of no use. On the other hand, succinylation of a small fraction of the residues introduces only a small charge alteration, leading to less resolution.

c. **Procedures.** Some proteins are appreciably dissociated and will, therefore, form hybrids even under essentially physiological conditions. With hemerythrin, for example, a 1:1 mixture of succinylated and native octamers was stored for 24 hr (pH 8) at 4°. Control solutions were also prepared containing native protein alone and succinylated protein alone.

Aliquots from each of these three solutions were placed on starch gels and exposed to the same electrophoretic conditions. The 1:1 mixture showed a diffuse but strong band which spanned a wide range of mobilities from that of the native protein to that of the succinylated octamer (Keresztes-Nagy *et al.*, 1965). Clearly the mixture contained a population of hybrids with a wide range of electrical charges.

Hemoglobin hybridization is usually achieved by exposure of the variant proteins to dilute acid (or base). For example, a 1:1 mixture of hemoglobin S and radioactively labeled hemoglobin A was dialyzed for 48 hr at 3° against 0.1 *M* sodium acetate buffer (pH 5). Chromatographic separation yielded radioactive hemoglobin S (Vinograd *et al.*, 1959). It is also possible to add to a mixture of hemoglobins in 0.2 *M* NaCl an equal volume of acetate buffer of 0.2 ionic strength (pH 4.7). Hybridization was observed when such a solution was allowed to stand for only 20 min at 23° (Singer and Itano, 1959).

A typical example of the use of denaturation by urea is provided by experiments with aldolase (Meighen and Schachman, 1970a). Unmodified enzyme and succinylated aldolase were mixed (in varying molar ratios) in buffer containing 4.0 *M* urea and 0.1 *M* dithiothreitol (pH 6.50) at 4°. After 30 min, the solution was dialyzed against buffer to remove the urea. Zone electrophoresis was then carried out on cellulose acetate strips, and the proteins were stained to make their positions visible.

All of the procedures described so far are *in vitro* techniques. In order to apply these techniques, one must have at the outset samples of two electrophoretically distinguishable variants, each constituted of a single subunit type, and hybridization must take place in at least one of the *in vitro* environments described. It has been discovered recently (Lew and Roth, 1971) that hybridization may also be instituted under *in vivo* conditions. For example, the enzyme 6-phosphogluconate dehydrogenase appears in both *E. coli* and *Salmonella typhimurium* but in electrophoretically distinct forms. It is possible to create a strain of *Salmonella* which carries two copies of the structural gene for this dehydrogenase enzyme, one being the normal chromosomal copy and the other being provided by an episome from *E. coli*. Each of these genes is functional in this special strain of *Salmonella* and produces its specific form of dehydrogenase subunits. Thus a mixed pool of subunits is generated in this bacterium, and enzymatically active oligomers of each of the possible hybrids are assembled by random association *in vivo*. These can be distinguished, as usual, by gel electrophoresis.

***d.* Applications.** Specific applications of hybridization methods to ascertain the number of subunits in an oligomeric protein have been largely limited to pairs of variants, each containing identical subunits. Thus Eq. (7) is the arithmetic relationship to apply.

Natural isozymes have been used widely. With lactic dehydrogenase, the hybridization of two individually homogeneous isozymes produced five electrophoretically distinct variants and thus established the tetrameric structure of the enzyme (Markert, 1963). Reversible dissociation and reassociation of aldolase isozymes (types A and C) also generated five electrophoretic bands (Penhoet *et al.*, 1967; Lebherz and Rutter, 1969) and hence resolved long-standing contradictory conclusions from sedimentation studies in favor of a tetrameric quaternary structure for aldolase A. Hybridization of creatine kinase isozymes has been accomplished in the presence of urea (Dawson *et al.*, 1967). After removal of the urea and subsequent electrophoresis, three protein bands were detected, which is compatible with the presence of two subunits. In some situations it is possible to hybridize electrophoretically distinct isozymes from widely divergent organisms. Thus it has been possible to hybridize creatine kinase from rabbit muscle with that from chicken muscle (Dawson *et al.*, 1967); the results indicate two subunits in each enzyme. Mixing beef and rattlesnake lactate dehydrogenase led to five bands of differing electrophoretic mobility, again indicating the presence of four subunits in each enzyme (Markert, 1968).

Succinylated proteins have also been used effectively. Native aldolase mixed with succinylated aldolase (Meighen and Schachman, 1970a) gave five components on electrophoresis, a result expected from the four-chain model of the enzyme. Similar experiments with glyceraldehyde-3-phosphate dehydrogenase have revealed the presence of five components (Meighen and Schachman, 1970b), a result consistent with the presence of four subunits. Electrophoresis of a mixture of the native ($\alpha\beta$) and succinylated ($\alpha_s\beta_s$) variants of bacterial luciferase, a dimeric protein, yielded three components: $\alpha\beta$, $\alpha\beta_s$, and $\alpha_s\beta$ in a single band, and $\alpha_s\beta_s$ (Meighen *et al.*, 1971).

Hybridization *in vivo* by the introduction of a second gene has been achieved with histidinol dehydrogenase as well as with 6-phosphogluconate dehydrogenase (Lew and Roth, 1971). Each of these enzymes has been shown to be dimeric.

With all of these procedures there may be cases where hybridization produces an unexpected number of electrophoretic variants. For example, Tate and Meister (1970) have hybridized *Alcaligenes* and *Pseudomonas* aspartate β-decarboxylase, a protein containing twelve subunits. Instead of the expected thirteen electrophoretic bands, they observed only seven. This can only be explained by hybridization of dimeric units of the enzyme from the two sources. Baker and Mintz (1969), investigating the hybridization of tetrameric mouse malate dehydrogenase isozymes, observed only three electrophoretically distinct bands instead of the expected five bands. Even after five repeated freeze–thaw cycles at high or low salt concentrations the three-banded pattern persisted. Urea treatment (0.25 *M*) similarly led to the production of only three electrophoretic variants. Higher concentrations of urea, without freezing and thawing, followed by dialysis, yielded no hybrid bands. In 4 *M* urea the enzyme was irreversibly inactivated. These results are consistent with hybridization involving dimeric units of malate dehydrogenase.

H. Table of Subunits

Many proteins have been examined for subunit composition by the methods described above. Table II contains a comprehensive list of such oligomeric macromolecules. Other less complete compilations have been published previously (Darnall and Klotz, 1972; Klotz and Darnall, 1969; Klotz, 1967).

In Table II, individual polypeptide chains that are held together by disulfide linkages have not been classified individually as subunits.[1] For example, the subunit weight listed for insulin is 5733, that of the A and B chains together, since the A and B chains are linked by disulfide bonds.

In many instances the subunit listed (Table II) may not be the minimal subunit obtainable, but instead the minimal subunit that has been unequivocally obtained under environmental conditions that eliminate cleavage of peptide bonds. For some proteins, two (or more) stages of dissociation can be clearly recognized; in such instances two (or more) entries specifying the relations between the different aggregates are given.

The most accessible references are given for each entry; they do not necessarily indicate the source most deserving of credit for establishing the subunit interrelations; these sources are mentioned in the cited works.

It should be pointed out that in many instances the number of subunits in a particular protein has been determined only after arduous efforts by several groups of investigators, often in serious disagreement for a long time. For myosin, almost ten years were required to reach the current state of understanding. For aldolase and aspartate transcarbamylase, careful characterizations by several groups of authoritative investigators over a period of years led to two seemingly divergent answers. The major pitfalls seem to lie not in the physicochemical methods used but rather in the complex association–dissociation patterns of protein oligomers, which may be perturbed by slight differences in components in the aqueous solvent environment.

TABLE II

Subunit Constitution of Proteins[a,b,c]

Protein	Molecular weight	Subunits No.	Subunits Molecular weight
Insulin (1)	11,466	2	5,733
Beef brain acidic protein (S-100 protein) (2)	21,300	3	7,000
E. coli mercaptopyruvate sulfur transferase (EC 2.8.1.2) (3)	23,800	2	12,000
Cytochrome (*cc'*) (4)	28,000	2	14,000
Nerve growth factor (5)	26,518	2	13,259
Rat pituitary luteinizing hormone (6)	31,000	2	15,500
Lactose specific factor III (7)	33,000	4	8,000
β-Lactoglobulin (8)	35,000	2	17,500
Human follicle-stimulating hormone (9)	35,000	2	17,500
β-Hydroxydecanoyl thioester dehydrase (10)	36,000	2	18,000
Rhodanese (EC 2.8.1.1) (11)	37,000	2	18,500
Chymotrypsin inhibitor I (12)	39,000	4	9,800
E. coli superoxide dismutase (13)	39,500	2	21,600
Catabolite gene-activator protein (14)	45,000	2	22,000
Neurospora cytoplasmic protein (15)	45,000	3	15,000
Bovine growth hormone (16)	48,000	2	25,000
α-Amylase (EC 3.2.1.1) (17)	50,000	2	25,000
Human red blood cell galactokinase (EC 2.7.1.6) (18)	53,000	2	27,000
Triosephosphate isomerase (EC 5.3.1.1) (19)	53,000	2	26,500
Neurospora malate dehydrogenase (EC 1.1.1.37) (20)	54,000	4	13,500
Hair follicle transglutaminase (21)	54,000	2	27,000
Azoferredoxin (22)	55,000	2	27,500
ω-Amidase (23)	58,000	2	27,000
NADP-Linked isocitrate dehydrogenase (EC 1.1.1.42) (24)	60,000	2	32,000
Staphylococcus deoxycytidylate deaminase (25)	60,000	2	29,000
Hydrogenase (EC 1.12.1.1) (26)	60,000	2	30,000
Bovine lactose synthetase (27)	60,000	1	44,000
		1	15,000
Aldose reductase (EC 1.1.1.21) (28)	61,000	1	22,300
		1	36,600
Human prealbumin (29)	62,000	4	15,500
Serine dehydratase (EC 4.2.1.13) (30)	64,000	2	34,000
Hemoglobin (31)	64,500	4	16,000
T_u–T_s Complex (32)	65,000	1	41,500
		1	28,500
Thiogalactoside transacetylase (EC 2.3.1.18) (33)	65,300	2	29,700

TABLE II (*Continued*)

Protein	Molecular weight	Subunits No.	Subunits Molecular weight
Muscle phosphoglycerate mutase (EC 2.7.5.3) (34)	66,000	2	33,000
Rat liver malate dehydrogenase (EC 1.1.1.37) (35)	66,300	2	37,500
O-Acetylserine sulfhydrylase A (36)	68,000	2	34,000
Tropomyosin B (37)	68,000	2	33,500
Transaldolase III (EC 2.2.1.2) (38)	68,000	2	34,000
Rat liver adenylate kinase (EC 2.7.4.3) (39)	68,000	3	23,000
Glycerol-3-phosphate dehydrogenase (EC 1.1.1.8) (40)	68,000	2	34,000
Human 17β-estradiol dehydrogenase (41)	68,000	2	33,000
Human hypoxanthine-guanine phosphoribosyltransferase (EC 2.4.2.8) (42)	68,000	2	34,000
Avidin (43)	68,300	4	18,000
Hydroxypyruvate reductase (EC 1.1.1.29) (44)	70,000	2	35,000
Thymidylate synthetase (EC 2.1.1.b) (45)	70,000	2	35,000
Pea nucleoside diphosphate kinase (EC 2.7.4.6) (46)	70,000	4	17,000
Nicotinamide–adenine dinucleotide glycohydrolase (EC 3.2.2.5) (47)	70,000	2	38,000
2-Keto-3-deoxy-6-phosphogluconate aldolase (EC 4.1.2.12) (48)	72,000	3	24,000
Bovine heart malate dehydrogenase (EC 1.1.1.37) (49)	72,000	2	37,000
E. coli tryptophanyl tRNA synthetase (EC 6.1.1.2) (50)	74,000	2	37,000
Candida sedoheptulose-1,7-diphosphatase (51)	75,000	2	35,000
Chorismate mutase-prephenate dehydrogenase (52)	76,000	2	40,000
Glycerol-1-phosphate dehydrogenase (EC 1.1.1.6) (53)	78,000	2	40,000
Hydroxyindole-*O*-methyltransferase (EC 2.1.1.4) (54)	78,000	2	39,000
Uridine diphosphogalactose 4-epimerase (EC 5.1.3.2) (55)	79,000	2	39,000
Bacterial luciferase (56)	79,000	1	42,000
		1	37,000
Veillonella phosphotransacetylase (EC 2.3.1.8) (57)	80,000	2	(40,000)
Alkaline phosphatase (EC 3.1.3.1) (58)	80,000	2	40,000
Creatine kinase (EC 2.7.3.2) (59)	80,000	2	40,000
Liver alcohol dehydrogenase (EC 1.1.1.1) (60)	80,000	2	41,000

(*Continued*)

TABLE II (*Continued*)

Protein	Molecular weight	Subunits No.	Subunits Molecular weight
Yeast aldolase (EC 4.1.2.13) (61)	80,000	2	40,000
Lombricine kinase (EC 2.7.3.5) (62)	80,000	2	40,000
Taurocyamine kinase (EC 2.7.3.4) (63)	80,000	2	40,000
Prephenoloxidase (64)	80,000	2	40,000
Troponin (65)	(80,000)	(2)	22,000
		1	40,000
Rabbit enolase (EC 4.2.1.11) (66)	82,000	2	41,000
ATP-Creatine transphosphorylase (67)	82,600	2	41,300
Haptoglobin 1-1 (68)	85,000	2	40,000
Acid phosphatase (EC 3.1.3.2) (69)	85,000	2	42,000
E. coli alkaline phosphatase (EC 3.1.3.1) (70)	86,000	2	43,000
Histidinol dehydrogenase (71)	87,000	2	40,000
Yeast enolase (EC 4.2.1.11) (72)	88,000	2	44,000
Bovine procarboxypeptidase A (73)	88,000	1	40,000
		2	23,000
Serum high density lipoprotein (74)	88,000	2	27,000
		2	17,000
Penicillium 2-deoxycitrate synthase (75)	90,000	2	45,000
Purine nucleoside phosphorylase (EC 2.4.2.1) (76)	90,000	2	47,000
		2	24,000
Pyruvate dehydrogenase (EC 1.2.4.1) (77)	90,000	2	45,000
Salycilate hydroxylase (EC 1.14.1.a) (78)	91,000	2	45,000
Firefly luciferase (79)	92,000	2	52,000
Succinate dehydrogenase (EC 1.3.99.1) (80)	97,000	1	70,000
		1	27,000
L-Ribulokinase (EC 2.7.1.16) (81)	98,000	2	50,000
Pseudomonas N-formimino-L-glutamate iminohydrolase (82)	100,000	2	50,000
D-Amino acid oxidase (EC 1.4.3.3) (83)	100,000	2	50,000
Diacetyl (acetoin) reductase (EC 1.1.1.5) (84)	100,000	4	25,000
Cysteamine oxygenase (EC 1.13.1.22) (85)	100,000	2	50,000
Galactokinase (EC 2.7.1.6) (86)	100,000	4	23,000
Citrate synthase (EC 4.1.3.7) (87)	100,000	2	50,000
Aspartate aminotransferase (EC 2.6.1.1) (88)	100,000	2	50,000
Seryl-tRNA synthetase (EC 6.1.1.11) (89)	100,000	2	50,000
Yeast hexokinase (EC 2.7.1.1) (90)	102,000	2	51,000
Rat liver 6-phosphogluconate dehydrogenase (EC 1.1.1.44) (91)	102,000	2	52,000
Yeast nucleoside diphosphokinase (EC 2.7.4.6) (92)	102,000	6	17,000
Bovine liver ornithine transcarbamylase (EC 2.1.3.3) (93)	108,000	3	36,000

TABLE II (*Continued*)

Protein	Molecular weight	Subunits No.	Subunits Molecular weight
Concanavalin A (94)	108,000	2	54,000
	54,000	2	27,000
Lipoxygenase (EC 1.99.2.1) (95)	108,000	2	54,000
Hemerythrin (96)	108,000	8	13,500
Rat liver glutamine transaminase (97)	110,000	2	54,000
Tubulin (98)	110,000	1	56,000
		1	53,000
Tryptophanyl-tRNA synthetase (EC 6.1.1.2) (99)	110,000	(2–4)	(58,000–27,000)
Yeast phosphoglycerate mutase (EC 2.7.5.3) (100)	110,000	4	27,900
Dihydrolipoyl dehydrogenase (101)	112,000	2	56,000
Tyrosine aminotransferase (EC 2.6.1.5) (102)	115,000	4	32,000
Malate-lactate transhydrogenase (EC 1.1.1.37) (103)	115,000	(3–4)	30,000
Aspartokinase (EC 2.7.2.4) (104)	116,000	(2)	(47,000)
		(2)	(17,000)
Human liver arginase (EC 3.5.3.1) (105)	118,000	4	30,000
Leucine-tRNA synthetase (EC 6.1.1.4) (106)	120,000	2	60,000
Pasteurella protein toxin B (107)	120,000	(5–6)	24,000
	24,000	2	12,000
Pyrophosphatase (EC 3.6.1.1) (108)	120,000	6	20,000
Spinach leaf aldolase (EC 4.1.2.13) (109)	120,000	4	30,000
DDT-Dehydrochlorinase (EC 4.5.1.1) (110)	120,000	4	30,000
Bacillus alkaline phosphatase (EC 3.1.3.1) (111)	121,000	2	55,000
Tryptophan oxygenase (EC 1.13.1.12) (112)	122,000	4	31,000
E. coli fatty acylthiokinase I (113)	122,000	4	30,000
Glutathione reductase (EC 1.6.4.2) (114)	124,000	2	56,000
T2-Bacteriophage-induced deoxycytidylate deaminase (115)	124,000	6	20,200
Tyrosinase (EC 1.10.3.1) (116)	128,000	4	32,000
Fructose diphosphatase (EC 3.1.3.11) (117)	130,000	2	29,000
		2	37,000
Mammary glucose-6-phosphate-dehydrogenase (EC 1.1.1.49) (118)	130,000	2	63,000
Ornithine aminotransferase (EC 2.6.1.13) (119)	132,000	4	33,000
E. coli L-asparaginase (EC 3.5.1.1) (120)	133,000	4	33,000
Phosphoglucose isomerase (EC 5.3.1.9) (121)	134,000	2	61,000
L-Amino acid oxidase (EC 1.4.3.2) (122)	135,000	2	70,000
Pig nucleoside diphosphokinase (123)	138,000	6	21,000
Yeast lysine-tRNA ligase (124)	138,000	2	69,000

(*Continued*)

TABLE II (*Continued*)

Protein	Molecular weight	Subunits No.	Subunits Molecular weight
Achromobacteraceae L-asparaginase (EC 3.5.1.1) (125)	138,000	4	35,000
Transketolase (EC 2.2.1.1) (126)	140,000	2	70,000
Protein phosphokinase (EC 2.7.1.37) (127)	140,000	1	80,000
		1	60,000
Succinyl-CoA synthetase (EC 6.2.1.5) (128)	140,000	2	38,500
		2	29,500
C-Reactive protein (129)	140,000	6	23,000
Lactate dehydrogenase (EC 1.1.1.27) (130)	140,000	4	35,000
L-Rhamnulose-1-phosphate aldolase (EC 4.1.2.b) (131)	140,000	4	35,000
Serratia anthranilate synthetase (132)	(141,000)	2	60,000
		2	21,000
Glyceraldehyde-3-phosphate dehydrogenase	144,000	2	72,000
(EC 1.2.1.12) (133)	72,000	2	37,000
Tartaric acid dehydrase (EC 4.2.1.c) (134)	145,000	4	39,000
Phosphofructokinase (EC 2.7.1.11) (135)	145,000	4	35,000
Bacillus malate dehydrogenase (EC 1.1.1.37) (136)	148,000	4	37,000
Tryptophan synthetase (EC 4.2.1.20) (137)	148,000	2	45,000
		2	28,700
Pyridoxamine pyruvate transaminase (EC 2.6.1.a) (138)	150,000	4	38,000
Yeast alcohol dehydrogenase (EC 1.1.1.1) (139)	150,000	4	37,000
Human platelet factor XIII (140)	(150,000)	2	81,000
Ceruloplasmin (141)	151,000	8	18,000
α-Acetylgalactosaminidase (142)	155,000	4	42,000
Muscle aldolase (EC 4.1.2.13) (143)	160,000	4	40,000
lac Repressor (144)	160,000	4	40,000
Salmonella cystathionine γ-synthetase (EC 4.2.1.21) (145)	160,000	4	40,000
Rat liver cystathionase (EC 4.2.1.15) (146)	160,000	8	20,000
Trimethylamine dehydrogenase (147)	160,000	2	80,000
Clostridium threonine deaminase (EC 4.2.1.16) (148)	160,000	4	40,000
Yeast pyruvate kinase (EC 2.7.1.40) (149)	161,000	8	20,000
Crotonase (EC 4.2.1.17) (150)	164,000	6	28,000
Beef liver carboxylesterase (EC 3.1.1.1) (151)	167,000	2	85,000
Pig liver esterase (EC 3.1.1.1) (152)	168,000	2	85,000
	85,000	2	42,000
Molybdoferredoxin (153)	168,000	2	59,000
		1	50,700

TABLE II (*Continued*)

Protein	Molecular weight	Subunits No.	Subunits Molecular weight
Aspartokinase II-homoserine dehydrogenase II (154)	169,000	4	43,000
E. coli carbamyl phosphate synthetase (EC 2.7.2.5) (155)	170,000	1	130,000
		1	42,000
E. coli aspartase (EC 4.3.1.1) (156)	170,000	4	45,000
Lysine-sensitive aspartokinase (EC 2.7.2.4) (157)	177,000	4	46,000
Methionyl-tRNA synthetase (EC 6.1.1.10) (158)	180,000	2	90,000
	90,000	2	45,000
Threonine dehydratase (EC 4.2.1.16) (159)	180,000	2	90,000
	90,000	2	45,000
Thetin homocysteine methylpherase (EC 2.1.1.10) (160)	180,000	3–4	50,000
Aminoacyl transferase I (161)	186,000	3	62,000
α-Dialkyl amino acid transaminase (162)	188,000	4	47,000
α-Isopropylmalate synthase (EC 2.3.1.c) (163)	190,000	4	47,500
Histidine decarboxylase (EC 4.1.1.22) (164)	190,000	5	9,000
		5	29,700
Rabbit phosphofructokinase (EC 2.7.1.11) (165)	190,000	2	78,000
Fumarase (EC 4.2.1.2) (166)	194,000	4	48,500
Salmonella threonine deaminase (EC 4.2.1.16) (167)	195,000	4	48,500
Dipeptidyl transferase (EC 3.4.4.9) (168)	197,000	2	100,000
	100,000	4	24,500
Salmonella phosphoenolpyruvate carboxylase (EC 4.1.1.31) (169)	198,000	4	49,200
Glutamine phosphoribosyl pyrophosphate amidotransferase (EC 2.4.2.14) (170)	200,000	2	100,000
	100,000	2	50,000
Polynucleotide phosphorylase (EC 2.7.7.8) (171)	200,000	2	95,000
Salmonella α-isopropylmalate synthase (172)	200,000	4	50,000
Sweet potato β-amylase (EC 3.2.1.2) (173)	201,000	4	50,000
Argininosuccinase (EC 4.3.2.1) (174)	202,000	2	100,000
	100,000	2	50,000
Human glucose-6-phosphate dehydrogenase (EC 1.1.1.49) (175)	204,800	2	101,400
	101,400	2	51,300
E. coli Qβ replicase (176)	205,000	1	70,000
		1	65,000
		1	45,000
		1	35,000
Isocitrate lyase (EC 4.1.3.1) (177)	206,000	4	48,200

(*Continued*)

TABLE II (*Continued*)

Protein	Molecular weight	Subunits No.	Subunits Molecular weight
Neurospora glucose-6-phosphate dehydrogenase (EC 1.1.1.49) (178)	206,000	2	104,000
	104,000	2	57,000
Human C5 (179)	(206,000)	1	123,000
		1	83,000
Bacillus tryptophanase (180)	208,000	4	50,500
Yeast pyruvate decarboxylase (EC 4.1.1.1) (181)	209,000	2	108,000
Invertase (EC 3.2.1.26) (182)	210,000	4	51,500
Cytidine triphosphate synthetase (EC 6.3.4.2) (183)	210,000	2	105,000
	105,000	2	50,000
Plasma high-density lipoprotein (184)	210,000	4	28,000
Phosphoribosyl adenosine triphosphate: pyrophosphate phosphoribosyltransferase (185)	215,000	6	36,000
Rabbit liver serine transhydroxymethylase (EC 2.1.2.1) (186)	215,000	4	47,000
Rat α-L-fucosidase (187)	216,000	2	47,000
		2	59,300
Glycerol kinase (EC 2.7.1.30) (188)	217,000	4	55,000
Acetol acetate-forming enzyme (189)	220,000	4	58,000
Tryptophanase (EC 4.2.1.e) (190)	220,000	2	110,000
	110,000	2	55,000
Paramyosin (191)	220,000	2	110,000
Salmonella chorismate mutase-prephenate dehydratase (192)	220,000	2	109,000
Arylamidase (193)	223,500	6	38,100
Streptococcus ornithine transcarbamylase (EC 2.1.3.3) (194)	223,000	3	74,000
	74,000	2	38,000
Glycyl-tRNA synthetase (EC 6.1.1.e) (195)	227,000	2	33,000
		2	80,000
Glyoxylate carboligase (EC 4.1.1.b) (196)	230,000	2	115,000
	115,000	2	61,000
Catalase (EC 1.11.1.6) (197)	232,000	4	57,500
Cytochrome b_2 (198)	235,000	4	57,000
Rabbit pyruvate kinase (EC 2.7.1.40) (199)	237,000	4	57,200
Clostridium formyltetrahydrofolate synthetase (EC 6.3.4.3) (200)	240,000	4	60,000
Rhodopsuedomonas δ-aminolevulinate dehydratase (EC 4.2.1.24) (201)	240,000	2	120,000
	120,000	3	40,000
Neurospora anthranilate synthetase complex (202)	240,000	6	40,000
Pasteurella protein toxin A (203)	240,000	10–12	24,000
	24,000	2	12,000
Protocollagen proline hydroxylase (204)	240,000	2	60,000
		2	65,000

TABLE II (*Continued*)

Protein	Molecular weight	Subunits No.	Subunits Molecular weight
2,5-Dihydroxypyridine oxygenase (205)	242,000	6	39,500
Horse liver aldehyde dehydrogenase (EC 1.2.1.3) (206)	245,000	4	57,000
δ-Aminolevulinate dehydratase (207)	280,000	2	140,000
	140,000	4	35,000
Rat liver cystathionine synthetase (EC 4.2.1.21) (208)	250,000	(2)	51,000
		(2)	73,000
Uridine diphosphate galactose 4-epimerase (EC 5.1.3.2) (209)	250,000	2	125,000
	125,000	2	60,000
Ascaris malic enzyme (EC 1.1.1.39) (210)	250,000	4	64,000
Phytochrome (211)	252,000	6	42,000
Leucine aminopeptidase (EC 3.4.1.1) (212)	255,000	4	63,500
Acetylcholinesterase (EC 3.1.1.7) (213)	259,000	6	42,000
Glycolate oxidase (EC 1.1.3.1) (214)	270,000	2	140,000
Phenylalanyl-tRNA synthetase (215)	276,000	2	75,000
		2	63,000
Muscle AMP-deaminase (EC 3.5.4.6) (216)	278,000	4	69,000
Bovine heart protein kinase (217)	280,000	(?)	42,000
		(?)	55,000
β-Glucuronidase (EC 3.2.1.31) (218)	280,000	4	75,000
Salmonella anthranilate synthetase complex (219)	280,000	2	62,000
		2	62,000
Lysine 2,3-aminomutase (220)	285,000	6	48,000
Yeast NAD-isocitrate dehydrogenase (EC 1.1.1.41) (221)	300,000	8	39,000
Cysteine synthetase (EC 4.2.1.22) (222)	309,000	1	160,000
		2	68,000
Human plasma factor XIII (223)	310,000	2	81,000
		2	81,000
Aspartyl transcarbamylase (EC 2.1.3.2) (224)	310,000	2	100,000
		3	34,000
	100,000	3	33,000
	34,000	2	17,000
Glutamate decarboxylase (EC 4.1.1.15) (225)	310,000	6	50,000
Horse serum cholinesterase (EC 3.1.1.8) (226)	315,000	4	77,300
Rat liver carbamoyl-phosphate synthase (EC 2.7.2.5) (227)	316,000	2	160,000
Glutamate dehydrogenase (EC 1.4.1.3) (228)	320,000	6	57,000
Chloroplast-coupling factor (229)	325,000	6	62,000
Bovine lens leucine aminopeptidase (EC 3.4.1.1) (230)	327,000	6	54,000
Potato L-phenylalanine ammonia lyase (EC 4.3.1.5) (231)	330,000	4	83,900

(*Continued*)

TABLE II (*Continued*)

Protein	Molecular weight	Subunits No.	Subunits Molecular weight
Hamster liver glutamine synthetase (EC 6.3.1.2) (232)	335,000	8	42,000
Acetoacetate decarboxylase (EC 4.1.1.4) (233)	340,000	6	62,000
	62,000	2	29,000
E. coli aspartokinase-homoserine dehydrogenase (234)	340,000	4	85,000
Maize phosphoenolpyruvate carboxylase (EC 4.1.1.31) (235)	340,000	2	160,000
Sheep red cell peptidase (236)	340,000	6	60,000
Arachin (237)	345,000	2	180,000
	180,000	6	30,000
N-Methylglutamate synthetase (238)	350,000	12	(30,000)
Rat liver glutamine synthetase (EC 6.3.1.2) (239)	352,000	8	44,000
Phycocyanin (240)	360,000	2	180,000
	180,000	6	30,000
L-Arabinose isomerase (EC 5.3.1.4) (241)	360,000	6	60,000
Neurospora glutamine synthetase (EC 6.3.1.2) (242)	360,000	4	90,000
Pseudomonas aspartate transcarbamylase (EC 2.1.3.2) (243)	360,000	2	180,000
Rat liver mitochondrial adenosine triphosphatase (EC 3.6.1.3) (244)	360,000	6	53,000
		1	28,000
		1	12,500
		1	9,000
Phosphorylase A (EC 2.4.1.1) (245)	370,000	4	92,500
Membrane adenosine triphosphatase (EC 3.6.1.3) (246)	385,000	12	33,000
E. coli RNA polymerase (EC 2.7.7.6) (247)	400,000	2	39,000
		1	155,000
		1	165,000
Rat liver RNA polymerase (EC 2.7.7.6) (248)	400,000	1	19,000
		1	150,000
		1	35,000
		1	25,000
Cholesterolesterase (EC 3.1.1.13) (249)	400,000	6	65,000
Lipovitellin (250)	400,000	2	200,000
Ovine glutamine synthetase (EC 6.3.1.2) (251)	400,000	8	49,000
E. coli phosphoenolpyruvate carboxylase (EC 4.1.1.31) (252)	402,000	4	99,600
Phosphoenolpyruvate carboxytransphosphorylase (EC 4.1.1.38) (253)	408,000	2	200,000
	200,000	2	100,000

TABLE II (*Continued*)

Protein	Molecular weight	Subunits No.	Subunits Molecular weight
Adenosine triphosphate sulfurylase (EC 2.7.7.4) (254)	440,000	8	56,000
Fatty acid synthetase (255)	450,000	2	230,000
Apoferritin (256)	460,000	24	18,500
Myosin (257)	468,000	2	212,000
		2–3	20,000
Urease (EC 3.5.1.5) (258)	480,000	2	240,000
	240,000	3	83,000
Uridine-diphosphate glucose pyrophosphorylase (EC 2.7.7.9) (259)	480,000	8	60,000
Hydrogenomonas ribulose-1,5-diphosphate carboxylase (EC 4.1.1.39) (260)	515,000	12–14	40,700
β-Galactosidase (EC 3.2.1.23) (261)	540,000	4	135,000
Plant ribulose diphosphate carboxylase (EC 4.1.1.39) (262)	550,000	8	52,000
		6	24,500
	24,500	2	12,000
E. coli L-malic enzyme (EC 1.1.1.40) (263)	550,000	8	67,000
Citrate lyase (EC 4.1.3.6) (264)	575,000	2	290,000
	290,000	2	137,000
	137,000	2	74,000
Chlorella ribulose-1,5-diphosphate carboxylase (EC 4.1.1.39) (265)	588,000	8	58,200
		8	15,300
Yeast phosphofructokinase (EC 2.7.1.11) (266)	590,000	6	100,000
E. coli glutamine synthetase (EC 6.3.1.2) (267)	592,000	12	48,500
Ovomacroglobulin (268)	650,000	2	325,000
Pyruvate carboxylase (EC 6.4.1.1) (269)	660,000	4	165,000
	165,000	4	45,000
Thyroglobulin (270)	669,000	2	335,000
Transcarboxylase (EC 2.1.3.1) (271)	670,000	1	320,000
		3	120,000
	320,000	3	115,000
	115,000	2	55,000
	120,000	2	12,000
		1	115,000
Bovine NAD-isocitrate dehydrogenase	670,000	2	330,000
(EC 1.1.1.41) (272)	330,000	8	41,000
Alcaligenes L-aspartate β-decarboxylase	675,000	6	112,000
(EC 4.1.1.12) (273)	112,500	2	57,000
NADP-Sulfite reductase (EC 1.8.1.2) (274)	700,000	4	53,000
		8	58,000
Propionyl carboxylase (EC 6.4.1.3) (275)	700,000	4	175,000
Azotobacter RNA polymerase (EC 2.7.7.6) (276)	782,000	2	391,000

(Continued)

TABLE II (*Continued*)

Protein	Molecular weight	Subunits No.	Subunits Molecular weight
Pseudomonas L-aspartate β-decarboxylase	800,000	2	400,000
(EC 4.1.1.12) (277)	400,000	4	100,000
α-Crystallin (278)	810,000	(30)	26,000
Arginine decarboxylase (EC 4.1.1.19) (279)	820,000	5	160,000
	160,000	2	82,000
RNA polymerase (EC 2.7.7.6) (280)	880,000	2	440,000
Dihydrolipoyl transacetylase (EC 2.3.1.12)	960,000	8	120,000
(281)	120,000	3	40,000
Hemocyanin (282)	300,000–		760,000
	9,000,000		380,000
			70,000
			35,000
Chlorocruorin (283)	2,750,000	12	250,000
Arenicola hemoglobin (284)	2,850,000	12	230,000
	230,000	4	54,000
Erythrocruorin (285)	3,000,000	162	18,500
Phage fII (286)	3,620,000	180	13,750
Pyruvate dehydrogenase complex (287)	4,000,000	1	960,000
		24	90,000
		12	112,000
	960,000	8	120,000
	120,000	3	40,000
	90,000	2	45,000
	112,000	2	56,000
Cowpea chlorotic mottle virus (288)	4,600,000	180	19,600
Bromegrass mosaic virus (289)	4,600,000	180	20,000
Broad bean mottle virus (290)	4,800,000	180	20,900
Turnip yellow mosaic virus (291)	5,000,000	150	21,000
Poliomyelitis virus (292)	5,500,000	130	27,000
Cucumber mosaic virus (293)	6,000,000	185	21,500
Alfalfa mosaic virus (294)	7,400,000	(140)	51,600
	51,600	2	24,500
Liver acetyl-CoA carboxylase (EC 6.4.1.2)	8,300,000	2	4,100,000
(295)	4,100,000	10	409,000
Rhinovirus 1A (296)	8,400,000	60	96,000
Bushy stunt virus (297)	9,000,000	120	60,000
Polyoma virus (298)	24,000,000	420	50,200
Potato virus X (299)	35,000,000	650	52,000
Tobacco mosaic virus (300)	40,000,000	2130	17,500

[a] Numbers in parentheses in columns other than 1 indicate doubt.
[b] Numbers in parentheses after entries in column 1 indicate reference source.
[c] This table indicates the status of these structures as of 1972.

TABLE II (*Continued*)

1. Crowfoot, D., *Proc. Roy. Soc., Ser. A* **164,** 580 (1938); Moody, L. S., Ph.D. Dissertation, University of Wisconsin, Madison (1944); Waugh, D. F., *Advan. Protein Chem.* **9,** 325 (1954).
2. Dannies, P. S., and Levine, L., *Biochem. Biophys. Res. Commun.* **37,** 587 (1969).
3. Vachek, H., and Wood, J. L., *Biochim. Biophys. Acta* **258,** 133 (1972).
4. Cusanovich, M. A., Tedro, S. M., and Kamen, M. D., *Arch. Biochem. Biophys.* **141,** 557 (1970).
5. Angeletti, R. H., and Bradshaw, R. A., *Proc. Nat. Acad. Sci. U. S.* **68,** 2417 (1971).
6. Ward, D. N., Reichart, L. E., Jr., Fitak, B. A., Nahm, H. S., Sweeney, C. M., and Neill, J. D., *Biochemistry* **10,** 1796 (1971).
7. Schrecker, O., and Hengstenberg, W., *FEBS (Fed. Eur. Biochem. Soc.) Lett.* **13,** 209 (1971).
8. Bull, H. B., *J. Amer. Chem. Soc.* **68,** 745 (1946); Townend, R., and Timasheff, S. N., *ibid.* **79,** 3613 (1957).
9. Ryan, R. J., Jiang, N.-S., and Hanlon, S., *Biochemistry* **10,** 1321 (1971); Saxena, B. B., and Rathnam, P., *J. Biol. Chem.* **246,** 3549 (1971).
10. Helmkamp, G. M., Jr., and Bloch, K., *J. Biol. Chem.* **244,** 6014 (1969).
11. Volini, M., DeToma, F., and Westley, J., *J. Biol. Chem.* **242,** 5220 (1967).
12. Melville, J. C., and Ryan, C. A., *Arch. Biochem. Biophys.* **138,** 700 (1970).
13. Keele, B. B., Jr., McCord, J. M., and Fridovich, I., *J. Biol. Chem.* **245,** 6176 (1970).
14. Riggs, A. D., Reiness, G., and Zubay, G., *Proc. Nat. Acad. Sci. U. S.* **68,** 1222 (1971).
15. Shannon, C. F., and Hill, J. M., *Biochemistry* **10,** 3021 (1971).
16. Edelhoch, H., Condliffe, P. G., Lippoldt, R. E., and Burger, H. G., *J. Biol. Chem.* **241,** 5205 (1966).
17. Connellan, J. M., and Shaw, D. C., *J. Biol. Chem.* **245,** 2845 (1970); Robyt, J. F., Chittenden, C. G., and Lee, C. T., *Arch. Biochem. Biophys.* **144,** 160 (1971).
18. Blume, K.-G., and Buetler, E., *J. Biol. Chem.* **246,** 6507 (1971).
19. Hartman, F. C., *Biochemistry* **10,** 146 (1971).
20. Munkres, K. D., *Biochemistry* **4,** 2180 and 2186 (1965).
21. Chung, S. I., and Folk, J. E., *Proc. Nat. Acad. Sci. U. S.* **69,** 303 (1972).
22. Nakos, G., and Mortenson, L., *Biochemistry* **10,** 455 (1971).
23. Hersh, L. B., *Biochemistry* **10,** 2881 (1971).
24. Magar, M. E., and Robbins, J. E., *Biochim. Biophys. Acta* **191,** 173 (1969).
25. Bessman, M. J., Diamond, G. R., Debeer, L. J., and Duncan, B. K., *Fed. Proc., Fed. Amer. Soc. Exp. Biol.* **30,** 1121 (1971).
26. Nakos, G., and Mortenson, L. E., *Biochemistry* **10,** 2442 (1971).
27. Trayer, I. P., and Hill, R. L., *J. Biol. Chem.* **246,** 6666 (1971); Magee, S. C., Mawal, R., and Ebner, K. E., *Fed. Proc., Fed. Amer. Soc. Exp. Biol.* **31,** 499 (1972).
28. Sheys, G. H., and Doughty, C. C., *Biochim. Biophys. Acta* **235,** 414 (1971).
29. Rask, L., Peterson, P. A., and Nilsson, S. F., *J. Biol. Chem.* **246,** 6087 (1971); Gonzalez, G., and Offord, R. E., *Biochem. J.* **125,** 309 (1971).
30. Inoue, H., Kasper, C. B., and Pitol, H. C., *J. Biol. Chem.* **246,** 2626 (1971).

31. Braunitzer, G., Hilse, K., Rudloff, V., and Hilschmann, N., *Advan. Protein Chem.* **19,** 1 (1964).
32. Hachmann, J., Miller, D. L., and Weissbach, H., *Arch. Biochem. Biophys.* **147,** 457 (1971).
33. Brown, J. L., Brown, D. M., and Zabin, I., *J. Biol. Chem.* **242,** 4254 (1967).
34. Scopes, R. K., and Penny, I. F., *Biochim. Biophys. Acta* **236,** 406 (1971).
35. Mann, K. G., and Vestling, C. S., *Biochemistry* **8,** 1105 (1969).
36. Becker, M. A., Kredich, N. M., and Tomkins, G. M., *J. Biol. Chem.* **244,** 2418 (1969).
37. Holtzer, A., Clark, R., and Lowey, S., *Biochemistry* **4,** 2401 (1965); Olander, J., Emerson, M., and Holtzer, A., *J. Amer. Chem. Soc.* **89,** 3058 (1967); Woods, E. F., *J. Biol. Chem.* **242,** 2859 (1967).
38. Tsolas, O., and Horecker, B. L., *Arch. Biochem. Biophys.* **136,** 303 (1970).
39. Criss, W. E., Sapico, V., and Litwack, G., *J. Biol. Chem.* **245,** 6346 (1970).
40. White, H. B., III, *Arch. Biochem. Biophys.* **147,** 123 (1971).
41. Burns, D. J. W., Engel, L. L., and Bethune, J. L., *Biochemistry* **11,** 2699 (1972); Jarabak, J., and Street, M. A., *ibid.* **10,** 3831 (1971); Burns, D. J. W., Engel, L. L., and Bethune, J. L., *Biochem. Biophys. Res. Commun.* **44,** 786 (1971).
42. Arnold, W. J., and Kelley, W. N., *J. Biol. Chem.* **246,** 7398 (1971).
43. Green, N. M., *Biochem. J.* **92,** 16c (1964); Green, N. M., and Ross, M. E., *ibid.* **110,** 59 (1968).
44. Utting, J. M., and Kohn, L. D., *Fed. Proc., Fed. Amer. Soc. Exp. Biol.* **30,** 1057 (1971).
45. Dunlap, R. B., Harding, N. G. L., and Huennekens, F. M., *Biochemistry* **10,** 88 (1971).
46. Edlund, B., *FEBS* (*Fed. Eur. Biochem. Soc.*) *Lett.* **13,** 56 (1971).
47. Green, S., and Dobrjansky, A., *Biochemistry* **10,** 4533 (1971).
48. Hammerstedt, R. H., Möhler, H., Decker, K. A., and Wood, W. A., *J. Biol. Chem.* **246,** 2069 and 2075 (1971).
49. Wolfenstein, C., Englard, S., and Listowsky, I., *J. Biol. Chem.* **244,** 6415 (1969).
50. Joseph, D. R., and Muench, K. H., *J. Biol. Chem.* **246,** 7610 (1971).
51. Traniello, S., Calcagno, M., and Pontremoli, S., *Arch. Biochem. Biophys.* **146,** 603 (1971).
52. Koch, G. L. E., Shaw, D. C., and Gibson, F., *Biochim. Biophys. Acta* **212,** 387 (1970); **229,** 805 (1971).
53. Pfleiderer, G., and Auricchio, F., *Biochem. Biophys. Res. Commun.* **16,** 53 (1964); Deal, W. C., and Holleman, W. H., *Fed. Proc., Fed. Amer. Soc. Exp. Biol.* **23,** 264 (1964).
54. Jackson, R. L., and Lovenberg, W., *J. Biol. Chem.* **246,** 4280 (1971).
55. Wilson, D. B., and Hogness, D. S., *J. Biol. Chem.* **244,** 2132 (1969).
56. Meighen, E. A., Nicoli, M. Z., and Hastings, J. W., *Biochemistry* **10,** 4062 (1971); Gunsalus-Miguel, A., Meighen, E. A., Nicoli, M. Z., Nealson, K. H., and Hastings, J. W., *J. Biol. Chem.* **247,** 398 (1972).
57. Whiteley, H. R., and Pelroy, R. A., *J. Biol. Chem.* **247,** 1911 (1972).
58. Garen, A., and Levinthal, C., *Biochim. Biophys. Acta* **38,** 470 (1960); Schlesinger, M. J., *Brookhaven Symp. Biol.* **17,** 66 (1964).
59. Dawson, D. M., Eppenberger, H. M., and Kaplan, N. O., *J. Biol. Chem.*

242, 211 (1967); Bayley, P. M., and Thomson, A. R., *Biochem. J.* **104,** 33c (1967).
60. Theorell, H., and Winer, A. D., *Arch. Biochem. Biophys.* **83,** 291 (1959); Li, T.-K., and Vallee, B. L., *Biochemistry* **3,** 869 (1964); Drum, D. E., Harrison, J. H., Li, T., Bethune, J. L., and Vallee, B., *Proc. Nat. Acad. Sci. U. S.* **57,** 1434 (1967); Castellino, F. J., and Barker, R., *Biochemistry* **7,** 2207 (1968); Weber, K., and Osborn, M., *J. Biol. Chem.* **244,** 4406 (1969); Green, R. W., and McKay, R. H., *ibid.* p. 5034.
61. Harris, C. E., Kobes, R. D., Teller, D. C., and Rutter, W. J., *Biochemistry* **8,** 2442 (1969).
62. Oriol, C., Landon, M. F., and Thoai, N. V., *Biochim. Biophys. Acta* **207,** 514 (1970).
63. Oriol, C., Landon, M. F., and Thoai, N. V., *Biochim. Biophys. Acta* **207,** 514 (1970).
64. Ashida, M., *Arch. Biochem. Biophys.* **144,** 749 (1971).
65. Ebashi, S., Wakabayashi, T., and Ebashi, F., *J. Biochem.* (*Tokyo*) **69,** 441 (1971).
66. Winstead, J. A., and Wold, F., *Biochemistry* **3,** 791 (1964); **4,** 2145 (1965); Cardenas, J. M., and Wold, F., *ibid.* **7,** 2736 (1968).
67. Yue, R. H., Palmieri, R. H., Olson, O. E., and Kuby, S. A., *Biochemistry* **6,** 3205 (1967).
68. Waks, M., and Alfsen, A., *Arch. Biochem. Biophys.* **123,** 133 (1968).
69. Jacobs, M. M., Nyc, J. F., and Brown, D. M., *J. Biol. Chem.* **246,** 1419 (1971).
70. Reynolds, J. A., and Schlesinger, M. J., *Biochemistry* **8,** 588 (1969); Schlesinger, M. J., and Barrett, K., *J. Biol. Chem.* **240,** 4284 (1965).
71. Loper, J., *J. Biol. Chem.* **243,** 3264 (1968); Yourno, J., *ibid.* p. 3277; Lew, K. K., and Roth, J. R., *Biochemistry* **10,** 204 (1971).
72. Mann, K. G., Castellino, F. J., and Hargrave, P. A., *Biochemistry* **9,** 4002 (1970).
73. Brown, J. R., Greenshields, R. N., Yamasaki, M., and Neurath, H., *Biochemistry* **2,** 867 (1963); Teller, D. C., *ibid.* **9,** 4201 (1970).
74. Scanu, A. M., Edelstein, C., and Lim, C. T., *Fed. Proc., Fed. Amer. Soc. Exp. Biol.* **31,** 829 (1972).
75. Måhlén, A., *Eur. J. Biochem.* **22,** 104 (1971).
76. Gilpin, R. W., and Sadoff, H. L., *J. Biol. Chem.* **246,** 1475 (1971).
77. Reed, L. J., *Curr. Top. Cell. Regul.* **1,** 233 (1969).
78. White-Stevens, R. H., and Kamin, H., *J. Biol. Chem.* **247,** 2358 (1972).
79. Travis, J., and McElroy, W. D., *Biochemistry* **5,** 2170 (1966).
80. Davis, K. A., and Hatefi, Y., *Biochemistry* **10,** 2509 (1971); Righetti, P., and Cerletti, P., *FEBS* (*Fed. Eur. Biochem. Soc.*) *Lett.* **13,** 181 (1971).
81. Lee, N., Patrick, J. W., and Barnes, N. B., *J. Biol. Chem.* **245,** 1357 (1970).
82. Wickner, R. B., and Tabor, H., *J. Biol. Chem.* **247,** 1605 (1972).
83. Fonda, M. L., and Anderson, B. M., *J. Biol. Chem.* **243,** 5635 (1968).
84. Hetland, Ø., Olsen, B. R., Christensen, T. B., and Størmer, F. C., *Eur. J. Biochem.* **20,** 200 (1971).
85. Federici, G., Barra, D., Fiori, A., and Costa, M., *Physiol. Chem. Phys.* **3,** 448 (1971); Cavallini, D., Cannella, C., Federici, G., Dupré, S., Fiori, A., and Del Grosso, E., *Eur. J. Biochem.* **16,** 537 (1970).

86. Rustum, Y. M., and Barnard, E. A., *Fed. Proc., Fed. Amer. Soc. Exp. Biol.* **30,** 1122 (1971).
87. Wu, J.-Y., and Yang, J. T., *J. Biol. Chem.* **245,** 212 (1970); Singh, M., Books, G. C., and Srere, P. A., *ibid.* p. 4636; Moriyama, T., and Srere, P. A., *ibid.* **246,** 3217 (1971).
88. Bertland, L. H., and Kaplan, N. O., *Biochemistry* **7,** 134 (1968).
89. Katze, J. R., and Konigsberg, W., *J. Biol. Chem.* **245,** 923 (1970).
90. Pringle, J. R., *Biochem. Biophys. Res. Commun.* **39,** 46 (1970); Schmidt, J. J., and Colowick, S. P., *Fed. Proc., Fed. Amer. Soc. Exp. Biol.* **29,** 334 (1970).
91. Procsal, D., and Holten, D., *Biochemistry* **11,** 1310 (1972).
92. Palmieri, R., Yue, R., Jacobs, H., Maland, L., Wu, L., and Kuby, S. A., *Fed. Proc., Fed. Amer. Soc. Exp. Biol.* **29,** 914 (1970).
93. Marshall, M., and Cohen, P. P., *J. Biol. Chem.* **247,** 1641 (1972).
94. Wang, J. L., Cunningham, B., and Edelman, G. M., *Proc. Nat. Acad. Sci. U. S.* **68,** 1130 (1971); Hardman, K. D., Wood, M. K., Schiffer, M., Edmundson, A. B., and Ainsworth, C. F., *ibid.* p. 1393.
95. Stevens, F. C., Brown, D. M., and Smith, E. L., *Arch. Biochem. Biophys.* **136,** 413 (1970).
96. Klotz, I. M., and Keresztes-Nagy, S., *Biochemistry* **2,** 445 and 923 (1963).
97. Cooper, A. J. L., and Meister, A., *Biochemistry* **11,** 661 (1972).
98. Feit, H., Slusarek, L., and Shelanski, M. L., *Proc. Nat. Acad. Sci. U. S.* **68,** 2028 (1971).
99. Preddie, E. C., *J. Biol. Chem.* **244,** 3958 (1969); Gros, C., Lemaire, G., Rapenbusch, R. V., and Labouesse, B., *ibid.* **247,** 2931 (1972).
100. Sasaki, R., Sugimoto, E., and Chiba, H., *Agr. Biol. Chem.* **34,** 135 (1970); Campbell, J. W., Hodgson, G. I., Watson, H. C., and Scopes, R. K., *J. Mol. Biol.* **61,** 257 (1971).
101. Reed, L. J., *Curr. Top. Cell. Regul.* **1,** 233 (1969).
102. Auricchio, F., Valeriote, F., Tomkins, G., and Riley, W. D., *Biochim. Biophys. Acta* **221,** 307 (1970).
103. Allen, S. H. G., and Patil, J. R., *J. Biol. Chem.* **247,** 909 (1972).
104. Biswas, C., Gray, E., and Paulus, H., *J. Biol. Chem.* **245,** 4900 (1970).
105. Carvajal, N., Venegas, A., Oestreicher, G., and Plaza, M., *Biochim. Biophys. Acta* **250,** 437 (1971).
106. Chirikjian, J. G., Wright, H. T., and Fresco, J. R., *Proc. Nat. Acad. Sci. U. S.* **69,** 1638 (1972).
107. Montie, T. C., and Montie, D. B., *Biochemistry* **10,** 2094 (1971).
108. Wong, S. C. K., Hall, D. C., and Josse, J., *J. Biol. Chem.* **245,** 4335 (1970).
109. Rapoport, G., Davis, L., and Horecker, B. L., *Arch. Biochem. Biophys.* **132,** 286 (1969).
110. Dinamarca, M. L., Levenbook, L., and Valdés, E., *Arch. Biochem. Biophys.* **147,** 374 (1971).
111. Hulett-Cowling, F. M., and Campbell, L. L., *Biochemistry* **10,** 1371 (1971).
112. Poillon, W. N., Maeno, H., Koike, K., and Feigelson, P., *J. Biol. Chem.* **244,** 3447 (1969).
113. Bonner, W. M., and Bloch, K., *J. Biol. Chem.* **247,** 3123 (1972).
114. Mavis, R. D., and Stellwagen, E., *J. Biol. Chem.* **243,** 809 (1968).
115. Maley, F., and Maley, G. F., *Fed. Proc., Fed. Amer. Soc. Exp. Biol.* **30,** 1113 (1971).
116. Zito, R., and Kertesz, D., *in* "Biological and Chemical Aspects of Oxy-

genases" (K. Bloch and O. Hayaishi, eds.), p. 290. Maruzen, Tokyo, 1966; Bouchilloux, S., McMahill, P., and Mason, H. S., *J. Biol. Chem.* **238,** 1699 (1963).
117. Sia, C. L., Traniello, S., Pontremoli, S., and Horecker, B. L., *Arch. Biochem. Biophys.* **132,** 325 (1969).
118. Levy, H. R., Raineri, R. R., and Nevaldine, B. H., *J. Biol. Chem.* **241,** 2181 (1966).
119. Peraino, C., Bunville, L. G., and Tahmisian, T. N., *J. Biol. Chem.* **244,** 2241 (1969).
120. Frank, B. H., Pekar, A. H., Veros, A. J., and Ho, P. P. K., *J. Biol. Chem.* **245,** 3716 (1970).
121. Carter, N. D., and Yoshida, A., *Biochim. Biophys. Acta* **181,** 12 (1969); Yoshida, A., and Carter, N. D., *ibid.* **194,** 151 (1969).
122. deKok, A., and Rawitch, A. B., *Biochemistry* **8,** 1405 (1969).
123. Hossler, F. E., and Rendi, R., *Biochem. Biophys. Res. Commun.* **43,** 530 (1971).
124. Rymo, L., Lundvik, L., and Lagerkvist, U., *J. Biol. Chem.* **247,** 3888 (1972); Lagerkvist, U., Rymo, L., Lindqvist, O., and Andersson, E., *ibid.* p. 3897.
125. Roberts, J., Holcenberg, J. S., and Dolowy, *J. Biol. Chem.* **247,** 84 (1972).
126. Heinrich, C. P., and Wiss, O., *FEBS (Fed. Eur. Biol. Soc.) Lett.* **14,** 251 (1971).
127. Tao, M., Salas, M. L., and Lipmann, F., *Proc. Nat. Acad. Sci. U. S.* **67,** 408 (1970); Miyamoto, E., Petzold, G. L., Harris, J. S., and Greengard, P., *Biochem. Biophys. Res. Commun.* **44,** 305 (1971).
128. Bridger, W. A., *Biochem. Biophys. Res. Commun.* **42,** 948 (1971); Leitzmann, C., Wu, J.-Y., and Boyer, P. D., *Biochemistry* **9,** 2338 (1970).
129. Kushner, I., and Somerville, J., *Biochim. Biophys. Acta* **207,** 105 (1970); Gotschlich, E. C., and Edelman, G. M., *Proc. Nat. Acad. Sci. U. S.* **54,** 558 (1965).
130. Castellino, F. J., and Barker, R., *Biochemistry* **7,** 2207 (1968); Heck, H., *J. Biol. Chem.* **244,** 4375 (1969); Schwert, G. W., Miller, B. R., and Peanasky, R. J., *ibid.* **242,** 3245 (1967); Adams, M. J., Ford, G. C., Koekoek, R., Lentz, P. J., Jr., McPherson, A., Jr., Rossmann, M. G., Smiley, I. E., Schevitz, R. W., and Wonacott, A. J., *Nature (London)* **227,** 1098 (1970).
131. Vance, D. E., and Feingold, D. S., *Fed. Proc., Fed. Amer. Soc. Exp. Biol.* **30,** 1057 (1971).
132. Zalkin, H., and Hwang, L. H., *J. Biol. Chem.* **246,** 6899 (1971).
133. Deal, W. C., and Holleman, W. H., *Fed. Proc., Fed. Amer. Soc. Exp. Biol.* **23,** 264 (1964); Harris, J. I., and Perham, R. N., *J. Mol. Biol.* **13,** 876 (1965); Harrington, W. F., and Karr, G. M., *ibid.* p. 885; Jaenicke, R., Schmid, D., and Knof, S., *Biochemistry* **7,** 919 (1968); Hoagland, V. D., Jr., and Teller, D. C., *ibid.* **8,** 594 (1969).
134. Hurlbert, R. E., and Jakoby, W. B., *J. Biol. Chem.* **240,** 2772 (1965).
135. Uyeda, K., and Kurooka, S., *Fed. Proc., Fed. Amer. Soc. Exp. Biol.* **29,** 399 (1970); Uyeda, K., and Kurooka, S., *J. Biol. Chem.* **245,** 3315 (1970).
136. Yoshida, A., *J. Biol. Chem.* **240,** 1113 (1965).
137. Henning, U., Helinski, D. R., Chao, F. C., and Yanofsky, C., *J. Biol. Chem.* **237,** 1523 (1962); Carlton, B. C., and Yanofsky, C., *ibid.* p. 1531; Wilson, D. A., and Crawford, I. P., *Bacteriol. Proc.* p. 92 (1964); Goldberg, M. E., Creighton, T. E., Baldwin, R. L., and Yanofsky, C., *J. Mol. Biol.* **21,** 71 (1966); Yanofsky, C., Drapeau, G. R., Guest, J. R., and Carlton, B. C.,

Proc. Nat. Acad. Sci. U. S. **57,** 296 (1967); Hathaway, G. M., and Crawford, I. P., *Biochemistry* **9,** 1801 (1970).

138. Kolb, H., Cole, R. D., and Snell, E. E., *Biochemistry* **7,** 2946 (1968).
139. Pfleiderer, G., and Auricchio, F., *Biochem. Biophys. Res. Commun.* **16,** 53 (1964); Harris, I., *Nature (London)* **203,** 30 (1964).
140. Schwartz, M. L., Pizzo, S. V., Hill, R. L., and McKee, P. A., *J. Biol. Chem.* **246,** 5851 (1971).
141. Kasper, C. B., and Deutsch, H. F., *J. Biol. Chem.* **238,** 2325 (1963); Poulik, M. D., *Nature (London)* **194,** 842 (1962); Poillon, W. N., and Bearn, A. G., *in* "Biochemistry of Copper" (J. Peisach, P. Aisen, and W. E. Blumberg, eds.), p. 525. Academic Press, New York, 1966.
142. Wang, C. T., and Weissmann, B., *Fed. Proc., Fed. Amer. Soc. Exp. Biol.* **29,** 333 (1970); Wang, C. T., and Weissmann, B., *Biochemistry* **10,** 1067 (1971).
143. Stellwagen, E., and Schachman, H. K., *Biochemistry* **1,** 1056 (1962); Deal, W. C., Rutter, W. J., and van Holde, K. E., *ibid.* **2,** 246 (1963); Schachman, H. K., and Edelstein, S. J., *ibid.* **5,** 2681 (1966); Penhoet, E., Kochman, M., Valentine, R., and Rutter, W. J., *ibid.* **6,** 2940 (1967); Sia, C. L., and Horecker, B. L., *Arch. Biochem. Biophys.* **123,** 186 (1968); Kawahara, K., and Tanford, C., *Biochemistry* **5,** 1578 (1966).
144. Adler, K., Beyreuther, K., Fanning, E., Geisler, N., Gronenborn, B., Klemn, A., Muller-Hill, B., Pfahl, M., and Schmitz, A., *Nature (London)* **237,** 322 (1972).
145. Kaplan, M. M., and Flavin, M., *J. Biol. Chem.* **241,** 5781 (1966).
146. Churchich, J. E., and Dupourque, D., *Biochem. Biophys. Res. Commun.* **46,** 524 (1972).
147. Colby, J., and Zatman, L. J., *Biochem. J.* **121,** 9P (1971).
148. Whiteley, H. R., *J. Biol. Chem.* **241,** 4890 (1966).
149. Ashton, K., and Peacocke, A. R., *FEBS (Fed. Eur. Biochem. Soc.) Lett.* **16,** 25 (1971).
150. Hass, G. M., and Hill, R. L., *J. Biol. Chem.* **244,** 6080 (1969).
151. Benöhr, V. H. C., and Krisch, K., *Hoppe-Seyler's Z. Physiol. Chem.* **348,** 1115 (1967).
152. Barker, D. L., and Jencks, W. P., *Biochemistry* **8,** 3879 (1969).
153. Dalton, H., Morris, J. A., Ward, M. A., and Mortenson, L. E., *Biochemistry* **10,** 2066 (1971).
154. Cohen, G. N., *Curr. Top. Cell. Regul.* **1,** 183 (1969).
155. Trotta, P. P., Burt, M. E., Haschmeyer, R. H., and Meister, A., *Proc. Nat. Acad. Sci. U. S.* **68,** 2599 (1971); Matthews, S. L., and Anderson, P. M., *Biochemistry* **11,** 1176 (1972).
156. Rudolph, F. B., and Fromm, H. J., *Arch. Biochem. Biophys.* **147,** 92 (1971).
157. Niles, E. G., and Westhead, E. W., *Fed. Proc., Fed. Amer. Soc. Exp. Biol.* **29,** 912 (1970).
158. Bruton, C. J., and Hartley, B. S., *Biochem. J.* **117,** 18P (1970); Cassio, D., and Waller, J.-P., *FEBS (Fed. Eur. Biochem. Soc.) Lett.* **12,** 309 (1971).
159. Simon, J. P., Schorr, J. M., and Phillips, A. T., *Fed. Proc., Fed. Amer. Soc. Exp. Biol.* **30,** 1057 (1971).
160. Durell, J., and Cantoni, G. L., *Biochim. Biophys. Acta* **35,** 515 (1959); Klee, W., *ibid.* **59,** 562 (1962).
161. McKeehan, W. L., and Hardesty, B., *J. Biol. Chem.* **244,** 4330 (1969).

162. Lamartiniere, C. A., and Dempey, W. B., *Fed. Proc., Fed. Amer. Soc. Exp. Biol.* **30,** 1121 (1971).
163. Bartholomew, J. C., and Calvo, J. M., *Biochim. Biophys. Acta* **250,** 568 and 577 (1971).
164. Riley, W. D., and Snell, E. E., *Biochemistry* **9,** 1485 (1970).
165. Uyeda, K., *Biochemistry* **8,** 2366 (1969); Paetkau, V. H., Younathan, E. S., and Lardy, H. A., *J. Mol. Biol.* **33,** 721 (1968).
166. Kanarek, L., Marler, E., Bradshaw, R. A., Fellows, R. E., and Hill, R. L., *J. Biol. Chem.* **239,** 4207 (1964); Penner, P. E., and Cohen, L. H., *ibid.* **246,** 4261 (1971).
167. Zarlengo, M. H., Robinson, C. W., and Burns, R. O., *J. Biol. Chem.* **243,** 186 (1968).
168. Metrione, R. M., Okuda, Y., and Fairclough, G. F., Jr., *Biochemistry* **9,** 2427 (1970).
169. Maeba, P., and Sanwal, B. D., *J. Biol. Chem.* **244,** 2549 (1969).
170. Rowe, P. B., and Wyngaarden, J. B., *J. Biol. Chem.* **243,** 6373 (1968).
171. Lehrach, H., Schafer, K., and Scheit, K. H., *FEBS* (*Fed. Eur. Biochem. Soc.*) *Lett.* **14,** 343 (1971).
172. Leary, T. R., and Kohlhaw, G. B., *J. Biol. Chem.* **247,** 1089 (1972).
173. Colman, P. M., and Matthews, B. W., *J. Mol. Biol.* **60,** 163 (1971); Spradlin, J., and Thoma, J. A., *J. Biol. Chem.* **245,** 117 (1970).
174. Schulze, I. T., Lusty, C. J., and Ratner, S., *J. Biol. Chem.* **245,** 4534 (1970).
175. Bonsignore, A., Cancedda, R., Lorenzoni, I., Cosulich, M. E., and DeFlora, A., *Biochem. Biophys. Res. Commun.* **43,** 94 (1971).
176. Kamen, R., *Nature* (*London*) **228,** 527 (1970); Kondo, M., Gallerani, R., and Weissman, C., *ibid.* p. 525.
177. McFadden, B. A., Rao, G. R., Cohen, A. L., and Roche, T. E., *Biochemistry* **7,** 3574 (1968).
178. Scott, W. A., *J. Biol. Chem.* **246,** 6353 (1971).
179. Nilsson, U., *Fed. Proc., Fed. Amer. Soc. Exp. Biol.* **31,** 740 (1972).
180. Hoch, S. O., and DeMoss, R. D., *J. Biol. Chem.* **247,** 1750 (1972).
181. Gounaris, A. D., Turkenkopf, I., Buckwald, S., and Young, A., *J. Biol. Chem.* **246,** 1302 (1971).
182. Meachum, Z. D., Jr., Calvin, H. J., Jr., and Braymer, H. D., *Biochemistry* **10,** 326 (1971).
183. Long, C. W., Levitzki, A., and Koshland, D. E., Jr., *J. Biol. Chem.* **245,** 80 (1970).
184. Cox, A. C., and Tanford, C., *J. Biol. Chem.* **243,** 3083 (1968); Scanu, A., Reader, W., and Edelstein, C., *Fed. Proc., Fed. Amer. Soc. Exp. Biol.* **26,** 435 (1967).
185. Voll, M. J., Appella, E., and Martin, R. G., *J. Biol. Chem.* **242,** 1760 (1967).
186. Martinez-Carrion, M., Critz, W., and Quashnock, J., *Biochemistry* **11,** 1613 (1972).
187. Carlsen, R. B., and Pierce, J. G., *J. Biol. Chem.* **247,** 23 (1972).
188. Thorner, J. W., and Paulus, H., *J. Biol. Chem.* **246,** 3885 (1971).
189. Huseby, N.-E., Christensen, T. B., Olsen, B. R., and Størmer, F. C., *Eur. J. Biochem.* **20,** 209 (1971).
190. Hoch, J. A., and DeMoss, R. D., *Biochemistry* **5,** 3137 (1966); Morino, Y., and Snell, E. E., *J. Biol. Chem.* **242,** 5591 (1967).
191. Lowey, S., Kucera, J., and Holtzer, A., *J. Mol. Biol.* **7,** 234 (1963); Olander,

J., Emerson, M., and Holtzer, A., *J. Amer. Chem. Soc.* **89,** 3058 (1967); McCubbin, W., and Kay, C. M., *Biochim. Biophys. Acta* **154,** 239 (1968).
192. Schmit, J. C., and Zalkin, H., *J. Biol. Chem.* **246,** 6002 (1971).
193. Little, G. H., Riley, P. S., and Behal, F. J., *Fed. Proc., Fed. Amer. Soc. Exp. Biol.* **30,** 1121 (1971).
194. Marshall, M., and Cohen, P. P., *J. Biol. Chem.* **247,** 1641 (1972).
195. Ostrem, D. L., and Berg, P., *Proc. Nat. Acad. Soc. U. S.* **67,** 1967 (1970).
196. Chung, S.-T., Tan, R. T. Y., and Suzuki, I., *Biochemistry* **10,** 1205 (1971).
197. Tanford, C., and Lovrien, R., *J. Amer. Chem. Soc.* **84,** 1892 (1962); Schroeder, W. A., Shelton, J. R., Shelton, J. B., and Olson, B. M., *Biochim. Biophys. Acta* **89,** 47 (1964); Weber, K., and Sund, H., *Angew. Chem.* **77,** 621 (1965); Schroeder, W. A., Shelton, J. R., Shelton, J. B., Robberson, B., and Apell, G., *Arch. Biochem. Biophys.* **131,** 653 (1969).
198. Monteilhet, C., and Risler, J. L., *Eur. J. Biochem.* **12,** 165 (1970); Lederer, F., and Simon, A. M., *Eur. J. Biochem.* **20,** 469 (1971).
199. Morawiecki, A., *Biochim. Biophys. Acta* **44,** 604 (1960); Steinmetz, M. A., and Deal, W. C., Jr., *Biochemistry* **5,** 1399 (1966).
200. Curthoys, N. P., Straus, L. D., and Rabinowitz, J. C., *Biochemistry* **11,** 345 (1972).
201. Heyningen, S. V., and Shemin, D., *Biochemistry* **10,** 4676 (1971).
202. Gaertner, F. H., and DeMoss, J. A., *J. Biol. Chem.* **244,** 2716 (1969).
203. Montie, T. C., and Montie, D. B., *Biochemistry* **10,** 2094 (1971).
204. Berg, R. A., Olsen, B. R., and Kivirikko, K. I., *Fed. Proc., Fed. Amer. Soc. Exp. Biol.* **31,** 479 (1972).
205. Gauthier, J. J., and Rittenberg, S. C., *J. Biol. Chem.* **246,** 3737 (1971).
206. Feldman, R. I., and Weiner, H., *J. Biol. Chem.* **247,** 260 (1972).
207. Doyle, D., *Fed. Proc., Fed. Amer. Soc. Exp. Biol.* **30,** 1230 (1971); van Heyningen, S., and Shemin, D., *ibid.* **29,** 937 (1970).
208. Kashiwamata, S., Kotake, Y., and Greenburg, D. M., *Biochim. Biophys. Acta* **212,** 501 (1970).
209. Darrow, R. A., and Rodstrom, R., *J. Biol. Chem.* **245,** 2036 (1970).
210. Fodge, D. W., Gracy, R. W., and Harris, B. G., *Biochim. Biophys. Acta* **268,** 271 (1972).
211. Correll, D. L., Steers, E., Jr., Towe, K. M., and Shropshire, W., Jr., *Biochim. Biophys. Acta* **168,** 46 (1968).
212. Melius, P., Moseley, M. H., and Brown, D. M., *Biochim. Biophys. Acta* **221,** 62 (1970).
213. Millar, D. B., and Grafius, M. A., *FEBS (Fed. Eur. Biochem. Soc.) Lett.* **12,** 61 (1970).
214. Frigerio, N. A., and Harbury, H. A., *J. Biol. Chem.* **231,** 135 (1958).
215. Schmidt, J., Wang, R., Stanfield, S., and Reid, B. R., *Biochemistry* **10,** 3264 (1971).
216. Boosman, A., Sammons, D., and Chilson, D., *Biochem. Biophys. Res. Commun.* **45,** 1025 (1971).
217. Rubin, C. S., Erlichman, J., and Rosen, O. M., *J. Biol. Chem.* **247,** 36 (1972).
218. Stahl, P. D., and Touster, O., *Fed. Proc., Fed. Amer. Soc. Exp. Biol.* **30,** 1121 (1971).
219. Henderson, E. J., and Zalkin, H., *J. Biol. Chem.* **246,** 6891 (1971).
220. Zappia, V., and Barker, H. A., *Biochim. Biophys. Acta* **207,** 505 (1970).

221. Barnes, L. D., Kuehn, G. D., and Atkinson, D. E., *Biochemistry* **10,** 3939 (1971).
222. Kredich, N. M., Becker, M. A., and Tomkins, G. M., *J. Biol. Chem.* **244,** 2428 (1969).
223. Schwartz, M. L., Pizzo, S. V., Hill, R. L., and McKee, P. A., *J. Biol. Chem.* **246,** 5851 (1971).
224. Gerhart, J. C., and Schachman, H. K., *Biochemistry* **4,** 1054 (1965); Schachman, H. K., and Edelstein, S. J., *ibid.* **5,** 2681 (1966); Changeux, J. P., Gerhart, J. C., and Schachman, H. K., *ibid.* **7,** 531 (1968); Weber, K., *Nature (London)* **218,** 1116 (1968); Wiley, D. C., and Lipscomb, W. N., *ibid.* p. 1119.
225. Strausbauch, P. H., and Fischer, E. H., *Biochemistry* **9,** 226 (1970).
226. Main, A. R., Tarkan, E., Aull, J. L., and Soucie, W. G., *J. Biol. Chem.* **247,** 566 (1972).
227. Virden, R., *Biochem. J.* **127,** 503 (1972).
228. Eisenberg, H., and Tomkins, G. M., *J. Mol. Biol.* **51,** 37 (1968); Cassman, M., and Schachman, H. K., *Biochemistry* **10,** 1015 (1971); Reisler, E., and Eisenberg, H., *ibid.* p. 2659; Josephs, R., *J. Mol. Biol.* **55,** 147 (1971).
229. Farron, F., *Biochemistry* **9,** 3823 (1970).
230. Melbye, S. W., and Carpenter, F. H., *J. Biol. Chem.* **246,** 2459 (1971).
231. Havir, E. A., and Hanson, K. R., *Fed. Proc., Fed. Amer. Soc. Exp. Biol.* **31,** 864 (1972).
232. Tiemeier, D. C., and Milman, G., *J. Biol. Chem.* **247,** 2272 (1972).
233. Tagaki, W., and Westheimer, F. H., *Biochemistry* **7,** 891 and 895 (1968).
234. Starnes, W. L., Munk, P., Maul, S. B., Cunningham, G. N., Cox, D. J., and Shive, W., *Biochemistry* **11,** 677 (1972); Clark, R. B., and Ogilvie, J. W., *ibid.* p. 1278.
235. Kerr, M. W., and Robertson, A., *Biochem. J.* **125,** 34P (1971).
236. Witheiler, J., and Wilson, D. B., *J. Biol. Chem.* **247,** 2217 (1972).
237. Tombs, M. P., and Lowe, M., *Biochem. J.* **105,** 181 (1967).
238. Pollock, R. J., and Hersh, L. B., *J. Biol. Chem.* **246,** 4737 (1971).
239. Tate, S. S., and Meister, A., *Proc. Nat. Acad. Sci. U. S.* **68,** 781 (1971).
240. MacColl, R., Lee, J. J., and Berns, D. S., *Biochem. J.* **122,** 421 (1971); Berns, D. S., *Biochem. Biophys. Res. Commun.* **38,** 65 (1970); Neufeld, G. J., and Riggs, A. F., *Biochim. Biophys. Acta* **181,** 234 (1969).
241. Patrick, J. W., and Lee, N., *J. Biol. Chem.* **244,** 4277 (1969).
242. Kapoor, M., Bray, D. F., and Ward, G. W., *Arch. Biochem. Biophys.* **134,** 423 (1969).
243. Adair, L. B., and Jones, M. E., *J. Biol. Chem.* **247,** 2308 (1972).
244. Lambeth, D. D., and Lardy, H. A., *Eur. J. Biochem.* **22,** 355 (1971); Catterall, W. A., and Pedersen, P. L., *J. Biol. Chem.* **246,** 4987 (1971).
245. Seery, V. L., Fischer, E. H., and Teller, D. C., *Biochemistry* **9,** 3591 (1970); Madsen, N. B., and Cori, C. F., *J. Biol. Chem.* **223,** 1055 (1956); Seery, V. L., Fischer, E. H., and Teller, D. C., *Biochemistry* **6,** 3315 (1967); DeVincenzi, D. L., and Hedrick, J. L., *ibid.* p. 3489.
246. Schnebli, H. P., Vatter, A. E., and Abrams, A., *J. Biol. Chem.* **245,** 1122 (1970).
247. Burgess, R. R., *J. Biol. Chem.* **244,** 6168 (1969); Johnson, J. C., DeBacker, M., and Boezi, J. A., *ibid.* **246,** 1222 (1971).

248. Weaver, R. F., Blath, S. P., and Rutler, W. J., *Proc. Nat. Acad. Sci. U. S.* **68,** 2994 (1971).
249. Hyun, J., Steinberg, M., Treadwell, C. R., and Vahouny, G. V., *Biochem. Biophys. Res. Commun.* **44,** 819 (1971).
250. Bernardi, G., and Cook, W. H., *Biochim. Biophys. Acta* **44,** 96 and 105 (1960); Burley, R. W., and Cook, W. H., *Can. J. Biochem. Physiol.* **40,** 363 (1962).
251. Tate, S. S., and Meister, A., *Proc. Nat. Acad. Sci. U. S.* **68,** 781 (1971).
252. Smith, T. E., *J. Biol. Chem.* **246,** 4234 (1971).
253. Haberland, M. E., Willard, J. M., and Wood, H. G., *Biochemistry* **11,** 712 (1972).
254. Tweedie, J. W., and Segel, I. H., *J. Biol. Chem.* **246,** 2438 (1971).
255. Butterworth, P. H. W., Yang, P. C., Bock, R. M., and Porter, J. W., *J. Biol. Chem.* **242,** 3508 (1967).
256. Bryce, C. F. A., and Crichton, R. R., *J. Biol. Chem.* **246,** 4198 (1971); Bjork, I., and Fish, W. W., *Biochemistry* **10,** 2844 (1971).
257. Holtzer, A., and Lowey, S., *J. Amer. Chem. Soc.* **81,** 1370 (1959); Mueller, H., *J. Biol. Chem.* **239,** 797 (1964); Tonomura, Y., Appel, P., and Morales, M., *Biochemistry* **5,** 515 (1966); Richards, E. G., Chung, C.-S., Menzel, D., and Olcott, H., *ibid.* **6,** 528 (1967); Gershman, L., Stracher, A., and Dreizen, P., *J. Biol. Chem.* **244,** 2726 (1969); Kominz, D., Carroll, W. R., Smith, E., and Mitchell, E., *Arch. Biochem. Biophys.* **79,** 191 (1959); Frederiksen, D., and Holtzer, A., *Biochemistry* **7,** 3935 (1968).
258. Contaxis, C. C., and Reithel, F. J., *J. Biol. Chem.* **246,** 677 (1971); Bailey, C. J., and Boulter, D., *Biochem. J.* **113,** 669 (1969); Creeth, J. M., and Nichol, L. W., *ibid.* **77,** 230 (1960); Reithel, F. J., Robbins, J. E., and Gorin, G., *Arch. Biochem. Biophys.* **108,** 409 (1964).
259. Levine, S., Gillett, T. A., Turnquist, M., and Hansen, R. G., *Fed. Proc., Fed. Amer. Soc. Exp. Biol.* **30,** 1121 (1971); Levine, S., Gillett, T. A., Hageman, E., and Hansen, R. G., *J. Biol. Chem.* **244,** 5729 (1969).
260. Kuehn, G. D., and McFadden, B. A., *Biochemistry* **8,** 2403 (1969).
261. Craven, G. R., Steers, E., Jr., and Anfinsen, C. B., *J. Biol. Chem.* **240,** 2468 (1965); Fowler, A. V., and Zabin, I., *ibid.* **245,** 5032 (1970).
262. Rutner, A. C., *Biochem. Biophys. Res. Commun.* **39,** 923 (1970); Kawashima, N., and Wildman, S. G., *ibid.* **41,** 1463 (1970); Trown, P. W., *Biochemistry* **4,** 908 (1965); Haselkorn, R., Fernández-Morán, H., Kieras, F. J., and van Bruggen, E. J. F., *Science* **150,** 1598 (1965).
263. Spina, J., Jr., Bright, H. J., and Rosenbloom, J., *Biochemistry* **9,** 3794 (1970).
264. Bowen, T. J., and Mortimer, M. G., *Biochem. J.* **117,** 71P (1970); Mahadik, S. P., and SivaRaman, C., *Biochem. Biophys. Res. Commun.* **32,** 167 (1968).
265. Sugiyama, T., Ito, T., and Akazawa, T., *Biochemistry* **10,** 3406 (1971).
266. Wilgus, H., Pringle, J. R., and Stellwagen, E., *Biochem. Biophys. Res. Commun.* **44,** 89 (1971).
267. Woolfolk, C. A., and Stadtman, E. R., *Arch. Biochem. Biophys.* **122,** 174 (1967); Valentine, R. C., Shapiro, B. M., and Stadtman, E. R., *Biochemistry* **7,** 2143 (1968).
268. Donovan, J. W., Mapes, C. J., Davis, J. G., and Hamburg, R. D., *Biochemistry* **8,** 4190 (1969).
269. Valentine, R. C., Wrigley, N. G., Scrutton, M. C., Irias, J. J., and Utter, M. F., *Biochemistry* **5,** 3111 (1966).

270. Steiner, R. F., and Edelhoch, H., *J. Amer. Chem. Soc.* **83,** 1435 (1961); Edelhoch, H., and de Crombrugghe, B., *J. Biol. Chem.* **241,** 4357 (1966).
271. Ahmad, F., Jacobson, B., and Wood, H. G., *J. Biol. Chem.* **245,** 6486 (1970); Jacobson, B., Gerwin, B. J., Ahmad, F., Waegell, P., and Wood, H. G., *ibid.* p. 6471; Ahmad, F., Jacobson, B., Wood, H. G., Valentine, R. C., Green, M., and Wrigley, N., *Fed. Proc., Fed. Amer. Soc. Exp. Biol.* **30,** 1058 (1971).
272. Giorgio, N. A., Jr., Yip, A. T., Fleming, J., and Plaut, G. W. E., *J. Biol. Chem.* **245,** 5469 (1970); Harvey, R. A., Giorgio, N. A., Jr., and Plaut, G. W. E., *Fed. Proc., Fed. Amer. Soc. Exp. Biol.* **29,** 532 (1970).
273. Bowers, W. F., Czubaroff, V. B., and Haschemyer, R. H., *Biochemistry* **9,** 2620 (1970); Tate, S. S., and Meister, A., *ibid.* p. 2626.
274. Siegel, L. M., and Kamin, H., *Fed. Proc., Fed. Amer. Soc. Exp. Biol.* **30,** 1261 (1971).
275. Kaziro, Y., Ochoa, S., Warner, R. C., and Chen, J., *J. Biol. Chem.* **236,** 1917 (1961).
276. Lee-Huang, S., and Warner, R. C., *J. Biol. Chem.* **244,** 3793 (1969).
277. Kakimoto, T., Kato, J., Shibatani, T., Nishimura, N., and Chibata, I., *J. Biol. Chem.* **245,** 3369 (1970).
278. Bloemendal, H., Bont, W. S., Jongkind, J. F., and Wisse, J. H., *Exp. Eye Res.* **1,** 300 (1962); Bloemendal, H., Bont, W. S., Benedett, E. L., and Wisse, J. H., *ibid.* **4,** 319 (1965).
279. Boeker, E. A., Fischer, E. H., and Snell, E. E., *J. Biol. Chem.* **244,** 5239 (1969); Boeker, E. A., and Snell, E. E., *ibid.* **243,** 1678 (1968).
280. Stevens, A., Emery, A. J., Jr., and Sternberger, N., *Biochem. Biophys. Res. Commun.* **24,** 929 (1966); Richardson, J. P., *Proc. Nat. Acad. Sci. U. S.* **55,** 1616 (1966).
281. Schwartz, E. R., and Reed, L. J., *J. Biol. Chem.* **244,** 6074 (1969); Reed, L. J., *Curr. Top. Cell. Regul.* **1,** 233 (1969).
282. DePhillips, H. A., Nickerson, K. W., Johnson, M., and Van Holde, K. E., *Biochemistry* **8,** 3665 (1969); Pickett, S. M., Riggs, A. F., and Larimer, J. L., *Science* **151,** 1005 (1966); Van Holde, K. E., and Cohen, L. B., *Biochemistry* **3,** 1803 (1964); Fernández-Morán, H., van Bruggen, E. J. F., and Ohtsuki, M., *J. Mol. Biol.* **16,** 191 (1966); Lontie, R., and Witters, R., *in* "The Biochemistry of Copper" (J. Peisach, P. Aisen, and W. E. Blumberg, eds.), p. 455. Academic Press, New York, 1966.
283. Guerritore, D., Bonacci, M. L., Brunori, M., Antonini, E., Wyman, J., and Rossi-Fanelli, A., *J. Mol. Biol.* **13,** 234 (1965).
284. Waxman, L., *J. Biol. Chem.* **246,** 7318 (1971).
285. Swaney, J. B., and Klotz, I. M., *Arch. Biochem. Biophys.* **147,** 475 (1971).
286. Hohn, T., and Hohn, B., *Advan. Virus Res.* **16,** 43 (1970).
287. Reed, L. J., *Curr. Top. Cell. Regul.* **1,** 233 (1969); Perham, R., and Thomas, J. O., *FEBS (Fed. Eur. Biochem. Soc.) Lett.* **15,** 8 (1971).
288. Bancroft, J. B., Hiebert, E., Rees, M. W., and Markham, R., *Virology* **34,** 224 (1968).
289. Bockstahler, L. E., and Kaesberg, P., *Biophys. J.* **2,** 1 (1962).
290. Miki, T., and Knight, C. A., *Virology* **25,** 478 (1965).
291. Markham, R., *Discuss. Faraday Soc.* **11,** 221 (1951); Harris, J. I., and Hindley, J., *J. Mol. Biol.* **3,** 117 (1961).
292. Anderer, F. A., and Restle, H., *Z. Naturforsch. B* **19,** 1026 (1964).
293. Yamazaki, H., and Kaesberg, P., *Biochim. Biophys. Acta* **53,** 173 (1961).

294. Kelley, J. J., and Kaesberg, P., *Biochim. Biophys. Acta* **55,** 236 (1962); *ibid.* **61,** 865 (1962); Kruseman, J., Kraal, B., Jaspars, E. M. J., Bol. J. F., Brederode, F. T., and Veldstra, H., *Biochemistry* **10,** 447 (1971).
295. Gregolin, C., Ryder, E., Warner, R. C., Kleinschmidt, A. K., and Lane, M. D., *Proc. Nat. Acad. Sci. U. S.* **56,** 1751 (1966).
296. Medappa, K. C., McLean, C., and Rueckert, R. R., *Virology* **44,** 259 (1971).
297. Hersh, R. T., and Schachman, H. K., *Virology* **6,** 234 (1958).
298. Fine, R., Mass, M., and Murakami, W. T., *J. Mol. Biol.* **36,** 167 (1968).
299. Reichmann, M. E., *J. Biol. Chem.* **235,** 2959 (1960); Reichmann, M. E., and Hatt, D. L., *Biochim. Biophys. Acta* **49,** 153 (1961).
300. Anderer, F. A., *Advan. Protein Chem.* **18,** 1 (1963); Caspar, D. L. D., *Advan. Protein Chem.* **18,** 37 (1963).

By far the majority of proteins possessing a quaternary structure are constituted of two or four subunits (Table III). Well over half of the 300 proteins listed in Table II have either two or four subunits. Nearly one-third of the proteins listed have only two subunits, and few proteins have more than twelve subunits with the exception of large enzyme complexes and viruses.

Most of the proteins listed in Table II have an even number of subunits. Only ten proteins of the nearly 300 examined have an odd number, and all of these have three subunits per oligomer. There have been no oligomers found which contain only five subunits, although there are three decameric oligomers which are composed of two five-membered rings.

Most of the proteins listed in Table II are composed of identical sub-

TABLE III

Distribution of Subunit Stoichiometries

Number of subunits	Number of proteins with designated number of identical subunits[a]	Number of proteins with designated number of nonidentical subunits[a]
2	96	13
3	7	3
4	77	17
5	0	0
6	26	1
7	0	0
8	11	2
9	0	0
10	2	1
11	0	0
12	7	3

[a] Summed from entries in Table II.

TABLE IV

Constitution of Proteins Composed of Nonidentical Subunits[a]

Protein	Molecular weight	Subunit composition
Rat pituitary leutinizing hormone	31,000	$\alpha\beta$
Neurospora malate dehydrogenase	54,000	$\alpha_3\beta$
Lactose synthetase	60,000	$\alpha\beta$
Aldose reductase	61,000	$\alpha\beta$
Hemoglobin	64,500	$\alpha_2\beta_2$
T_u–T_s Complex	65,000	$\alpha\beta$
Bacterial luciferase	79,000	$\alpha\beta$
Troponin	(80,000)	$\alpha_2\beta$
Bovine procarboxypeptidase A	88,000	$\alpha\beta\gamma$
Serum high density lipoprotein	88,000	$\alpha_2\beta_2$
Succinate dehydrogenase	97,000	$\alpha\beta$
Tuberlin	110,000	$\alpha\beta$
Aspartokinase	116,000	$\alpha_2\beta_2$
Pasteurella protein toxin B	120,000	$(\alpha_6\beta_6)$
Fructose diphosphatase	130,000	$\alpha_2\beta_2$
Protein kinase	140,000	$\alpha\beta$
Succinyl-CoA synthetase	140,000	$\alpha_2\beta_2$
Serratia anthranilate synthetase	(141,000)	$\alpha_2\beta_2$
Tryptophan synthetase	148,000	$\alpha_2\beta_2$
Molybdoferredoxin	168,000	$\alpha_2\beta$
E. coli carbamyl phosphate synthetase	170,000	$\alpha\beta$
Histidine decarboxylase	190,000	$\alpha_5\beta_5$
Dipeptidyl transferase	197,000	$(\alpha_6\beta_2)$
E. coli Qβ replicase	205,000	$\alpha\beta\gamma\delta$
Human C5	(206,000)	$\alpha\beta$
Rat α-L-fucosidase	216,000	$\alpha_2\beta_2$
Glycyl-tRNA synthetase	227,000	$\alpha_2\beta_2$
Neurospora anthranilate synthetase complex	240,000	$\alpha_2\beta_4$
Pasteurella protein toxin A	240,000	$(\alpha_{12}\beta_{12})$
Protocollagen proline hydroxylase	240,000	$\alpha_2\beta_2$
Cystathionine synthetase	250,000	$\alpha_2\beta_2$
Phenylalanine-tRNA synthetase	276,000	$\alpha_2\beta_2$
Bovine protein kinase	280,000	$(\alpha\beta)$
Salmonella anthranilate synthetase complex	280,000	$\alpha_2\beta_2$
Cysteine synthetase	309,000	$\alpha_2\beta$
Aspartate transcarbamylase	310,000	$\alpha_6\beta_6$
Human plasma factor XIII	310,000	$\alpha_2\beta_2$
Arachin	345,000	$\alpha_2\beta_2\gamma_8$
		$\alpha_4\gamma_8$
		$\beta_4\gamma_8$
Rat liver mitochondrial ATPase	360,000	$(\alpha\beta\gamma_8)$
Membrane ATPase	385,000	$\alpha_6\beta_6$

(*Continued*)

TABLE IV (*Continued*)

Protein	Molecular weight	Subunit composition
E. coli RNA polymerase	400,000	$\alpha_2\beta\beta'$
RNA polymerase-sigma factor	486,000	$\alpha_2\beta\beta'\delta$
Chlorella ribulose-1,5 diphosphate carboxylase	588,000	$\alpha_8\beta_8$
Transcarboxylase	670,000	$(\alpha\beta)$
TPNH-Sulfite reductase	700,000	$\alpha_4\beta_8$

[a] Literature references to these entries may be obtained by reference to Table II. As in Table II, entries enclosed by parentheses are somewhat uncertain.

units, although in recent years the number of oligomeric proteins with nonidentical subunits has increased to approximately 15% of the known oligomeric proteins. Table IV lists the proteins that are composed of nonidentical subunits. In general, proteins that are composed of nonidentical subunits have only two different subunits, as all but seven of the entries in Table IV show. The most prevalent oligomers that contain nonidentical subunits are the dimers and tetramers (Tables III and IV).

III. GEOMETRIC ARRANGEMENTS OF IDENTICAL SUBUNITS

Since most proteins are constituted of a small number of subunits, there is a finite number of spatial arrangements which the oligomeric ensembles may assume. We shall now describe the different geometries possible and, when known, cite the architectural details deduced directly from electron microscopy or X-ray diffraction.

When identical protein monomers associate, the macromolecular assembly may possess one of several different geometries. The number of arrangements increases with the number of subunits but can nevertheless be limited if we impose certain restrictions at the outset. First, we shall assume that all subunits in an oligomeric protein are in equivalent (or pseudoidentical) environments. This assumption seems valid, since with only one or two exceptions, the 30–40 proteins that have been examined by electron microscopy or X-ray diffraction fulfill this specification. Thus we shall disregard linear arrays of subunits and variations of the linear structure. Second, since most protein oligomers are composed of a small number of subunits, the bonding potential or binding regions of the subunits must be saturated or else higher protein aggregates would be encountered. Hence we shall assume that arrangements of subunits in the oligomeric ensemble are restricted to closed sets. For example, if a protein is composed of four subunits, the formation of a hexamer or an octamer would be unlikely. This means the monomers of the tetramer must have their interface strong-binding regions saturated.

These two restrictions require that the subunits be regularly packed around a central point in the oligomer. Therefore only point group symmetry is possible. Although some virus structures possess line symmetry, i.e., helical structures, such an arrangement is not possible for small protein aggregates because the subunits at the end are necessarily in a different environment from those in the interior. Such a situation is ruled out by our first restriction.

Point groups contain rotation axes and inversion axes. However, for proteins composed of L-amino acids, inversion symmetry is not possible, since a protein is necessarily different from its mirror image. Therefore only rotation axes passing through a point need be considered. Monod *et al.* (1965) and Hanson (1966, 1968) have described some of the possible geometries and point groups that are found in protein oligomers, and Klug (1967) has discussed those that pertain particularly to virus structure. Here we shall consider point groups pertinent to small protein oligomers. These point groups may be categorized as (1) cyclic, (2) dihedral, and (3) cubic.

A. Cyclic Symmetry

In cyclic symmetry, the simplest symmetry possible, the subunits are so disposed that there is an n-fold rotational axis, where n is the number of subunits. Thus the subunits are arranged in a head-to-tail fashion around a circle. A dimer ($n = 2$) has a single twofold rotational axis. The dimer is a member of the cyclic point group[3] having C_2 symmetry (rotation of 360°/2 or 180° transposes the dimer into itself). A trimer ($n = 3$) has a single threefold axis of rotation and C_3 symmetry (rotation of 360°/3 or 120° transposes the trimer into itself). These are the only two geometries possible if the subunits of the dimers or trimers are in identical environments (see Fig. 3). A tetramer ($n = 4$) possesses C_4 symmetry, the subunits being arranged at the corners of a square (Fig. 4). Alternatively, the subunits of a tetramer can have a tetrahedral arrangement (see Fig. 4 and Section III,B). A pentamer ($n = 5$) has C_5 symmetry and the subunits are arranged at the corners of a pentagon (Fig. 3). Within our restrictions, this is the only symmetry possible for

[3] In this chapter we use the Schönfliess notation for symmetry operations. For conversion to the crystallographic notation, see Jaffé and Orchin (1965).

Fig. 3 Cyclic symmetry in arrangements of subunits in a protein dimer, trimer, and pentamer.

a pentamer. If arranged in cyclic symmetry, the subunits of each higher aggregate simply would occupy the corners of a regular polygon.

When two protomers associate to form a dimer in which the protomers are sterically equivalent, the dimer possesses a single twofold axis of symmetry (Fig. 3). Such association is said to be *isologous* (Monod *et al.*, 1965), i.e., the contact sites are identical in both subunits. Isologous bonding is the only type possible for a dimer if the subunits are assumed to be equivalent and a closed structure is formed (Fig. 5). However, it appears that some proteins, e.g., hexokinase B (Steitz, 1971) and insulin (Adams *et al.*, 1969), are exceptions. X-Ray diffraction studies (Steitz, 1971) have shown that the two subunits of hexokinase B are

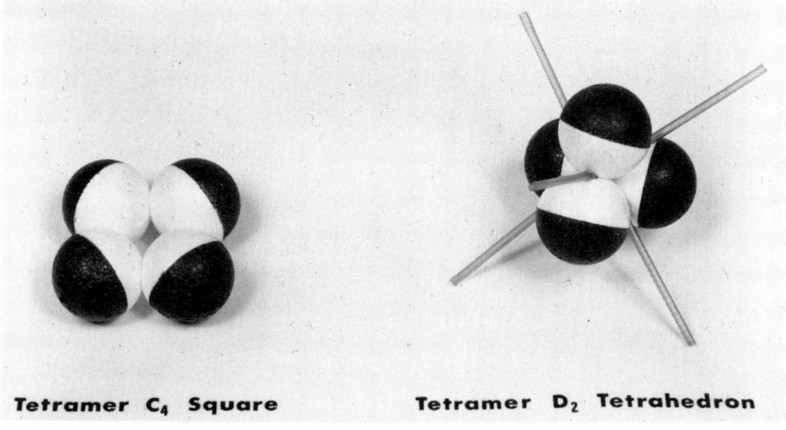

Fig. 4 Cyclic and dihedral symmetry in arrangements of subunits in a protein tetramer. The three twofold axes of symmetry are indicated in the tetrahedral ensemble.

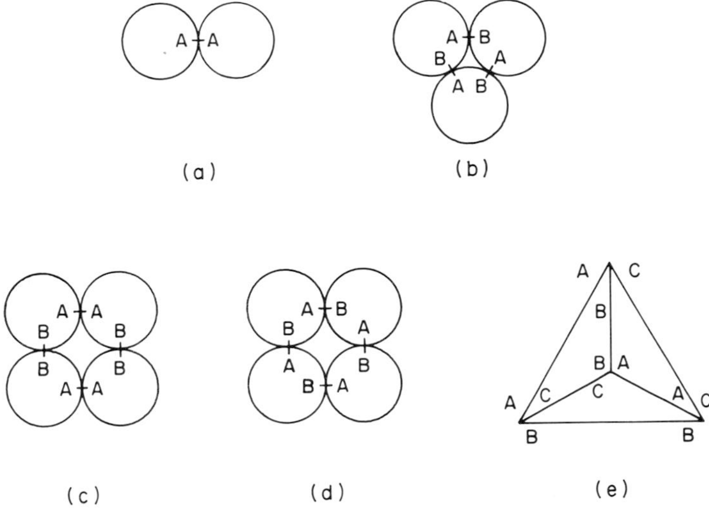

Fig. 5 Isologous and heterologous bonding between protein subunits. The letters indicate the bonding domain on the protein: A—A represents isologous bonding while A—B or A—C indicates heterologous bonding. (a) Dimer, isologous bonding, C_2 symmetry; (b) trimer, heterologous bonding, C_3 symmetry; (c) tetramer, isologous bonding, D_2 symmetry; (d) tetramer, heterologous bonding, C_4 symmetry; and (e) tetramer, isologous bonding, D_2 symmetry. The protein subunits would be placed at each of the four corners of the tetrahedron in (e), but have been omitted to simplify the representation.

related by an approximate rather than by a true dyad axis, i.e., one subunit is translated relative to the other by 3.6 Å along the molecular symmetry axis. X-Ray analysis of insulin (Adams *et al.*, 1969) has also shown that a distorted axis is present.

If three subunits associate to form a trimer in which the protomers are sterically equivalent, the trimer necessarily possesses a single threefold axis of symmetry (Fig. 3 and Table V). Since isologous association is not possible in this case, the bonding is said to be *heterologous* (Monod *et al.*, 1965), i.e., the subunit contact regions between any two monomers are nonidentical (see Fig. 5b). In fact, in any oligomer that has an odd number of subunits, only heterologous subunit interactions are possible. On the other hand, the subunit association in an oligomer with an even number of subunits may be either isologous or heterologous (Fig. 5c,d).

Included among proteins with C_2 symmetry are rhodanese, a dimer whose structure has been revealed by X-ray crystallography (Drenth and Smit, 1971), and the acetolactate-forming enzyme, which has been

TABLE V

Symmetry and Bonding in Protein Oligomers

Number of subunits	Geometry	Symmetry	Total number of binding regions between subunits	Binding regions per subunit	Can have both isologous and heterologous binding regions?[a]	Types of binding regions[a]
2	Linear	C_2	1	1	No	Isologous
3	Triangle	C_3	3	2	No	Identical heterologous
4	Square	C_4	4	2	No	Identical heterologous
4	Square	D_2	4	2	No	Two different isologous
4	Tetrahedron	D_2	6	3	No	Three different isologous
5	Pentagon	C_5	5	2	No	Identical heterologous
6	Hexagon	C_6	6	2	No	Identical heterologous
6	Hexagon	D_3	6	2	No	Two different isologous
6	Trigonal prism	D_3	9	3 (1 isologous, 2 heterologous)	Yes	Isologous and heterologous
6	Octahedron	D_3	12	4 (2 isologous, 2 heterologous)	Yes	Isologous and heterologous
7	Heptagon	C_7	7	2	No	Identical heterologous
8	Octagon	C_8	8	2	No	Identical heterologous
8	Octagon	D_4	8	2	No	Two different isologous
8	Cube	D_4	12	3 (1 isologous, 2 heterologous)	Yes	Isologous and heterologous
8	Square antiprism	D_4	16	4 (2 isologous, 2 heterologous)	Yes	Isologous and heterologous

[a] Assuming all subunits are in identical environments.

examined by electron microscopy (Huseby *et al.*, 1971). As mentioned above, both hexokinase B (Steitz, 1971) and insulin (Adams *et al.*, 1969) have been shown to possess distorted dyad axes of symmetry.

Electron microscopy has shown that the subunits of the tetrameric enzymes tryptophanase (Morino and Snell, 1967a), pyruvate carboxylase (Valentine *et al.*, 1966), and diacetyl reductase (Hetland *et al.*, 1971) are arranged at the corners of a square. An unequivocal distinction between C_4 and D_2 symmetry cannot be made, however, on the basis of the electron micrographs alone, except possibly for tryptophanase which appears to have D_2 symmetry (see below). A decameric protein, arginine decarboxylase, is actually composed of two rings containing five subunits each. Each of these rings may possess C_5 symmetry (Boeker and Snell, 1968). Hemocyanin also shows pentagonal (or decagonal) subunit disposition (Fernández-Morán *et al.*, 1966). Phycocyanin, with six subunits, possesses sixfold symmetry with its subunits arranged at the corners of a hexagon (Berns and Edwards, 1965; Neufeld and Riggs, 1969). Hexagonal organization has definitely been identified in electron micrographs of erythrocruorin (Levin, 1963), protocollagen proline hydroxylase (Olsen *et al.*, 1970a), and the 12 S subunit of transcarboxylase (Ahmad *et al.*, 1971). Electron micrographs also show a planar hexagonal arrangement of the subunits in ATPase (Schnebli *et al.*, 1970).

B. Dihedral Symmetry

If a subunit assembly has n twofold axes at right angles to any single n-fold axis, then the oligomer possesses dihedral symmetry. In this case, the number of subunits must be $2n$ (Klug, 1967). Hence only oligomers with an even number of subunits may possess dihedral symmetry. Dihedral symmetry is designated as D_n.

A tetrahedral array of four subunits (Fig. 4) provides the simplest example of D_2 symmetry ($n = 2$). Normally a tetrahedral arrangement of identical atoms exhibits cubic symmetry, i.e., four threefold axes and three twofold axes (see below). However, since protein subunits are necessarily asymmetric, the threefold axes are nonexistent and only three twofold axes are present (see Fig. 4). In principle, it should be possible to arrange the subunits of a tetramer at the corners of a square and still obtain D_2 symmetry. For example, the square arrangement of subunits with all isologous bonds (Fig. 5c) would have three twofold axes of symmetry which would be mutually perpendicular. Nevertheless, the square array is not common.

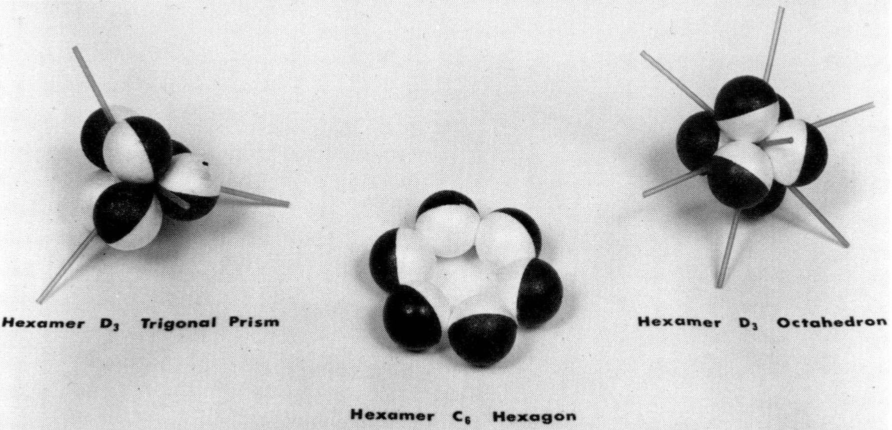

Fig. 6 Cyclic and dihedral symmetry in arrangements of subunits in a protein hexamer. The threefold and three twofold axes are indicated for each D_3 ensemble.

For an assembly of six subunits ($n = 3$), two noncyclical geometries, both D_3, are possible: the trigonal prism and the octahedron (Fig. 6). Here again, an octahedral arrangement of atoms would normally be cubic in symmetry, but since protein subunits are asymmetric, the symmetry of the oligomer is reduced to D_3.

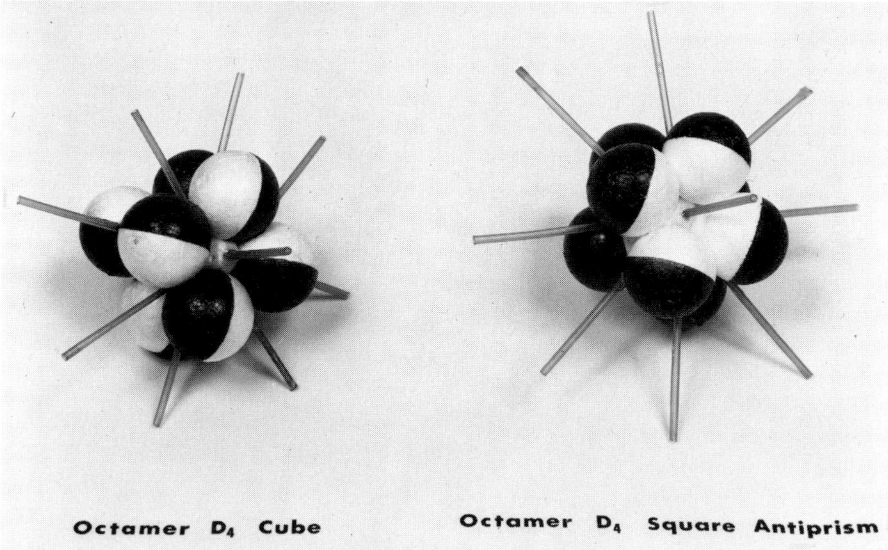

Fig. 7 Dihedral symmetry in arrangements of subunits in a protein octamer. The fourfold and three twofold axes of symmetry are indicated.

For an ensemble of eight subunits ($n = 4$), the two possible geometries are the cube and the square antiprism (Fig. 7). As in the case of $n = 2$ or 3, a cubic arrangement of these molecules normally has cubic symmetry, but because the subunits are asymmetric, the symmetry of the oligomer is lowered.

Structures with D_5, D_6, or D_n symmetry exist for proteins with 10, 12, or $2n$ subunits, respectively.

Hanson (1966) has correctly pointed out that it is also possible for dihedral symmetry to prevail when isologous ring structures are formed of $2n$ subunits (see, for example, Fig. 5c).

The possible geometries and types of bonding between subunits of oligomers exhibiting dihedral symmetry are summarized in Table V.

At least eleven different tetrameric proteins examined by electron microscopy or X-ray diffraction have been found to have D_2 symmetry (Table VI). These are isocitrate lyase (McFadden *et al.*, 1968), aldolase (Penhoet *et al.*, 1967; Lebherz and Rutter, 1969), hemoglobin (Perutz *et al.*, 1960), lactic dehydrogenase (Rossmann *et al.*, 1967), glyceraldehyde-3-phosphate dehydrogenase (Watson and Banaszak, 1964), prealbumin (Blake *et al.*, 1971), concanavalin A (Hardman *et al.*, 1971; Quiocho *et al.*, 1971), phosphoglycerate mutase (Campbell *et al.*, 1971), phosphorylase (Kiselev and Lerner, 1971), L-asparaginase (Epp *et al.*, 1971), and avidin (Green *et al.*, 1971). All have their four subunits placed at the vertices of a tetrahedron. As judged by electron microscopy, one protein, tryptophanase (Morino and Snell, 1967a), appears to have D_2 symmetry with its subunits arranged at the corners of a square.

Only one protein, aspartate transcarbamylase, has been shown to possess D_3 symmetry (Wiley and Lipscomb, 1968). A single protein, arginine decarboxylase, exhibits D_5 symmetry (Boeker and Snell, 1968) with two rings of five subunits each associating to form a decamer. On the other hand, four proteins, glutamine synthetase (Valentine *et al.*, (1968), chlorocruorin (Cucrritore *et al.*, 1965), erythrocruorin (Levin, 1963), and *Arenicola* hemoglobin (Waxman, 1971) have been found to possess D_6 symmetry. Electron microscopy has shown that the subunits in each of these oligomers are arranged at the corners of a hexagonal prism; two cyclic hexamers make up the oligomer. Since a cyclic hexamer is a closed structure, complete dissociation of the dodecamer may produce observable hexamers as an intermediate step.

C. Cubic Symmetry

The remaining combinations of rotational axes are the cubic point groups. These point groups may not be of great importance in describ-

TABLE VI

Some Symmetries Found in Proteins

Dimer	Trimer	Tetramer	Pentamer	Hexamer	Octamer
C_2	C_3	C_4	C_5	C_6	C_8
β-Lactoglobulin Hexokinase B Insulin Rhodanese		Tryptophanase (Pyruvate carboxylase)[a] (Diacetyl reductase)[a]	Arginine decarboxylase Hemocyanin	Phycocyanin Erythrocruorin Adenosine triphosphatase Transcarboxylase Protocollagen proline hydroxylase	
		D_2		D_3	D_4
		(Hemoglobin)[b] Aldolase Lactic dehydrogenase Glyceraldehyde-3-phosphate dehydrogenase *i*-Citrate lyase Prealbumin Concanavalin A Phosphoglycerate mutase Phosphorylase Asparaginase Avidin		Aspartate transcarbamylase	Hemerythrin

[a] These may have D_2 symmetry rather than C_4.

[b] To place hemoglobin in this column, one must ignore the actual differences between α and β subunits.

ing the smaller proteins since only quaternary structures with $12n$ subunits possess cubic symmetry and most oligomeric proteins have less than 12 subunits (see Table III). On the other hand, cubic symmetry appears to be of great importance in describing the multisubunit spherical viruses (Klug, 1967).

Cubic point groups characteristically possess more than one rotational axis higher than twofold. Geometric arrangements that fulfill this requirement are the tetrahedron, the cube, and the icosahedron.

The common tetrahedron (imagine each ball in Fig. 4 to be a uniform white sphere) possesses four threefold axes (one from each vertex to the opposite face) and three twofold axes (those through the midpoints of opposite edges). This symmetry is designated as *T*. As mentioned above, the threefold symmetry is lost when asymmetric protein subunits are at the vertices of a tetrahedron. However, threefold symmetry is preserved if a group of three subunits (in a triangular array) is placed at each of the four vertices of the tetrahedron. For this structure the minimum number of asymmetric subunits is 12 (see Klug, 1967). Likewise, it can be shown that if groups of three subunits are placed at each of the vertices of a cube, cubic symmetry designated *O* (three fourfold, four threefold, and six twofold axes) is created. For this symmetry to be achieved, the oligomer must contain at least 24 subunits. Correspondingly, for icosahedral symmetry at least 60 subunits must be present (Klug, 1967).

Since most subunit proteins fall into the cyclic or dihedral classes of symmetry, there are at present only a few examples of protein oligomers possessing *T* or *O* symmetry. However, electron micrographs of L-aspartate-β-decarboxylase, an enzyme with 12 subunits, seem to indicate a structure with *T* symmetry, i.e., three subunits reside at each of the four corners of a tetrahedron (Bowers *et al.*, 1970). Electron microscopy and X-ray diffraction of dihydrolipoyl transsuccinylase (24 subunits) show, as expected, that *O* symmetry prevails, with three subunits disposed at each of the eight corners of a cube (DeRosier *et al.*, 1971). There is also circumstantial evidence that ferritin and dihydrolipoyl transacetylase possess *O* symmetry, with three subunits at each of the eight vertices of a cube. Recent hydrodynamic and chemical studies (Crichton, 1972) indicate that there are 24 subunits in apoferritin instead of the 20 previously believed present. Similarly the transacetylase enzyme, constituted of 24 identical polypeptide chains, appears in the electron microscope to be arranged in eight morphological ensembles, each containing three subunits (Henney *et al.*, 1967).

D. General Comments

Table V lists the types of bonding and geometric arrangements in proteins containing up to eight subunits. When $n = 2$, 3, or any odd number, only one type of geometry is possible, namely a cyclic structure.

When $n = 4$, 6, or 8, there are four possible geometries, one cyclic and two dihedral for the tetramer, and one cyclic and three dihedral for the hexamer and octamer. Table V also shows that it is possible for an oligomer with 4, 6, or 8 subunits to have all isologous, all heterologous, or mixed isologous and heterologous binding regions. Monod *et al.* (1965) have suggested that isologous bonding should be the most prevalent since most oligomers are dimers and tetramers, and closed dimers cannot be formed by heterologous association. Certainly dimers are the most prevalent oligomers; however, there are large numbers of oligomers with 4, 6, or 8 subunits (Table III). Hence there is no *a priori* reason why isologous bonding should prevail over heterologous bonding. Thermodynamic calculations also indicate that there is no reason why isologous bonding should be preferred over heterologous bonding (Cornish-Bowden and Koshland, 1971).

Under some circumstances it may be possible to predict the geometry most likely to appear. To do so we must look closely at the association phenomenon. The driving force for association of isolated monomers must come from increased stabilization due to the formation of interface bonds between subunits. Other things being equal, the spatial arrangement with the largest number of subunit contacts or regions of bonding should be the most stable. On this basis, a tetrahedral geometry should be the most stable for a tetramer, an octahedral arrangement should be the most stable for a hexamer, and a square antiprism the most stable for an octamer (Table V). In actuality, of course, an energy term contains an intensity factor as well as a quantity factor. Thus not only the number of bonds formed but also the strength of these bonds play a role in determining the stability of a particular geometric arrangement of the subunits in an oligomer; and these bond strengths have not been well established. Nevertheless, it is interesting to note that of fourteen tetrameric proteins scrutinized at the Å unit level, eleven are in a tetrahedral arrangement.

In quaternary structures with cyclic symmetry the intersubunit contact regions (bonds) are all identical, while in those with dihedral or cubic symmetry there may be up to four different bonding regions (Table V). (To confirm this, one may need to examine three-dimensional models.) Bonds of the same class form a closed cycle centered on the rotational symmetry axis of this class. Hence, if the symmetry of a protein ensemble is known, one may be able to predict the type of intermediate aggregates produced upon dissociation. For example, a tetramer with tetrahedral symmetry (D_2), containing one type of dimer bond and two types of tetramer bond (bonds where two dimers meet to form the tetramer), would be expected to produce a stable dimer under selective dissociating conditions. Stable dimers have indeed been found for hemoglobin (Rosemeyer and Huehns, 1967) and for glyceraldehyde-3-phosphate dehydrogenase (Hoagland and Teller, 1969), both of which are tetrahedral. Similarly, a hexamer with dihedral symmetry (D_3) should form trimers under suitable dissociating conditions. Trimers have been observed in aspartate transcarbamylase (Rosenbusch and Weber, 1971a).

Conversely, if the symmetry of the subunits is not known but intermediate dissociation stages have been identified, it may be possible to predict the symmetry of the oligomer. For example, examination of acetoacetate decarboxylase (Tagaki and Westheimer, 1968), which has 12 subunits, shows the presence of intermediate dimers. These are most likely closed structures with C_2 symmetry. Six closed-structured isologous dimers must then associate to form the dodecamer, with each subunit in an identical environment. This means that the dodecamer would most likely have D_6 symmetry. By similar reasoning, one might predict that a protein with eight subunits which intermediately dissociates into tetramers or dimers would have D_4 symmetry, with the subunits arranged at the corners of a cube or square antiprism.

Most conclusive methods for determining the geometry of an oligomeric protein have been based on either X-ray diffraction or electron microscopy. Electron microscopy is, however, rarely able to reveal the geometric arrangement of the subunits of oligomeric proteins having molecular weights below 100,000. A clever method for determining the symmetry of a small oligomer has nevertheless been developed by Green and co-workers (1971) by combining electron microscopy with chemical methods. Green *et al.* (1971) investigated the subunit structure of avidin, a tetrameric protein (MW 68,000) that binds biotin very strongly. They synthesized a series of bisbiotin derivatives in which the ureido rings were joined by hydrocarbon chains of different lengths so that the number of bonds between the biotin carboxyl groups varied between nine and twenty-five. When the number of bonds was twelve, bifunctional binding between two tetrameric avidins occurred, as judged by the combining ratios and the appearance of polymers in the electron microscope. These polymers took the form of linear chains whose width corresponded to that of a single avidin molecule and showed almost no branch points. If the subunits in avidin are all in equivalent environments, then only C_4 and D_2 symmetry would be possible (heterologous or isologous association). Only if the subunits are arranged in D_2 symmetry can polymers be obtained that are one molecule thick. An arrangement of C_4 symmetry would lead to two-dimensional networks or to a unidimensional repeating system in which each tetramer is associated in a zigzag manner to two others. Thus, although the subunits are too small to be resolved within the tetramer by electron microscopy, the electron microscope nevertheless discriminates between the possibilities and thereby establishes the symmetry relation between the subunits.

The available evidence, which in the absence of crystallographic data must necessarily remain inconclusive, supports the concept that most oligomeric proteins whose subunits are in identical environments possess maximum or near maximum symmetry. There are, however, a few disturbing exceptions. Insulin (Adams *et al.*, 1969) and hexokinase (Steitz, 1971) have been mentioned already. Tables III and IV also show that there are a number of proteins with nonidentical subunits. Those dimeric proteins with nonidentical subunits necessarily have their subunits in nonequivalent environments. The tetrameric proteins of the $\alpha_2\beta_2$ type nevertheless may possess pseudosymmetry axes, as does hemoglobin. Hemocyanin (Fernández-Morán *et al.*, 1966) has ten subunits arranged in a circle with tenfold symmetry, but the decamers are stacked one on top of the other with up to twelve decamers per stack. The end decamers are then in different environments from those in the interior of the stack. Perhaps hemocyanin would be more accurately described by line symmetry, as have the helical viruses (Klug, 1967).

One other potential contradiction of our basic structural assumption seems to have evaporated under the heat of improved data. Originally it seemed that ferritin, bovine kidney dihydrolipoyl transacetylase, and some other proteins might be oligomers with 20 subunits arranged at the (20) vertices of a pentagonal dodecahedron. In such an arrangement, the identical units are not actually in identical environments[4] (Hanson, 1968; Cornish-Bowden and Koshland, 1971). One would be obliged to conclude, therefore, that in an oligomer identical subunits can exist in two different environments and perhaps even in two different conformations. However, recent chemical and physical measurements (Bjork and Fish, 1971; Crichton, 1972) indicate that there are 24 subunits in ferritin and in the other anomalous proteins in purported dodecahedral arrays. As mentioned previously, 24 subunits can be placed in an en-

[4] This is perhaps best seen in the dodecahedral clathrate hydrates (see Klotz, 1965). A pentagonal dodecahedron has 20 (identical) H_2O molecules with the oxygen atoms at the 20 vertices. The dodecahedron also has 30 edges, each with a hydrogen bond between two oxygen vertices. Thus, of the 40 hydrogens in $H_{40}O_{20}$, 30 are within the edges and 10 stick out from 10 of the 20 oxygens at the vertices of the polyhedron. The remaining 10 oxygens each have an electron pair pointing out of the dodecahedron. Thus there are two different types of orientations of identical water molecules at the vertices of a dodecahedron.

semble with *O* symmetry in which every subunit is in an identical environment.

IV. ENERGETICS OF SELF-ASSEMBLY

Having delineated the stoichiometric and geometric constitution of a complete quaternary ensemble, we proceed to examine the degree of association or dissociation at different concentrations of the protein. The accretion of monomers to build the completed quaternary structure may proceed as a gradual stepwise transition or as an extremely sharp transition from the subunits to the final organized ensemble. In either event, a true *self*-assembly is subject to thermodynamic constraints, i.e., the energetics of the successive steps in the construction of the quaternary structure. At each stage in the aggregation process, the succeeding subunit may be attached with the same intrinsic free energy change as its predecessor or successor; in this case we say that there are *no interactions* between subunits in the self-assembly. On the other hand, the addition of successive subunits may be facilitated by their predecessors in the intermediate oligomer; in these circumstances *cooperative*[5] interactions are being manifested. Alternatively, the addition of successive monomers may be hindered, in which case *antagonistic* or *obstructive*[5] (or, as they are often called, *anticooperative*[5]) interactions are taking place.

In a noninteracting system, the intrinsic free energies may be represented as in Fig. 8a; each step is at the same level as its neighbors. In contrast, when cooperative interactions exist the free energy levels drop progressively (Fig. 8b). Correspondingly, when antagonistic interactions prevail, the free energy changes are less deep as successive monomers are inserted into the aggregate (Fig. 8c).

In all of these situations the only available avenue for evaluating the free energies starts with equilibrium constants. To obtain these constants one must measure equilibrium concentrations of all species in the association–dissociation process. The most widely used procedure for such measurements is equilibrium centrifugation, and this will be described in some detail. Nevertheless, it should be mentioned that light-scattering,

[5] A *cooperative* interaction implies a mutually "helping" effect. Anticooperative literally implies a nullifying of the "help" or "cooperativity" but not the antonym or contrary of "helpful." The contrary action is to "hinder, impede, *obstruct,* block, thwart."

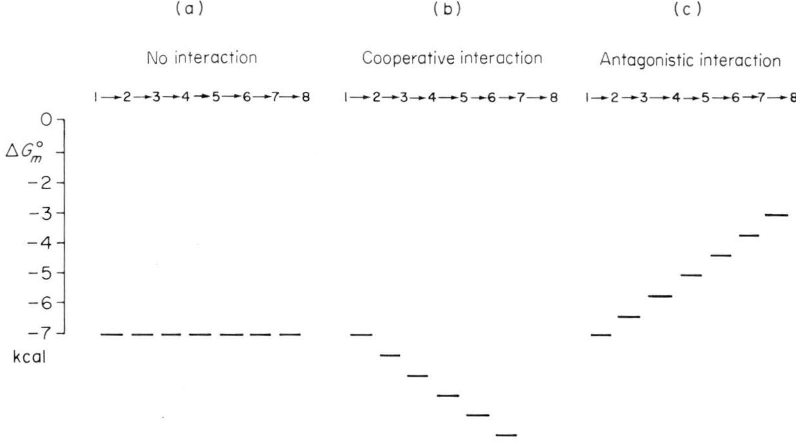

Fig. 8 Step (intrinsic) free energy changes, $\Delta G_m{}^\circ$, for the stepwise addition of monomer in the assembly of an octamer. (a) $\Delta G_m{}^\circ$ for monomer → dimer equals that for dimer → trimer, etc. (b) $\Delta G_m{}^\circ$ becomes more favorable in a stepwise fashion. (c) $\Delta G_m{}^\circ$ becomes progressively less favorable. These illustrations are for a hypothetical system in which the equilibrium constant for the first step, formation of dimer, is 10^5.

osmotic pressure, and gel filtration methods have also been utilized for many protein oligomers (Nichol *et al.*, 1964; Ackers, 1970).

A. Equilibrium Constants from Ultracentrifugation

Sedimentation equilibrium experiments provide data from which one can evaluate average molecular weights as a function of concentration of the protein. From this information (if it covers a wide enough concentration range) one can determine by extrapolation the molecular weight of the monomeric subunit and of the complete quaternary ensemble, and one can calculate the relative amounts of each of the macromolecular species in association equilibrium in solution.

We shall limit this discussion to proteins whose behavior as solutes is ideal. In these circumstances each species of solute in solution at sedimentation equilibrium will be distributed exponentially along the radial axis of the centrifugal field. Specifically, the concentration, c_i, of each species i is given by the following function of its distance r from the axis of rotation:

$$c_i = B_i \exp \frac{M_i(1 - \bar{v}_i\rho)\omega^2}{RT}\frac{r^2}{2} \tag{11}$$

In this equation B_i is an integration constant which can be evaluated from appropriate boundary conditions (e.g., the concentration at the meniscus), M_i and $\bar{v}_i$ are the molecular weight and partial specific volume of species i, respectively, ρ is the density of the solution, ω is the angular velocity of the rotor, R is the gas constant, and T is the absolute temperature. For convenience this expression is usually condensed to

$$c_i = B_i e^{\sigma_i r^2/2} \tag{12}$$

and σ_i is referred to as the "reduced molecular weight." Equation (12) serves as the starting point for deducing the mode of association and the associated energy quantities.

The most direct treatment of the ultracentrifugal observations subjects the concentration data to least-squares analysis based on the function described by Eq. (12). In essence, one determines two parameters, B_i and σ_i, for each macromolecular species participating in the association–dissociation equilibrium. Such an approach has been applied to aspartate β-decarboxylase (Haschemeyer and Bowers, 1970; Bowers *et al.*, 1970). This procedure must be used with caution, however, since a large number of adjustable parameters (all the B_i's and σ_i's) may fit the concentration distribution curve.

An alternative treatment in terms of the curves described by Eq. (11) has been proposed by Dyson and Isenberg (1971). For n species, each distributed exponentially with radial position, the total concentration, $c(z)$, may be expressed as a function of the relative position $z = r_0^2 - r^2 (r \leq r_0)$, by the equation

$$c(z) = \sum_{1}^{n} c_i(0) \exp - \frac{M_i(1 - v_i\rho)\omega^2}{2RT} z \tag{13}$$

where $c_i(0)$ is the concentration of species i at the reference radial position r_0. The $2n$ moments

$$\mu_k = \int_0^Z c(z)z^k \, dz \qquad (k = 0, 1, \ldots, 2m - 1) \tag{14}$$

are then defined and suitably minimized in order to replace the original data with a smaller number of more precise representations. This transform has been used to analyze the equilibrium distribution of a mixture of ovalbumin and serum albumin in a centrifugal field. Predicted and calculated values for the respective molecular weights and for $c_i(0)$ agree very closely (Dyson and Isenberg, 1971). The determination of n, the number of species, has not been thoroughly examined yet. Nevertheless, the use of moments, each formed from the entire set of original data, seems to offer the distinct advantage of exhibiting much less noise than the exponential representation.

The most common treatment of sedimentation equilibrium data relates average molecular weights to the equilibrium constants of stepwise association reactions. As early as 1943 a very generalized analysis of multiple association, focusing on the aggregation of small molecules, was published by Kreuzer. The pioneering paper treating macromolecular associations in terms of average molecular weights is by Steiner (1952).

For a stepwise association of monomers A_1 to give oligomeric species A_i

$$iA_1 \rightleftharpoons A_i \tag{15}$$

we may define an association constant K_i by

$$K_i = (A_i)/(A_1)^i \tag{16}$$

Simultaneously, if we express the weight-average molecular weight by

$$M_w = \Sigma(A_i)M_i^2/\Sigma(A_i)M_i \tag{17}$$

then combination of Eqs. (16) and (17), followed by appropriate algebraic manipulation, leads to the relation

$$M_w = (M_1/c_t)\Sigma i^2 K_i(A_1)^i \tag{18}$$

where c_t is the total stoichiometric protein concentration expressed in terms of monomer units. Equation (18) provides an expression of M_w in terms of (A_1), K_i, M_1, and c_t. As Steiner (1952) showed, further algebraic manipulation then leads to

$$\ln \frac{c_1}{c_t} = \int_0^{c_t} \left(\frac{M_1}{M_w} - 1\right) d \ln c_t \tag{19}$$

which relates the fraction of monomer, c_1/c_t, to the weight-average molecular weight, M_w, and to c_t. Equation (19) can be used to evaluate c_1 at a series of values of c_t, using M_w data derived from the sedimentation experiments. Corresponding equations also exist for other moments of the molecular weight distribution.

The Steiner procedure has been used in studies of the aggregation of α-chymotrypsin (Rao and Kegeles, 1958) and in an analysis of the monomer–octamer equilibrium of hemerythrin (Langerman and Klotz, 1969). The major uncertainty in this approach lies in the extrapolation to zero protein concentration demanded by Eq. (19). A reliable extrapolation depends upon precise data at very low protein concentration, the region where experimental difficulties are most severe.

Adams and co-workers (Adams and Williams, 1964; Adams, 1967,

1969; Adams and Lewis, 1968) have extended the procedure of Steiner (1952) and developed methods for treating a variety of types of self-association of identical monomers [Eq. (15)] and mixed monomers whose aggregation may be represented by

$$A + B = AB; \qquad K_{AB} = (AB)/(A)(B) \tag{20}$$

This treatment also starts with Eq. (11), a statement of the conservation of mass in the form

$$c_t = c_1 + \sum_{i=2}^{n} K_i c_1^{\,i} \tag{21}$$

and an expression defining the weight-average molecular weight, e.g., Eq. (17). Assuming the specific type of association that is presumed to occur, one then writes n equations, one for each of the species participating in the association. By proper combination of two of these equations, one can eliminate the term for the nth species, the completed oligomer. Successive combinations of the resultant equations will successively remove the next highest term representing the next highest intermediate oligomer. By this procedure, one can ultimately generate an equation containing one unknown, c_1. To evaluate c_1, one must still depend upon Eq. (19) or an equivalent relationship; hence data of high precision at low concentration are still essential. Once c_1 is known, one can work backward to obtain K_1, K_2, etc. If a consistent set of equilibrium constants is obtained, then the assumed mode of association is presumed to be a correct description of the oligomeric system. If not, one can consider other modes of association and generate a corresponding set of algebraic relationships between the successive equilibrium constants.

The Adams procedure also lends itself readily to inclusion of nonideality corrections. Adams and Lewis (1968) have applied the procedure to the study of β-lactoglobulin. A similar approach has been used by Millar *et al.* (1969)[6] to evaluate the association constants for lactate dehydrogenase.

A very different treatment of sedimentation equilibrium data utilizes more than one type of molecular weight average to deduce the associating species and constants (Roark and Yphantis, 1969; Roark, 1970; Hoagland and Teller, 1969; see also Sophianopoulos and Van Holde, 1964). Moments of the molecular weight distribution may be determined readily and accurately at sedimentation equilibrium, provided that

[6] Disagreement with the conclusions of these investigators has recently been expressed by Huston *et al.* (1972) and by Fosmire and Timasheff (1972).

the concentration of macromolecules at the meniscus in the ultracentrifuge cell is low enough. Since the exact method to be used depends upon the mode of association assumed, we cannot deal with all possible cases. Therefore, we shall describe only the so-called "two-species plot."

If an ideal solution consists of two species of molecular weights M_1 and M_2 and of weight fraction $[1 - \alpha(r)]$ and $\alpha(r)$ at point r in the centrifugal field, then the number-average (M_n) and weight-average (M_w) molecular weights are given by

$$1/M_n(r) = 1 - \alpha(r)/M_1 + \alpha(r)/M_2 \tag{22}$$

$$1/M_w(r) = [1 - \alpha(r)]M_1 + \alpha(r)M_2 \tag{23}$$

Combining these two equations to eliminate $\alpha(r)$, we obtain

$$M_w(r) = -M_1M_2[1/M_n(r)] + M_1 + M_2 \tag{24}$$

Thus, if only two species are present, a plot of the weight-average molecular weight against the reciprocal number-average molecular weight will give a straight line whose slope is $-M_1M_2$ and whose y intercept is $M_1 + M_2$.

This analysis has been generalized to include other moments of the molecular weight distribution. The general relationship is

$$M_k(r) = -M_1M_2[1/M_{k-1}(r)] + M_1 + M_2 \tag{25}$$

where $k = 0, 1, 2, 3$, corresponding respectively to M_n, M_w, M_z, and M_{z+1}.

After the "two-species plot" has been used to deduce the species present, various equations may be used to evaluate the relative concentrations of each species or the equilibrium constants. The most straightforward approach is to fit Eq. (11) by a least-squares procedure similar to that described above (Haschemeyer and Bowers, 1970). This time, however, only one adjustable parameter (B_i) is necessary, and hence the danger of computational noise is greatly reduced. Alternatively, the number-, weight-, and z-average molecular weights at each point in the centrifugal field may be used directly. The appropriate relationships are

$$\frac{2K_{2,n}}{M_1} = \frac{2M_{n,r}(M_{n,r} - M_1)}{c(2M_1 - M_{n,r})^2} \tag{26}$$

$$\frac{2K_{2,w}}{M_1} = \frac{M_1(M_{w,r} - M_1)}{c(2M_1 - M_{w,r})^2} \tag{27}$$

$$\frac{2K_{2,z}}{M_1} = \frac{(M_{z,r} - M_1)(3M_1 - M_{z,r})}{4c(2M_1 - M_{z,r})^2} \tag{28}$$

where M_1 is the molecular weight of the smallest species entering into the association and c is the total concentration in grams per liter. The $K_{2,i}$'s are molar association constants, i referring to the molecular weight moment upon which K is based. Equation (26) was derived by Nichol *et al.* (1964), Eq. (27) by Rao and Kegeles (1958), and Eq. (28) by Hoagland and Teller (1969).

These methods have been used in the study of the dimer–tetramer association of glyceraldehyde-3-phosphate dehydrogenase (Hoagland and Teller, 1969), the dimerization of yeast aldolase (Curtis *et al.*, 1969), and the self-association of β-lactoglobulin (Roark and Yphantis, 1969). The procedures have been tested carefully by Roark and Yphantis (1969) and Hoagland and Teller (1969) and found to be relatively insensitive to experimental noise. The success of these methods, nevertheless, depends upon the evaluation of the various molecular weight moments as a function of concentration over a fairly broad concentration range and thus requires highly efficient and computationally stable methods of data analysis.

The final method to be discussed in this section is that based on multinomial theory (Derechin, 1968, 1969a,b, 1971). This treatment involves the solution of the general "multinomial" expression for the specific constants describing the assumed mode of association. For example, for a monomer–dimer–tetramer system, the association constants are given by

$$K_2 = -\left[\frac{d(M_1/M_w)}{dc}\right]_{c=0} \tag{29}$$

$$K_4 = -\frac{8}{3}\left[\frac{d(M_1/M_w)}{dc}\right]^3_{c=0} + \left[\frac{d(M_1/M_w)}{dc}\right]_{c=0}\left[\frac{d^2(M_1/M_w)}{dc^2}\right]_{c=0} - \frac{1}{18}\left[\frac{d^3(M_1/M_w)}{dc^3}\right]_{c=0} \tag{30}$$

It is obvious that the use of these equations requires very accurate data of exceptionally low noise in the region of $c = 0$. This method has been tested by Derechin (1971) for lysozyme and for cytidine (using data of Van Holde *et al.*, 1969), and the conclusions reached are not the same as those based on alternative methods.

B. Association Constants and Thermodynamic Quantities

Table VII lists equilibrium constants and related energy quantities for a number of associating systems. Included in this table are also some aggregation reactions. "Subunit" is perhaps an inappropriate term for the A_1 species in these aggregates since the proteins (e.g., chymo-

TABLE VII

Thermodynamic Parameters of Subunit Association

System	K_i (molar scale)	$\Delta G_i^{\circ a}$ (kcal/mole)	ΔH_i° (kcal/mole)	ΔS_i° (eu)	$-\Delta G_m^{\circ b}$ (kcal/mole subunit)	Reference
D-Amino acid oxidase, pH 8.5, 10°	1.2×10^5	−6.5	0	23	−3.25	Henn and Ackers (1969)
$2A_1 \rightleftarrows A_2$ pH 8.5, 20°	3.1×10^5	−7.3	0	25	−3.65	
α-Amylase, pH 7.20° (*Bacillus subtilis*)						Kakiuchi *et al.* (1965)
$2A_1 + Zn^{3+} \rightleftarrows A_2(Zn^{2+})$	2×10^9	−12.5		−6.3		
Apoferritin, pH 4.6, 25°						Richter and Walker (1967)
$2A_1 \rightleftarrows A_2$	4.5×10^5	−7.72			−3.86	
$A_1 + A_2 \rightleftarrows A_3$	3.6×10^5	−7.59				
Arginosuccinase, pH 7.5 (0.1 *M* PO_4), 24.5°						Schulze *et al.* (1970)
$2A_1 \rightleftarrows A_2$	3.2×10^7	−10.2	46	189	−5.1	
Chymotrypsinogen A, pH 7.9, 25°						Hancock and Williams (1969)
$A_1 + A_{i-1} = A_i$ (isodesmic)	1.27×10^3	−4.2			−2.1	
Enolase, pH 7.4, 40° (yeast)						Keresztes-Nagy and Orman (1971)
$2A_1 \rightleftarrows A_2$	3.7×10^6	−9.4	−80	−221	−4.7	
Glutamate dehydrogenase, pH 7 25°						Reisler and Eisenberg (1971)
$A_1 + A_{i-1} = A_i$ (isodesmic)	1.07×10^5	−7.8	0	25	−3.9	
Glucagon, pH 10, 25°						Swann and Hammes (1969),
$2A_1 \rightleftarrows A_2$	3.3×10^3	−4.78			−2.39	Gratzer and Beaven (1969)
$6A_1 \rightleftarrows A_6$	10.9×10^{17}	−24.5			−4.08	
Rabbit muscle D-glyceraldehyde-3-phosphate dehydrogenase, pH 7.5°						Hoagland and Teller (1969)
$2A \rightleftarrows A_4$	2×10^6	−8.0			−4.0	

Hemerythrin, pH 7.0, 5° (*Golfingia gouldii*)						Langerman and Klotz (1969)
$8A_1 \rightleftarrows A_8$	3.4×10^{36}	−46.3	0	+22	−5.8	
Hemoglobin, pH 7.0, 20°						Kellett (1971), Evans *et al.* (1970)
$2\alpha\beta \rightleftarrows (\alpha\beta)_2$	8×10^4 $(M\ heme)^{-1}$	−6.5	−7	−1.5	−3.25	
Hb(CO), pH 6.8, 4°						Ackers and Thompson (1965)
$2\alpha\beta \rightleftarrows (\alpha\beta)_2$	2.5×10^5	−6.8			−3.4	
Hemoglobin, pH 5.6, 20°, deoxy (*Petromyzon marinus*)						Anderson (1971)
$2A_1 \rightleftarrows A_2$	2.5×10^6	−8.5			−4.25	
$2A_2 \rightleftarrows A_4$	1.7×10^3	−4.3			2.15	
Insulin, pH 2, 25°						Jeffrey and Coates (1966)
$2A_1 \rightleftarrows A_2$	10.2×10^3	−5.47	−7.1	−5.5	−2.73	
$2A_2 \rightleftarrows A_4$	0.90×10^3	−3.95	−16.3	−41		
$A_2 \rightleftarrows A_4 \rightleftarrows A_6$	0.58×10^3	−3.85	+49.0	+177		
Isocitrate dehydrogenase, pH 7.2, 5° (bovine heart)						Giorgio *et al.* (1970)
$2A_1 \rightleftarrows A_2$	1.3×10^6	−7.75			−3.87	
α-Isopropylmalate synthase, pH 7.0, 5°						Leary and Kohlhaw (1972)
$4A_1 \rightleftarrows A_4$	9×10^{16}	−21.5			−5.4	
β-Lactoglobulin, pH 4.55, 4.6°						Kumosinski and Timasheff (1964), Roark and Yphantis (1969)
$4A_2 \rightleftarrows A_8$	2.4×10^{11}	−15.2	−56	−151	−3.8	
β-Lactoglobulin, pH 5.6, 25°	7.3×10^7	−10.0				Townend *et al.* (1960), Kelly and Reithel (1971)
$2A_1 \rightleftarrows A_2$	(4.3×10^4)	(−8.08)	(−6.9)	(+3.6)	(−4.04)	
β-Lactoglobulin A, pH 6.9, 20°						Zimmerman *et al.* (1970)
$2A_1 \rightleftarrows A_2$	4.88×10^4	−6.3			−3.1	
β-Lactoglobulin B, pH 6.9, 20°						
$2A_1 \rightleftarrows A_2$	1.42×10^5	−6.9			−3.4	
β-Lactoglobulin, pH 8.8, 20°						Georges *et al.* (1962)
$2A_1 \rightleftarrows A_2$	4.1×10^3	−4.8			−2.4	

(*Continued*)

TABLE VII (*Continued*)

System	K_i (molar scale)	$\Delta G_i^{\circ a}$ (kcal/mole)	ΔH_i° (kcal/mole)	ΔS_i° (eu)	$-\Delta G_m^{\circ b}$ (kcal/mole subunit)	Reference
Myosin, pH 7.3 (0.2 *M* PO_4) 6.0°						Herbert and Carlson (1971)
$2A_1 \rightleftarrows A_2$	4.8×10^5	−7.2			−3.6	
C-Phycocyanin, pH 6, 20°						MacColl *et al.* (1971b)
$6A_1 \rightleftarrows A_6$	1.0×10^{30}	−40			−6.7	
Procarboxypeptidase A, pH 7.5, 5°, Fraction II (bovine)						Teller (1970)
$A_1 + A_{i-1} = A_i$ (isodesmic)	8.6×10^3	−4.9			−2.45	
Trypsin inhibitor, pH 7, 25°, (soybean)						Harry and Steiner (1969)
$2A_1 \rightleftarrows A_2$	3.65×10^3	−4.85	0	19	−2.42	
Tryptophanase, pH 8, 5°						Morino and Snell (1967a)
$2A_2 \rightleftarrows A_4$	1.2×10^4	−5.17	+50	+198	−2.58	
Tryptophan synthetase, β chain, pH 7.3, 20° (*E. coli*)						Hathaway and Crawford (1970)
$2A_1 = A_2$	4.6×10^5	−7.6	0	+25	−3.8	
$A_1 + A_2 = A_3$	6.2×10^4	−6.4				

[a] ΔG_i° is computed from $-RT \ln K_i$.
[b] ΔG_m° is $-\Delta G_i^\circ/n$, where n is the stoichiometric coefficient of the reactant in the reaction $nA_1 = A_n$.

trypsinogen A), as normally isolated in the pure state, exist as monomers in solution. A comparison of thermodynamic parameters, however, reveals no precise distinction between structures usually viewed as associating–dissociating quaternary systems and those usually considered as aggregating systems of normally self-sufficient macromolecular particles. As in many other categorizations, it is clear that at the extremes there are unequivocal examples of precisely arranged, very stable quaternary structures of *n*-mers (e.g., hemoglobin, hemerythrin, viruses) and, in contrast, there are many proteins that remain predominantly in the monomeric state in solutions near physiological pH (e.g., ribonuclease, lysozyme, and serum albumin). On the other hand, there are numerous examples of systems whose state of aggregation is highly variable in concentration ranges readily accessible to experimental measurement. It is not always obvious what range of concentration, and hence which aggregate, corresponds to that found in the native state in tissues or to that in the crystal. One of the more interesting examples in this regard is β-lactoglobulin. Over a wide range of pH near neutrality, β-lactoglobulin is unquestionably a dimer in which $M_1 = 18{,}500$. At pH values near 4.5 and at low temperatures, however, the dimer associates to form an octamer (Timasheff and Townend, 1964; Roark and Yphantis, 1969). A model for the octamer has been proposed (Timasheff and Townend, 1964) which describes it as a square antiprism having D_4 symmetry, one of the structures likely for an octameric ensemble (see Section III). A study by Georges *et al.* (1962) at high pH (8.8, 20°) indicates that the protein is a dimer in equilibrium with the monomer. At low pH (<3) the protein also exists as a monomer (18,500 gm $mole^{-1}$). Thus the state of association may be very dependent upon the solution environment.

C. Interactions in Assembly of Quaternary Ensemble

Even with clearly recognizable quaternary structures constituted of subunits linked noncovalently, disaggregation must occur at some dilution (Le Chatelier's principle). From the complementary reference point, monomeric species predominate at very low concentrations and assemble themselves into a specific quaternary ensemble with increasing concentration of protein. We recognize two different classes of behavior. In some association–dissociation systems, significant concentrations of intermediate species exist between the monomer and the final quaternary structure. In lactic dehydrogenase, for example, dimers are important intermediates between monomers and tetramers (Millar *et al.*, 1969).[6]

In hemoglobin, dimers may provide a stable intermediate stage between monomers and tetramers (Kirshner and Tanford, 1964; Kawahara *et al.*, 1965; Schachman and Edelstein, 1966). In contrast, in some systems only the initial and final states are manifested. In hemerythrin, for example, no intermediate is detected between the monomeric and octameric stages (Langerman and Klotz, 1969).

In descriptive terms it is apparent that the difference between stepwise and all-or-none association to n-mer depends upon the degree of cooperativity between successively bound subunits. When these considerations are analyzed quantitatively, it becomes apparent that a change from a stepwise to a very sharp transition may follow from a very minor increment in the degree of cooperativity.

It is instructive to examine an elementary quantitative treatment of subunit interactions in which we take no explicit note of statistical factors and assume that the degree of cooperativity is the same for each associating step. We may then define a cooperativity parameter by

$$\alpha = \frac{k_{i+1}}{k_i} \tag{31}$$

where k_i is the equilibrium constant for the addition of a monomeric A molecule to oligomer A_{i-1} containing $(i - 1)$ monomeric subunits:

$$A + A_{i-1} = A_i; \qquad k_i = \frac{A_i}{(A)(A_{i-1})} \tag{32}$$

If one explicitly writes out the individual steps in the successive associations represented by Eq. (32), then one can readily show that

$$\frac{A_i}{(A)^i} = \prod_2^i k_j \tag{33}$$

Making use of α from Eq. (31) we may write

$$\begin{aligned} k_3 &= \alpha k_2 \\ k_4 &= \alpha k_3 = \alpha^2 k_2 \\ \vdots\ & \quad\ \vdots \qquad\quad \vdots \\ k_i &= \alpha k_{i-1} = \alpha^{i-2} k_2 \end{aligned} \tag{34}$$

Combining the relations of Eq. (34) with Eq. (33) we find

$$\begin{aligned} A_i &= (\Pi k_j)(A)^i = k_2^{i-1}(\alpha \cdot \alpha^2 \ . \ . \ . \ \alpha^{i-2})A^i \\ &= k_2^{i-1}\alpha^\sigma A^i \end{aligned} \tag{35}$$

where

$$\sigma = 1 + 2 + \cdot \cdot \cdot + (i - 2) = \frac{(i - 1)(i - 2)}{2} \tag{36}$$

The experimentally fitted single association constant for a monomer i-mer equilibrium would be written as

$$K_i = \mathrm{A}_i/(\mathrm{A})^i \tag{37}$$

From Eqs. (35) and (37) it follows that

$$k_2{}^{i-1}\alpha^\sigma = K_i \tag{38}$$

The influence of degree of cooperativity on extent of association and on concentrations of intermediate species may be illustrated in terms of a hypothetical model system consisting of subunits having molecular weights of 10,000. Considering, as an example, an octameric quaternary structure and assuming a k_2 of 1×10^5, we can compute M_w as a function of total protein concentration for values of α equal to 0.01, 1, and 3, respectively. As Fig. 9 illustrates, a noninteracting assembly ($\alpha = 1$) produces a gradual increase in M_w. An obstructive interaction slows down the buildup of oligomers; when $\alpha = 0.01$, this hindrance is so large that it would be hard to exceed an M_w of 20,000, the molecular weight of the dimer. In contrast, a cooperative interaction builds up higher oligomers quickly; when $\alpha = 3$, the octameric molecular weight is approached very rapidly with increasing protein concentration, and

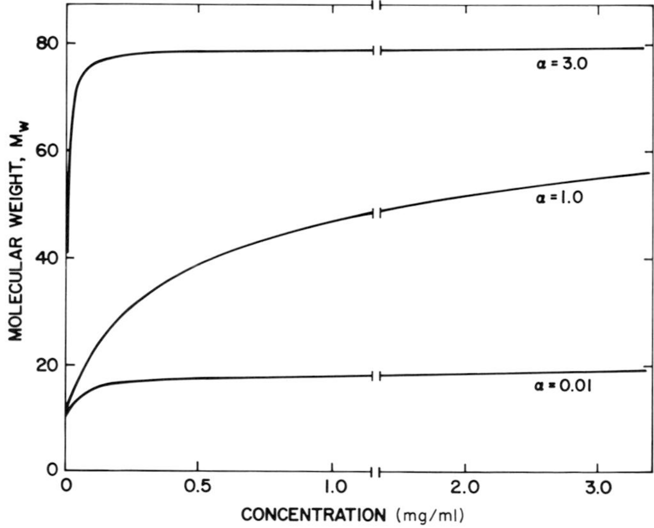

Fig. 9 Weight-average molecular weight, M_w, as a function of total protein concentration for an octamer. $M_w = 10{,}000$; $k_2 = 1 \times 10^5$, and parameter α is varied from 0.01 (obstructive interaction) to 3.0 (cooperative interaction).

values near 80,000 are attained even at protein concentrations of 0.02% (Fig. 9).

The macromolecular distributions which correspond to the M_w values illustrated in Fig. 9 can be computed readily and the results can then be visualized in terms of charts such as those given in Figs. 10–12. In each chart the height of the vertical bar is a relative measure of the mole fraction (in moles of monomer content) of each oligomeric species (monomer = 1, dimer = 2, etc.) at a given total protein concentration, c_t. Figure 10 traces the course of the relative concentrations of species when there is no interaction between subunits, i.e., $\alpha = 1$. At extreme dilutions all of the protein exists as monomer. With increasing total concentration of protein, the assembly proceeds through dimer, trimer, etc., each species appearing in appreciable concentration. Octamer becomes noticeable near $c_t = 0.2$ mg/ml and continues to increase until it dominates the distribution above $c_t = 2$ mg/ml. Thus there is a stepwise transition from the smallest species to the final structure.

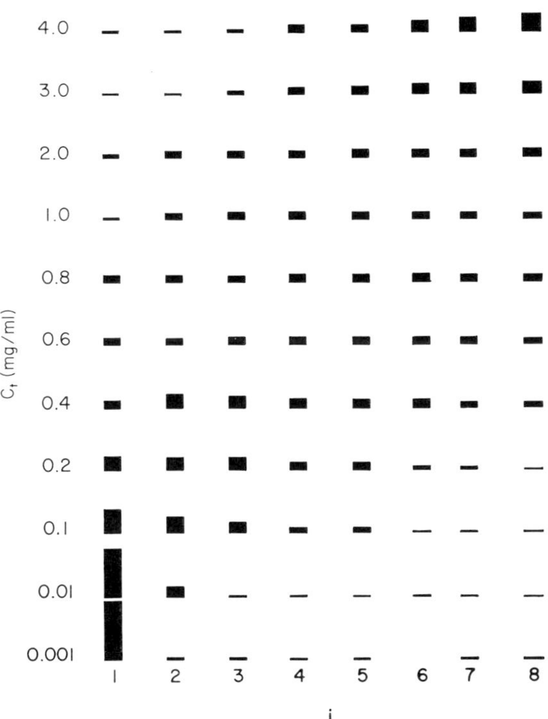

Fig. 10 Relative mole fractions (in terms of content of moles of monomer units) at different total protein concentrations c_t, of various oligomeric species 1, 2, . . . , 8 for an octameric assembly in which $k_2 = 10^5$, $M_1 = 10^4$, and $\alpha = 1.0$.

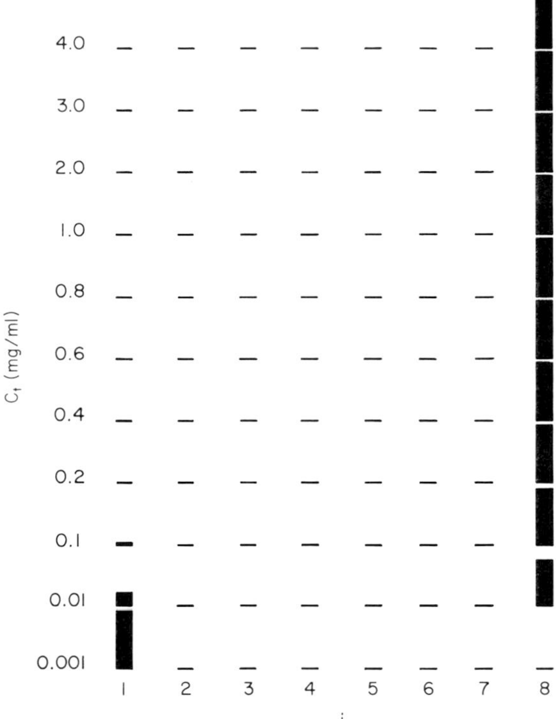

Fig. 11 Relative mole fractions (in terms of content of moles of monomer units) at different total protein concentrations c_t, of various oligomeric species, 1, 2, . . . , 8 for an octameric assembly in which $k_2 = 10^5$, $M_1 = 10^4$, and $\alpha = 3$.

In marked contrast is the behavior of an ensemble in which $\alpha = 3.0$ (Fig. 11). Again, at extreme dilutions, only monomer is present. When a modest cooperativity exists, however, octamer appears almost exclusively as the total concentration of protein is increased. The transition is thus extremely sharp, and only negligible concentrations of intermediate species are evident at any stage.

An alternative type of behavior is exhibited by a system with obstructive interactions, e.g., $\alpha = 0.01$ (Fig. 12). Again, monomer is predominant at very low concentrations. As total protein is increased, monomer is almost exclusively converted to dimer. Only at very high protein concentrations, $c_t > 3$ mg/ml, do small amounts of trimer appear in solutions which are overwhelmingly populated by dimers.

Each vertical column in Figs. 10–12 gives the relative concentration of each species of oligomer (A_i) as a function of total protein concentration (c_t). It is also possible to derive from the plots in Figs. 10–12

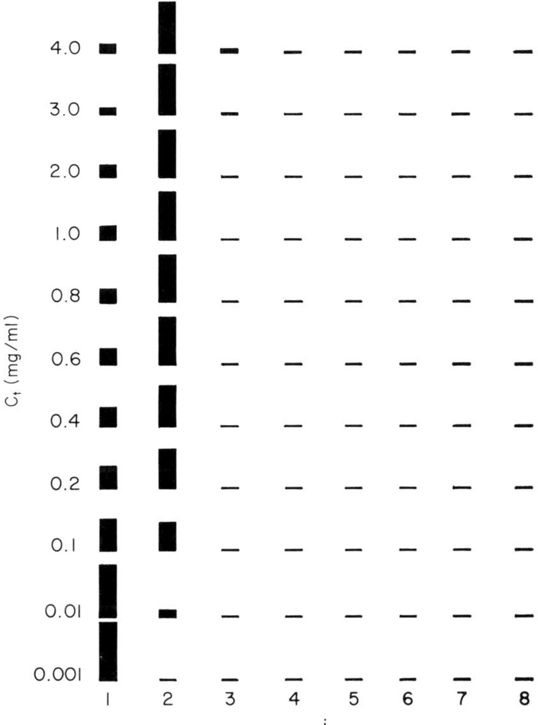

Fig. 12 Relative mole fractions (in terms of content of moles of monomer units) at different total protein concentrations c_t, of various oligomeric species 1, 2, . . . , 8 for an octameric assembly in which $k_2 = 10^5$, $M_1 = 10^4$, and $\alpha = 0.01$.

an algebraic relation between the concentration of a particular intermediate species and the total protein concentration. Such a relation can be used to find the protein concentration at which the mole fraction of any particular intermediate (dimer, trimer, etc.) reaches a maximum (Pranis, 1972; N. R. Langerman, unpublished analysis). It can be shown that with increasing cooperativity, this maximum is progressively shifted to lower protein concentrations.

It is apparent from Eq. (35) that the concentration of any species (A_i) depends not only upon the parameter α but also upon the association constant k_2, which essentially is a measure of the intrinsic interaction between the subunits. Distribution charts analogous to those in Figs. 10–12 can be assembled for alternative values of k_2, and M_w distributions as a function of total protein concentration can be calculated from such data. A plot of M_w vs. c_t is shown in Fig. 13 for a system in which $k_2 = 10^7$ and should be compared to the plot in Fig. 9 ($k_2 = 10^5$). A

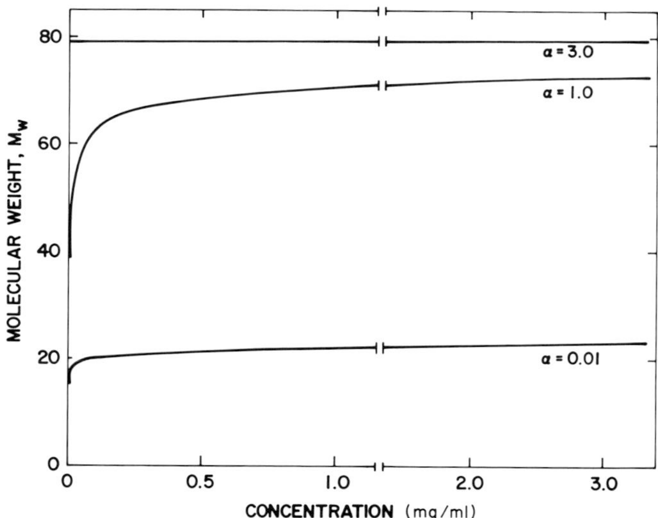

Fig. 13 Weight-average molecular weight, M_w, as a function of total protein concentration for an octamer ($M_1 = 10^4$) when $k_2 = 10^7$ and α is varied as shown.

hundredfold increase in k_2 corresponds to an increase in intrinsic affinity of approximately 2.5 kcal/mole. When $\alpha = 3$, this is sufficient to maintain the ensemble in the octameric state throughout the observable concentration range. When $\alpha = 0.01$, the system is uniformly in the dimeric state. Some intermediate species would be observable if the subunits did not interact.

The effect of an increase in k_2 (from 10^5 to 10^7) on the distribution of intermediate species is illustrated in the "pie" charts in Fig. 14. In each "pie," the shaded area of the sector (or the angle subtended at the center) indicates the relative amount of each species. In each pair of circles, the first represents the concentration when $k_2 = 10^5$, the second when $k_2 = 10^7$. It may be seen that at any concentration, c_t, the distribution is shifted toward the more highly assembled intermediates in the system with the higher k_2.

It is obvious from these illustrations that a small increment in the cooperativity parameter (e.g., from $\alpha = 1.0$ to $\alpha = 3.0$) has a large effect on the concentration dependence of M_w. In terms of free energy changes, this increment is given by Eq. (39):

$$\Delta(\Delta G^\circ) = RT \ln \frac{k_{i+1}}{k_i} = RT \ln \alpha = RT \ln 3 = 0.65 \text{ kcal} \qquad (39)$$

A small increase in the area of contact between monomer and oligomer would be sufficient to produce such an increment in the free energy of

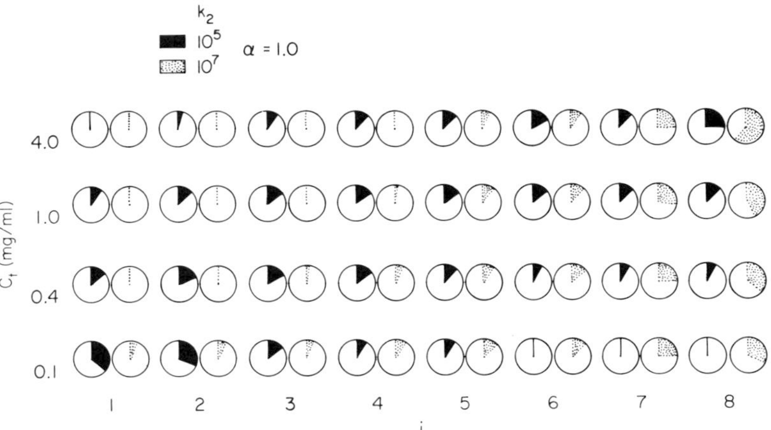

Fig. 14 Comparison of concentrations of various assembled species i, when k_2 is varied. In each pair of circles, the first represents the concentration when $k_2 = 10^5$ and the second when $k_2 = 10^7$. For all sets, $\alpha = 1$, $M_1 = 10^4$.

cooperativity and would be manifested as a steeply rising association curve.

The crucial role that a slight change in structure may play in the association–dissociation energetics can also be viewed in the following way.

For comparison of proteins having different numbers of subunits, the free energy of association of the highest oligomer may be equally distributed among the monomers, and the free energy of formation of monomers, $\Delta G_m{}^\circ$, may be defined as follows:

$$nA = A_n; \qquad \frac{\Delta G^\circ}{n} = -\Delta G_m{}^\circ = -\frac{1}{n} RT \ln K_n \tag{40}$$

$\Delta G_m{}^\circ$ may be viewed as the difference in free energy levels between the monomeric and oligomeric states of a subunit. In hemerythrin, for example (Table VII),

$$\Delta G_m{}^\circ = \tfrac{1}{8} RT \ln K_8 = 5.8 \text{ kcal/mole monomer} \tag{41}$$

We may now examine the effect of small changes in $\Delta G_m{}^\circ$ on the association–dissociation equilibrium. Some typical calculations for hemerythrin are summarized in Fig. 15. A change in $\Delta G_m{}^\circ$ of only 2 kcal (from -6.8 to -4.8 in Fig. 15) would convert this quaternary structure into one that would appear completely monomeric in the range below 0.5 mg/ml, and overwhelmingly dissociated (instead of its true state, which is almost completely oligomeric) up to very high concen-

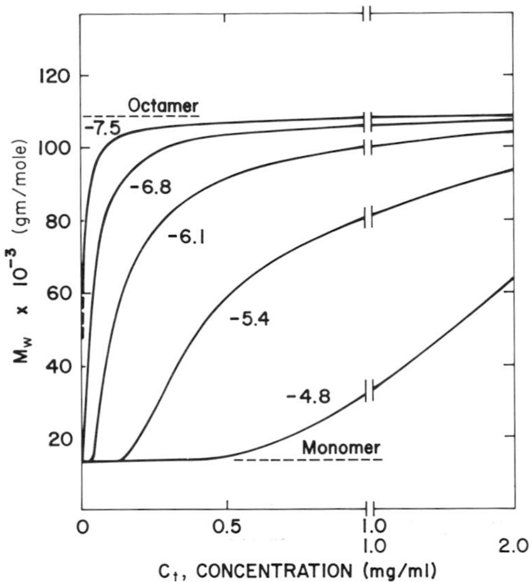

Fig. 15 Dependence of association on $\Delta G°$. Curves of weight-average molecular weight, M_w, are plotted against concentrations of protein computed for different values of $-\Delta G_m°$ (in kcal/mole), the difference in free energy levels between the monomeric and octameric state of the subunit.

trations of protein. Even a change of 0.8 kcal would markedly affect the dependence of M_w upon protein concentration (Fig. 15).

Since $\Delta G_m°$ represents a difference in free energy, its magnitude can be changed by varying $G°$ of either the initial (oligomeric) or the final (dissociated) state. Within the oligomer, such variations may arise from small perturbations of side chain interactions of neighboring monomers. Within the dissociated monomer, such variations may arise from perturbations of interactions of side chains with each other or with the surrounding solvent.

Energetic quantities in themselves cannot establish the molecular origin of the interactions or the changes in interactions. On the other hand, it is useful to keep in mind that minor changes in structure or orientation of small groups are accompanied by substantial changes in free energy. For example, replacement of an asparagine residue by an aspartic acid residue would be accompanied by a substantial free energy change due to the ionization of a new carboxyl group. As a first approximation, the free energy change, $\Delta G°$, is that of the ionization of a typical COOH group (Harned and Owen, 1958):

$$\text{—COOH} = \text{—COO}^- + \text{H}^+; \qquad \Delta G° = 6 \text{ kcal} \tag{42}$$

A somewhat smaller free energy change would be associated with the transfer of a COOH group from an aqueous environment to an apolar one. This has been observed in lysozyme (Parsons and Raftery, 1972), where the p*K* of Glu 35 was shifted by approximately 1.5 units, from its normal value of 4.5 (Nozaki and Tanford, 1967) to near 6. This change in p*K* corresponds to a free energy change of 2 kcal. Conversely, immobilization of an amide residue, according to Eq. (43),

$$\mathrm{CHRCONH_{flexible}} = \mathrm{CHROCONH_{immobile}}; \qquad \Delta S° = -5 \text{ eu} \quad (43)$$

should be accompanied by an entropy loss of 3–7 eu (Schellman, 1958) corresponding to a free energy change of 1.5 kcal. Figure 15 shows that changes in $\Delta G_m°$ as small as 1–2 kcal are accompanied by marked perturbations of the oligomeric state of a protein.

V. INTERSUBUNIT COMMUNICATION

The mechanism of long-range interactions between distant parts of a protein molecule, as well as between its subunit entities, has long been a subject of thoughtful speculation. Electronic conduction, particularly through organized hydrogen-bonded arrays, was a provocative early suggestion (Szent-Györgyi, 1946; Evans and Gergely, 1949). Long-range dispersion interactions, particularly between prosthetic groups such as hemes with a large electronic polarizibility, have also been invoked (Minton and Libby, 1968). With the current progress in delineating the detailed atomic structure of protein molecules and their rearrangements in response to environmental changes, it has become clear that subunit interactions involve specific residues at the interfaces and that ligand-linked conformational changes follow displacements of these interface residues.

A. Molecular Structure at Subunit Interfaces

Since each protein is a special case, no generalizations can be made about residues and interactions at contact interfaces. Because the most extensive and detailed studies have been made with hemoglobin (Perutz *et al.*, 1968a,b; Perutz and Ten Eyck, 1972), these studies will be summarized here.

The four subunits in hemoglobin, two α's and two β's, are in a tetrahedral array. A specific subunit forms three different types of interfacial contact with its neighbors (see Table V and Fig. 5). Since there are two nonidentical monomers, α and β, four different subunit interfaces must be present: $\alpha_1\beta_1$, $\alpha_1\beta_2$, $\alpha_1\alpha_2$, and $\beta_1\beta_2$. Contacts between unlike subunits (heteromeric) involve chiefly apolar residues, whereas those between like subunits (homomeric), if present, involve polar residues. Heteromeric interfaces, i.e., between α and β monomers, are of two types: $\alpha_1\beta_1$ and $\alpha_1\beta_2$. The residues involved in these contacts can be conveniently shown in chart form (Fig. 16).

As is apparent from Fig. 16, the $\alpha_1\beta_1$ interface contains the more numerous contacts. In the deoxy form, 32 residues or about 98 atoms come within a distance of 4 Å of each other, and in the oxy form, 34 residues or 110 atoms are in contact. These interactions, primarily apolar, arise largely from contacts between sections of B, G, and H helices of the α chain with H, G, and B helices of the β chain. The four or five interchain hydrogen bonds are all exposed to solvent and remain constant upon oxygenation.

At the $\alpha_1\beta_2$ interface (Fig. 16), 20 residues or 69 atoms are in contact in deoxyhemoglobin as compared to 19 residues or 80 atoms in oxyhemoglobin. At this interface the CD polypeptide region of one chain is dovetailed into the FG region of the other.

The $\alpha_1\beta_1$ contact remains relatively unperturbed during oxygenation. In contrast, the $\alpha_1\beta_2$ interface changes drastically. The CDα and the FGβ polypeptide sections shift substantially, although the dovetailing of CDβ and FGα is not appreciably affected. The shifts in relative positions of interface residues can be visualized from Fig. 17, which illustrates contacts before and after oxygenation. One of the most striking changes is the replacement of the Tyr 42α . . . Asp 99β hydrogen bond by an alternative bond between Asp 94α and Asn 102β. Marked conformational shifts also appear in the carboxyl-terminal regions. In the met (and presumably also oxy) state, the carboxyl-terminal residues Arg 141α and His 146β are free to rotate, i.e., not fixed in position, and the penultimate tyrosine residues in each chain also can assume alternative positions. In contrast, in deoxyhemoglobin the tyrosine groups occupy pockets between the F and H helices, and their —OH groups form a hydrogen bond with the C=O of the corresponding valine residue in the FG 5 region of the polypeptide chain. The carboxyl-terminal residue Arg 141α_1 forms a double salt bridge with the other α subunit: the free COO^- is linked to the ^+H_3N of amino-terminal Val 1α_2 and the —$NHC(NH_2)_2^+$ to the ^-OOC of Asp 126α_2. At the same time His 146β forms a pair of hydrogen bonds: the free COO^- is linked to the ^+H_3N of Lys 40α_2 and the charged imidazole moiety, ImH^+, to the ^-OOC of Asp 94β_1.

These residue shifts, which appear when hemoglobin combines with

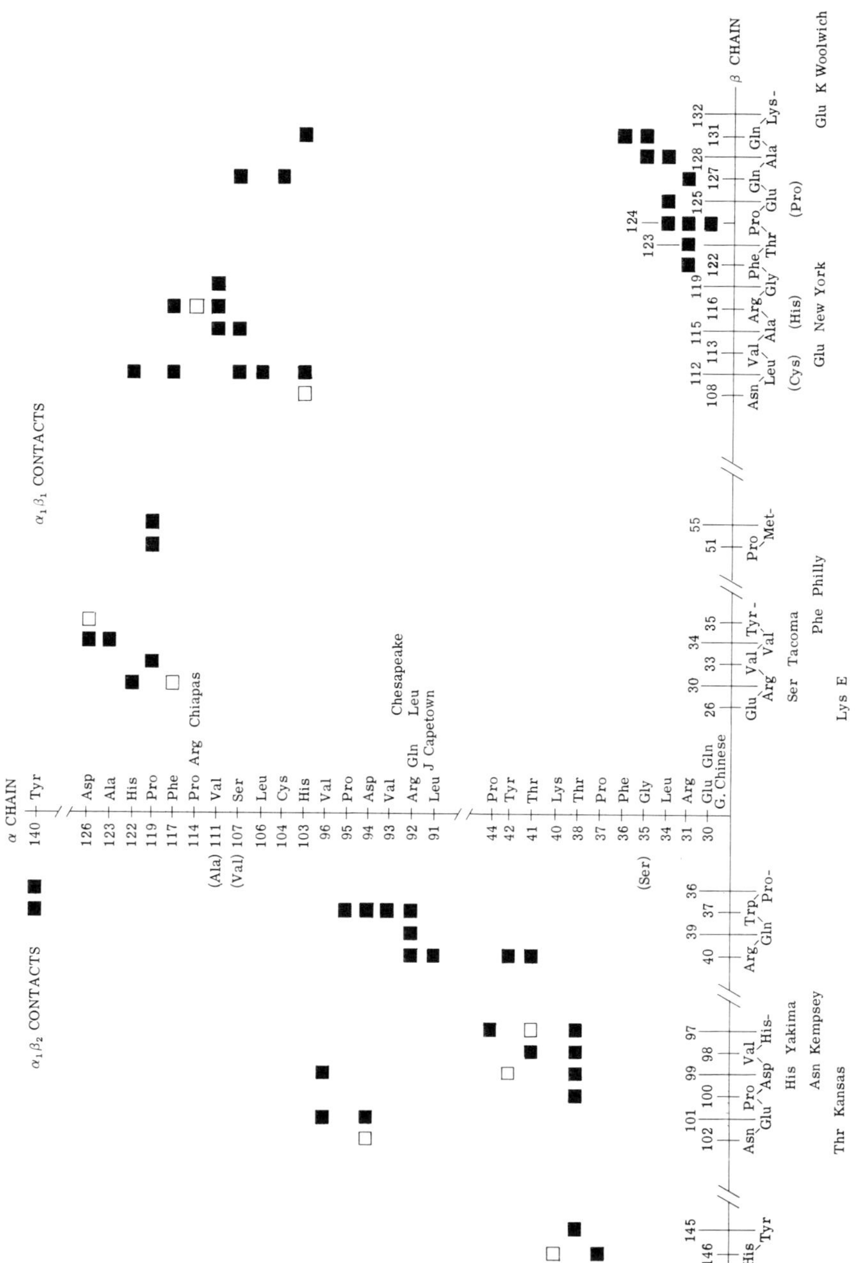

a ligand, produce a concomitant displacement of the helical regions of 2–3 Å and a change in the tilt of the heme plane relative to the polypeptide chain. These changes are also linked to a displacement of the Fe atom from a position approximately 0.75 Å out of the plane of the porphyrin ring in deoxyhemoglobin to a coplanar position in oxyhemoglobin. Simultaneously in the β subunit the γ-methyl groups of Val E11, which block the pocket of oxygen in the deoxy state, move aside to create a gap adequate for insertion of O_2. No corresponding change appears to occur in the α subunit. In both subunits, the distance from the coordinated histidine to the Fe atom is increased by 0.75–0.95 Å.

It is thus apparent that the electronic changes at the iron locus are intimately interlinked with the stereochemical displacements of the subunit interfaces. This linkage provides a basis for "communication" between subunits, even in their interior regions.

Some progress has also been made in delineating the residue contacts at the interface of insulin subunits (Blundell *et al.*, 1972) and lactate dehydrogenase subunits (Rossmann *et al.*, 1972). It is also apparent from the recent Cold Spring Harbor Symposium (1972) that several other protein structures, particularly those of multimeric regulatory enzymes, are approaching a level of resolution that will reveal subunit contacts in atomic detail. From the data already available it is evident that each protein has a pattern specific to itself.

B. Factors Modifying Subunit Interactions

We shall examine a number of factors that can perturb quaternary structure. These may be conveniently classified as (1) intrinsic and (2) extrinsic.

Fig. 16 Residues at contact interfaces between subunits in horse oxyhemoglobin. The residues of the α chain involved in subunit contacts are shown on the ordinate and the residues of the β chain are shown on the abscissa. Filled squares indicate van der Waals contacts; open squares indicate that the contact includes a hydrogen bond. The right quadrant shows the residues in the $\alpha_1\beta_1$ interface, while the left shows those residues in the $\alpha_1\beta_2$ interface. Thirty-four residues participate in the $\alpha_1\beta_1$ contacts. Twenty-one of these are common to all normal mammalian hemoglobins so far examined. Nineteen residues participate in $\alpha_1\beta_2$ contacts. With one exception, all are common to mammalian hemoglobins.

The amino acid residues in parentheses represent the substitutions in human hemoglobin A as compared to horse hemoglobin. Where there are no parentheses, the residues are identical in both proteins. Mutant human hemoglobins (in which only one mutation has occurred) are also indicated. Mutations involving the α chain are shown to the right of the α axis and mutations of the β chain are shown below the β axis. See the text and Table VIII for the effects of these mutations.

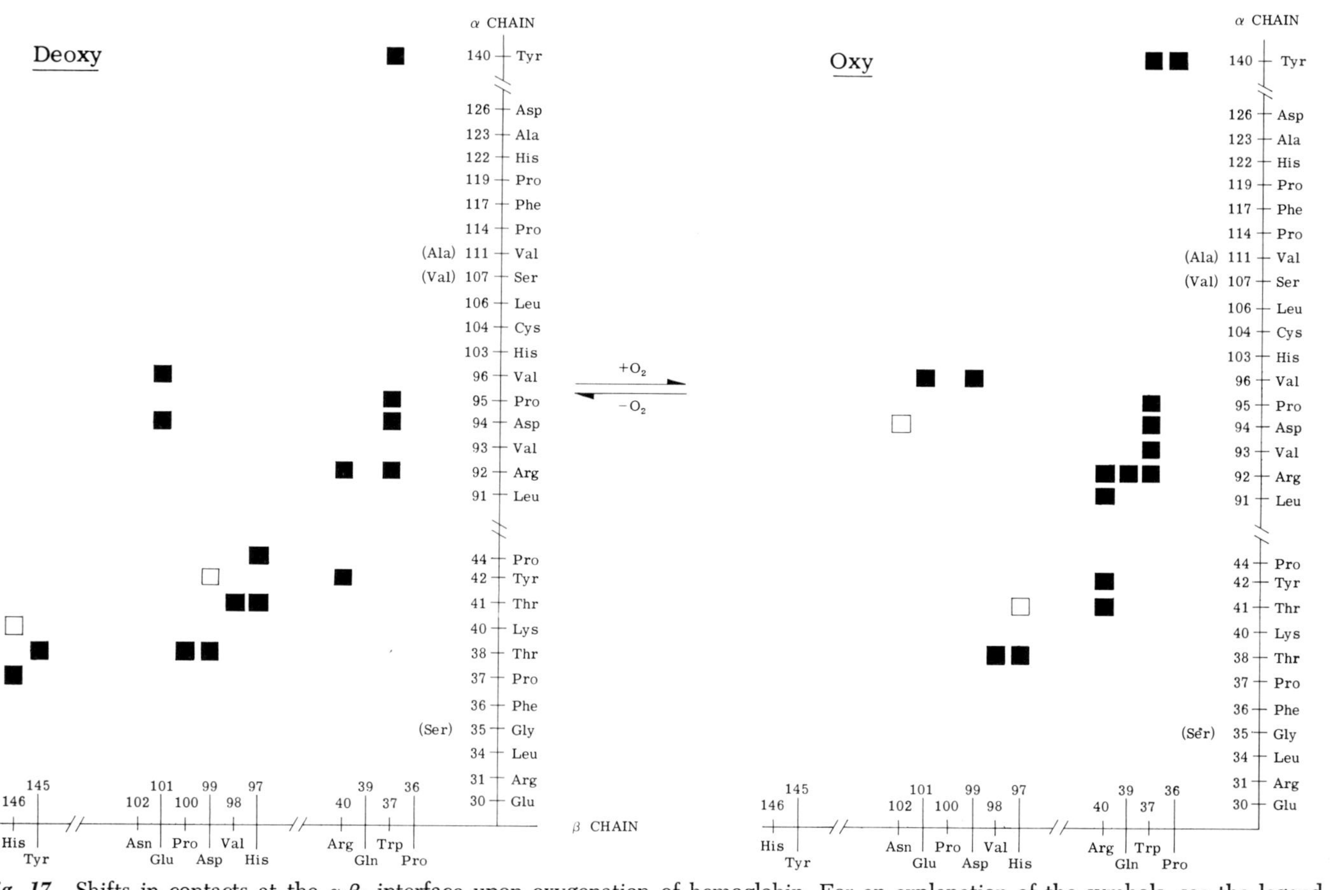

Fig. 17 Shifts in contacts at the $\alpha_1\beta_2$ interface upon oxygenation of hemoglobin. For an explanation of the symbols, see the legend to Fig. 16.

1. *Intrinsic Factors*

Intrinsic factors that modify subunit interactions are those caused by variations in the primary structure of a protein. Once again the most extensively studied protein is hemoglobin. Over one hundred abnormal hemoglobins arising from point mutations are known (Dayhoff, 1969). In all of these, hemoglobin exists as a tetramer. When residue substitutions occur at or near the contact interfaces, one can expect subunit interactions to be affected.

When hemoglobin ($\alpha_2\beta_2$) dissociates into two symmetrical dimers, $\alpha\beta$, the separation probably occurs at the $\alpha_1\beta_2$ and $\alpha_2\beta_1$ interfaces (Perutz *et al.*, 1968b). Symmetrical dissociation requires breaking only one of the apolar contacts and both polar contacts. In contrast, asymmetrical dissociation depends upon splitting both apolar contacts but no polar contacts. Since dissociation into dimers is facilitated by concentrated electrolyte solutions, it is the polar bonds that are most likely to be cleaved. The $\alpha_1\beta_2$ interface, being smaller in area, should be the first to open.

Perutz and Lehmann (1968) have scrutinized the relationships between point mutations and subunit interactions in hemoglobin. Table VIII contains a list of some abnormal hemoglobins with their associated clinical symptoms and abnormal biochemical properties. The residue replacement for each of these mutants can be seen in Fig. 16, as can the contact region affected by this interchange.

With the structural information now at hand it is often possible to rationalize the effects of mutations on the quaternary structure of hemoglobin (Perutz and Lehmann, 1968). Hemoglobin E, for example, carries a lysine residue in place of Glu 26β. The latter in normal hemoglobin probably forms hydrogen bonds with basic residues in its vicinity, and these bonds would be disrupted by the substitution of a lysine residue. On this basis one can account for the displacement of β^E chains from $\alpha_2{}^A\beta_2{}^E$ by normal β^A chains. Since splitting of hemoglobin dimers to monomers occurs at the $\alpha_1\beta_1$ interface, one would also expect increased dissociation of $\alpha\beta$ dimers of hemoglobin E. On related structural grounds Perutz and Lehmann (1968) predict that hemoglobin Tacoma and hemoglobin New York should show increased dissociation of dimers into monomers, since the mutant replacements (see Fig. 16) should weaken contacts at the $\alpha_1\beta_1$ interface.

In hemoglobin Philly, replacement of Tyr 35β by phenylalanine is accompanied by increased dissociation to monomers. This tyrosine, in helix C of the β chain, projects into the interior of the tetramer and forms a hydrogen bond with Asp 126α (see Fig. 16). Phenylalanine can-

TABLE VIII

Effects of Mutations at Contact Interfaces of Subunits of Hemoglobin[a]

Designation	Residue No.	Residue substitution: From	Residue substitution: To	Clinical symptoms	Biochemical Properties[b]: Oxygen affinity	Heme–heme interaction	Bohr effect	Other
A. *α-Chain Mutations*								
G Chinese	30	Glu	Gln	None	0	0	0	
J Capetown	92	Arg	Gln	(Polycythemia)	+	−	(0)	
Chesapeake	92	Arg	Leu	Polycythemia	+	−	0	
Chiapas	114	Pro	Arg	None	0	0	0	
M Iwate[c]	87	His	Tyr	Cyanosis	−	−	−	
B. *β-Chain Mutations*								
E	26	Glu	Lys	None or anemia	−	0		Diminished affinity for α chains
Tacoma	30	Arg	Ser	None	+			Unstable at 50°
Philly	35	Tyr	Phe	Mild anemia				All 6 Cys react with mercuribenzoate
Yakima	99	Asp	His	Polycythemia	+	−	0	
Kempsey	99	Asp	Asn	Polycythemia	+	−	(0)	
New York	113	Val	Glu	None				No increased electrophoretic mobility despite extra charge
K Woolwich	132	Lys	Glu	None	0	0	0	
Kansas	102	Asn	Thr	Cyanosis	−	−	(+)	Unstable at 50°; easily autoxidized
Ranier	145	Tyr	Cys	Polycythemia	+	−	−	
Richmond	102	Asn	Lys	None	0	(−)	0	
M Hyde Park[c]	92	His	Tyr	Methemoglobinemia	0	−	0	

[a] Modified from the table of Perutz and Lehmann (1968) and Greer (1972).

[b] Key: +, increase compared to normal hemoglobin; −, decrease; 0, no change.

[c] These mutations are at positions not really at the interface. Rather the heme-linked His is replaced by a Tyr. Nevertheless, marked changes in tertiary structure ensue.

not form such a hydrogen bond because it lacks the —OH substituent, and hence a contact in the $\alpha_1\beta_1$ interface is weakened.

Hemoglobin Kansas is a particularly interesting mutant (see Fig. 16) in that it dissociates into $\alpha\beta$ dimers more readily than does hemoglobin A. The interface hydrogen bond between Asn 102β and Asp 94α of hemoglobin A disappears with the substitution of Thr 102β in the mutant (although an alternate hydrogen bond to an aspartic acid residue *within* the β chain could be formed). The dissociation constant changes from 1.2×10^{-6} for hemoglobin A to 2×10^{-4} for hemoglobin Kansas. This corresponds to a change of about 3 kcal in $\Delta G°$ of dissociation. This free energy change is the net result of entropic contributions from localized molecular rearrangements (such as that of the released Asp 94α side chain) as well as of enthalpic contributions from hydrogen bond energies and van der Waals interactions. As was illustrated earlier (Fig. 15), a change of 3 kcal in free energy should be accompanied by very large changes in the degree of dissociation of an oligomeric structure.

Hemoglobin Richmond is closely related to Kansas in that the mutation appears at Asp 102β, but in the former the replacement is a lysine residue. This substitution also leads to anomalously high dissociation of tetramer into dimer, a reflection of the perturbation at the contact interface, even though the other functional properties of hemoglobin Richmond are normal (Table VIII).

Hemoglobins Chesapeake, J Capetown, Yakima, Kempsey, and Kansas also illustrate the importance of specific contacts at the $\alpha_1\beta_2$ interface for the expression of heme–heme interactions. Diminished heme–heme interaction has been found in each of these mutant hemoglobins (see Table VIII). On the other hand, all of these hemoglobins show a Bohr effect. Nuclear magnetic resonance (Shulman *et al.*, 1972; Huestis and Raftery, 1972; Lindstrom *et al.*, 1972) and electron paramagnetic resonance studies (Ogata and McConnell, 1972) may provide some of the insights needed to rationalize the molecular detail provided by X-ray diffraction.

Mutations that affect subunit interface contacts may also be accompanied by clinical symptoms (see Table VIII). In a sense the most striking manifestation of this is the original "molecular disease" protein, hemoglobin S, in which the replacement of Glu 6β (hemoglobin A) by Val 6β (hemoglobin S) causes an anemia which is clearly a consequence of the much reduced solubility of the mutant pigment (Murayama, 1962).

Detailed diffraction studies have been made with some of the mutant hemoglobins, particularly those with drastically altered functional properties (Greer, 1972). With hemoglobin Rainier, for example, the replacement of Tyr 145β by Cys 145β, accompanied by formation of an intrachain disulfide linkage, Cys 145β—S—S—Cys 93β, displaces the carboxyl-terminal His 146β from its normal position and breaks up the salt bridges to Lys 40α and to Asp 94β (see Section V,A). The new disulfide bond accounts for the strong resistance of hemoglobin Rainier to denaturation by alkali and to digestion by carboxypeptidase. The disruption of the salt bridges, responsible for the Bohr effect, accounts for the decreased Bohr effect observed in hemoglobin Rainier (Perutz and Ten Eyck, 1972).

The vital role played by residues at or near the carboxyl terminus is also seen when mutations occur at position 146β. In hemoglobin Hiroshima, His 146β is replaced by Asp 146β (Imai *et al.*, 1972). Clearly this would perturb one of the normal salt linkages in this region (see Fig. 16). Ultraviolet spectra suggest that during oxygenation, normal conformational changes in the $\alpha_1\beta_2$ interface are impaired and cooperativity is diminished.

The important role played by the salt bridge contributed by His 146β is also demonstrated by the behavior of Des-His hemoglobin, in which the carboxyl-terminal residue has been specifically removed from the β chain by digestion with carboxypeptidase (Kilmartin and Wootton, 1970; Kilmartin and Hewitt, 1972). This hemoglobin shows a substantially reduced Bohr effect. Its oxygen affinity and Hill coefficient are practically identical with that of hemoglobin Hiroshima (His 146β replaced by Asp 146β).

Corresponding removal of the carboxyl-terminal residue of the α chain, Arg 141α, to produce Des-Arg 141α hemoglobin is also followed by marked changes in functional properties. Particularly interesting is the enhanced dissociation into dimers in the presence of salt (Kilmartin and Hewitt, 1972; Hewitt *et al.*, 1972).

Other examples of marked changes in oligomeric state are provided by hemoglobins G Georgia and Rampa (Smith *et al.*, 1972). Both of these variants are relatively stable proteins in which Pro 95α has been replaced by leucine or by serine, respectively. Clearly the interface contact with Trp 37β (see Fig. 16) has been ruptured. In both of these variants, the liganded form shows a much greater tendency to dissociate into dimers under conditions in which normal hemoglobin is tetrameric (e.g., 0.1 *M* NaCl, pH 7, 25°). In the deoxygenated state, both Georgia and Rampa hemoglobins remain tetramers under the same conditions, as well as in much more concentrated salt solutions. Both variants show increased oxygen affinities and decreased heme–heme interactions.

Table IX summarizes the tetramer → dimer dissociation constants for hemoglobins. These vary over more than a thousandfold range. Thus a single amino acid substitution can cause a dramatic change in the oligomeric state. Nevertheless, the corresponding free energy change is only

TABLE IX

Dissociation Constants ($\alpha_2\beta_2 \rightarrow 2\alpha\beta$) of Liganded Hemoglobins in Dilute Salt Solution

Variant	Residue substitution: From	Residue substitution: To	Ligand	$K \times 10^5$ (M)[a]
A	—	—	O_2	(0.1)[b]
Richmond	Asn 102β	Lys 102β	CN	3
Kansas	Asn 102β	Thr 102β	O_2, CO	20
Bibba	Leu 136α	Pro 136α	CN	50
Rampa	Pro 95α	Ser 95α	CN	165
G Georgia	Pro 95α	Leu 95α	CN	225

[a] These measurements are summarized in Smith *et al.* (1972), which also lists references to the original investigators.

[b] See Kellett and Schachman (1971) and Kellett (1971).

of the order of a few kilocalories. If the subunit aggregate is on the verge of dissociating, a small change in $\Delta G°$ is sufficient to lead to an observable disaggregation. In contrast, it is of interest to note that deoxyhemoglobins tend to maintain the tetrameric structure. Clearly the additional ionic interactions of the salt bridges in the deoxygenated state provide an appreciable free energy of stabilization, more than enough to compensate for the few kilocalories of destabilization introduced by an amino acid substitution that is critical in the oxy form.

Studies of the effects of residue replacements are also being carried out with other proteins, but on a much less extensive scale. For example, a study has been made comparing the ease of dissociation of β-lactoglobulin A with the dissociation of β-lactoglobulin B (Zimmerman *et al.*, 1970). These two proteins differ in only two residues (per subunit) in primary structure, and the tertiary structure is still unknown. The dissociation constants of dimer into monomer (Table VII) differ by a factor of about 3, corresponding to a free energy change of only 0.6 kcal. Again we see that small changes in interaction energy can produce a marked effect on the state of aggregation of subunits.

In addition to covalently linked residues in the polypeptide chain, there are occasionally other substituents, such as metal ions, which play a crucial role in maintaining quaternary structure. In many such cases a substituent is so tightly bound that it cannot be removed by dialysis even against solutions with chelators, and hence it seems appropriate to classify its contributions to structure as intrinsic. A striking example is provided by aspartate transcarbamylase which contains six moles of Zn^{2+}

(Rosenbusch and Weber, 1971b; Nelbach *et al.*, 1972) per oligomer (molecular weight 310,000; Table II), that is, one per protomer (constituted of one catalytic and one regulatory monomer). This Zn^{2+} is not removed by prolonged dialysis against solutions containing 1,10-phenanthroline, nor is it exchanged perceptibly for free radioactive $^{65}Zn^{2+}$, even in an exposure time of 40 days (Nelbach *et al.*, 1972).

When aspartate transcarbamylase is dissociated into its constituent subunits, Zn^{2+} can be easily removed by addition of chelators (Nelbach *et al.*, 1972). The isolated catalytic subunit is free of Zn^{2+} and still possesses catalytic activity. Clearly the metal is not a constituent of the active site. Regulatory subunits containing one Zn^{2+} per monomer (MW 17,000) can be obtained. If Zn^{2+} is removed from isolated regulatory subunits, this apoprotein is unable to associate with catalytic subunits to regenerate aspartate transcarbamylase. However, if Zn^{2+} is added to the mixture, complete reconstitution of the oligomeric enzyme (MW 310,000) is achieved, and this quaternary ensemble contains firmly bound zinc. Related metals in the IIb group of the Periodic Table, such as Hg^{2+} and Cd^{2+}, also facilitate reconstitution of the enzyme. It is evident, then, that Zn^{2+} (and related metal) ions play a key role in stabilizing the quaternary architecture of aspartate transcarbamylase, apparently by enhancing interface interactions between heteromeric subunits. The stabilization is so great that no exchange can be observed between aspartate transcarbamylase and radioactively labeled subunits, even after prolonged incubation (Nelbach *et al.*, 1972).

Since the apocatalytic subunit of aspartate transcarbamylase maintains a trimeric structure even in the absence of Zn^{2+}, the metal is not involved in interchain bonding of catalytic subunits. In contrast, Zn^{2+} promotes association of regulatory monomers to the dimeric form, and hence the metal does seem to act at this interface. It is also possible that Zn^{2+} plays a role at the catalytic–regulatory interface, but its contribution here is more difficult to establish. Spectroscopic observations point to —SH groups as being ligands for the Zn^{2+}. This seems substantiated by the fact that mercurials displace Zn^{2+} from aspartate transcarbamylase and simultaneously split the enzyme into its catalytic and regulatory subunits.

Zinc has also been assigned a role in cementing the quaternary structure of α-amylase (Stein and Fischer, 1960; Isemura and Kakiuchi, 1962; Robyt and Ackerman, 1972). Apparently one zinc ion binds two subunits of α-amylase together. Removal of zinc dissociates the enzyme into monomers which are nevertheless still active.

Zinc is also involved in the stabilization of the quaternary structure of alkaline phosphatase (Simpson and Vallee, 1968) and alcohol dehydrogenase (Drum *et al.*, 1969). In these enzymes, however, the cementing role is not clearly established because the metal is also directly involved in catalytic activity.

2. *Extrinsic Factors*

Extrinsic factors may be analyzed in terms of the effects of (1) substrate, (2) other ligands, and (3) third components in the solvent.

In many proteins, quaternary structure is modified when the substrate is bound. The example with the longest history and the greatest molecular detail is, of course, hemoglobin. It was originally noted by Haurowitz (1938) that oxygenation produces a drastic change in the shape of hemoglobin crystals. From this observation of macroscopic changes Haurowitz prophetically predicted that the deoxyhemoglobin molecules themselves must undergo a transition in shape with uptake of oxygen. Thermodynamic arguments in the same direction were provided by Eley (1943), Wyman and Allen (1951), and St. George and Pauling (1951). These indirect indications were confirmed by the direct demonstration from X-ray diffraction that oxygenation produces marked rearrangement of the β subunits in the tetramer (see Section V,A). Even the rate of rearrangement has now been measured by following the change in reactivity of the 93β sulfhydryl group upon oxygenation of hemoglobin; the conformational adaptation occurs in less than a millisecond (Antonini and Brunori, 1969).

Studies with tryptophanase (Morino and Snell, 1967a) provide a striking example of the capabilities of hydrodynamic methods in following structural metamorphoses. The binding of the coenzyme, pyridoxal phosphate, to tryptophanase changes the sedimentation coefficient from 9.5 S to 10.5 S and the intrinsic viscosity from 5.5 to 3.8, with no change in molecular weight. These changes in hydrodynamic behavior provide a definite indication of a conformational adaptation of the tetramer to a more symmetrical, compact form. The quaternary structure of tryptophan oxygenase is also stabilized by the binding of tryptophan or α-methyl tryptophan (Koike *et al.*, 1969).

Even smaller changes in sedimentation coefficient have been detected using differential methods (Gerhart and Schachman, 1968; Kirschner and Schachman, 1971a,b). For example, with these techniques it has been found that the enzyme aspartate transcarbamylase sediments 3.5% slower in the presence of succinate and carbamyl phosphate than in the absence of these ligands. This change corresponds, in a spherical protein, to an 11% increase in hydrodynamic volume in the protein with ligands attached, i.e., to a substantial swelling of the enzyme. If the macromolecule is not spherical, binding of ligands must produce a more asymmetric quaternary structure. A general opening of the structure is also evident from the much increased reactivity of the sulfhydryl groups of the enzyme toward chloromercuribenzoate. Comparative studies have also been made of the effects of succinate and carbamyl phosphate on the isolated catalytic subunit.

Other ligands besides substrate may also transform a quaternary structure (for examples, see Table I). Again, an especially interesting case in point is provided by hemoglobin. Chanutin and co-workers (Sugita and Chanutin, 1963; Chanutin and Curnish, 1965) originally observed that a number of phosphates form complexes with hemoglobin. Subsequently it was discovered by Benesch and Benesch (1967) and by Chanutin and Curnish (1967) that certain organic phosphates, particularly 2,3-diphosphoglycerate, markedly decrease the oxygen affinity of hemoglobin. It has now been clearly demonstrated (Chanutin and Hermann, 1969; Benesch *et al.*, 1969; Benesch and Benesch, 1969) that this effect arises from a much stronger binding of 2,3-diphosphoglycerate by deoxyhemoglobin than by oxyhemoglobin, with a consequent shift in protein equilibrium to the deoxy conformation in the presence of ligand. It was originally proposed by Benesch *et al.* (1968) that this preferential binding by deoxyhemoglobin occurs in the central cavity of the tetramer along the dyad axis of the quaternary structure. This cavity has a shape somewhat like that of two adjoining boxes, each about 20 Å long (perpendicular to the dyad axis), 8–10 Å wide, and 25 Å deep (along the dyad axis) (Perutz *et al.*, 1968b). One box separates the α chains, the other the β chains. The internal cavity provided by these boxes is lined with polar residues and hence offers a very suitable environment for an organic phosphate ion. Bunn and Briehl (1970) have shown that the α-NH_3^+ group of Val 1β and the imidazole groups of His 143β are necessary for the lowering of oxygen affinity on addition of 2,3-diphosphoglycerate. It was therefore suggested by Perutz and Ten Eyck (1972) that the four nitrogens and the two ϵ-amino groups of Lys 82β form salt linkages with the acidic groups of diphosphoglycerate. These expectations, deduced from chemical evidence, have now been amply confirmed and extended by X-ray diffraction studies of crystals of deoxyhemoglobin containing 2,3-diphosphoglycerate (Arnone, 1972). These show that diphosphoglycerate does indeed plug the entrance to the central cavity at the amino-terminal region of the β subunit and that it forms salt linkages with seven cationic groups of the β chains: two Val 1β; two His 2β; two His 143β; and one Lys 82β (Fig. 18). The combination with ligand is accompanied by a movement of the A helices and the two β subunits closer together, but the H helices are displaced from each other, and the

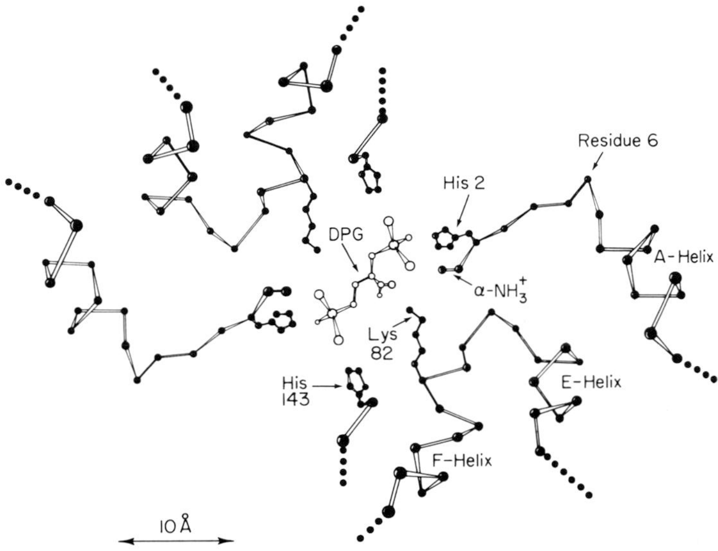

Fig. 18 Diagram showing salt bridges formed by 2,3-diphosphoglycerate (DPG) with cationic side chains of hemoglobin (from Arnone, 1972). The diphosphoglycerate fills the channel in the tetramer at the β chain portal. The binding of this anion is accompanied by a movement of the A helix and the residue mutation site of hemoglobin S toward the E helix and the EF bend.

β subunits as a whole may also move apart. Thus diphosphoglycerate binding is linked to large changes in tertiary structure and perhaps to some displacements in quaternary arrangement.

There are numerous other examples of the effects of ligands on quaternary structure, but none so well delineated in the molecular sense. Tryptophanase undergoes substantial changes in sedimentation coefficient in the presence of K^+ ions (Morino and Snell, 1967a). In the presence of ClO_4^- ions, the reactivity of the sulfhydryl group of hemerythrin (Garbett *et al.*, 1971) is markedly attenuated. In contrast, in the presence of iron-coordinating ligands such as N_3^-, rates of reaction of mercurials with the mercaptan group of hemerythrin are greatly enhanced (Keresztes-Nagy and Klotz, 1965). The change in reactivity of —SH groups in aspartate transcarbamylase with the binding of substrate (Gerhart and Schachman, 1968) has already been alluded to. Evidence of conformational changes in this enzyme, promoted by the nucleotide activator and inhibitor, ATP and CTP, respectively, as well as by other ligands, is also provided by observations of marked decreases in susceptibility to tryptic digestion (McClintock and Markus, 1968) and changes in immunologic reactivity (von Fellenberg *et al.*, 1968). In the presence of various additives, glutamine synthetase undergoes a variety of conformational adaptations, including dissociation into subunits (Shapiro and Ginsburg, 1968) as is evident from the hydrodynamic behavior and from the reactivity of its mercaptan groups. Differential reactivities of other side chains such as lysine and tyrosine have also been observed in some proteins, e.g., tobacco mosaic virus (Fraenkel-Conrat and Colloms, 1967), and these, too, reflect conformational adaptations.

Finally, additives such as salts or denaturants added to the protein solution in large quantity may also affect conformation. For example, at moderate salt concentration (0.4 M ionic strength), RNA polymerase (MW 782,000) is dissociated into two subunits (Richardson, 1966; Lee-Huang and Warner, 1969). At higher salt concentrations, hemoglobin separates into dimers (Rossi-Fanelli *et al.*, 1961; Kirshner and Tanford, 1964). Many examples of the effects of high concentrations of urea or other denaturants on quaternary structure were cited in Section II.

When a third solute (besides protein and solvent water) is added to a solution in very low concentration, we can be reasonably sure that any perturbing effect on the quaternary structure of the protein is a direct one, since the conformational adaptation can be traced to preferential binding of the small molecule by one of the conformations in equilibrium. At high concentrations, also, the third solute may produce conformational rearrangement by direct combination. However, it is essential to remember that the chemical potentials of solute and solvent in a solution are inextricably intertwined (Klotz, 1966). This is a necessary consequence of the two laws of classical thermodynamics and is expressed concisely in the Gibbs–Duhem equation

$$n_1\, d\mu_1 + n_2\, d\mu_2 = 0 \tag{44}$$

where n_i represents the moles of each component and μ_i its chemical potential. From this relationship it follows that if we change the chemical potential of the solvent (e.g., by adding a third solute), we must simultaneously alter the chemical potential, and hence the behavior, of the (macromolecular) solute. In other words, the conformation of a macromolecule may be modified because the additive to the aqueous solution changes the interactions of the water molecules. The solvent environment may thus participate in the molding of the conformation of a dissolved macromolecule.

Summarizing this survey of different types of extrinsic factors that affect conformation, we find that a wide gamut of substances may perturb quaternary structures. It seems evident, therefore, that there can be multiple conformations for a particular protein in aqueous solution, even in the native state.

VI. FUNCTIONAL ASPECTS

A. Why Subunits?

Large proteins almost universally are constituted of an ensemble of subunits linked noncovalently. Why are these not composed of single, giant, covalently linked macromolecules, with peptide threads tying together the various functional regions? Conformational adaptations and adjustments are just as feasible in the latter arrangement as in the former. The following factors delineate some of the advantages of an assembly of subunits.

1. *Economy in Genetic Information*

A significant feature of oligomeric ensembles is that their architecture, stability, and function are governed completely by the properties of the constituent monomers. Once the monomers have been synthesized on the ribosome, they possess all the information necessary for self-assembly and function. A readily apparent biologic advantage of a structure that can be self-assembled is that it is completely defined by the genetic information required to specify the monomers. Clearly less DNA is needed to code for several identical subunits in an ensemble than for a single giant macromolecule.

2. *Efficiency in Eliminating Errors*

Given a small frequency of defects in the biosynthesis of a polypeptide, e.g., 1 in 10^3, it is clear that the construction of a large polypeptide chain of 10^3 residues runs a very high risk of being defective. In contrast, the construction of 10 subunits each with 10^2 residues will result in an overwhelming proportion of effective subunits. In a large pool of subunits, groups of ten could then assemble themselves into functional ensembles of 10^3 residues.

Similarly in oligomers containing in every ensemble 50% defective subunits that abolish function of the entire macromolecule, disproportionation could rearrange the distribution of subunits so that as much as half of the new ensembles would be functionally perfect. This type of phenomenon has been demonstrated *in vitro* in chemically injured glyceraldehyde-3-phosphate dehydrogenase (Smith and Schachman, 1971), a tetrameric ensemble (see Table II). All of the enzyme molecules can react with mercuribenzoate to the extent that fully *inactivated* material results. When this inactivated enzyme is allowed to sit for one hour, 70% of its initial activity is regained. Inactive molecules disproportionate to yield active native tetramers and, in this particular case, fully mercaptan-blocked monomers.

It seems very likely that many oligomeric systems are in dynamic equilibrium. This has been shown by the hybridization experiments, described in Section II,G. Wherever hybridization can occur, disproportionation will always be feasible, since it corresponds to the complementary process.[7]

3. *Entropic Factors*

There is another way in which the behavior of an oligomeric structure is fundamentally different from that of a single, giant molecule. In the former, an association–dissociation equilibrium [Eq. (1)] will exist under certain conditions. Thus at low enough concentration, an oligomeric ensemble will dissociate. Likewise, as the concentration of monomers is increased from an extremely low value, oligomers will form. The state of the assembly is thus concentration-dependent in an oligomeric structure, whereas it is independent of concentration in a single giant molecule. This gives an extra dimension of functional control to quaternary structures which may be particularly useful at the site of biosynthesis. For example, if the subunits themselves are nonfunctional as is true for many enzymes (Frieden, 1971), they could be synthesized at one locus, diffuse away, and accumulate elsewhere until they reach a concentration high enough to assemble into the functional quaternary ensemble. Disturbances of metabolic pathways at the site of biosynthesis would be minimized. Furthermore, subunits, being smaller than oligomers, would diffuse away more rapidly than oligomers from the site of biosynthesis, and hence leave the site open for continued production of new copies.

4. *Molecular Evolution*

Monod *et al.* (1965) first pointed out the advantages of an oligomeric ensemble in terms of evolution. Consider a protein monomer that has many functional groups spaced over its surface, many of which may

[7] It may be essential for the function of some proteins that they do *not* undergo hybridization or disproportionation. The possibility of hybridization or disproportionation could be eliminated by cross-linking the constituent units with covalent S—S bonds. In the IgG immunoglobulins, for example, the two haptenic binding sites must have the same specificity if a network is to form between a specific (multivalent) antigen and (divalent) antibody. This may be the reason that the constituent heavy and light chains are linked together by S—S bonds instead of being assembled into a quaternary structure containing two (noncovalently linked) heavy monomers and two light monomers.

possess inherent chemical affinity for one another. Since the distance between two such groups on the surface of two individual monomers is necessarily the same, an antiparallel association of two monomers would create a dimer with two bonds and a symmetry axis. The association between two monomers at this point might be weak because of mutual repulsion forces between monomers, but consider the mutation of one functional group in a monomer which might produce the capacity to form another bond with its partner. This would result in the formation of *two* new bonds at the subunit contact sites. Any such mutational event would symmetrically and cooperatively affect the functional properties of the two monomers due to the changes in subunit interactions. Any mutation that produces a slight change in the tertiary structure of the protein might be amplified in the same manner through the subunit interactions in the ensemble. Although we have described the situation for dimers, these effects would be even more magnified in higher oligomers. Thus the structural and functional effects of a single mutation in an oligomer which possesses axes of symmetry should be greatly amplified as compared to the same mutation in a monomeric protein. Oligomeric proteins should, therefore, constitute particularly sensitive targets for molecular evolution allowing much stronger selective pressures to prevail in the random selection of structures that are functionally more efficient.

B. Distribution of Subunit Ensembles within Metabolic Pathways

The subunit status of enzymes in two metabolic pathways has been examined to see if any pattern emerges. Figure 19 summarizes the quaternary structures of the enzymes in the glycolytic pathway, and Fig. 20 presents corresponding information for enzymes in the tricarboxylic acid cycle. The striking feature of these charts is the overwhelming presence of enzymes constituted of subunits. The number of subunits in a specific enzyme often varies with its biologic source (even though the metabolic reactions catalyzed are the same), but if one source produces a subunit ensemble, so do the others. It is apparent that regulatory processes are intertwined throughout a metabolic sequence. This must lead to a dampening of the effect of any specific perturbation. Extensive dampening networks would provide a very effective mechanism for homeostasis at the cellular level.

C. Modes of Metabolic Regulation

Let us examine first the various options intrinsically available to the quaternary ensemble (Fig. 21).

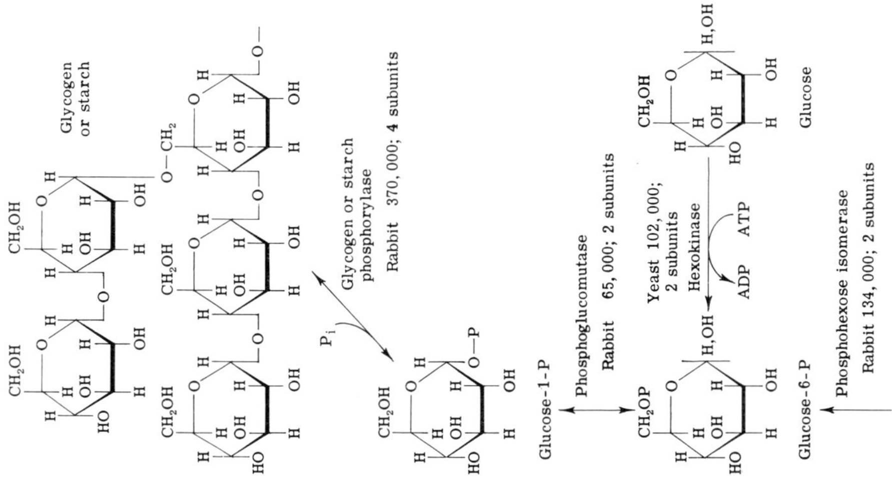
Glycogen or starch
Glycogen or starch phosphorylase
Rabbit 370,000; 4 subunits
P_i
Glucose-1-P
Phosphoglucomutase
Rabbit 65,000; 2 subunits
Glucose
Yeast 102,000; 2 subunits
Hexokinase
ATP
ADP
Glucose-6-P
Phosphohexose isomerase
Rabbit 134,000; 2 subunits

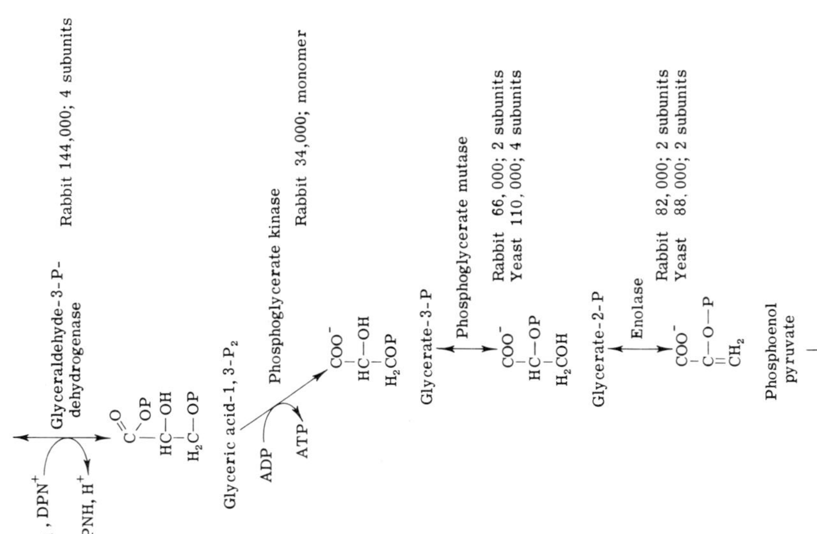
P_i, DPN^+
DPNH, H^+
Glyceraldehyde-3-P-dehydrogenase
Rabbit 144,000; 4 subunits
Glyceric acid-1, 3-P_2
ADP
ATP
Phosphoglycerate kinase
Rabbit 34,000; monomer
Glycerate-3-P
Phosphoglycerate mutase
Rabbit 66,000; 2 subunits
Yeast 110,000; 4 subunits
Glycerate-2-P
Enolase
Rabbit 82,000; 2 subunits
Yeast 88,000; 2 subunits
Phosphoenol pyruvate

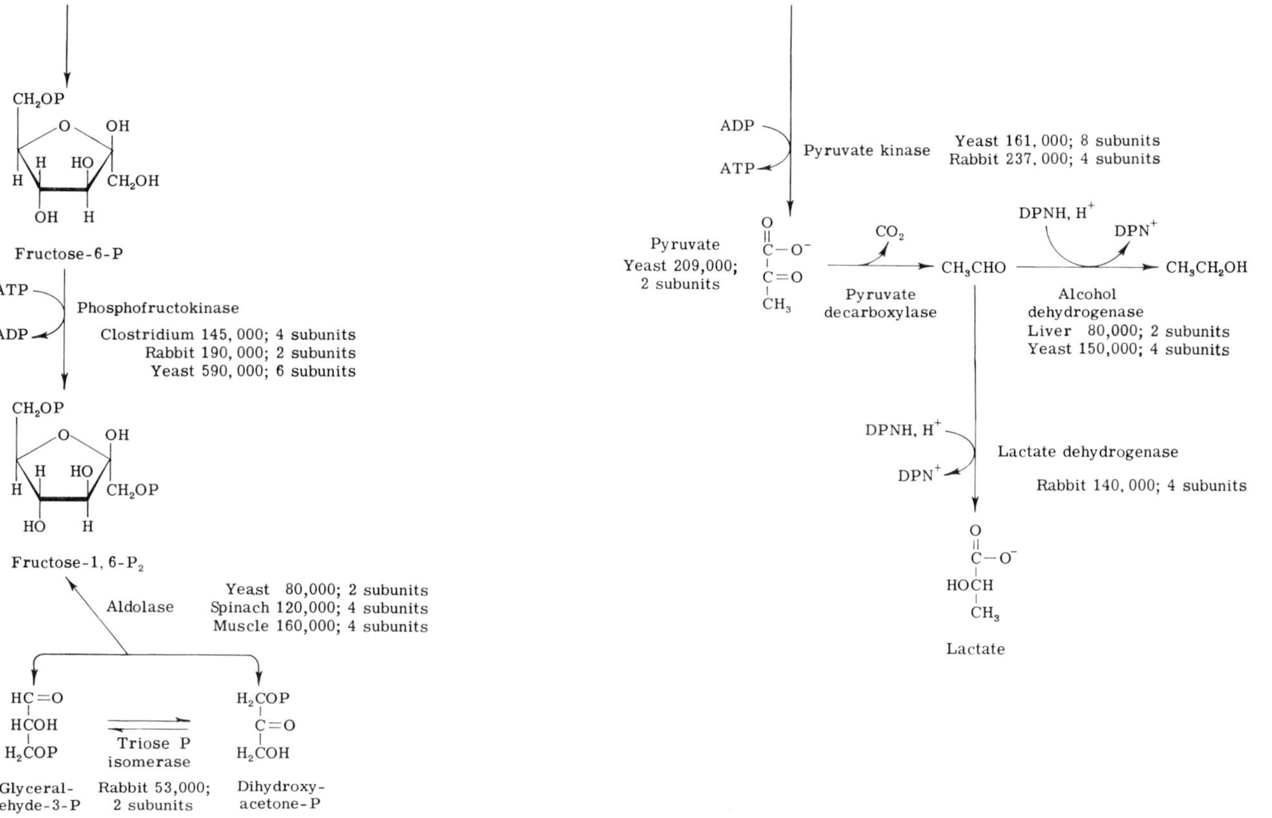

Fig. 19 Quaternary structures of enzymes in the glycolytic pathway.

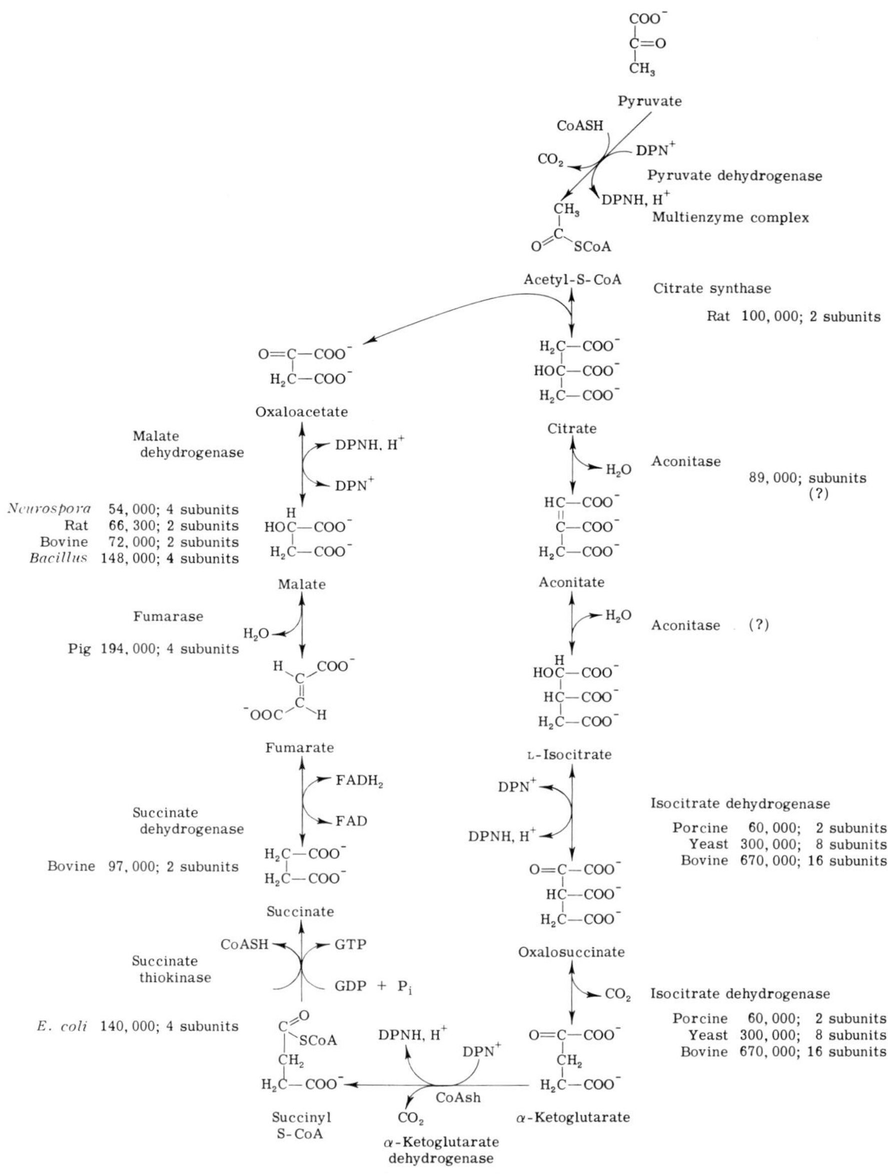

Fig. 20 Quaternary structures of enzymes in the tricarboxylic acid cycle.

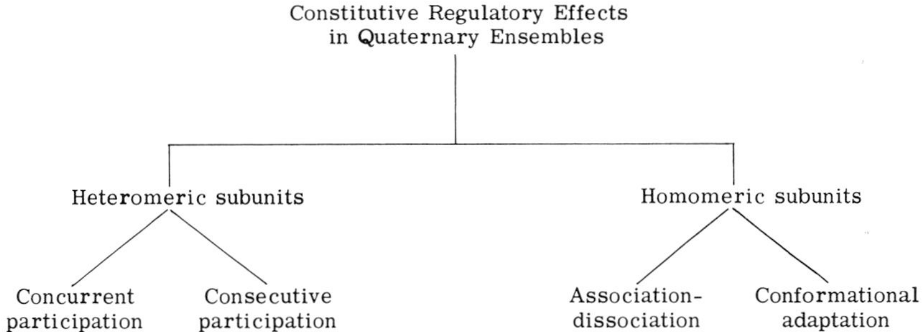

Fig. 21 Modes of metabolic regulation in a quaternary ensemble.

1. Heteromeric Subunits

In a quaternary ensemble constituted of nonidentical subunits, different types of subunits may interact either *concurrently* or *consecutively*. In essence, the first category encompasses those quaternary structures that are constituted of a catalytic and a regulatory subunit. The most extensively studied example of this type of interaction is aspartate transcarbamylase (Gerhart and Schachman, 1968; Nelbach *et al.*, 1972). In this superstructure, six catalytic subunits (C) are combined with six regulatory monomers (R) to give an oligomer (C_6R_6) possessing D_3 symmetry (Wiley and Lipscomb, 1968; Wiley *et al.*, 1972). As Fig. 6 indicates, there are two alternative hexameric arrangements with D_3 symmetry. Electron microscopy combined with physicochemical evidence (Richards and Williams, 1972; Cohlberg *et al.*, 1972) strongly suggests the eclipsed trigonal prism arrangement (Fig. 22) for aspartate transcarbamylase. Since the optical rotatory properties of the enzyme differ from the sum of the contributions of the catalytic and regulatory subunits, it seems clear that the assembly of the quaternary structure is accompanied by changes in conformation. The regulatory subunit binds certain ligands (e.g., CTP), and this binding in the intact C_6R_6 ensemble leads to inhibition of enzymatic activity. The interplay of inhibiting and activating ligands in regulating the enzymatic functions of aspartate transcarbamylase has been described extensively in the literature and can be interpreted in terms of changes in the spatial orientation of the subunits (Gerhart, 1970; Cohlberg *et al.*, 1972).

In aspartate transcarbamylase, the catalytic subunit possesses normal enzymatic specificity (and high activity) in the absence of the regulatory chains. However, in some instances it is also possible for the second type of subunit to modify the specificity of the catalytic chain. This has been illustrated with lactose synthetase (Brew *et al.*, 1968), in which the B

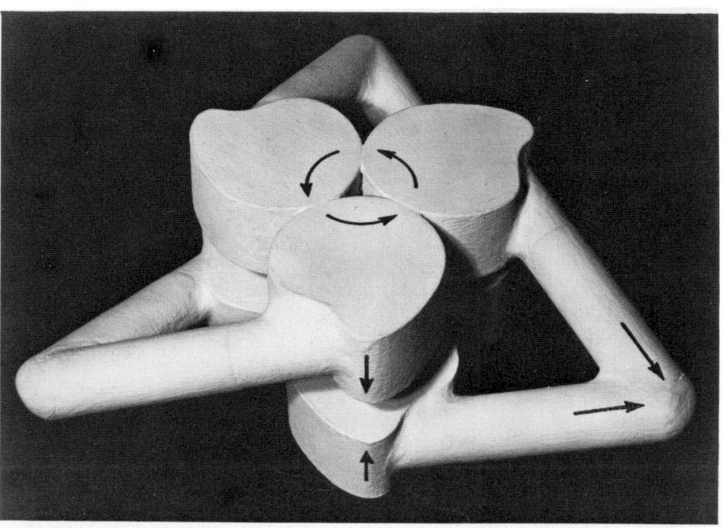

Fig. 22 A model for the structure of aspartate transcarbamylase. Each catalytic monomer is shown as a slightly oblate entity associated in a heterologous fashion with two others to give a trimeric subunit with a threefold axis of symmetry. Each regulatory dimeric subunit is represented by a pair of cylinders meeting at an acute angle in isologous association, consistent with a twofold axis of symmetry. The three dimeric regulatory subunits interconnect the two trimeric catalytic subunits. Evidence for the eclipsed arrangement of catalytic subunits and for the diagonal interconnections by regulatory subunits has been presented by Richards and Williams (1972) and by Cohlberg *et al.* (1972).

protein (identical with α-lactalbumin) modifies the substrate specificity of the isolated A protein so that, in the presence of glucose, synthesis of lactose proceeds.

An example of extreme modification of the specificity of the catalytic C subunit by the regulatory R subunit is provided by the cyclic AMP-dependent protein kinases (Gill and Garren, 1970; Tao *et al.*, 1970; Reimann *et al.*, 1971). In these enzymes, the C·R complex is completely inactive. The binding of cyclic AMP to R is accompanied by dissociation of the complex into the free, active C subunit and the R–AMP subunit.

A regulatory relationship between subunits appears among hormonal proteins as well as among enzymes. Interstitial cell-stimulating hormone is composed of two chemically dissimilar subunits, ICSH-α and ICSH-β. Studies of the ovulation-inducing activity of the intact hormone and of the α and β subunits, respectively, show that the intact hormone and the β subunit induce ovulation, whereas the α subunit is unable to do so

(Yang *et al.*, 1972). Nevertheless, the β subunit is substantially less effective than the intact hormone. Evidently the α subunit, combined with the β, enhances the ovulation-inducing activity.

A striking example of a quaternary ensemble in which the dissimilar subunits perform *consecutive* functions is provided by tryptophan synthetase (Crawford and Yanofsky, 1958; Henning *et al.*, 1962; Goldberg *et al.*, 1966). This enzyme is composed of two pairs of nonidentical subunits, $\alpha_2\beta_2$. Highly purified preparations of each of the separated subunits catalyze different reactions:

$$\text{Indoleglycerol phosphate} \overset{\alpha}{\rightleftarrows} \text{indole} + \text{glyceraldehyde-3-phosphate} \qquad (45a)$$

$$\text{Indole} + \text{L-serine} \overset{\beta}{\rightleftarrows} \text{L-tryptophan} \qquad (45b)$$

The native enzyme, $\alpha_2\beta_2$, catalyzes a third reaction which is actually the net result of combining reactions (a) and (b) in consecutive order. In the presence of $\alpha_2\beta_2$, indole, the product of (a) and reactant of (b), is not detected as a free intermediate. Furthermore, the $\alpha_2\beta_2$ ensemble increases the rates of reactions (a) and (b) one hundredfold over those observed with the separated subunits. This enhancement of rates has been shown to be an effect of each type of subunit on its complementary partner. Nevertheless, the key accomplishment of the $\alpha_2\beta_2$ ensemble is the linkage in sequence of the two biochemical steps, reactions (a) and (b).

Carbamyl phosphate synthetase (from *E. coli*) appears to be another example of a heteromeric ensemble whose monomers function by consecutive participation. The protomer of this enzyme, $\alpha\beta$ (see Table IV), contains one heavy chain (MW 130,000) and one light chain (MW 42,000) (Trotta *et al.*, 1971) and catalyzes the reaction

$$\text{L-Gln} + 2\,\text{ATP} + \text{HCO}_3^- \rightarrow \text{H}_2\text{N—C}\begin{matrix}\diagup\!\!\diagup \text{O} \\ \diagdown \text{OPO}_3^{2-}\end{matrix} + 2\,\text{ADP} + 2\,\text{H}^+ + \text{P}_i + \text{L-Glu} \qquad (46)$$

Neither subunit alone catalyzes glutamine-dependent carbamyl phosphate synthesis. Nevertheless, the heavy subunit by itself does catalyze carbamyl phosphate synthesis if ammonia is available as the source of nitrogen. It also can catalyze partial reactions, e.g., bicarbonate-dependent ATP cleavage and synthesis of ATP from carbamyl phosphate and ADP. Furthermore, the various positive effectors (ornithine, NH_3, IMP) and the negative effector (UMP) of the intact enzyme also act on the isolated heavy chain, which therefore must have binding sites for these small molecules. In contrast, the light subunit by itself shows only glutaminase activity. The addition of light chain to heavy chain leads to a heteromeric combination capable of using glutamine for production of carbamyl phosphate. These observations can all be accounted for if the light chain, acting as a glutaminase, generates the $—NH_2$ that is used by the (regulated) heavy chain to feed into the active intermediate (produced from ATP and HCO_3^-) to give the final product, carbamyl phosphate (Trotta *et al.*, 1971). The appearance of such a heteromeric enzyme would also be understandable from an evolutionary viewpoint. If the heavy chain was the original enzyme, using ammonia for its nitrogen, and if in time ammonia was replaced by glutamine as the principal source of nitrogen, then association with the second chain, endowed with glutaminase activity, would provide a distinct selective advantage for the organism.

An example of a heteromeric oligomer whose subunits behave in both a concurrent and a consecutive fashion is the anthranilate synthetase complex from *Salmonella typhimurium* (Henderson and Zalkin, 1971). The enzyme is an $\alpha_2\beta_2$ system in which each of the subunits has a molecular weight of 62,000 (see Table IV). The first two reactions of tryptophan biosynthesis in *Salmonella typhimurium* are catalyzed by this complex which is composed of the products of the *trp* A and *trp* B genes of the *trp* operon. The *trp A* gene product, anthranilate synthetase (α subunit) catalyzes the reaction:

$$\text{Chorismate} + NH_3 \xrightarrow{Mg^{2+}} \text{anthranilate} + \text{pyruvate} \tag{47a}$$

Aggregation of anthranilate synthetase (α chain) with the *trp B* gene product, anthranilate-5-phosphoribosylpyrophosphate phosphoribosyltransferase (β subunit), alters the specificity of the α chain so that either ammonia or glutamine can be utilized for anthranilate synthesis. Evidence indicates that the glutamine binding site of the aggregate is located on the β chain. In addition, the aggregated or unaggregated *trp B* gene product (β chain) catalyzes the second reaction in the pathway of tryptophan biosynthesis as shown below:

$$\text{Anthranilate} + \text{PP-ribose-P} \xrightarrow{Mg^{2+}} N\text{-(5}'\text{-phosphoribosyl)anthranilate} + PP_i \tag{47b}$$

Thus the β chain of the complex is bifunctional, i.e., its function is concerned with (a) glutamine binding and (b) phosphoribosyltransferase activity.

Another regulatory relationship between subunits is found in the interaction of the T_u and T_s protein subunits which are the elongation factors important in protein biosynthesis. The current view of the function of T_u and T_s in protein biosynthesis is shown below (Lucas-Lenard and Lipmann, 1971):

$$T_u\text{-GDP} + T_s \rightleftarrows T_u\text{-}T_s + \text{GDP} \tag{48a}$$

$$T_u\text{-}T_s + \text{GTP} + \text{Aminoacyl-tRNA} \rightleftarrows \text{aminoacyl-tRNA-}T_u\text{-GTP} + T_s \tag{48b}$$

$$\text{Aminoacyl-tRNA-}T_u\text{-GTP} \xrightarrow[\text{mRNA}]{\text{ribosome}} (\text{aminoacyl-tRNA-mRNA-ribosome}) + T_u\text{-GDP} + P_i \tag{48c}$$

Very recently the T_u–T_s complex has also been shown to function in an entirely different reaction catalyzed by Qβ replicase. The enzyme Qβ replicase, responsible for the replication of the RNA of *E. coli* phage Qβ, is composed of four nonidentical subunits designated as I, II, III, and IV. Subunits I, III, and IV are coded for by the bacterial genome, while the other subunit (II) is phage-specific. The four polypeptides of Qβ replicase can be separated into complexes of subunits I + II and subunits III + IV. Kamen (1970) found that neither complex alone shows activity in the poly(C)-dependent assay, but activity could be restored after the two were mixed. Blumenthal *et al.* (1972) have now found that subunits III and IV of Qβ replicase are identical with T_u and T_s, respectively, the two elongation factors identified as part of the mechanism of protein biosynthesis. Travers *et al.* (1970a,b) proposed that subunits III + IV (termed $\psi_r^{Q\beta}$) enable DNA-dependent RNA polymerase from *E. coli* to synthesize ribosomal RNA. If Travers' model is correct, T_u and T_s (subunits III and IV) are functional in three different biosynthetic processes: protein synthesis, stable RNA synthesis, and phage RNA synthesis. This is one of the very few cases where protein subunits are known to be multifunctional.

In some proteins the specific function of each heteromer may not be clear. For example, in bacterial luciferase ($\alpha\beta$) it has been shown that alterations in the α subunit profoundly affect catalytic steps in the bioluminescent reaction (Meighen *et al.*, 1971). Corresponding modifications of the β subunit are without effect on the luminescence lifetime or the Michaelis constant for the binding of flavin. Evidently flavin binding and light-emitting reactions occur on the α subunit. It appears that only the α subunit contributes residues to the active site and that the β subunit is required to maintain the active conformation of the catalytic subunit α (Cline and Hastings, 1972).

2. *Homomeric Subunits*

Turning to homomeric quaternary ensembles, we find many enzymes whose state of aggregation is highly sensitive to the presence of substrates or effectors (see Table I and Frieden, 1971). It is therefore likely that the activity of these systems *in vivo* would depend upon the state of aggregation. In practice, however, few studies have been detailed and extensive enough to establish that an association–dissociation reaction functions in a regulatory manner *in vivo*. With isocitrate dehydrogenase, for example, it has been shown that the active form has a molecular weight of 110,000 and that an aggregated ensemble, favored by high concentrations of DPN as well as by slightly acid pH, is inactive (LeJohn *et al.*, 1969). In this enzyme system, disaggregation to the active form is promoted by citrate or isocitrate. Only circumstantial evidence has been obtained, however, to show that disaggregation is important in the living organism. With cytosine triphosphate synthetase it has been observed that binding of ATP and UTP is accompanied by association of dimer to tetramer, and that the tetramer has an appreciably higher turnover number (Long *et al.*, 1970; Levitzki and Koshland, 1972).

Rabbit muscle phosphorylase has been shown to exist in two molecular forms: phosphorylase *b*, a dimer (MW 185,000) that is essentially inactive in the absence of AMP, and phosphorylase *a*, a tetramer (MW 370,000) that is active in the absence of any effector, provided it is fully saturated with substrate. Phosphorylation of phosphorylase *b* is accompanied by the transition of the predominantly dimeric *b* state to the predominantly tetrameric *a* state. Dimeric phosphorylase *b* is also shifted primarily toward active tetrameric phosphorylase *b* by AMP. Since both active forms of phosphorylase have a strong tendency to tetramerize, it was first believed that the activity depended upon this change in quaternary structure. This view is now thought to be questionable because of the following reasons. Formation of tetramers of phosphorylase *b* by AMP does not occur under assay conditions (high dilution) and hence is not a prerequisite for activity (Ullmann *et al.*, 1964). In addition, rabbit muscle phosphorylase *a* dissociates from tetramer into dimer in the presence of glycogen, glucose, or high salt concentration, with a concomitant increase in activity (Metzger *et al.*, 1967; Wang and Graves, 1964). It is not known whether any of these changes in quaternary structure are of physiological importance, since conditions under which the enzyme exists *in vivo,* i.e., apparently bound to glycogen particles (Meyer *et al.*, 1970), differ considerably from those that obtain *in vitro* (Haschke *et al.*, 1970).

Regulation by conformational adaptation, or allosteric effects, pervades all cellular phenomena. It has long been evident that a specific protein can exist in multiple conformational states. Shifts in equilibria between conformational peers can be produced by a variety of perturbations, of which ligand binding is the most important from a functional, regulatory viewpoint. Such responses to ligand binding were originally called "conformational adaptability" by Karush (1950) and, on recognition of their role in cellular regulation, "allosteric effects" by Monod *et al.* (1963).

Hemoglobin is the allosteric protein whose conformational adaptations have been most clearly delineated at the atomic level (see Section V). The specific structural states favored by oxygen, the homotropic ligand, and by 2,3-diphosphoglycerate, one of the heterotropic ligands, have been determined to angstrom unit resolution.

With hemoglobin, furthermore, there is now emerging a causal thread linking molecular structure to physiological behavior, at least in hemoglobin S. As has been described in Section V,B,2, X-ray diffraction has clearly established that 2,3-diphosphoglycerate plugs the channel of the tetramer near the amino-terminal portal of the β chain. The specific arrangement of residues and of the anionic ligand is shown in Fig. 18 (Arnone, 1972). Furthermore, it has been found that cyanate reacts with hemoglobin at the amino-terminal valine residues and is an effective inhibitor of the gelation of hemoglobin S on deoxygenation (Cerami and Manning, 1971). It is well known that cyanate forms carbamylates when it reacts with amino groups of proteins (Stark *et al.*, 1960; Kilmartin and Rossi-Bernardi, 1969):

$$\mathrm{P{-}NH_2 + H{-}NCO \rightarrow P{-}\overset{\displaystyle H}{\overset{|}{N}}{-}\overset{\displaystyle O}{\overset{\|}{C}}{-}NH_2} \qquad (49)$$

It is clear from Fig. 18 that carbamylated Val 1 residues would project into the diphosphoglycerate channel and distort the binding of this ligand. Furthermore, it is apparent from detailed examination of models (Arnone, 1972) that in the presence of diphosphoglycerate, the Val $6\beta_1$ and Val $6\beta_2$ residues of hemoglobin S are closer together, and this may account for solubility changes.

Alternative approaches for blocking the α-NH_3^+ groups of Val 1β residues can be suggested. There is a wide variety of metabolically functional acylating agents that could be introduced directly into blood. Those which penetrate the membrane of the erythrocyte should acylate the amino-terminal valine residues of hemoglobin and may be effective anti-sickling agents. A particularly interesting synthetic acylating agent is aspirin, acetylsalicylic acid. This reagent has been shown to acylate amino groups of plasma albumin *in vivo* (Hawkins *et al.*, 1968, 1969). Although the pK_a of its COOH is well below 7, some of the nonionized form should be in equilibrium with the anionic state, and the nonionized molecule should be able to penetrate the erythrocyte, although larger dosages of aspirin than normally used may be necessary. Alternatively, it would not be difficult to prepare acetylsalicylate esters or related derivatives that would be uncharged and hence capable of entering red blood cells with ease. Thus aspirin or some of its derivatives may be promising anti-sickling agents (Klotz and Tam, 1973). Since acetylsalicylate has a long history of safe clinical use and its side effects are well known, it may be more attractive for this purpose than other potential anti-sickling agents. In any event it is clear that a detailed understanding of the quaternary structure and chemistry of hemoglobin is providing a rational basis for interpreting its biologic properties and for controlling its physiological functions.

As judged by a variety of physicochemical methods, many enzymes also undergo ligand-linked conformational adaptations. The role of these adaptations in the regulation of enzymatic activity has been reviewed frequently (Atkinson, 1966; Koshland and Neet, 1968; Kurganov, 1968).

There are also quaternary systems that combine modes of control. A particularly interesting example is glutamate dehydrogenase, a key enzyme that links the tricarboxylic acid cycle with the pathways of biosynthesis of amino acids. In this system, metabolic regulation depends upon association–dissociation and conformational adaptation (Tomkins *et al.*, 1963; Stadtman, 1966; Eisenberg, 1971; Frieden, 1971). Our reference state for this enzyme is the hexameric oligomer (MW 320,000) composed of six identical subunits (Table II). Electron microscopy (Josephs, 1971) indicates that the subunits are arranged octahedrally (see Fig. 6) with D_3 symmetry. In solution, however, the hexamer can exist in two different conformations as well as participate in an aggregation process (Fig. 23). Conformation B has alanine hydrogenase activity; conformation A and polymer have glutamate dehydrogenase

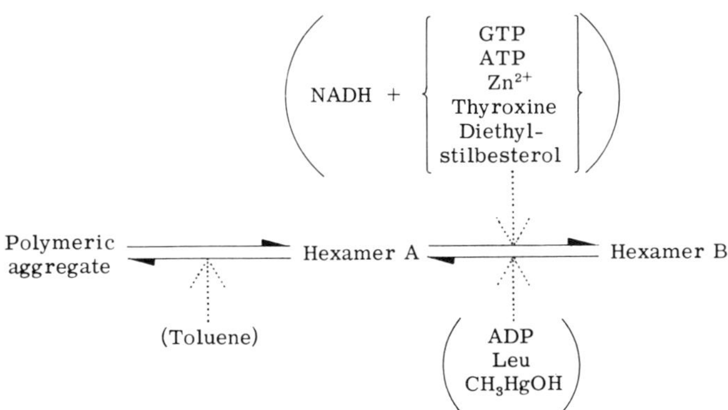

Fig. 23 Conformational and aggregational equilibria of glutamate dehydrogenase, and some of the ligands which effect these quaternary ensembles.

activity. Addition to the enzyme of NADH and GTP, for example, inhibits glutamate dehydrogenase activity and simultaneously depolymerizes the aggregate, by shifting the equilibrium (Fig. 23) to hexamer B. In contrast, reagents that promote aggregation of the enzyme (e.g., ADP) simultaneously increase its glutamate dehydrogenase activity. Curiously enough, toluene has also been found to promote association, evidently by being bound to the aggregating form of the enzyme (Eisenberg, 1971). This may be a hint that the lipid environment of a mitochondrion, where the enzyme is localized, plays a role in the regulation of quaternary structure and function. Since the concentration of glutamate dehydrogenase is quite high in some mitochondria (e.g., liver), it seems very likely that the aggregation reaction occurs *in vivo*.

The linkage between the various quaternary ensembles participating in the regulatory process is provided by the small molecules that are bound preferentially to specific macromolecules. The interlocking dynamic rearrangements for a simplified situation were presented schematically at the outset of this chapter, in Fig. 1. As is illustrated in this figure, binding of a small molecule (A) by one conformation, represented as a square, will shift the conformational equilibria shown at the upper left, and this shift in turn must perturb the aggregation equilibria, one of which is illustrated at the upper right. It is the interplay within such a general manifold of interactions that provides the fine modulation in functional regulation.

Once these effects are delineated in descriptive form, it is interesting to rephrase them in analytic terms to provide quantitative expressions to account for observed kinetic and thermodynamic behavior.

For systems involving association–dissociation reactions the linkage with ligand binding may be represented by

$$\begin{array}{lll} P_n & \rightleftarrows P_{n-1} \cdots & \rightleftarrows nP \\ + & \quad + & \\ A & \quad A & \\ \upharpoonleft\downharpoonright & \quad \upharpoonleft\downharpoonright & \\ P_nA & \rightleftarrows P_{n-1}A & \\ \cdot & \quad \cdot & \\ \cdot & \quad \cdot & \\ \cdot & \quad \cdot & \\ \upharpoonleft\downharpoonright & & \\ P_nA_n & & \end{array} \tag{50}$$

The relationships between the equilibrium constants involved and the extent of ligand binding have been described by several groups of investigators (Frieden, 1967; Nichol *et al.*, 1967; Klapper and Klotz, 1968; Czerlinski, 1968; Kurganov, 1968; Noble, 1969). From these relationships and appropriate assumptions about oligomeric forms that possess enzymatic activity, expressions for kinetic consequences can also be obtained (Frieden, 1971).

The quantitative treatment of conformational adaptations within a quaternary ensemble has been dominated by the allosteric models proposed by Monod *et al.* (1965) and by Koshland *et al.* (1966). The formalism of the former is based on the equilibria

$$\begin{array}{ccc} P_n & \rightleftarrows & P_n' \\ + & & + \\ A & & A \\ \upharpoonleft\downharpoonright & & \upharpoonleft\downharpoonright \\ P_nA & & P_n'A \\ \cdot & & \cdot \\ \cdot & & \cdot \\ \cdot & & \cdot \\ \upharpoonleft\downharpoonright & & \upharpoonleft\downharpoonright \\ P_nA_n & & P_n'A_n \end{array} \tag{51}$$

where P_n and P_n' are two complementary states of the subunit ensemble, also called the relaxed (R) and the taut (T) forms. If one of the states, for example P_n', binds the substrate preferentially, the presence of substrate will pull the equilibrium mixture toward the P_n' state. Mathematical analysis shows that the model will account for sigmoidal kinetic or thermodynamic saturation curves and other characteristics of allosteric behavior. Although the model is not necessarily limited to two states, its emphasis is on the symmetry inherent in *concerted* transitions between boundary states.

The sequential model of Koshland *et al.* (1966) interprets the kinetic and cooperative properties of enzymes on the basis of *stepwise* changes in conformation involving each subunit consecutively in the quaternary ensemble. In equilibria corresponding to Eq. (51), the Koshland model explicitly includes all of the intermediates of the type $P_x'\ P_{n-x}$:

$$P_n \rightleftarrows P'P_{n-1} \rightleftarrows \cdot\cdot\cdot \rightleftarrows P_x'P_{n-x} \rightleftarrows \cdot\cdot\cdot \rightleftarrows P_n' \tag{52}$$

The conformational change of each subunit may in turn affect the conformation of adjacent subunits. Again the conformational adaptation is ligand-linked, and all of the associated ligand equilibria of $P_x'\ P_{n-x}$ must be added to Eq. (52). At any stage in ligand binding, the strength of the subunit interactions may be increased, decreased, or remain the same. The binding of a ligand will depend not only upon the parameters characteristic of a monomeric subunit but also upon the changes in subunit interactions. The sequential model analyzes these effects in terms of parameters that account for the cooperativity and kinetic effects in molecular terms. These molecular parameters are then used to calculate the concentrations of intermediate molecular species present. The existence of intermediate states, in which some of the protomers have conformations different from others, is the essential distinction between the sequential and concerted models.

Much effort has been expended in the past few years in an attempt to prove that allosteric proteins conform to one model or the other. However, as Frieden (1971) has observed, binding or kinetic data alone will not prove a particular mechanism correct, and frequently these are the only data available.

Weber (1972) has recently suggested that the small number of thermodynamically defined protein conformations implicit in the concerted and sequential models is unrealistic. Weber bases his stand on the following argument. Consider a single polypeptide chain capable of binding two different ligands at two different sites. If there are three allowed conformations, i.e., one without ligand, one with a high affinity for one ligand, and one with high affinity for the other ligand, then the presence of either of the ligands will tend to pull the equilibrium conformation toward that liganded form. If, however, both ligands are present in saturating concentrations, neither ligand displaces the other, and the protein must assume a *new* conformational form intermediate between the two liganded forms. This analysis indicates where the other models may be unrealistic.

Weber (1972) proposes an alternative, more realistic approach. Instead of fitting observed binding curves with equilibrium constants between various conformational states, he describes the system in terms of free energies of the various liganded states. The measured free energy change upon binding the first ligand includes the free energy of conformational adjustment of the protein. A similar free energy change for binding the second ligand to free protein can also be observed. The measured free energy change upon binding of both ligands, however, will be still different. The difference between the free energies of the latter case and the sum of the free energies of binding either specific ligand in the absence of the other is called the "coupling free energy." The coupling free energy is negative if the two ligands assist each other in binding, positive if they oppose each other in binding, and zero if the respective bindings are independent of one another. The free energy change upon binding one ligand in the presence of saturating concentrations of the other is termed "conditional free energy." In oligomeric protein systems, it is reasonable to presume that any conformational change in a protomer linked with binding of a ligand can be transmitted to other parts of the oligomer. The subunit contact sites are particularly sensitive to this type of conformational change; it is much simpler to disrupt the quaternary structure of an oligomer than to denature the protein. If, upon binding a ligand to one subunit, a greater portion of the free energy change is transmitted to other subunits, then the binding of the second ligand will be changed. The result will be a cooperative binding.

Thermodynamic and kinetic models of conformational adaptation provide a very general conceptual framework for interpreting biochemical behavior of quaternary ensembles. These models will be subject to further modification as the molecular structural features of these macromolecules are revealed in ever-increasing detail.

VII. CONCLUSION

Quaternary ensembles of subunits provide a plan of macromolecular architecture with unique advantages in structure and in potential for flexibility in behavior and regulation. Thus they play a key role in a broad spectrum of cellular and physiological activities which couple metabolic transformations with the productive processes necessary for biologic functions.

REFERENCES

Ackers, G. K. (1970). *Advan. Protein Chem.* **24**, 343.

Ackers, G. K., and Thompson, T. E. (1965). *Proc. Nat. Acad. Sci. U. S.* **53**, 342.

Adams, E. T., Jr. (1967). *Biochemistry* **6**, 1864.

Adams, E. T., Jr. (1969). *Ann. N. Y. Acad. Sci.* **164**, 226.

Adams, E. T., Jr., and Lewis, M. S. (1968). *Biochemistry* **7**, 1044.

Adams, E. T., Jr., and Williams, J. W. (1964). *J. Amer. Chem. Soc.* **86**, 3454.

Adams, M. J., Blundell, T. L., Dodson, E. J., Dodson, G. G., Vijayan, M., Baker, E. N., Harding, M. M., Hodgkin, D. C., Rimmer, B., and Sheat, S. (1969). *Nature (London)* **224**, 491.

Ahmad, F., Jacobson, B., Wood, H. G., Valentine, R. C., Green, M., and Wrigley, N. (1971). *Fed. Proc., Fed. Amer. Soc. Exp. Biol.* **30**, 1058.

Allen, S. H. G., and Patil, J. R. (1972). *J. Biol. Chem.* **247**, 909.

Anderson, M. E. (1971). *J. Biol. Chem.* **246**, 4800.

Anderson, P. M., and Marvin, S. V. (1970). *Biochemistry* **9**, 171.

Antonini, E., and Brunori, M. (1969). *J. Biol. Chem.* **244**, 3909.

Antonini, E., Brunori, M., Bruzzesi, M. R., Chiancone, E., and Massey, V. (1966). *J. Biol. Chem.* **241**, 2358.

Arnone, A. (1972). *Nature (London)* **237**, 146.

Atkinson, D. E. (1966). *Annu. Rev. Biochem.* **35**, 85.

Baker, W. W., and Mintz, B. (1969). *Biochem. Genet.* **2**, 351.

Barker, D. L., and Jencks, W. P. (1969). *Biochemistry* **8**, 3879.

Barnes, L. D., Kuehn, G. D., and Atkinson, D. E. (1971). *Biochemistry* **10**, 3939.

Benesch, R., and Benesch, R. E. (1967). *Biochem. Biophys. Res. Commun.* **26**, 162.

Benesch, R., and Benesch, R. E. (1969). *Nature (London)* **221,** 618.

Benesch, R., Benesch, R. E., and Enoki, T. (1968). *Proc. Nat. Acad. Sci. U. S.* **61,** 1102.

Benesch, R. E., Benesch, R., and Yu, C. I. (1969). *Biochemistry* **8,** 2567.

Bernal, J. D. (1958). *Discuss. Faraday Soc.* **25,** 7.

Berns, D. S., and Edwards, M. R. (1965). *Arch. Biochem. Biophys.* **110,** 511.

Bjork, I., and Fish, W. W. (1971). *Biochemistry* **10,** 2844.

Blake, C. C. F., Swan, I. D. A., Rerat, C., Berthou, J., Laurent, A., and Rerat, B. (1971). *J. Mol. Biol.* **61,** 217.

Blumenthal, T., Landers, T. A., and Weber, K. (1972). *Proc. Nat. Acad. Sci. U. S.* **69,** 1313.

Blundell, T. L., Cutfield, J. F., Dodson, E. J., Dodson, G. G., Hodgkin, D. C., and Mercola, D. A. (1972). *Cold Spring Harbor Symp. Quant. Biol.* **36,** 233.

Boeker, E. A., and Snell, E. E. (1968). *J. Biol. Chem.* **243,** 1678.

Boeker, E. A., Fischer, E. H., and Snell, E. E. (1969). *J. Biol. Chem.* **244,** 5239.

Bonsignore, A., Cancedda, R., Nicolini, A., Damiani, G., and DeFlora, A. (1971). *Arch. Biochem. Biophys.* **147,** 493.

Bowers, W. F., Czubaroff, V. B., and Haschemeyer, R. H. (1970). *Biochemistry* **9,** 2620.

Braunitzer, G., Beyreuther, K., Fujiki, H., and Schrank, B. (1968). *Hoppe-Seyler's Z. Physiol. Chem.* **349,** 265.

Brew, K., Vanaman, T. C., and Hill, R. L. (1968). *Proc. Nat. Acad. Sci. U. S.* **59,** 491.

Brewer, J. M., Fairwell, T., Travis, J., and Lovins, R. E. (1970). *Biochemistry* **9,** 1011.

Bunn, H. F., and Briehl, R. W. (1970). *J. Clin. Invest.* **49,** 1088.

Burford, G. D., Ginsburg, M., and Thomas, P. J. (1971). *Biochim. Biophys. Acta* **229,** 730.

Butler, P. J. G., Harris, J. I., Hartley, B. S., and Leberman, R. (1969). *Biochem. J.* **112,** 679.

Campbell, J. W., Hodgson, G. I., Watson, H. C., and Scopes, R. K. (1971). *J. Mol. Biol.* **61,** 257.

Carlsen, R. B., and Pierce, J. G. (1972). *J. Biol. Chem.* **247,** 23.

Carvajal, N., Venegas, A., Oestreicher, G., and Plaza, M. (1971). *Biochim. Biophys. Acta* **250,** 437.

Cerami, A., and Manning, J. M. (1971). *Proc. Nat. Acad. Sci. U. S.* **68,** 1180.

Chanutin, A., and Curnish, R. R. (1965). *Proc. Soc. Exp. Biol. Med.* **120,** 291.

Chanutin, A., and Curnish, R. R. (1967). *Arch. Biochem. Biophys.* **121,** 96.

Chanutin, A., and Hermann, E. (1969). *Arch. Biochem. Biophys.* **131,** 180.

Chignell, D. A., Gratzer, W. B., and Valentine, R. C. (1968). *Biochemistry* **7,** 1082.

Chilson, O. P., Costello, L. A., and Kaplan, N. O. (1964). *J. Mol. Biol.* **10,** 349.

Chilson, O. P., Costello, L. A., and Kaplan, N. O. (1965). *Biochemistry* **4,** 271.

Clark, R. B., and Ogilvie, J. W. (1972). *Biochemistry* **11,** 1278.

Cline, T. W., and Hastings, J. W. (1972). *Biochemistry* **11,** 3359.

Cohlberg, J. A., Pigiet, V. P., Jr., and Schachman, H. K. (1972). *Biochemistry* **11,** 3396.

Cold Spring Harbor Symposia on Quantitative Biology. (1972). Volume 36.

Cooper, A. J. L., and Meister, A. (1972). *Biochemistry* **11,** 661.

Cornish-Bowden, A. J., and Koshland, D. E., Jr. (1971). *J. Biol. Chem.* **246,** 3092.

Cottam, G. L., Hollenberg, P. F., and Coon, M. J. (1969). *J. Biol. Chem.* **244,** 1481.
Crawford, I. P., and Yanofsky, C. (1958). *Proc. Nat. Acad. Sci. U. S.* **44,** 1161.
Crichton, R. R. (1972). *Biochem. J.* **126,** 761.
Curthoys, N. P., Straus, L. D., and Rabinowitz, J. C. (1972). *Biochemistry* **11,** 345.
Curtis, E. H., Kobes, R. D., Teller, D. C., and Rutter, W. J. (1969). *Biochemistry* **8,** 2442.
Czerlinski, G. (1968). *Curr. Mod. Biol.* **2,** 219.
Darnall, D. W., and Klotz, I. M. (1972). *Arch. Biochem. Biophys.* **149,** 1.
Davies, G. E., and Stark, G. R. (1970). *Proc. Nat. Acad. Sci. U. S.* **66,** 651.
Dawson, D. M., Eppenberger, H. M., and Kaplan, N. O. (1967). *J. Biol. Chem.* **242,** 210.
Dayhoff, M. O. (1969). "Atlas of Protein Sequence and Structure," pp. 61–73. Nat. Biomed. Res. Found., Silver Spring, Maryland.
Derechin, M. (1968). *Biochemistry* **7,** 3253.
Derechin, M. (1969a). *Biochemistry* **8,** 921.
Derechin, M. (1969b). *Biochemistry* **8,** 927.
Derechin, M. (1971). *Biochemistry* **10,** 4981.
DeRosier, D. J., Oliver, R. M., and Reed, L. J. (1971). *Proc. Nat. Acad. Sci. U. S.* **68,** 1135.
Dinamarca, M. L., Levenbook, L., and Valdés, E. (1971). *Arch. Biochem. Biophys.* **147,** 374.
Dixon, H. B. F., and Perham, R. N. (1969). *Biochem. J.* **109,** 312.
Drenth, J., and Smit, J. D. G. (1971). *Biochem. Biophys. Res. Commun.* **45,** 1320.
Drum, D. E., Li, T.-K., and Vallee, B. L. (1969). *Biochemistry* **8,** 3873.
Dyson, R. D., and Isenberg, I. (1971). *Biochemistry* **10,** 3233.
Efremov, G. D., Huisman, T. H. J., Smith, L. L., Wilson, J. B., Kitchens, J. L., Wrightstone, R. N., and Adams, H. R. (1969). *J. Biol. Chem.* **244,** 6105.
Eisenberg, H. (1971). *Accounts Chem. Res.* **4,** 379.
Eley, D. D. (1943). *Trans. Faraday Soc.* **39,** 172.
Epp, O., Steigemann, W., Formanek, H., and Huber, R. (1971). *Eur. J. Biochem.* **20,** 432.
Evans, M. G., and Gergely, J. (1949). *Biochim. Biophys. Acta* **3,** 188.
Evans, W. J., Forlani, L., Brunori, M., Wyman, J., and Antonini, E. (1970). *Biochim. Biophys. Acta* **214,** 64.
Fernández-Morán, H., van Bruggen, E. F. J., and Ohtsuki, M. (1966). *J. Mol. Biol.* **16,** 191.
Fosmire, G. J., and Timasheff, S. N. (1972). *Biochemistry* **11,** 2455.
Fowler, A. V., and Zabin, I. (1970). *J. Biol. Chem.* **245,** 5032.
Fraenkel-Conrat, H., and Colloms, M. (1967). *Biochemistry* **6,** 2740.
Frieden, C. (1963). *J. Biol. Chem.* **238,** 3286.
Frieden, C. (1967). *J. Biol. Chem.* **242,** 4045.
Frieden, C. (1971). *Annu. Rev. Biochem.* **40,** 653.
Garbett, K., Darnall, D. W., and Klotz, I. M. (1971). *Arch. Biochem. Biophys.* **142,** 455.
Gazith, J., Schulze, I. T., Gooding, R. H., Womack, F. C., and Colowick, S. P. (1968). *Ann. N. Y. Acad. Sci.* **151,** 307.
Georges, C., Guinand, S., and Tonnelat, J. (1962). *Biochim. Biophys. Acta* **59,** 737.
Gerhart, J. C. (1970). *Curr. Top. Cell. Regul.* **2,** 275.
Gerhart, J. C., and Schachman, H. K. (1965). *Biochemistry* **4,** 1054.
Gerhart, J. C., and Schachman, H. K. (1968). *Biochemistry* **7,** 538.

Gilbert, G. A., and Jenkins, R. C. L. (1959). *Proc. Roy. Soc., Ser. A* **253**, 420.
Gill, G. N., and Garren, L. D. (1970). *Biochem. Biophys. Res. Commun.* **39**, 335.
Giorgio, N. A., Jr., Yip, A. T., Fleming, J., and Plaut, G. W. E. (1970). *J. Biol. Chem.* **245**, 5469.
Goldberg, M. E., Creighton, T. E., Baldwin, R. L., and Yanofsky, C. (1966). *J. Mol. Biol.* **21**, 71.
Gonzalez, G., and Offord, R. E. (1971). *Biochem. J.* **125**, 309.
Gounaris, A. D., Turkenkoff, I., Buckwald, S., and Young, A. (1971). *J. Biol. Chem.* **246**, 1302.
Gratzer, W. B., and Beaven, G. H. (1969). *J. Biol. Chem.* **244**, 6675.
Green, N. M., Knoieczny, L., Toms, E. J., and Valentine, R. C. (1971). *Biochem. J.* **125**, 781.
Greer, J. (1972). *Cold Spring Harbor Symp. Quant. Biol.* **36**, 315.
Gregolin, C., Ryder, E., Warner, R. C., Kleinschmidt, A. K., Chang, H-C., and Lane, M. D. (1968). *J. Biol. Chem.* **243**, 4236.
Guerritore, D., Bonacci, M. L., Brunori, M., Antonini, E., Wyman, J., and Rossi-Fanelli, A. (1965). *J. Mol. Biol.* **13**, 234.
Haberland, M. E., Willard, J. M., and Wood, H. G. (1972). *Biochemistry* **11**, 712.
Hancock, D. K., and Williams, J. W. (1969). *Biochemistry* **8**, 2598.
Hanson, K. R. (1966). *J. Mol. Biol.* **22**, 405.
Hanson, K. R. (1968). *J. Mol. Biol.* **38**, 133.
Hardman, K. D., Wood, M. K., Schiffer, M., Edmundson, A. B., and Ainsworth, C. F. (1971). *Proc. Nat. Acad. Sci. U. S.* **68**, 1393.
Harned, H. S., and Owen, B. B. (1958). "The Physical Chemistry of Electrolytic Solutions," 3rd ed. Van Nostrand-Reinhold, Princeton, New Jersey.
Harris, J. I., and Hindley, J. (1965). *J. Mol. Biol.* **13**, 894.
Harry, J. B., and Steiner, R. F. (1969). *Biochemistry* **8**, 5060.
Haschemeyer, R. H., and Bowers, W. F. (1970). *Biochemistry* **9**, 435.
Haschke, R. H., Meyer, F., Heilmeyer, L., Jr., and Fischer, E. H. (1970). *J. Biol. Chem.* **245**, 6657.
Hathaway, G. M. (1972). *J. Biol. Chem.* **247**, 1440.
Hathaway, G. M., and Crawford, I. P. (1970). *Biochemistry* **9**, 1801.
Haurowitz, F. (1938). *Hoppe-Seyler's Z. Physiol. Chem.* **254**, 266.
Hawkins, D., Pinckard, R. N., and Farr, R. S. (1968). *Science* **160**, 780.
Hawkins, D., Pinckard, R. N., Crawford, I. P., and Farr, R. S. (1969). *J. Clin. Invest.* **48**, 536.
Helmreich, E., Michaelides, M. C., and Cori, C. F. (1967). *Biochemistry* **6**, 3695.
Henderson, E. J., and Zalkin, H. (1971). *J. Biol. Chem.* **246**, 6891.
Henn, S. W., and Ackers, G. K. (1969). *Biochemistry* **8**, 3829.
Henney, H. R., Willms, C. R., Muramutsu, M., Mukherjee, B. B., and Reed, L. J. (1967). *J. Biol. Chem.* **242**, 898.
Henning, U., Helinski, D. R., Chao, F. C., and Yanofsky, C. (1962). *J. Biol. Chem.* **237**, 1523.
Herbert, T. J., and Carlson, F. D. (1971). *Biopolymers* **10**, 2231.
Hetland, Ø., Olsen, B. R., Christensen, T. B., and Størmer, F. C. (1971). *Eur. J. Biochem.* **20**, 200.
Hewitt, J. A., Kilmartin, J. V., Ten Eyck, L. F., and Perutz, M. F. (1972). *Proc. Nat. Acad. Sci. U. S.* **69**, 203.
Hoagland, V. D., Jr., and Teller, D. C. (1969). *Biochemistry* **8**, 594.
Hoch, S. O., and DeMoss, R. D. (1972). *J. Biol. Chem.* **247**, 1750.

Huestis, W. H., and Raftery, M. A. (1972). *Biochemistry* **11**, 1648.
Huseby, N.-E., Christensen, T. B., Olsen, B. R., and Størmer, F. C. (1971). *Eur. J. Biochem.* **20**, 209.
Huston, J. S., Fish, W. W., Mann, K. G., and Tanford, C. (1972). *Biochemistry* **11**, 1609.
Imai, K., Hamilton, H. B., Miyaji, T., and Shibata, S. (1972). *Biochemistry* **11**, 114.
Ioppolo, C., Chiancone, E., Antonini, E., and Wyman, J. (1969). *Arch. Biochem. Biophys.* **132**, 249.
Isemura, T., and Kakiuchi, K. (1962). *J. Biochem.* (*Tokyo*) **51**, 385.
Ishihama, A. (1972). *Biochemistry* **11**, 1250.
Itano, H. A., and Singer, S. J. (1958). *Proc. Nat. Acad. Sci. U. S.* **44**, 522.
Iwatsuki, N., and Okazaki, R. (1967). *J. Mol. Biol.* **29**, 139.
Jacq, C., and Lederer, F. (1972). *Eur. J. Biochem.* **25**, 41.
Jaenicke, R., Schmid, D., and Knof, S. (1968). *Biochemistry* **7**, 919.
Jaffé, H. H., and Orchin, M. (1965). "Symmetry in Chemistry." Wiley, New York.
Jarabak, J., and Street, M. A. (1971). *Biochemistry* **10**, 3831.
Jeffrey, P. D., and Coates, J. H. (1966). *Biochemistry* **5**, 489.
Jolley, R. L., Robb, D. A., and Mason, H. S. (1969). *J. Biol. Chem.* **244**, 1593.
Jones, R. T., and Schroeder, W. A. (1963). *Biochemistry* **2**, 1357.
Josephs, R. (1971). *J. Mol. Biol.* **55**, 147.
Kakiuchi, K., Hamaguchi, K., and Isemura, T. (1965). *J. Biochem.* (*Tokyo*) **57**, 167.
Kamen, R. (1970). *Nature* (*London*) **228**, 527.
Karush, F. (1950). *J. Amer. Chem. Soc.* **72**, 2705.
Kastenschmidt, L. L., Kastenschmidt, J., and Helmreich, E. (1968). *Biochemistry* **7**, 4543.
Kawahara, K., Kirshner, A. G., and Tanford, C. (1965). *Biochemistry* **4**, 1203.
Kellett, G. L. (1971). *J. Mol. Biol.* **59**, 401.
Kellett, G. L., and Schachman, H. K. (1971). *J. Mol. Biol.* **59**, 387.
Kelly, M. J., and Reithel, F. J. (1971). *Biochemistry* **10**, 2639.
Keresztes-Nagy, S., and Klotz, I. M. (1963). *Biochemistry* **2**, 923.
Keresztes-Nagy, S., and Klotz, I. M. (1965). *Biochemistry* **4**, 919.
Keresztes-Nagy, S., and Orman, R. (1971). *Biochemistry* **10**, 2506.
Keresztes-Nagy, S., Lazer, L., Klapper, M. H., and Klotz, I. M. (1965). *Science* **150**, 357.
Kilmartin, J. V., and Hewitt, J. A. (1972). *Cold Spring Harbor Symp. Quant. Biol.* **36**, 311.
Kilmartin, J. V., and Rossi-Bernardi, L. (1969). *Nature* (*London*) **222**, 1243.
Kilmartin, J. V., and Wootton, J. F. (1970). *Nature* (*London*) **228**, 766.
Kirschner, M. W., and Schachman, H. K. (1971a). *Biochemistry* **10**, 1900.
Kirschner, M. W., and Schachman, H. K. (1971b). *Biochemistry* **10**, 1919.
Kirshner, A. G., and Tanford, C. (1964). *Biochemistry* **3**, 291.
Kiselev, N. A., and Lerner, F. Ya. (1971). *J. Mol. Biol.* **62**, 537.
Klapper, M. H., and Klotz, I. M. (1968). *Biochemistry* **7**, 223.
Klapper, M. H., Barlow, G. H., and Klotz, I. M. (1966). *Biochem. Biophys. Res. Commun.* **25**, 116.
Klotz, I. M. (1965). *Fed. Proc., Fed. Amer. Soc. Exp. Biol.* **24**, No. 2, Part III, S-24.
Klotz, I. M. (1966). *Arch. Biochem. Biophys.* **116**, 92.
Klotz, I. M. (1967). *Science* **155**, 697.
Klotz, I. M., and Darnall, D. W. (1969). *Science* **166**, 126.
Klotz, I. M., and Keresztes-Nagy, S. (1962). *Nature* (*London*) **195**, 900.

Klotz, I. M., and Tam, J. W. O. (1973). *Proc. Nat. Acad. Sci. U. S.* **70,** 1313.
Klug, A. (1967). *Symp. Int. Soc. Cell Biol.* **6,** 1.
Kohlhaw, G., and Boatman, G. (1971). *Biochem. Biophys. Res. Commun.* **43,** 741.
Koike, K., Poillon, W. N., and Feigelson, P. (1969). *J. Biol. Chem.* **244,** 3457.
Konings, W. N., Dijk, J., Wichertjes, T., Beuvery, E. C., and Gruber, M. (1969). *Biochim. Biophys. Acta* **188,** 43.
Koshland, D. E., Jr., and Neet, K. E. (1968). *Annu. Rev. Biochem.* **37,** 359.
Koshland, D. E., Jr., Nemethy, G., and Filmer, D. (1966). *Biochemistry* **5,** 365.
Kreuzer, J. (1943). *Z. Phys. Chem., Abt. B* **53,** 273.
Kuczenski, R. T., and Suelter, C. H. (1970). *Biochemistry* **9,** 2043.
Kumar, S., and Porter, J. W. (1971). *J. Biol. Chem.* **246,** 7780.
Kumosinski, T. F., and Timasheff, S. N. (1964). *J. Amer. Chem. Soc.* **88,** 5653.
Kurganov, B. I. (1968). *Mol. Biol.* **2,** 430.
Kvamme, E., Tveit, B., and Svenneby, G. (1970). *J. Biol. Chem.* **245,** 1871.
Labeyrie, F., and Baudras, A. (1972). *Eur. J. Biochem.* **25,** 33.
Langerman, N. R., and Klotz, I. M. (1969). *Biochemistry* **8,** 4746.
Lazarus, N. R., Derechin, M., and Barnard, E. A. (1968). *Biochemistry* **7,** 2390.
Leary, T. R., and Kohlhaw, G. (1972). *J. Biol. Chem.* **247,** 1089.
Lebherz, H. G., and Rutter, W. J. (1969). *Biochemistry* **8,** 109.
Lee-Huang, S., and Warner, R. C. (1969). *J. Biol. Chem.* **244,** 3793.
LeJohn, H. B., McCrea, B. E., Suzuki, I., and Jackson, S. (1969). *J. Biol. Chem.* **244,** 2484.
Levin, Ö. (1963). *J. Mol. Biol.* **6,** 95.
Levitzki, A., and Koshland, D. E., Jr. (1972). *Biochemistry* **11,** 247.
Lew, K. K., and Roth, J. R. (1971). *Biochemistry* **10,** 204.
Linderstrøm-Lang, K. U. (1952). "Lane Medical Lectures: Proteins and Enzymes." Stanford Univ. Press, Stanford, California.
Lindstrom, T. R., Noren, I. B. E., Charache, S., Lehmann, H., and Ho, C. (1972). *Biochemistry* **11,** 1677.
Long, C. W., Levitzki, A., and Koshland, D. E., Jr. (1970). *J. Biol. Chem.* **245,** 80.
Lucas-Lenard, J., and Lipmann, F. (1971). *Annu. Rev. Biochem.* **40,** 409.
McClintock, D. K., and Markus, G. (1968). *J. Biol. Chem.* **243,** 2855.
MacColl, R., Berns, D. S., and Koven, N. L. (1971a). *Arch. Biochem. Biophys.* **146,** 477.
MacColl, R., Lee, J. J., and Berns, D. S. (1971b). *Biochem. J.* **122,** 421.
McFadden, B. A., Rao, G. R., Cohen, A. L., and Roche, T. E. (1968). *Biochemistry* **7,** 3574.
MacKenzie, R. E., and Rabinowitz, J. C. (1971). *J. Biol. Chem.* **246,** 3731.
Madsen, N. B., and Cori, C. F. (1956). *J. Biol. Chem.* **223,** 1055.
Maley, G. F., and Maley, F. (1968). *J. Biol. Chem.* **243,** 4506.
Mankovitz, R., and Segal, H. L. (1969). *Biochemistry* **9,** 3757.
Mann, K. G., and Vestling, C. S. (1970). *Biochemistry* **9,** 3020.
Markert, C. L. (1963). *Science* **140,** 1329.
Markert, C. L. (1968). *Ann. N. Y. Acad. Sci.* **151,** 14.
Markert, C. L., and Whitt, G. S. (1968). *Experientia* **24,** 977.
Marshall, M., and Cohen, P. P. (1972). *J. Biol. Chem.* **247,** 1641.
Meighen, E. A., and Schachman, H. K. (1970a). *Biochemistry* **9,** 1163.
Meighen, E. A., and Schachman, H. K. (1970b). *Biochemistry* **9,** 1177.
Meighen, E. A., Pigiet, V., and Schachman, H. K. (1970). *Proc. Nat. Acad. Sci. U. S.* **65,** 234.

Meighen, E. A., Nicoli, M. Z., and Hastings, J. W. (1971). *Biochemistry* **10,** 4062.
Metzger, B. E., Helmreich, E., and Glaser, L. (1967). *Proc. Nat. Acad. Sci. U. S.* **57,** 994.
Meyer, F., Heilmeyer, L., Jr., Haschke, R. H., and Fischer, E. H. (1970). *J. Biol. Chem.* **245,** 6642.
Millar, D. B., Frattali, V., and Willick, G. E. (1969). *Biochemistry* **8,** 2416.
Minton, A., and Libby, W. F. (1968). *Proc. Nat. Acad. Sci. U. S.* **61,** 1191.
Monod, J., Changeux, J. P., and Jacob, F. (1963). *J. Mol. Biol.* **6,** 306.
Monod, J., Wyman, J., and Changeux, J. P. (1965). *J. Mol. Biol.* **12,** 88.
Morimoto, K., and Kegeles, G. (1971). *Arch. Biochem. Biophys.* **142,** 247.
Morino, Y., and Snell, E. E. (1967a). *J. Biol. Chem.* **242,** 5591.
Morino, Y., and Snell, E. E. (1967b). *J. Biol. Chem.* **242,** 5602.
Murayama, M. (1962). *Nature (London)* **194,** 933.
Nelbach, M. E., Pigiet, V. P., Jr., Gerhart, J. C., and Schachman, H. K. (1972). *Biochemistry* **11,** 315.
Neufeld, G. J., and Riggs, A. F. (1969). *Biochim. Biophys. Acta* **181,** 234.
Nichol, L. W., Bethune, J. L., Kegeles, G., and Hess, E. L. (1964). *In* "The Proteins" (H. Neurath, ed.), 2nd ed., Vol. 2, p. 305. Academic Press, New York.
Nichol, L. W., Jackson, W. J. H., and Winzor, D. J. (1967). *Biochemistry* **6,** 2449.
Noble, R. W. (1969). *J. Mol. Biol.* **39,** 479.
Nowak, T., and Himes, R. H. (1971). *J. Biol. Chem.* **246,** 1285.
Nozaki, Y., and Tanford, C. (1967). *J. Biol. Chem.* **242,** 4731.
Ogata, R. T., and McConnell, H. M. (1972). *Cold Spring Harbor Symp. Quant. Biol.* **36,** 325.
Olsen, B. R., Jimenez, S. A., Kivirikko, K. I., and Prockop, D. J. (1970a). *J. Biol. Chem.* **245,** 2649.
Olsen, B. R., Svenneby, G., Kvamme, E., Tveit, B., and Eskeland, T. (1970b). *J. Mol. Biol.* **52,** 239.
Page, M., and Godin, C. (1969). *Can. J. Biochem.* **47,** 401.
Palacian, E., and Neet, K. E. (1969). *Fed. Proc., Fed. Amer. Soc. Exp. Biol.* **28,** 536.
Parsons, S. M., and Raftery, M. A. (1972). *Biochemistry* **11,** 1623.
Penhoet, E., Kochman, M., Valentine, R., and Rutter, W. J. (1967). *Biochemistry* **6,** 2940.
Perutz, M. F., and Lehmann, H. (1968). *Nature (London)* **219,** 902.
Perutz, M. F., and Ten Eyck, L. F. (1972). *Cold Spring Harbor Symp. Quant. Biol.* **36,** 295.
Perutz, M. F., Rossmann, M. G., Cullis, A. F., Muirhead, H., Will, G., and North, A. C. T. (1960). *Nature (London)* **185,** 416.
Perutz, M. F., Muirhead, H., Cox, J. M., Goaman, L. C. G., Mathews, F. S., McGandy, E. L., and Webb, L. E. (1968a). *Nature (London)* **219,** 29.
Perutz, M. F., Muirhead, H., Cox, J. M., and Goaman, L. C. G. (1968b). *Nature (London)* **219,** 131.
Pranis, R. A. (1972). Ph.D. Dissertation, Northwestern University, Evanston, Illinois.
Pringle, J. R. (1970). *Biochem. Biophys. Res. Commun.* **39,** 46.
Procsal, D., and Holten, D. (1972). *Biochemistry* **11,** 1310.
Quiocho, F. A., Reeke, G. N., Becker, J. W., Lipscomb, W. N., and Edelman, G. D. (1971). *Proc. Nat. Acad. Sci. U. S.* **68,** 1853.
Rao, M. S. N., and Kegeles, G. (1958). *J. Amer. Chem. Soc.* **80,** 5724.
Reimann, E. M., Brostrom, C. O., Corbin, J. D., King, C. A., and Krebs, E. G. (1971). *Biochem. Biophys. Res. Commun.* **42,** 187.

Reisler, E., and Eisenberg, H. (1971). *Biochemistry* **10,** 2659.

Reynolds, J. A., and Schlesinger, M. J. (1969). *Biochemistry* **8,** 4278.

Richards, K. E., and Williams, R. C. (1972). *Biochemistry* **11,** 3393.

Richardson, J. P. (1966). *Proc. Nat. Acad. Sci. U. S.* **55,** 1616.

Richter, G. W., and Walker, G. F. (1967). *Biochemistry* **6,** 2871.

Roark, D. E. (1970). Ph.D. Thesis, SUNY, Buffalo, New York.

Roark, D. E., and Yphantis, D. A. (1969). *Ann. N. Y. Acad. Sci.* **164,** 245.

Robyt, J. F., and Ackerman, R. (1972). *Fed. Proc., Fed. Amer. Soc. Exp. Biol.* **31,** 474.

Rosemeyer, E. R., and Huehns, E. R. (1967). *J. Mol. Biol.* **25,** 253.

Rosenbusch, J. P., and Weber, K. (1971a). *J. Biol. Chem.* **246,** 1644.

Rosenbusch, J. P., and Weber, K. (1971b). *Proc. Nat. Acad. Sci. U. S.* **68,** 1019.

Rossi-Fanelli, A., Antonini, E., and Caputo, A. (1961). *J. Biol. Chem.* **236,** 397.

Rossmann, M. G., Jeffrey, B. A., Main, P., and Warren, S. (1967). *Proc. Nat. Acad. Sci. U. S.* **57,** 515.

Rossmann, M. G., Adams, M. J., Buchner, M., Ford, G. C., Hackert, M. L., Lentz, P. J., Jr., McPherson, A., Jr., Schevitz, R. W., and Smiley, I. E. (1972). *Cold Spring Harbor Symp. Quant. Biol.* **36,** 179.

St. George, R. C. C., and Pauling, L. (1951). *Science* **114,** 629.

Scanu, A. M., Edelstein, C., and Lim, C. T. (1972). *Fed. Proc., Fed. Amer. Soc. Exp. Biol.* **31,** 829.

Schachman, H. K., and Edelstein, S. J. (1966). *Biochemistry* **5,** 2681.

Schellman, J. (1958). *C. R. Trav. Lab. Carlsberg, Ser. Chim.* **30,** 363.

Schmit, J. C., and Zalkin, H. (1971). *J. Biol. Chem.* **246,** 6002.

Schnebli, H. P., Vatter, A. E., and Abrams, A. (1970). *J. Biol. Chem.* **245,** 1122.

Schulze, I. T., Lusty, C. J., and Katner, S. (1970). *J. Biol. Chem.* **245,** 4534.

Scott, W. A. (1971). *J. Biol. Chem.* **246,** 6353.

Shapiro, A. L., Viñuela, E., and Maizel, J. V., Jr. (1967). *Biochem. Biophys. Res. Commun.* **28,** 815.

Shapiro, B. M., and Ginsburg, A. (1968). *Biochemistry* **7,** 2153.

Shaw, C. R. (1964). *Brookhaven Symp. Biol.* **17,** 117.

Shulman, R. G., Ogawa, S., and Hopfield, J. J. (1972). *Cold Spring Harbor Symp. Quant. Biol.* **36,** 337.

Sia, C. L., and Horecker, B. L. (1968). *Biochem. Biophys. Res. Commun.* **31,** 731.

Sia, C. L., Traniello, S., Pontremoli, S., and Horecker, B. L. (1969). *Arch. Biochem. Biophys.* **132,** 325.

Siegel, L. M., and Kamin, H. (1971). *Fed. Proc., Fed. Amer. Soc. Exp. Biol.* **30,** 1261.

Simpson, R. T., and Vallee, B. L. (1968). *Biochemistry* **7,** 4343.

Singer, S. J., and Itano, H. A. (1959). *Proc. Nat. Acad. Sci. U. S.* **45,** 174.

Smith, G. D., and Schachman, H. K. (1971). *Biochemistry* **10,** 4576.

Smith, G. P., Hood, L., and Fitch, W. M. (1971). *Annu. Rev. Biochem.* **40,** 969.

Smith, L. L., Plese, C. F., Barton, B. P., Charache, S., Wilson, J. B., and Huisman, T. H. J. (1972). *J. Biol. Chem.* **247,** 1433.

Sophianopoulos, A. J., and Van Holde, K. E. (1964). *J. Biol. Chem.* **239,** 2516.

Stadtman, E. R. (1966). *Advan. Enzymol.* **28,** 41.

Stancel, G. M. (1969). *Fed. Proc., Fed. Amer. Soc. Exp. Biol.* **28,** 345.

Stancel, G. M., and Deal, W. C., Jr. (1968). *Biochem. Biophys. Res. Commun.* **31,** 398.

Stark, G. R., Stein, W. H., and Moore, S. (1960). *J. Biol. Chem.* **235,** 3177.

Steers, E., Jr., Craven, G. R., Anfinsen, C. B., and Bethune, J. L. (1965). *J. Biol. Chem.* **240,** 2478.
Stein, E. A., and Fischer, E. H. (1960). *Biochim. Biophys. Acta* **39,** 287.
Steiner, R. F. (1952). *Arch. Biochem. Biophys.* **39,** 133.
Steitz, T. A. (1971). *J. Mol. Biol.* **61,** 695.
Strausbauch, P. H., and Fischer, E. H. (1970). *Biochemistry* **9,** 226.
Sugita, Y., and Chanutin, A. (1963). *Proc. Soc. Exp. Biol. Med.* **112,** 72.
Svedberg, T. (1929). *Nature (London)* **123,** 871.
Svedberg, T., and Fahraeus, R. (1926). *J. Amer. Chem. Soc.* **48,** 430.
Svedberg, T., and Hedenius, A. (1934). *Biol. Bull.* **66,** 191.
Svedberg, T., and Heyroth, F. F. (1929). *J. Amer. Chem. Soc.* **51,** 550.
Svedberg, T., and Pedersen, K. O. (1940). "The Ultracentrifuge." Oxford Univ. Press (Clarendon), London and New York.
Swann, J. C., and Hammes, G. G. (1969). *Biochemistry* **8,** 1.
Szent-Györgyi, A. (1946). *Nature (London)* **157,** 875.
Tagaki, W., and Westheimer, F. H. (1968). *Biochemistry* **7,** 891.
Tao, M., Salas, M. L., and Lipmann, F. (1970). *Proc. Nat. Acad. Sci. U. S.* **67,** 408.
Tate, S. S., and Meister, A. (1970). *Biochemistry* **9,** 2626.
Teipel, J. W., and Hill, R. L. (1971). *J. Biol. Chem.* **246,** 4859.
Teller, D. C. (1970). *Biochemistry* **9,** 4201.
Timasheff, S. N., and Townend, R. (1964). *Nature (London)* **203,** 517.
Tomkins, G. M., Yielding, K. L., Talal, N., and Curran, J. F. (1963). *Cold Spring Harbor Symp. Quant. Biol.* **26,** 331.
Townend, R., Weinberger, L., and Timasheff, S. N. (1960). *J. Amer. Chem. Soc.* **82,** 3175.
Travers, A. A., Kamen, R. I., and Schleif, R. F. (1970a). *Nature (London)* **228,** 748.
Travers, A. A., Kamen, R. I., and Cashel, M. (1970b). *Cold Spring Harbor Symp. Quant. Biol.* **35,** 415.
Trotta, P. P., Burt, M. E., Haschemeyer, R. H., and Meister, A. (1971). *Proc. Nat. Acad. Sci. U. S.* **68,** 2599.
Ullmann, A., Vagelos, P. R., and Monod, J. (1964). *Biochem. Biophys. Res. Commun.* **17,** 86.
Uyeda, K. (1969). *Biochemistry* **8,** 2366.
Valentine, R. C., Wrigley, N. G., Scrutton, M. C., Irias, J. J., and Utter, M. F. (1966). *Biochemistry* **5,** 3111.
Valentine, R. C., Shapiro, B. M., and Stadtman, E. R. (1968). *Biochemistry* **7,** 214.
van Heyningen, S., and Shemin, D. (1971). *Biochemistry* **10,** 4676.
Van Holde, K. E., Rossetti, G. P., and Dyson, R. D. (1969). *Ann. N. Y. Acad. Sci.* **164,** 297.
Vinograd, J. R., Hutchinson, W. D., and Schroeder, W. A. (1959). *J. Amer. Chem. Soc.* **81,** 3168.
von Fellenberg, R., Bethell, M. R., Jones, M. E., and Levine, L. (1968). *Biochemistry* **7,** 4322.
Wampler, D. E., and Westhead, E. W. (1968). *Biochemistry* **7,** 1661.
Wang, C. T., and Weissmann, B. (1971). *Biochemistry* **10,** 1067.
Wang, J. H., and Graves, D. J. (1964). *Biochemistry* **3,** 1437.
Wang, J. H., Kwok, S.-C., Wirch, E., and Suzuki, I. (1970). *Biochem. Biophys. Res. Commun.* **40,** 1340.
Watson, H. C., and Banaszak, L. J. (1964). *Nature (London)* **204,** 918.

Waxman, L. (1971). *J. Biol. Chem.* **246,** 7318.
Weber, G. (1972). *Biochemistry* **11,** 864.
Weber, K., and Osborn, M. (1969). *J. Biol. Chem.* **244,** 4406.
Weisenberg, R. C., and Timasheff, S. N. (1970). *Biochemistry* **9,** 4110.
Welch, W. H., Buttlaire, D. H., Hersh, R. T., and Himes, R. H. (1971). *Biochim. Biophys. Acta* **236,** 599.
Wetlaufer, D. B. (1961). *Nature (London)* **190,** 113.
Whanger, P. D., Phillips, A. T., Rabinowitz, K. W., Piperno, J. R., Shada, J. D., and Wood, W. A. (1968). *J. Biol. Chem.* **243,** 167.
Whiteley, H. R., and Pelroy, R. A. (1972). *J. Biol. Chem.* **247,** 1911.
Wickner, R. B., and Tabor, H. (1972). *J. Biol. Chem.* **247,** 1605.
Wiley, D. C., and Lipscomb, W. N. (1968). *Nature (London)* **218,** 1119.
Wiley, D. C., Evans, D. R., Warren, S. G., McMurray, C. H., Edwards, B. P. F., Franks, W. A., and Lipscomb, W. N. (1972). *Cold Spring Harbor Symp. Quant. Biol.* **36,** 285.
Wilk, S., Meister, A., and Haschemeyer, R. H. (1969). *Biochemistry* **8,** 3168.
Williams-Ashman, H. G., Notides, A. C., Pabalan, S. S., and Lorand, L. (1972). *Proc. Nat. Acad. Sci. U. S.* **69,** 2322.
Wyman, J., and Allen, D. W. (1951). *J. Polym. Sci.* **7,** 499.
Yang, S. T., and Deal, W. C., Jr. (1969). *Biochemistry* **8,** 2806.
Yang, W. H., Sairam, M. R., Papkoff, H., and Li, C. H. (1972). *Science* **175,** 638.
Yue, R. H., Palmieri, R. H., Olson, O. E., and Kuby, S. A. (1967). *Biochemistry* **6,** 3204.
Zimmerman, J. K., Barlow, G. H., and Klotz, I. M. (1970). *Arch. Biochem. Biophys.* **138,** 101.

6

Electron Microscopy of Proteins

J. T. FINCH

I. INTRODUCTION

With the relative ease of operation of present day instruments, electron microscopy has become very popular in the investigation of biologic structures. In many cases, one can obtain images that are fairly faithful records of the detail in the specimen within only a few minutes. The two main drawbacks of the method are (1) the relative transparency of proteins to electrons and (2) the disruption of protein structure as the specimens dry out in the evacuated microscope and are bombarded by the electron beam. The usual method of increasing the contrast is by adding heavy metals in one way or another. The simplest way of doing this, the method of negative staining, is fortunately also the most successful in preserving specimen order. Even so, the images obtained by negative staining show that structural detail is normally preserved to only about 20 Å resolution. If care is taken to minimize electron beam damage, the resolution can be extended to around 10 Å, although modern microscopes are capable of resolving details to 2 or 3 Å.

With the effective resolution limited to 20 Å in electron micrographs of proteins, little, even in terms of the shape of the molecule, can be deduced from images of average-sized, isolated monomers. Electron microscopy has therefore been most successful in the determination of the quaternary structures of assemblies of protein molecules, and the main concern of this chapter is a description of the results obtained from studies of enzymes, muscle proteins, and viruses. Since the electron microscopy of immunoglobulins has been reviewed in detail by Green (1969), work on collagen by Traub and Piez (1971) and work on keratin and other fibrous proteins by Fraser and MacRae (1973), further reviews here would be superfluous. Before describing the results, a brief review is given of the various techniques that have been used in the electron microscopy of proteins and in the analysis of the resulting images. First, however, the relationship between the specimen and the image is discussed, since this is the basis for all of the structural interpretations.

II. RELATIONSHIP BETWEEN THE ELECTRON MICROSCOPE IMAGE AND THE SPECIMEN

The determination of the structure of a specimen from electron microscope images is based on the assumption that each image corresponds to a projection of the structure in a particular direction. This assumption may be felt to be justified if the images are recorded with the objective lens of the microscope close to focus, since the electron lenses of modern microscopes are virtually perfect to the 20 Å effective limit of resolution. However, since images taken exactly in focus show little contrast, micrographs are usually taken somewhat under focus in order to show detail more clearly. Erickson and Klug (1971) have shown that this procedure is valid, provided that the underfocusing is not excessive.

Erickson and Klug (1971) investigated the effects of defocusing on the transfer function, which is the ratio of the amplitudes of the periodicities, or Fourier components, in the image to those in the specimen. For an in-focus image, the transfer function decreases for smaller periodicities, and the image is produced purely by amplitude contrast arising from the elastic scattering of electrons by the specimen outside the aperture of the objective lens. When this lens is underfocused, the phase differences between the electron waves which have passed through different parts of the specimen give rise to corresponding amplitude differences across the image. This phase contrast effect is first appreciable for smaller periodicities and reinforces their contributions to the image, so that at 800 Å under focus (using the experimental conditions described by Erickson and Klug, 1971), a fairly constant transfer function is obtained. At about 5000 Å under focus, the phase contrast effect is sufficient to increase the transfer function for periodicities in the range of 20 to 50 Å by about five times that of the in-focus image. Since this range of

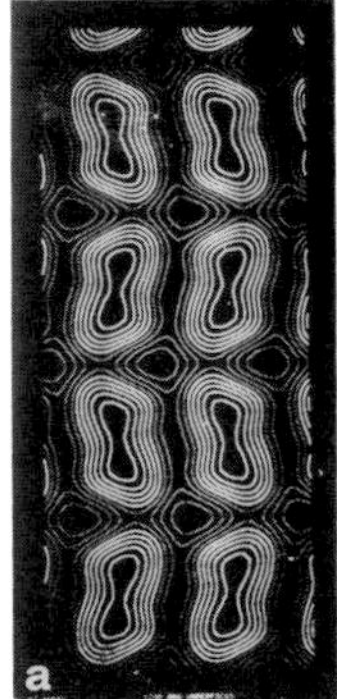

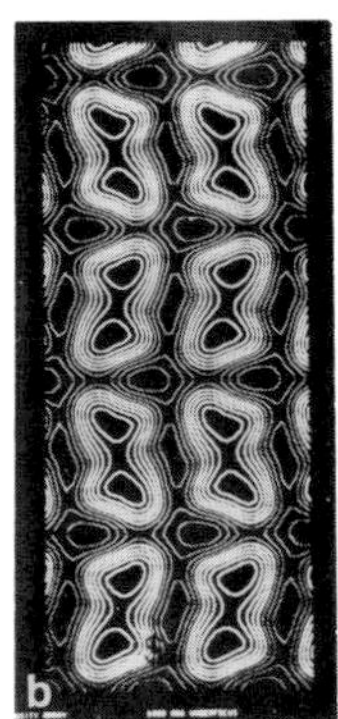

Fig. 1 The effect of underfocusing on the detail in images of catalase molecules (Erickson and Klug, 1971). Contour plots are shown of the density in filtered images obtained computationally from electron micrographs of negatively stained thin catalase crystals. (a) The true projection of the specimen, which would be obtained with a constant transfer function relating the details in the specimen to those in the image; this essentially occurs for images at about 800 Å under focus. (b) Calculated projection obtained from an image 5400 Å under focus; details corresponding to periodicities in the 20 to 50 Å range are enhanced yet present with the correct phase. (c, d) Calculated projections obtained from images underfocused by 14,500 and 19,500 Å, respectively; phase reversal effects are now present in these images because of excessive underfocusing.

detail is of particular interest in protein studies, 5000 Å was considered to be the optimum under-focus setting for such work. With further underfocusing, the transfer function decreases for the smaller periodicities and changes sign so that false details (i.e., periodicities with the wrong phase) are introduced into the image. This effect can be readily diagnosed by optical diffraction from an electron micrograph. The zero value of the transfer function is revealed as a ring of minimum diffracted intensity which should be well beyond the spacing of the minimum detail under investigation to ensure that no phase reversal artifacts occur.

These effects were tested using a thin negatively stained crystal of catalase as a specimen. Erickson and Klug (1971) demonstrated that it was possible to computationally correct the measured amplitudes of periodicities obtained from a series of electron micrographs taken at various focus settings and in this way obtain the true projection of the catalase molecules shown in Fig. 1a. The calculated projections from a 5400 Å under-focus image (Fig. 1b) show that detail in the region of spacings of 20 to 50 Å is somewhat exaggerated. The effect of excessive underfocusing in introducing periodicities with the wrong phase is shown in Figs. 1c and 1d.

III. SPECIMENS

A. Types of Specimens

Many attempts have been made to characterize isolated molecules by electron microscopy. While this has been useful in the case of exceptionally large proteins (e.g., myosin), it has been less successful with more normally sized molecules, since it is often impossible to determine whether the different images observed arise from different degrees of distortion, etc., or from different views of the molecule (Mellema *et al.*, 1968). In these cases it is more useful to study regular aggregates in which the component molecules are in the same or symmetrically related orientations. Fortunately, many proteins occur naturally in such aggregates with the individual subunits arranged with point group symmetry (as in the shells of spherical viruses and many enzymes) or helical symmetry (as in the rod-shaped viruses and muscle protein filaments). Also, in many cases proteins can be induced to form regular arrays with helical symmetry, in two-dimensional sheets or three-dimensional crystals. The great advantage of these arrays is the possibility of minimizing the effects of local distortions and background detail by averaging the data from a large number of subunits (see Section IV).

B. Specimen Preparation

The simplest method of mounting a specimen for electron microscopy is to apply a solution of the specimen to a thin support film (usually carbon) attached to a metal grid. The solution can be applied as a drop or sprayed onto the grid. Because protein specimens are relatively transparent to electrons, they must be reinforced in some way with heavy atoms to provide sufficient contrast in the electron micrographs. This may be done either by evaporating metal onto the protein (shadowing) or by allowing a solution of an electron-dense compound to dry around the protein specimen (negative staining). More elaborate techniques are normally required for thicker specimens such as protein crystals.

1. Shadowing

In this procedure a metal is evaporated from a hot filament held at a fixed distance from the specimen at a fixed angle. The evaporated metal collects on surfaces exposed to the source, leaving the far sides clear. Since the shadowed portion of the specimen scatters the electrons whereas the protected portion is practically transparent to the electron beam, the detail in the specimen can be deduced from the variations in metal thickness. At best, the resolution of detail is limited to about 20 Å in the (projected) direction of shadowing and is considerably worse in other directions. Knowing the angle of shadowing, an estimate of particle size can be made from the dimensions of the metal-clear shadows, but these may be exaggerated by the buildup of metal on the other side. Also, the specimen may well be distorted or collapse as it dries in the vacuum chamber before shadowing. To minimize this damage, freeze-drying (Williams, 1953) and critical point drying (Anderson, 1951) techniques have been developed.

Shadowing has been extended to the study of the surfaces of thick crystals by using replication techniques. The crystals can be air-dried or, to reduce disorder upon drying, frozen, and the surface solution etched away by approaching it with a colder object (Steere, 1957). For example, a replica of the crystal surface can be made by shadowing it with a metal, then coating it with a thin layer of evaporated carbon to stabilize the surface. The crystals can then be dissolved away or the replica floated off and placed on a grid for examination in the microscope.

Internal crystal planes can be revealed by fracturing the crystal while frozen, before etching.

2. *Sectioning*

Protein crystals have also been studied by the standard method of thin sectioning. The usual procedure involves fixing the crystal (for example, with glutaraldehyde) to enable it to withstand subsequent drying in alcohol, and embedding in a resin which is then hardened. If the crystals are large and well shaped, sections can be cut in chosen crystallographic directions, placed on a grid, and examined in the electron microscope. Stains can be applied both to the crystal during the drying procedure and also to the individual sections. Preparations of microcrystals will yield many small sections in random orientations which must then be correlated from the patterns in the micrographs.

Considerable shrinkage of unit cell dimensions can occur upon drying and further distortion may be introduced by compression during sectioning. The thinnest sections obtainable at present are about 200 to 300 Å, and optical diffraction patterns show that the detail within these sections extends to spacings of about 50 Å at best.

3. *Negative Staining*

Most of the results described later in this chapter were obtained by negative staining. Besides being a simple and quick procedure, it yields specimens that often show well-preserved order. The effect was first observed by Farrant (1954), Hall (1955), and Huxley (1957) in electron micrographs of specimens containing residual staining solution. Particles embedded in stain show up in electron micrographs because their density is lower than that of the stain, thus producing a difference in contrast. Particle substructure is revealed by the penetration of stain into holes and crevices. The clarity of detail depends upon the extent to which the stain remains amorphous as it dries. Neutralized phosphotungstic acid, uranyl acetate, and related compounds have been the stains most commonly employed, but many others have also been used successfully.

Probably the safest way of preparing a negatively stained specimen is that described by Huxley and Zubay (1960a). A drop of the specimen solution is applied to the support film on the grid, the excess specimen not attached to the film is washed off, and a few drops of the stain solution are applied. The excess stain is then withdrawn by touching it with the edge of a filter paper. The intermediate step of washing not only removes any solution which may react with the stain but also removes any specimen still in solution, thus reducing the possibility of a change in structure being induced by the stain. Specimens stable in a particular stain can be mixed with it and the resulting mixture then dropped or sprayed onto grids (Brenner and Horne, 1959).

The support film should be thin in order to maximize the contrast between the stain and specimen; it should also be as uniform as possible so that its substructure contributes minimally to the detail seen in the specimen images. Huxley and Zubay (1960a) showed that the clarity of the images was greatly improved by examining the specimen area over holes in the support film; specimens of sufficient concentration will spread out over the holes and become suspended only in a film of stain.

C. Specimen Preservation in Negative Stain

In the course of investigations into the structures of viruses, especially in experiments in which the changes in image were recorded as the specimen was tilted in the microscope, much information has been obtained on the preservation of the structure of individual particles in negative stain. If the stain has spread evenly and has completely enveloped the particles on the support film, the symmetry of the particles can be well preserved, particularly when the particles are in close-packed arrays (Klug and Finch, 1968). However, if the particles are only partially embedded in regions of thin stain, they appear to be appreciably flattened against the substrate. If the particles are embedded in stain over holes in the substrate, particle structure can also be well preserved (Finch and Klug, 1966; Josephs, 1971; Josephs and Borisy, 1972). Although films of stain over holes tend to split and contract in the electron beam, the particle structure remains intact in regions where the contraction is uniform rather than sheared to one side. In the study of long polymers or crystalline arrays it can, however, prove difficult to find large, undistorted areas over holes.

The images of negatively stained specimens usually contain details to a resolution of about 20 Å; optical diffraction patterns from electron micrographs of negatively stained periodic specimens fade out at these spacings. Recent work, however, has shown that the order preserved by the stain is considerably better than this, and that deterioration occurs as a result of the action of the electron beam. For example, electron diffraction patterns from negatively stained thin crystals of catalase, which had not been previously irradiated, extend to spacings of about 8 Å (Glaeser, 1971), but subsequent exposure for a few seconds to the higher beam current required to focus and take electron micrographs of images decreases the order to about 25 Å. In order to reduce the effect of beam damage, Williams and Fisher (1970) have developed a method of taking electron micrographs of specimens minimally exposed to the electron beam. The helical periodicities in tobacco mosaic virus show up very clearly by this method, and fine features in aspartate transcarbamylase have been observed which are absent in normal images (Richards and Williams, 1972). Another method of minimizing beam damage, applicable only to large periodic arrays, has been developed by H. P. Erickson (unpublished work). In this case images are recorded at low magnification and low intensity, and although there is little local detail visible because of electron noise, optical diffraction patterns from the images extend to spacings of 15 Å. Images up to this resolution have been reconstructed using optical or computational filtering techniques.

IV. INTERPRETATION AND ANALYSIS OF IMAGES

A. Symmetry Requirements of Molecular Assemblies

Electron microscopy is usually applied to the investigation of the structure of molecular aggregates, since the resolution is usually insufficient to give much dependable detail in images of isolated monomers, unless they are large and appreciably aspherical. In order to derive the structure of an aggregate from its image, a knowledge of the possible ways in which aggregation can occur is of great value. If the individual molecules constituting the aggregate are identical, it is likely that the same chemical bonds at the same sites link each molecule to its neighbors. Every molecule is then structurally equivalent to every other molecule and as a result the aggregate will possess symmetry, i.e., any one molecule can be superposed exactly on any other in the aggregate by rotation about symmetry axes, or by a translation, or by a combination of these operations. The structurally equivalent units need not necessarily be individual molecules or chemical subunits but can be groups of molecules. To include these cases, the terms *structure unit* (Caspar and Klug, 1962) and *protomer* (Monod *et al.*, 1965) will be used in this discussion.

If the structure units are related solely by rotation about symmetry axes, the aggregate will be closed and will contain a fixed number of units as, for example, in the case of the oligomeric enzymes. Since the symmetry axes all pass through one particular point at the center of the aggregate, the possible symmetries are called point groups. There are three types of point groups: cyclic, dihedral, and cubic. In the cyclic

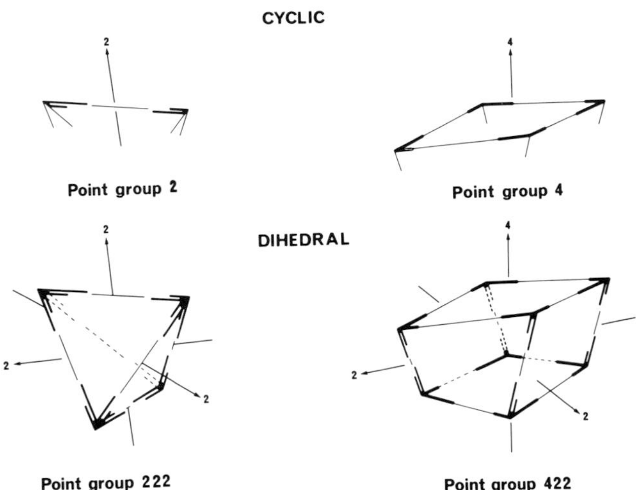

Fig. 2 Arrangements of identical subunits (structure units) with the cyclic point group symmetries 2 and 4 and the dihedral point group symmetries 222 and 422. The directions of the rotation axes are marked by arrows. The structure units are represented by the spiky objects, the spikes representing bond sites. The cyclic groups possess only a single class of bonds; the dihedral (and cubic) point groups require at least two classes of bonds to define the structure, although three distinct classes of bonds are possible. Taken from Klug (1968).

point group n (or C_n), n structure units are arranged in a circle and are related by an n-fold rotational symmetry axis through the center; all the contacts between neighboring structure units around the circle are identical (Fig. 2). If two such rings of structure units associate face-to-face, they are related by a twofold rotational symmetry axis between the rings, perpendicular to their common n-fold axis; the $2n$ structure units are now arranged with $n2$ (or D_n) dihedral symmetry. In dihedral arrangements there must be at least two types of bonds or contacts between the structure units, one between the units in each ring related by the n-fold axis, and one between units of different rings related by the twofold axis (Fig. 2). Since these bonds are likely to be of different characters, the dissociation of a dihedral arrangement is usually a two-stage process, first into dimers or n-mers depending upon the relative strengths of the bonds, and then into individual monomers. A cyclic arrangement will dissociate directly into monomers, provided no change in bonding occurs during dissociation. The scheme of dissociation can thus often differentiate between cyclic and dihedral symmetries.

The cubic point groups are so called because they have at least one set of threefold axes distributed as the body diagonals of a cube (Fig. 3).

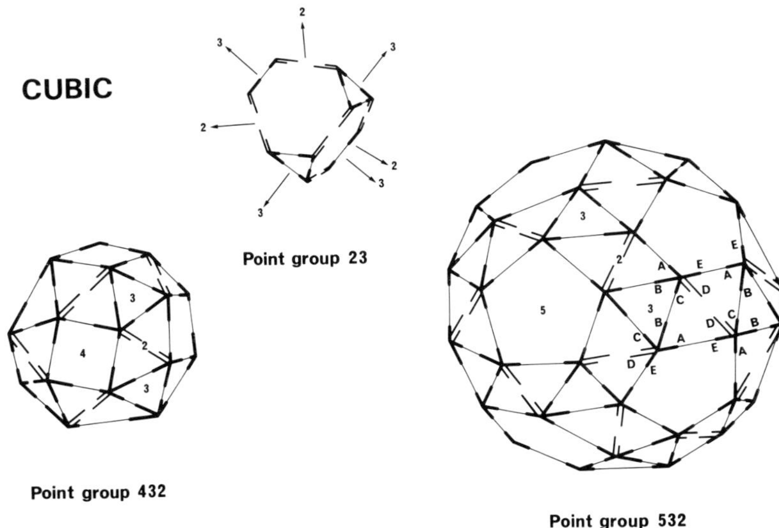

Fig. 3 Arrangements of structure units with the cubic point group symmetries 23, 432, and 532.

The tetrahedral point group, 23, has twofold and threefold rotation axes relating 12 structure units; the octahedral point group, 432, has fourfold, threefold, and twofold axes relating 24 structure units; the icosahedral point group, 532, has fivefold, threefold, and twofold axes relating 60 structure units. Arrangements with cubic point group symmetry must be fairly isometric since the structure units lie on a sphere.

If the relationship between structure units in an aggregate involves a translation, the number of structure units is not fixed. Translations in one, two, or three dimensions, combined with rotational symmetry axes, yield the line, plane, and space groups, respectively (see the International Tables for X-Ray Crystallography, 1952). Included in the line groups are helical arrangements in which the structure units are related by a rotation about a fixed axis, plus a translation parallel to that axis (the screw symmetry operation); if the helical axis is also an n-fold rotational symmetry axis, the result is a stack of rings of n structure units. The plane groups describe the possible relationships between the structure units in two-dimensional sheets, and the space groups describe the possible arrangements in three-dimensional crystals.

A regular aggregate in which identical subunits occupy structurally equivalent positions must fall into one of the categories described above. However, these categories can be extended if the rule of strict structural equivalence is relaxed somewhat. Such an extension has been found necessary to describe the symmetries of the protein shells of the small isometric viruses (Caspar and Klug, 1962). Since the shells are isometric, one would expect the structure units within them to be arranged with one of the cubic point group symmetries. However, in several cases the numbers of identical chemical subunits have been found to be far greater than 60, which is the largest number of structural units that can be arranged equivalently with cubic symmetry in the icosahedral point group. An arrangement of 60 equivalent units, represented by commas, is shown in Fig. 4a, together with the edges of an icosahedron which define a spherical surface lattice for the arrangement; the commas are

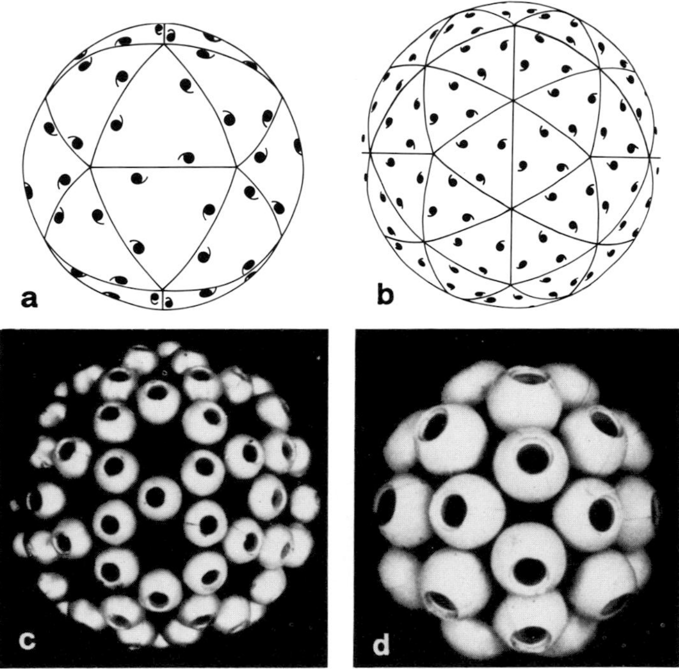

Fig. 4 (a) Sixty structure units, represented by commas, arranged with icosahedral (532) symmetry on the surface of a spherical icosahedron. This is the largest number of structure units that can be arranged isometrically, in strictly identical environments, three in each of the twenty triangles of the spherical surface lattice formed by the edges of the icosahedron. (b) An arrangement of 180 structure units on the $T = 3$ icosahedral surface lattice. This lattice results from subdividing a sphere with icosahedral symmetry into 3×20 triangles. With this type of design, the protein shells of some viruses are built from 3×60 identical molecules using the same bonds between neighbors. Although the molecules occur in three different types of environments, these are very nearly or *quasi*-equivalent. (c, d) Two of the striking surface patterns that arise by particular choices for the positions of the structure units in (b). If the units are located near the centers of the edges of the surface lattice, they cluster into dimers (c). Clustering into hexamers and pentamers (d) occurs when the units are located close to vertices of the triangles.

in equivalent environments and distributed three per triangle of the lattice. Caspar and Klug (1962) showed that more than 60 structure units could be accommodated by subtriangulating the sphere further. Provided the surface lattice so formed has icosahedral symmetry, the structure units lie in quasi-equivalent environments, thus preserving identical bonding patterns between them. Caspar and Klug described the possible icosahedral surface lattices in terms of the triangulation numbers T required to accommodate $60T$ structure units. The possible values of T are given by $T = h^2 + hk + k^2$, where h and k are any integers. An arrangement of 180 structure units (commas) on the $T = 3$ icosahedral surface lattice is shown in Fig. 4b. Although the structure units occur in three (i.e., T) different environments, these are quasi-equivalent.

B. Model Building and Image Simulation

Besides being the most convenient way of displaying the results of a structure determination, models can play an important part in that determination by predicting the types of images to be expected from a particular arrangement. A few clear images of different aspects of a small, closed molecular assembly can be sufficient to limit its symmetry to that of one particular point group. If this is cyclic, the radius and shape of the structure units must be known in order to build a model. For all other point groups, the relative positions of the structure units are not fixed by the symmetry and so these must also be determined from the images. (The location of one structure unit relative to the symmetry axes is sufficient to fix the locations of the remainder in any particular point group arrangement.) The shapes and positions of the structure units within a model can be adjusted to account for the detail in the images.

One of the most useful ways of investigating the plausibility of a particular proposed structure is to simulate the images arising from a model of the structure. There are, however, several limitations when this method is applied to real physical models. First, the density in a physical model must be projected in some way onto a plane to simulate the electron microscope image. Although the shadows cast by a skeletal model in a parallel beam of light have been used with some success (Klug and Finch, 1965), this method suffers from the fact that overlapping parts of the structure do not give darker shadows, and extra shadows appear from any components used to hold the model together; with a realistic model, it is quite likely that no light would get through

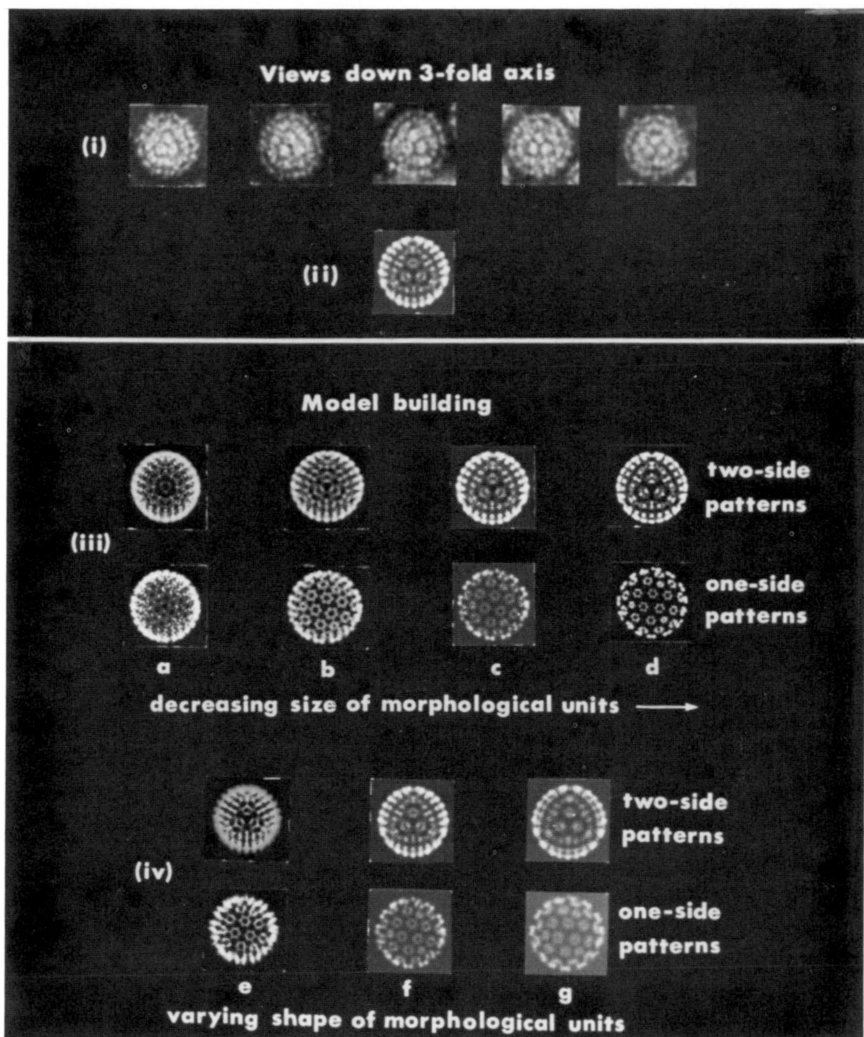

Fig. 5 Model building by computer. (i) Images of negatively stained particles of human wart virus seen in directions close to a threefold axis. (ii) Simulated image which gives the best agreement with (i). (iii, iv) Projections of various models based on the $T = 7$ icosahedral surface lattice. In (iii) the effect of clustering the structure units into hexamer and pentamer morphological units is shown. In (iv) the structure units in (c) are varied in radial extent.

the model at all. X-Radiographs of cork models can yield much better results, and Caspar (1966) has simulated the effect of negative stain by coating a cork model with Plaster of Paris. The radiographs obtained in this way were strikingly similar to the electron microscope images of the negatively stained virus particles under investigation (see Section V,C,2,*a*).

However, for all but the simplest arrangements, the construction of a physical model can be time-consuming and, since each refinement requires a new model, the whole procedure can be rather tedious. A method of model building by computer has therefore been developed and has proved very useful in work on viruses (Finch and Klug, 1967; Klug and Finch, 1968). In the case of human wart virus, illustrated in Fig. 5, the images suggested that the structure units were associated in rings of five and six to form large morphological units on the surface of a particular icosahedral surface lattice ($T = 7$). Coordinates of points on this lattice were therefore chosen as the location of the structure units and fed into a computer. The computer generated the symmetrically related points and calculated the projection of the overall arrangment in the direction of the threefold symmetry axis of the lattice, since a particularly striking electron microscope image was obtained for this aspect of the virus particle. The computed projection was photographed from a display screen for comparison with the images. By representing the structure units as a series of points and by varying the coordinates, radii, and sizes of these points, the best agreement between the computed model projections and the images was found. The agreement was best at the centers of the patterns. Less detail was visible around the periphery of the image than is shown in the projection, since the electron microscope records the projection of the stain distribution around the particle rather than the projection of the particle itself; however allowance can easily be made for this discrepancy. The method of computing projections has also been applied to helical structures (Kiselev and Klug, 1969).

C. Tilting Experiments

Correspondence between the images simulated from a model structure and those obtained from electron micrographs is a good indication that the model is correct. Tilting experiments can, however, provide a much more definitive test of a model by determining whether the image of a particular particle changes, as it is tilted in the electron microscope, as predicted by the model. The particle structure must, of course, be well preserved for such work.

In addition to testing a model, tilting experiments can give information on the handedness of a structure. At sufficiently good resolution all structures will have an element of skewness, i.e., they will lack mirror planes of symmetry, although this will not always be evident in the case of small molecular assemblies seen at the normal 20 Å resolution. Skewness, however, is often very noticeable in larger assemblies, for example, in the isometric virus particles, and of course in helical aggregates, where the various families of helices are either right- or left-handed. On the basis of isolated images alone it is not possible to differentiate between a particular model structure and its enantiomorph because both structures yield identical projections and, if stained uniformly, produce identical electron microscope images. However, a decision can be made in favor of one of the structures by tilting the specimen, since the rotation required to change from one particular image to another will be of opposite sense for the two enantiomorphs.

Tilting experiments with the isometric human wart virus (Klug and Finch, 1968) have shown that the particle symmetry is well preserved in negatively stained arrays where the particles are in contact and are covered by a fairly uniform sheet of stain. The changes in images of individual particles produced by tilting correlate with those predicted by the model shown in Fig. 5. The icosahedral surface lattice upon which this model is based can exist in two enantiorphic forms, *dextro* ($T = 7d$) and *levo* ($T = 7l$). From the sense of the tilt it was shown that the structure of this virus is based upon the $T = 7d$ version of the lattice.

In the case of a helical structure, it is again theoretically possible to determine the handedness from two images taken before and after tilting through a known angle in a known sense. In practice the difficulty lies in finding the same point on the particle axis in the two images. If this can be done (for example, with a stacked ring arrangement), the handedness can be deduced by tilting about the particle axis: the super position pattern (i.e., the projection of both sides of the structure) will appear to have moved along the particle, and the handedness can be deduced by correlating the direction of movement with the sense of tilt. With a helical arrangement of close-packed identical structure units, as, for example, in tobacco mosaic virus, it is virtually impossible to find the same point on the particle axis in two images and there is no obvious superposition pattern. In this case the hand of the helix has been determined by tilting about an axis perpendicular to the particle axis and looking for the side of the image in which a cusplike periodicity occurs as the helix on that side becomes parallel to the direction of view (Finch, 1972; see Section V,C,*1*,*a*).

D. Optical Analysis and Optical Filtering

The fine detail in electron microscope images of identical molecules can vary considerably from image to image due to differences in stain distribution and specimen preparation and to the nonuniformity of the substrate. For this reason the consistent features of a given type of image are selected from many examples before a structural interpretation is attempted. When the specimen has an extended periodic structure, the same factors contribute a nonperiodic "noise" which can obscure the local structural detail and occasionally make even the basic periodicities difficult to see. The difficulties are multiplied when two periodic layers overlap, as, for example, the two sides of a helical arrangement. Markham *et al.* (1964) introduced the technique of enhancing the periodic detail by superposing a number of prints of an image displaced successively by a vector of the specimen lattice and averaging out the nonperiodic contributions. To apply this technique with confidence, one must know the unit cell of the lattice, and this may not be obvious by visual inspection.

The investigation of an image for possible periodicities can be most conveniently made by optical diffraction directly from the electron microscope plate (Klug and Berger, 1964). For example, a two-dimensional periodic array yields a lattice of discrete spots in the Fraunhofer diffraction pattern from which the unit cell of the array can be determined (Fig. 6). A helical array will give rise to a helical diffraction pattern on layer lines appropriate to the helical parameters. Nonperiodic noise in the images contributes diffracted intensity which is not confined to spots or layer lines.

The difference in character between diffraction from the periodic and non-periodic components of a micrograph is the basis of the method of optical filtering by which an image of the periodic component alone can be obtained (Klug and DeRosier, 1966). The diffracted rays from this component are isolated by an appropriate mask and recombined with a lens to form the filtered image. In a similar way, one periodic layer can be filtered from an image of several superposed layers, provided they lie at different angles to each other, as, for example, the two sides of a helical structure. Local variations in detail are preserved in optically filtered images because the substructure of each diffraction spot contains

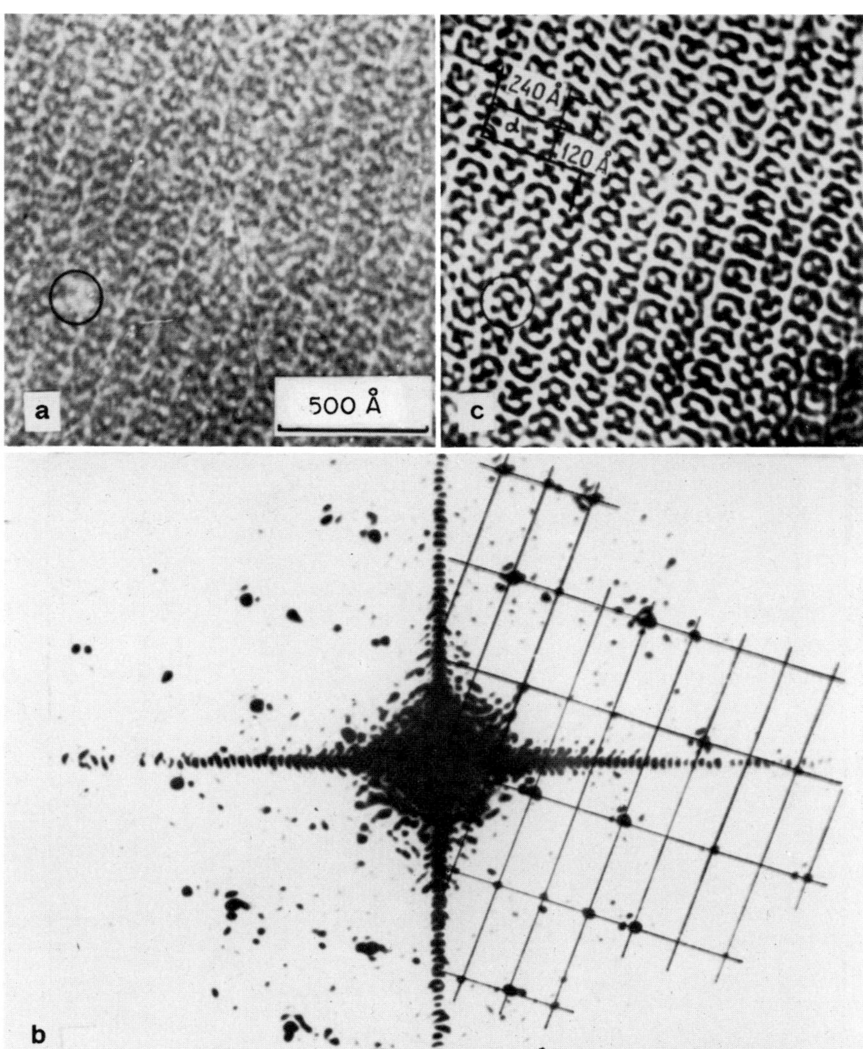

Fig. 6 Optical filtering of periodic images. (a) Image of a negatively stained thin crystal of muscle phosphorylase *b* (Kiselev *et al.*, 1971), printed so that the protein is black. (b) Optical diffraction pattern of (a). The diffracted rays from the crystal fall on the lattice drawn on the right, while nonperiodic noise in the image contributes diffracted intensity between the lattice points. (c) Filtered image obtained by recombining only the diffracted rays on the lattice in (b). The projections of the molecules in the crystal can be seen quite clearly.

the information on the areas of the original image which incorporate the corresponding periodicity.

The considerable increase in clarity of the image which can result from optical filtering is demonstrated in Fig. 6, taken from Kiselev *et al.* (1971). The original electron micrograph (Fig. 6a) is of a negatively stained thin crystal of muscle phosphorylase *b*, and although periodicities are apparent, little detail in the images of individual molecules can be seen. The optical diffraction pattern (Fig. 6b) shows a clear lattice of diffracted rays arising from the periodicities in the image. If these rays are isolated by a mask from the rest of the diffracted intensity which is due to the nonperiodic background noise and brought to a focus, the filtered image shown in Fig. 6c is obtained, in which a regular molecular image, with the appearance of a semicircle over a cross, is clearly visible (in two orientations). This image can be interpreted on the basis of the proposed molecular model of the enzyme (see Section V,A,*4* for discussion).

Further image analysis and processing by optical methods can become extremely laborious and is better left to computational techniques such as those described in the next section. In comparison to optical methods, computational techniques are more flexible and quantitative and hence provide a more precise indication of the reliability of the results.

E. Computational Analysis and Processing

The first step in computational work is the conversion of the optical density in the image to an array of numbers by densitometry. The computational equivalent of optical diffraction from the image is the calculation of the Fourier transform of the array, i.e., the distribution of sinusoidal density variations (Fourier components) which make up the image. Whereas a photograph of the Fraunhofer diffraction pattern is a record of the amplitudes of the Fourier transform only, the calculated transform yields both the amplitudes and the phases with which the Fourier components are combined in the image. A knowledge of the phases can be useful in the determination of the symmetry of a helical structure; a distinction between odd- and even-start families of helices can be made by comparing the phases of the contributions made to the Fourier transform from the near and far sides of these helices (Finch and Klug, 1971). The extent to which these phases correlate is also an excellent guide to the state of preservation of the particle structure.

As is the case with the optical methods, it is possible to isolate the contributions to the transform from the periodic component of an image, recombine these periodicities in their correct phase relationship, and build up a filtered image (Amos and Klug, 1972).

The computed transform has also been used in the investigation of images for possible rotational symmetry (Crowther and Amos, 1971a). The weight of a specified n-fold rotational harmonic in the image is calculated for the best image center consistent with n-fold rotational symmetry. By repeating this for a range of values of n, a series of rotational power spectra is built up which shows any dominant rotational symmetry in the image. By combining only rotational components consistent with this symmetry, a rotationally filtered image can be produced. The advantage of this method over the rotational version of the optical

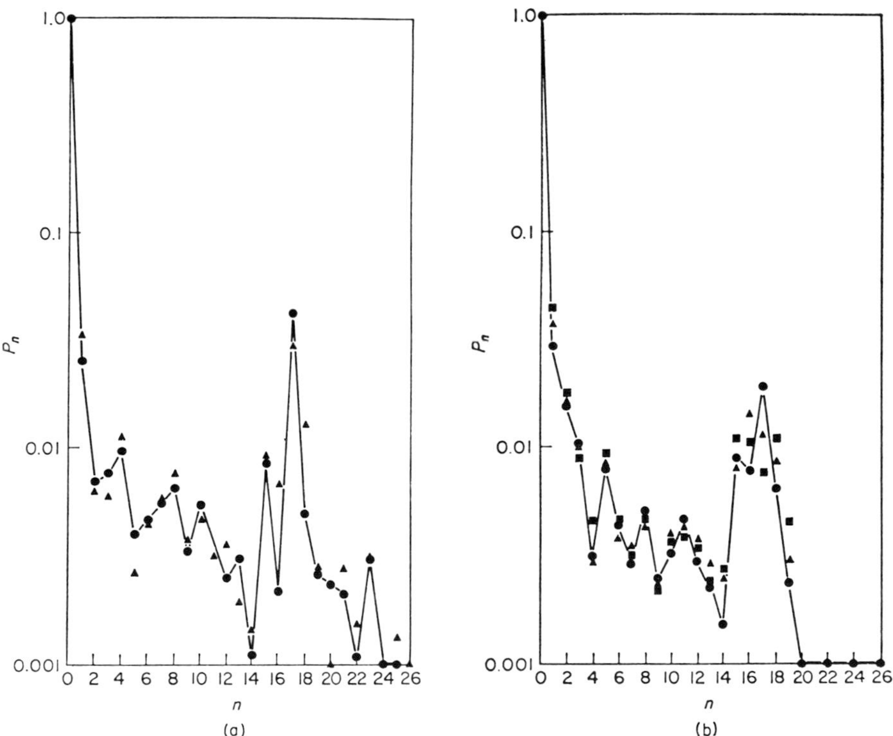

Fig. 7 The rotational power spectra from two images of the disk aggregate of tobacco mosaic virus protein (Crowther and Amos, 1971a). (a) A well-preserved disk in which one harmonic (17-fold) is dominant. The curve (—●—) refers to the origin which maximizes the 17-fold component. The points (▲) refer to an origin which maximizes the 16- and 18-fold components. (b) A poorly preserved disk in which no rotational harmonic is dominant. The points (▲), (●), and (■) refer to origins which maximize the 16-, 17- and 18-fold components, respectively.

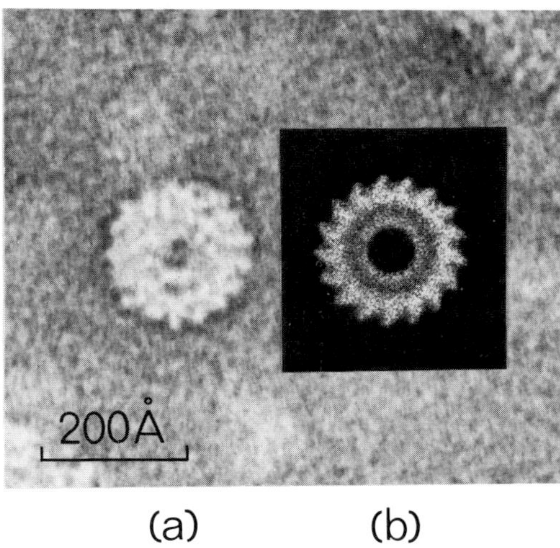

Fig. 8 (a) Image of a disk of the protein of tobacco mosaic virus. The rotational power spectrum shows a dominant seventeenfold component. (b) The seventeenfold rotationally filtered image obtained from (a).

superposition method (Markham *et al.*, 1963) is the possibility of numerically assessing the dominance of one particular symmetry on the basis of the power spectrum rather than choosing the most striking rotationally averaged image. The rotational power spectra from two images of the disk aggregate of tobacco mosaic virus protein (see Section V,C,*1*,*a*) are shown in Fig. 7. The spectrum in Fig. 7a shows a clear maximum for seventeenfold rotational symmetry, and the corresponding rotationally filtered image is shown in Fig. 8. The spectrum in Fig. 7b does not have such a dominant maximum; for different choices of image center, the rotational harmonics from fifteen to eighteen all become present with about the same weight, making this image a poor one from which to judge the symmetry. The detail in the latter image was distorted by the presence of a close, neighboring disk.

F. Reconstruction of Particles from Images

With specimens which are periodic in one plane only and which are photographed normal to that plane, image processing is limited to two-dimensional filtering. However, if the periodicity is not confined to one plane, it is possible to extend the Fourier methods to three dimensions and recreate the structure of the specimen. Helical structures are probably the subjects most suited to this treatment, provided micrographs of reasonably straight stretches of well-ordered specimens are available (DeRosier and Klug, 1968; DeRosier and Moore, 1970), but the three-dimensional structures of several isometric viruses have also been re-created (Crowther *et al.*, 1970; Crowther and Amos, 1971b).

As in the two-dimensional case, the density within a three-dimensional object can be represented as the sum of sinusoidal Fourier density components, the amplitudes and phases of which are mapped in the three-dimensional Fourier transform of the object. An electron microscope image corresponds to a projection of the density of the object in the direction of view, and the two-dimensional Fourier transform of this projection is the central section of the three-dimensional transform perpendicular to the direction of view. With an asymmetric object, it is necessary to collect many images from different directions of view in order to build up the complete three-dimensional transform to the required resolution. With present electron microscope techniques it is not possible to do this from only one asymmetric specimen. The best one might hope to do is to collect images from several equivalent specimens and try to correlate the different directions of view. However, the presence of symmetry within the specimen gives rise to corresponding symmetry in the three-dimensional transform, enabling the transform and hence the structure of the object to be completely determined, to a given resolution, from fewer sampled sections (i.e., images) as compared to asymmetric specimens.

In the case of helical particles, the symmetry of the transform is known from the helical parameters. If these parameters are favorable (so that there are many different views of the component subunits), the transform can be determined from one image at the normal 20 Å resolution. In less favorable cases, more than one view may be necessary to resolve ambiguities in the outer parts of the transform. After the transform has been determined, the particle structure can then be synthesized from the Fourier components. In practice, cylindrical coordinates are used in the computational techniques, and the structure is built up from sets of helical density fluctuations by Fourier–Bessel inversion of the transform data.

The procedure is more complex when applied to the isometric viruses. The icosahedral symmetry of several of these viruses has been found to be preserved to a resolution of 30 to 40 Å. To determine the three-dimensional Fourier transform to this limit requires only three or four images in the case of the smaller viruses, since each image will, in general, yield sixty equivalent, symmetrically related sections of the transform. The direction of view for each image relative to the symmetry axes must first be determined; usually this can be found to within 1 or

2° by image simulation and then refined by testing the transform of the image for consistency with the symmetry. The various image transforms are then scaled to each other to build up the complete three-dimensional transform from which the particle structure can then be synthesized.

V. RESULTS

A. Subunit Arrangements in Enzymes and Enzyme Complexes

1. Glutamic Dehydrogenase

Physicochemical studies have established that bovine liver glutamic dehydrogenase has a molecular weight of 312,000 and is composed of six identical subunits, each having a molecular weight of 52,000 (Eisenberg and Tomkins, 1968; Reisler *et al.*, 1970; Cassman and Schachman, 1971). The first electron micrographs of the enzyme showed many triangular-shaped images (Horne and Greville, 1963), and on the basis of this type of image, Valentine (1968) and Eisenberg and Reisler (1970) proposed a model for the enzyme in which the subunits were arranged with 32 symmetry at the vertices of a short triangular prism. The model has been disproved by Josephs (1971), who followed changes in the images which occurred when the electron microscope grid was tilted through angles up to ±40° in fields of particles suspended in stain over the holes in a carbon substrate. As the tilt angle was increased, none of the extra detail of the triangular images predicted from the earlier model appeared. The only changes observed were those consistent with tilted views of a ring of three subunits, and it was thus concluded that these images represented half-molecules.

In addition to the triangular images, Josephs (1971) observed three other distinct types of images. The first of these (Fig. 9a) has a doughnutlike appearance with a hollow center and an overall diameter of about 80 Å. These images tend to have a polygonal outline with about five or six vertices. The second type of image (Fig. 9b) is a crosslike arrangement of globular "units" with the arms of the cross about 100 Å long. The third image, which was frequently observed (Fig. 9c), consists of two parallel layers of density, each about 50 Å wide and 80 Å long, separated by a distance of 40 to 50 Å.

Of the two possible point group symmetries for an arrangement of six identical subunits, 6 and 32, cyclic sixfold symmetry was ruled out since a ring of six subunits cannot give rise to the images shown in Figs.

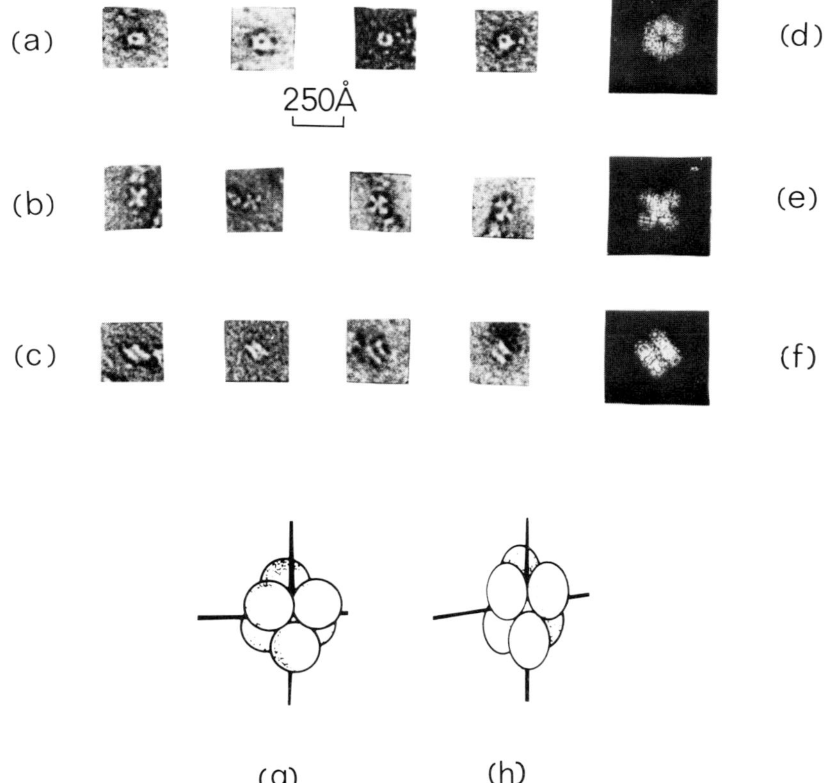

Fig. 9 Electron micrographs of glutamic dehydrogenase (Josephs, 1971). (a, b, c) The three striking types of images obtained from negatively stained molecules photographed over holes in the support film. (d, e, f) The corresponding computed superposition patterns from the model shown in (g). (g) A model of the molecule consisting of six spherical subunits arranged with 32 symmetry at the vertices of a triangular antiprism. (h) A model similar to (g) in which the subunits are elongated in the direction of the threefold axis. This model was derived from images of polymers.

9b and 9c. However, all three kinds of images can be accounted for by an arrangement with 32 symmetry, shown schematically in Fig. 9g, in which six globular subunits lie at the vertices of a triangular antiprism. Confirmation of this type of arrangement was obtained by following the changes that occurred in the images of individual molecules as the grid was tilted through known angles. The magnitude of the tilt angle required to convert one recognizable image into another and the orientations of the images relative to the tilt axis could be predicted from this model.

From the electron micrographs of single molecules it was not possible to determine more than the arrangement of the centers of the subunits, but Josephs (1971) was able to obtain some information about the molecular shape from electron micrographs of polymers of the molecules. According to physicochemical measurements (Olsen and Anfinsen, 1952), the reversible polymerization of glutamic dehydrogenase occurs linearly (Eisenberg and Tomkins, 1968; Sund *et al.*, 1969), and since the molecule has only one unique axis—the threefold axis—one would expect this to be the direction of polymerization. The distance between molecules along the polymers was found to be 100 to 120 Å, while the width of the chains was about 80 Å. The molecules are therefore elongated along the threefold axis with an axial ratio of about 1.5/1. This implies a similar elongation of the individual subunits, depicted in the model in Fig. 9h.

The tubular structures which form in solutions of glutamic dehydrogenase when the ionic strength is increased have been investigated by Josephs and Borisy (1972). These structures may well represent the storage arrangement in mitochondria where the enzyme is present in appreciable concentrations. Filamentous particles having dimensions similar to those of the tubes have been observed in mitochondria by Munn (1971). The images of the tubes were analyzed by optical diffraction and by computation and the tube symmetry was determined from images of particles tilted about their axes.

Complete tubes were found to be of three types, each formed by the association of four helically coiled linear polymers. In the first type (zero-start), the individual polymer chains are related by a fourfold axis of rotation so that the molecules of each chain are in register along the tube axis. In the second type (one-start), the four polymer chains are staggered by the length of one-quarter of a molecule and so are related by a fourfold screw axis. The third type of tube has a structure which is a compromise between those of the other two; the molecules of three of the chains are in register while those of the fourth are displaced by the length of one-half of a molecule. Various incomplete tubes were also found. These tubes followed the same structural patterns but lacked one or sometimes two of the polymer chains.

The structures of the tubes reveal a high degree of flexibility in the bonding between the enzyme molecules, not only in different chains (leading to the three different arrangements) but also within the chains. When these are in helical configurations, the angles between the threefold axes of adjacent molecules is about 19°.

2. *Catalase*

Physicochemical studies on catalase have provided strong evidence that the molecule consists of four identical subunits, each having a molecular weight of about 60,000 (Spitzberg, 1966; Sund *et al.*, 1967). The fact that the enzyme can be dissociated into half-molecules (Tanford and Lovrien, 1962; Samejima and Yang, 1963) suggests that there is more than one type of bond between subunits, thus making a cyclic arrangement of subunits unlikely. Therefore, of the two possible point group symmetries for four subunits, 4 and 222, it is likely that catalase has 222 symmetry. No detailed analysis of the electron microscope images of isolated molecules has been published, but several laboratories have investigated the various ordered arrangements in which catalase can be aggregated.

The first low resolution picture of the catalase substructure was obtained from an optical diffraction study of tubular aggregates by Kiselev *et al.* (1968). Catalase from human erythrocytes polymerizes into tubes at pH 6.5 to 7, and that from ox liver polymerizes at pH 5 to 6 (Kiselev *et al.*, 1967). In both cases, the unit of packing in the tubes is a single catalase molecule, the wall of the tube being a monomolecular layer. The tubes of erythrocyte catalase are about 420 Å in diameter, while those of ox liver catalase are narrower with a diameter of about 310 Å. Optical diffraction from images of the two types of tubes indicated that the local packing of catalase molecules is the same in both, although the helical parameters are different. An electron microscope image of a negatively stained tube of erythrocyte catalase is shown in Fig. 10, together with the filtered images of the near and far sides. The catalase molecules consist of two distinct elongated parts, separated by a low density region into which stain has penetrated. In some regions of the filtered images, each of the two halves of a molecule subdivided further into two smaller units. A model of the molecule which is consistent with the detail visible in the filtered images is shown in Fig. 11; the four subunits are arranged with 222 symmetry at the vertices of an elongated tetrahedron. The subunits have been elongated in the model to account for the fact that the two halves of the molecule display fairly constant elongation across the images despite the change in the azimuth of view around the tube.

In addition to the tubes, Kiselev *et al.* (1967) observed crystalline sheets, usually several layers thick, in their preparations of ox liver catalase. Since the molecules in successive layers were not in register below one another, the images obtained in electron micrographs were complicated superposition patterns of partially overlapping molecules and so were not easy to interpret directly in terms of molecular struc-

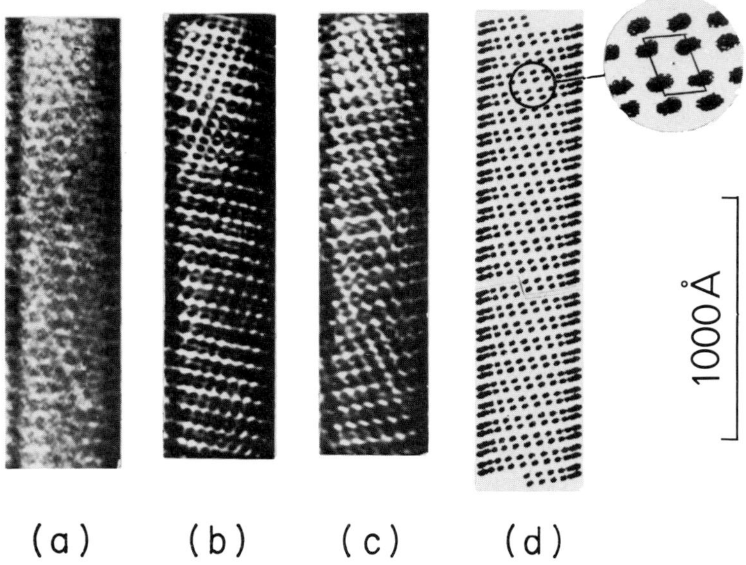

Fig. 10 Filtered images of tubes of erythrocyte catalase (Kiselev *et al.*, 1968). (a) Electron micrograph of a negatively stained tube of catalase printed so that the protein is black. (b) and (c) Optically filtered images of the near and far sides of the tube in (a). (d) Computer-generated projection of one side of a model structure with the unit cell indicated.

ture. Occasionally, however, monomolecular layers were observed. In these layers the same molecular shape could be seen as in the filtered images of tubes. The local arrangement of molecules is, however, different in the two cases since the molecules are present in two orientations in the sheets. As described earlier (Section II), these two-dimensional crystals were used as test specimens by Erickson and Klug (1971) in their investigation of the effects of defocusing and lens aberrations. The best images of individual catalase molecules, corrected for these effects, averaged over a crystal, and filtered from background noise, are shown in Fig. 1. The images are in good agreement with the model deduced from the optically filtered images of tubes (Fig. 11).

Several combined X-ray diffraction and electron microscope studies have been made on the hexagonal crystals of ox liver catalase grown in the pH range of 6 to 7.5. For example, Longley (1967) has obtained good agreement between the spacings observed in sections of fixed, dried, and embedded crystals and the unit cell of wet crystals measured by X-ray diffraction. The approximate packing of the molecules in the

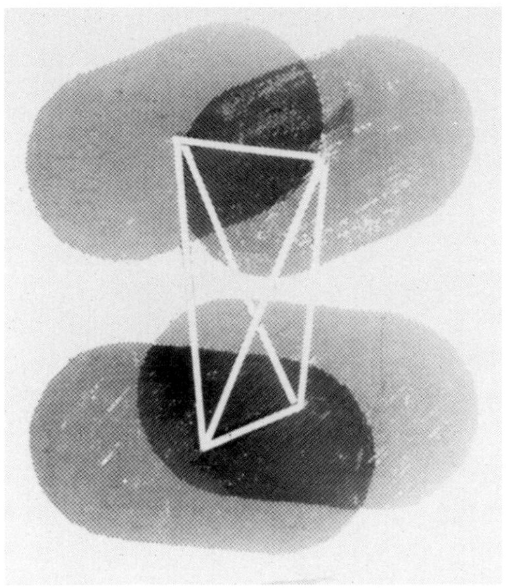

Fig. 11 A model of the catalase molecule deduced from the filtered images of tubes (Kiselev *et al.*, 1968). The four elongated globular subunits are arranged with 222 symmetry.

crystal was deduced from the electron micrographs but the order in the sections was insufficient to give information on the substructure of the molecules. More detail was obtained from electron micrographs of negatively stained thin crystals (Vainshtein *et al.*, 1968; Barynin and Vainshtein, 1972). These images were used to assign phases to the low angle X-ray reflections, enabling a Fourier map of the molecule to be calculated to a resolution of about 30 Å (Gurskaya *et al.*, 1972). The Fourier map shows a tetrahedral arrangement of subunits similar to that shown in Fig. 11. Similar maps were constructed from the electron micrographs alone, using images corresponding to different views of the crystals.

3. *Glutamate Decarboxylase*

While the physicochemical studies on this enzyme showed that it is composed of identical subunits, each having a molecular weight of 50,000, the number of subunits per molecule could be placed only between the limits of six and eight (Strausbauch and Fischer, 1970). However, from electron micrographs, To (1971) has deduced that glutamate decarboxylase from *E. coli* contains six subunits arranged with 32 symmetry at the vertices of an octahedron. Arrangements of seven and eight subunits were ruled out since no images with sevenfold, eightfold, or fourfold axes (the latter from a possible 42 arrangement) were observed. Since the dominant types of images are very similar to those of glutamic dehydrogenase, the proposed arrangement of subunits is also very similar, except that there are no indications that the decarboxylase molecule is elongated. The images of particles seen in the direction of a threefold axis do not show the sixfold symmetry one would expect from a projection of the model. Instead, the six subunits alternate in contrast. To (1971) explained this on the basis of the nonuniform distribution of stain around the molecule; because of the attraction of stain toward the supporting carbon film, the three subunits in contact with the film are more heavily stained than the other three subunits. The corresponding images of glutamic dehydrogenase given in Fig. 9 do not show this effect, since the molecules were photographed over holes in the substrate and the stain distribution is thus more uniform.

4. *Muscle Phosphorylase b*

In solution at neutral pH, phosphorylase *b* exists as a dimer having a molecular weight of 185,000 (Seery *et al.*, 1967). The dimer can be dissociated into monomers which are similar in size and properties (Madsen and Cori, 1956; Fischer and Krebs, 1966) and are presumably identical. The tetrameric form of phosphorylase *b* occurs by the association of dimers (for example, in the presence of protamine; Madsen and Cori, 1954) and thus is likely to have 222 symmetry.

Kiselev *et al.* (1971) have carried out an electron microscope study of the tetramers formed in the presence of protamine and the regular arrays in which they tend to aggregate at low temperature as the protein concentration is raised. When the specimens were prepared under conditions of minimal association, two distinctive images were observed. The first of these (Fig. 12a) has the appearance of a semicircle on one side and a cross or "V" on the other. The length of these images is about 120 Å. The second type of image (Fig. 12b) has the appearance of a figure 8, a length of about 120 Å, and a width of about 80 Å. As the conditions were changed to favor aggregation, various types of regular aggregates were formed, ranging from planar sheets and single- and double-walled tubes to three-dimensional crystals. Optically filtered images obtained from planar arrays showed the same semicircle and cross pattern recognized in the images of single tetramers (Fig. 12c). The filtered images from tubes had a more rectangular appearance (Fig. 12d). All of these images can be accounted for by the model shown in

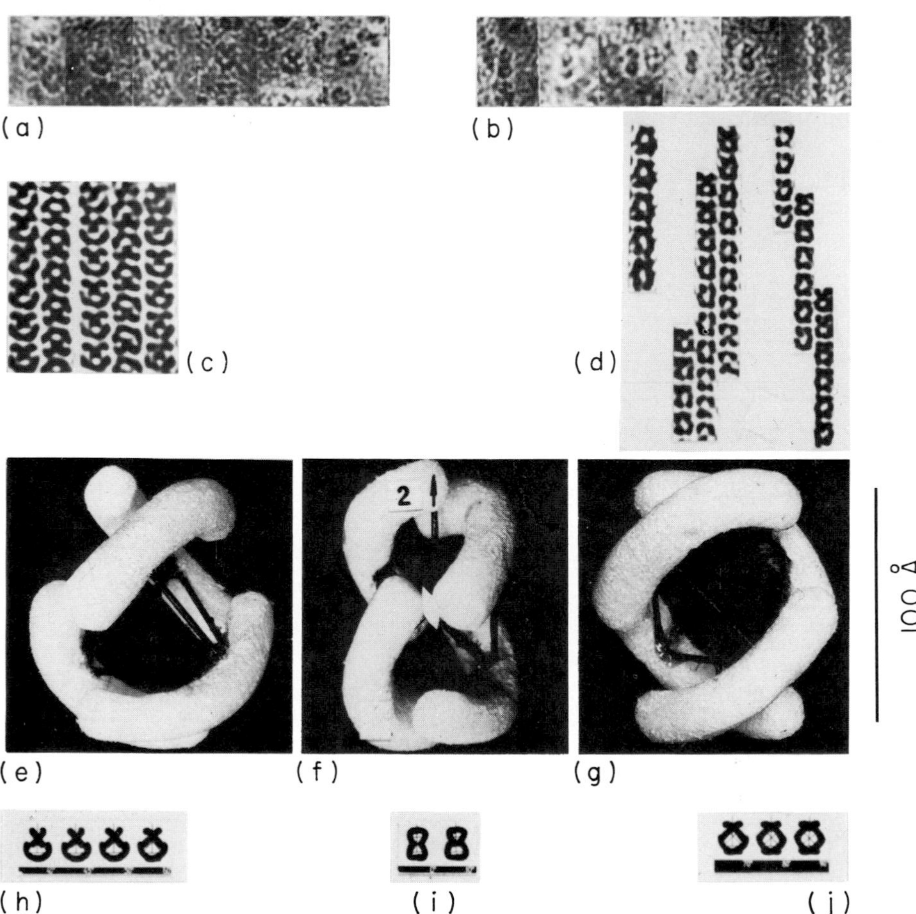

Fig. 12 Electron micrographs of muscle phosphorylase *b* (Kiselev *et al.*, 1971). (a) Images of isolated molecules having the appearance of a semicircle and cross. (b) Images of isolated molecules having the appearance of a figure 8. (c) Optically filtered image of a thin crystal (see Fig. 6) in which the molecules have a similar appearance to those in (a). (d) Optically filtered images of tubular aggregates. (e, f, g) Photographs and (h, i, j) corresponding projections of the proposed model of the structure which account for the images in (a), (b), and (d), respectively. The model has four banana-shaped subunits arranged with 222 symmetry.

Fig. 12 (e, f, g) which consists of four banana-shaped units arranged with 222 symmetry. There is good correspondence between the shadows of particular facets of this model and the various types of images, as can be seen in Fig. 12 (h, i, j).

In earlier electron micrographs of crystals of phosphorylase *b*, published by Valentine and Chignell (1968), the molecular shape appeared to be rhombic. Since this form is not consistent with 222 symmetry, it was suggested that the tetramer contained two different types of subunits. However, in a more recent study by Eagles and Johnson (1972), the filtered images were rectangular in shape, like those of Kiselev *et al.* (1971), and are thus consistent with 222 symmetry. The rhombic shape observed in the earlier work may have arisen from the incomplete staining of thicker crystals.

5. *Glucose-6-phosphate Dehydrogenase*

Physicochemical studies have shown that in the active state, human erythrocyte glucose-6-phosphate dehydrogenase is a mixture of dimers and tetramers derived from an inactive monomer having a molecular weight of 51,000 (Cohen and Rosemeyer, 1969; Bonsignore *et al.*, 1971). This scheme of association suggests that the tetramer has 222 symmetry and this was confirmed in an electron microscope study by Wrigley *et al.* (1972). When specimens were prepared under conditions favoring the formation of tetramers, the images of the enzyme showed that the monomers were arranged approximately at the vertices of a regular tetrahedron. Measurements of these images and those of the dimers suggested that the monomers in these aggregates are elongated with an axial ratio of about 1.4/1. A greater axial ratio (2.0/1), derived from measurements of images of monomers, was interpreted as indicating a change of conformation of the monomer upon association. It may be, however, that monomers are less stable under the conditions of microscopy. The apparent difference in shape could also be due to the difficulty of defining a molecular boundary when measuring images of such small molecules in negative stain.

6. *Pyruvate Carboxylase*

Pyruvate carboxylase isolated from chicken liver mitochondria has a molecular weight of 660,000 and contains bound biotin and manganese, each in the ratio of 4 moles/mole of enzyme (Scrutton and Utter, 1965). The enzyme dissociates into four apparently identical subunits upon cooling or treatment with maleic anhydride. Since there is no indication of dissociation into an intermediate dimer, one would expect the molecule to have cyclic 4 symmetry. This symmetry was confirmed by the electron micrographs of Valentine *et al.* (1966). The images almost exclusively show four globular subunits in a square arrangement, the distance between subunits being about 72 Å. Similar images have been obtained with enzymes from turkey liver and bovine liver (Valentine, 1969).

In the case of pyruvate carboxylase from baker's yeast, however, the four globular units, seen in the electron microscope images are at the vertices of a rhombus having sides of 72 Å (Valentine, 1969). The dissociation of the yeast enzyme with maleic anhydride is also peculiar because it proceeds through a dimer stage (Young *et al.*, 1971). The rhombic images are not, however, consistent with an arrangement having 222 symmetry which would be expected from a simple tetramer–dimer–monomer dissociation scheme. A likely explanation is that the yeast enzyme is built from two types of subunits and has only twofold symmetry. There are indications in the electron micrographs that the subunits are of two different sizes.

7. *Glutamine Synthetase*

Glutamine synthetase from *E. coli* has a molecular weight of 592,000 (Shapiro and Ginsburg, 1968) and is composed of twelve apparently identical subunits (Woolfolk *et al.*, 1966; Shapiro and Stadtman, 1967). As normally isolated, the enzyme is in a fairly stable ("taut") state. It becomes less stable ("relaxed") following the removal of divalent cations, but can be again stabilized ("tightened") when divalent cations are replaced. In an electron microscope study by Valentine *et al.* (1968), the arrangement of subunits was determined and attempts were made to detect differences in the structure corresponding to the taut, relaxed, and tightened forms of the enzyme. No differences were in fact found except for the tendency of the tightened form to polymerize. The ability of the enzyme to tolerate the conditions of negative staining was markedly improved by fixing it (in 0.5% glutaraldehyde, 0.1 *M* potassium phosphate buffer, pH 7.3 for 20 min) before applying the stain.

Three characteristic images were observed in the electron micrographs. The first has the appearance of a ring of six globular "units," consistent with the interpretation that the twelve subunits are arranged with 62 symmetry at the vertices of a hexagonal prism. The other images correspond to views of this arrangement in directions parallel and perpendicular to the edges of the hexagon. The distance between the centers of subunits both within a hexagon and between hexagons is about 45 Å.

The main difference between the taut and tightened states of the enzyme lies in the relative insolubility of the latter. Electron micrographs indicated that this was due to the polymerization of the molecules in the direction of their sixfold axes and side-to-side aggregation of the resulting polymers.

8. *β-Galactosidase*

β-Galactosidase from *E. coli* has a molecular weight close to 500,000 (Sund and Weber, 1963) and can be dissociated into subunits having a molecular weight of about 130,000 (Weber *et al.*, 1963; Cohn, 1957). The tetrameric nature of the enzyme indicated by these values has been confirmed by hybridization experiments (Zipser, 1963).

In an electron microscope study by Karlsson *et al.* (1964), the most uniform and detailed images were obtained by fixing the specimen with formaldehyde and negatively staining it with uranyl acetate. The dominant image shows four units arranged at the vertices of a square. This arrangement and the lack of any discrete dimer stage in the dissociation of the tetramer suggest that the tetramer has cyclic 4 symmetry. Other images having the appearance of a "notched ellipse" were interpreted as views of the tetramer perpendicular to the symmetry axis, and their asymmetry confirmed the polar nature of the tetramer. The polarity is preserved in stacks of the tetramers along their axes, the notches all pointing in one direction.

However, in addition to polymers of tetramers, sedimentation patterns of the enzyme also reveal aggregates corresponding to six and twelve monomers (Marchesi *et al.*, 1969). Such arrangements are difficult to account for on the basis of the structural scheme described above.

9. *Aspartate Transcarbamylase*

Aspartate transcarbamylase (ATCase) from *E. coli* has a molecular weight of about 300,000. The two functions of the enzyme can be attributed to the two dissociation products known as the catalytic and regulatory subunits (Gerhart and Holoubek, 1967). The catalytic subunit is a trimer of polypeptide chains, each having a molecular weight of 33,500, and the regulatory subunit is a dimer of polypeptide chains, each having a molecular weight of 17,000 (Weber, 1968; Rosenbusch and Weber, 1971; Meighen *et al.*, 1970). From the above values it was concluded that the enzyme contains six of both types of polypeptide chains, and this was confirmed by X-ray diffraction. In different crystal forms the ATCase molecules were located on twofold and threefold axes of rotation and hence had to possess both twofold and threefold symmetry (Wiley and Lipscomb, 1968). The existence of catalytic trimers and regulatory dimers is a strong indication that the point group symmetry of ATCase is 32. The electron micrographs of Richards and Williams (1972) have confirmed this choice of symmetry, and the model in Fig. 13 (Cohlberg *et al.*, 1972) shows the arrangement of subunits derived from these images.

An interesting and useful feature in the work of Richards and Williams (1972) was their ability to obtain different views of the molecules of ATCase by applying the specimen and stain in different ways to the

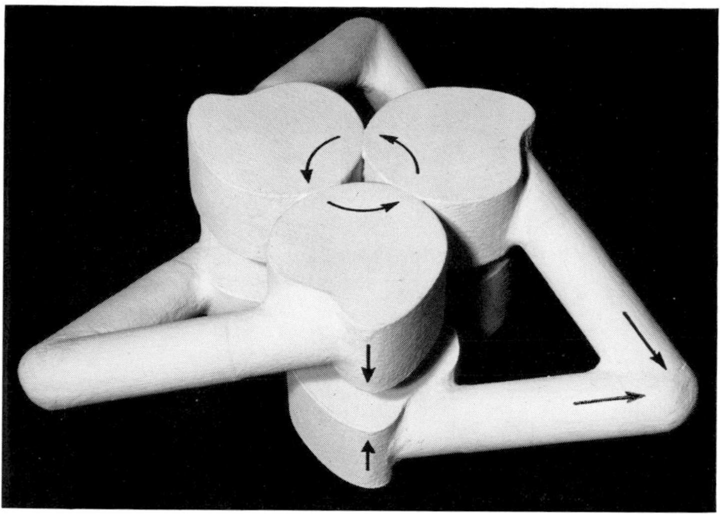

Fig. 13 A model of the structure of aspartate transcarbamylase derived from electron microscopy and consistent with the physicochemical data. The six large central units represent the catalytic polypeptide chains to which the V-shaped regulatory dimers are connected. Reprinted with permission from Cohlberg, J. A., Pigiet, U. P., and Schachman, H. K. (1972). *Biochemistry* **11**, 3396. Copyright by the American Chemical Society.

electron microscope grid. When the stain was applied first to a hydrophobic substrate, allowed to dry, and then sprayed with the enzyme solution, the triangular-shaped images shown in Fig. 14a,b were obtained. These images are bounded by a triangle of fine lines 145 Å long with usually prominent vertices. Almost enclosed by this triangle is another more substantial triangle, rotated by 60° with respect to the first, which often appears to be formed by three large "units." However, when the enzyme and stain are mixed and sprayed together onto the substrate, the dominant image consists of two identical lune-shaped parts with their slightly concave sides innermost and separated by a distance of 20 to 40 Å (Figs. 14c). The length of the lunes is 90 Å. Fine material can be seen beyond this distance, sometimes connecting the lunes.

Electron microscope images of the negatively stained catalytic trimer have the appearance of an equilateral triangle similar to the inner part of the triangular images of the complete molecule. The two catalytic trimers in ATCase must therefore lie over each other in fairly close register in the direction of the threefold axis, i.e., the six catalytic polypeptide chains lie at the vertices of a triangular prism. The double-line image corresponds to a view of this arrangement in a direction perpen-

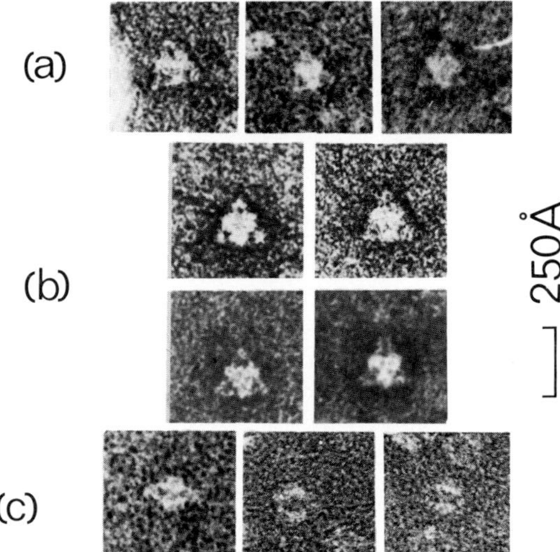

Fig. 14 Electron micrographs of aspartate transcarbamylase (Richards and Williams, 1972). The particles were negatively stained with potassium phosphotungstate and electron micrographs were taken by the minimum exposure technique of Williams and Fisher (1970). (a) Images of particles seen in the direction of their threefold axes, showing a faint triangular outline enclosing a more solid triangular center. (b) Averaged images obtained by superposing five images of the type shown in (a). These were selected from the same electron micrograph. (c) Images of particles seen perpendicular to their threefold axes, showing two fairly dense "units" connected by faint outer material. Reprinted with permission from Richards, K. E., and Williams, R. C. (1972). *Biochemistry* **11,** 3393. Copyright by the American Chemical Society.

dicular to the threefold axis. The fainter detail forming the outer triangle of the ATCase images is presumed to be due to the regulatory dimers. The model shown in Fig. 13 thus represents the most likely arrangement of subunits consistent with the physicochemical data and the electron microscope images.

Very similar results have been obtained from the recent X-ray diffraction investigation by Wiley *et al.* (1971). In the crystal form studied, the ATCase molecules occupy positions of 32 symmetry, confirming that this is indeed the point group symmetry. An electron density map, calculated to a resolution of 5.5 Å, indicates a subunit arrangement which is in close agreement with the electron microscope results.

10. Multienzyme α-Ketoacid Dehydrogenase Complexes

A typical member of this family of large enzyme complexes is the α-ketodehydrogenase complex (KGDC) from *E. coli* (Fig. 15a). It con-

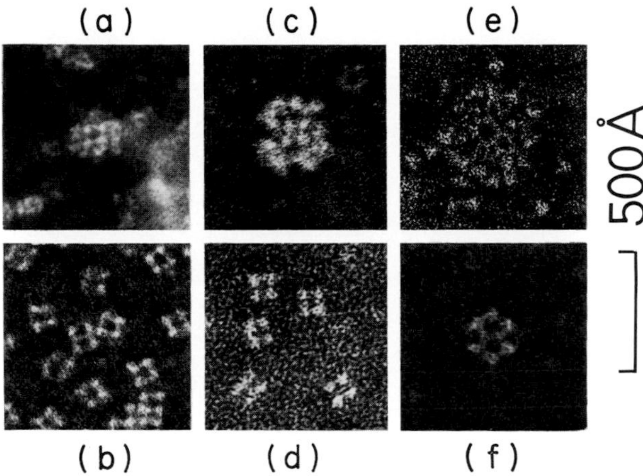

Fig. 15 Electron micrographs of multienzyme complexes. (a) The α-ketoglutarate dehydrogenase complex (KGDC) from *E. coli.* (b) The core of KGDC, lipoyl transsuccinylase (LTS). (c) The pyruvate dehydrogenase complex (PDC) from *E. coli* and (d) its core, lipoyl transacetylase (LTA). (e) PDC from beef kidney and (f) its LTA core. Taken from DeRosier and Oliver (1971).

tains three enzymes, lipoyl transuccinylase (LTS), dihydrolipoyl dehydrogenase, and α-ketoglutarate dehydrogenase.

The LTS component forms the core of the complex and can be isolated from it as a single molecule. It is made up of 24 similar if not identical polypeptide chains (Pettit *et al.*, 1973). Electron micrographs of LTS (Fig. 15b) show square and H-shaped images characteristic of views in the directions of the fourfold and twofold axes of a cube. It was concluded from these images that the twenty-four subunits are arranged with octahedral (432) symmetry and this was confirmed by X-ray diffraction of LTS crystals. The space group is F432, with one LTS molecule per lattice point (DeRosier *et al.*, 1971).

Electron micrographs have also been obtained from small negatively stained LTS crystals. The optical diffraction patterns from these micrographs show intensity distributions similar to those in the corresponding X-ray diagrams. A three-dimensional map of LTS has been reconstructed from these electron micrographs (DeRosier and Oliver, 1971). Because of the high symmetry of the crystals, the computed Fourier transforms from the images of crystals seen in the direction of twofold and fourfold axes yield all of the structure factors to a resolution of 40 Å. A model derived from the density map is shown in Fig. 16a. The resolution is insufficient to reveal the individual subunits, but their approximate arrangement is indicated in the model in Fig. 16b. This model has been

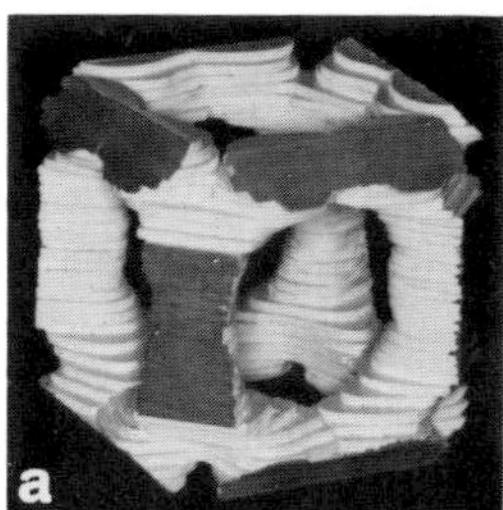

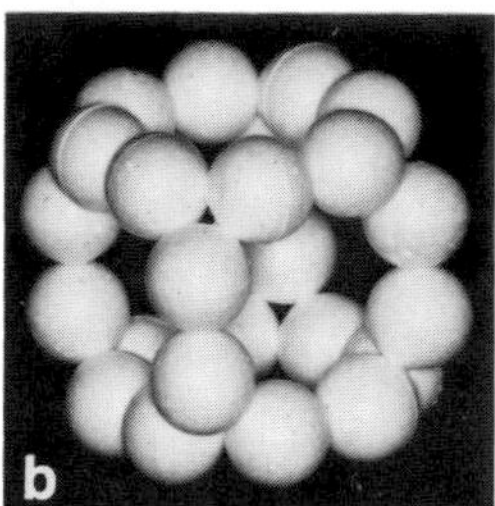

Fig. 16 (a) Model of a particle of LTS reconstructed from electron micrographs of thin crystals (DeRosier and Oliver, 1971). The flat gray surfaces correspond to planes cutting midway between neighboring particles in the crystal and define the volume of one molecule. (b) A model having essentially the same density distribution as (a), built from twenty-four spherical subunits at the vertices of a truncated cube.

used as a basis for allotting signs to the reflections in the X-ray diagrams out to a spacing of 18 Å. The refined structure produced in this way shows that the subunits are clustered in trimers at an inner radius on the cube diagonals.

The LTS core accepts only twelve copies each of the other two proteins in KGDC (Pettit *et al.*, 1973) so that the overall symmetry is probably 23.

The pyruvate dehydrogenase complex (PDC) from *E. coli* (Fig. 15c) has a structure similar to that of KGDC. It has a cube-shaped core of lipoyl transacetylase (LTA) to which molecules of pyruvate dehydrogenase and dihydrolipoyl dehydrogenase are attached. However, in the pyruvate dehydrogenase complex from beef kidney the LTA core has the shape of a pentagonal dodecahedron (Fig. 15f). In this case the overall complex has a structure that is an icosahedral version of that of the complex from *E. coli.*

11. Transcarboxylase

Transcarboxylase, isolated from propionic acid bacteria, is a biotin-containing enzyme having a molecular weight of about 790,000 and a sedimentation constant of 18 S. A study of the enzyme and its dissociation products has been made by Green *et al.* (1972) using electron microscopy, sedimentation equilibrium, and gel electrophoresis.

The interpretation of the electron microscope images of the enzyme was complicated by the presence of partially dissociated forms and various breakdown products. The most common particles appear to have a dense body about 100 Å in diameter to which smaller subunits are attached on one side, rather like feathers on a shuttlecock (Fig. 17a). On the basis of size, the dense unit was identified with a 12 S dissociation

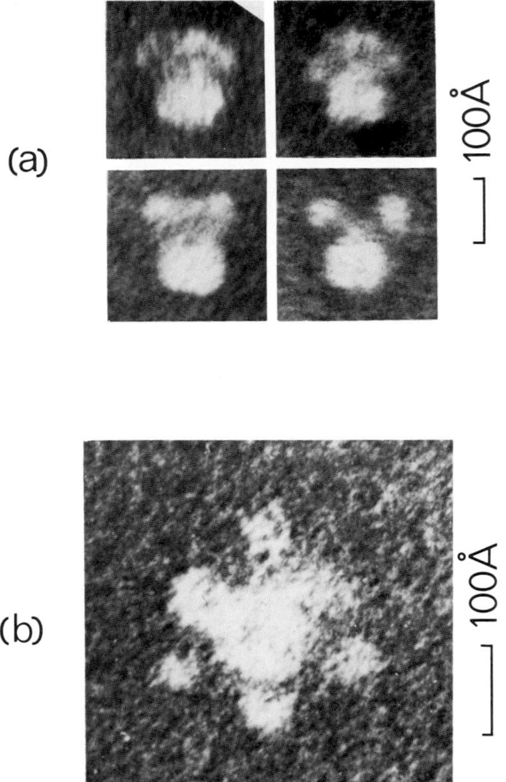

Fig. 17 Electron micrographs of transcarboxylase (Green *et al.*, 1972). (a) Images of particles showing a large dense body to which smaller units are attached on one side. (b) Image of a particle to which avidin is attached, showing threefold symmetry. Three of the outer units are believed to be avidin and the other three the smaller units visible in (a).

product and the "feathers" with a 6 S fragment containing biotin. The symmetry was not evident from the images because the enzyme molecules did not sit on-end on the specimen grid. They could be induced to do so, however, by adding the biotin-binding protein, avidin, to the preparation prior to microscopy. The end-on views then obtained showed threefold symmetry: six morphological units were attached to the central 12 S core, alternating in density and radius (Fig. 17b). Three of these units were identified with the 6 S subunits and three with the attached avidin.

The polarity of attachment of the 6 S subunits to the 12 S core limits the enzyme to cyclic threefold symmetry. There are, however, indications that the 12 S core has some remnants of 32 symmetry. Electron microscope images of the core show a pronounced equatorial groove, and the three subunits into which it dissociates under mild conditions themselves dissociate into pairs of similar, if not identical, subunits in urea or sodium dodecyl sulfate. Under certain conditions reaggregation led to the formation of nonpolar molecules in which three of the 6 S subunits are attached to each side of the core. The normal enzyme thus appears to be in a symmetrically degenerate form.

12. *Other Enzymes*

Aldolase has been studied by Penhoet *et al.* (1967). The electron microscope images showed the appearance of four subunits, consistent with the results of hybridization experiments, but the subunit arrangement was not determined. The X-ray diffraction study by Eagles *et al.* (1969) has since indicated that the subunits are arranged at the vertices of a tetrahedron, with 222 symmetry.

Tryptophanase images show four subunits at the vertices of a square (Morino and Snell, 1967a). The dissociation scheme, tetramer → dimer → monomer, suggests that the tetramer has 222 symmetry, although the monomers themselves are built from two identical or nearly identical halves (Morino and Snell, 1967b).

B. Muscle Proteins

Electron microscopy of sections of striated muscle fibers (so called because of their appearance in the light microscope) has shown them to consist of alternating bands containing arrays of thick and thin filaments with their axes parallel to the length of the fiber. These arrays interpenetrate each other to an extent which depends upon the state of contraction of the muscle, and cross-bridges can be seen which link the two types of filaments. The current theory of muscle contraction is that it is the swinging of these cross-bridges, induced by an increase in the concentration of calcium ions, that forces the arrays of filaments to slide past each other and thus contracts the fiber.

Besides its use in investigating the structure of intact muscle fibers, electron microscopy has been widely used in studying the constituent muscle proteins and their interactions. The technique is particularly valuable since these proteins are not easy to study individually by X-ray diffraction.

The two main structural proteins in striated muscle are myosin (MW about 500,000) and actin (MW 45,000), which are associated with the thick and thin filaments, respectively. Tropomyosin is probably present in both structural and regulatory capacities, in the latter cases in conjunction with another protein, troponin. Paramyosin is one of the major components of unstriated muscle. The presence of other proteins has also been demonstrated.

1. Myosin

Electron micrographs of shadowed preparations of individual myosin molecules (Fig. 18) show that they are rod-shaped structures about 1700 Å long and 20 Å wide with a thickened "head" about 200 Å long and 40 Å wide at one end (Rice, 1961, 1964; Huxley, 1963). By rotary shadowing (i.e., from no fixed direction but at one fixed angle) Slayter and Lowey (1967) have demonstrated that the head consists of two equally sized lobes.

By lowering the ionic strength of purified solutions of myosin, the molecules can be made to form large rod-shaped aggregates which appear to be very similar to the thick filaments of muscle fibers (Fig. 19). Both show projections all the way along their length except for a short, bare region about 2000 Å long in the center. Huxley (1963) has suggested that this appearance results from the stacking of myosin molecules with their bulbous heads (which appear as the protuberances) pointing away from the center, leaving the bare zone consisting only of the thin tails. The myosin molecules at the two ends of the thick filaments are thus arranged with opposite polarity. It follows from this interpretation that

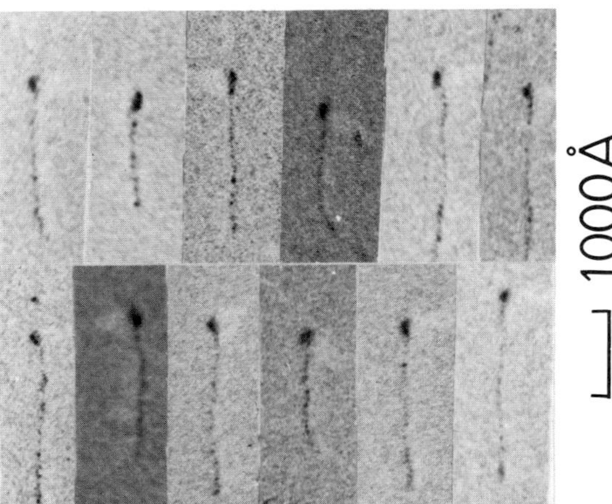

Fig. 18 Images of shadowed myosin molecules showing the "head and tail" structure. Taken from Huxley (1963).

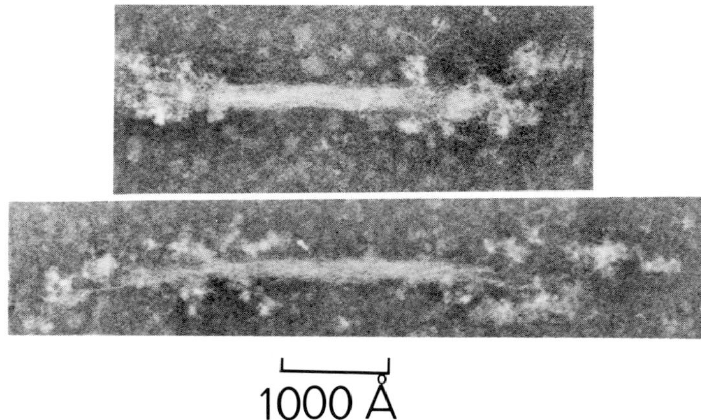

Fig. 19 Images of negatively stained filaments of reaggregated myosin showing the characteristic bare central shaft and irregular projections along the rest of the filament. Taken from Huxley (1963).

the protuberances, which form cross-bridges linking the thin and thick filaments in the muscle fibers, are the myosin heads.

The myosin molecule can be cleaved approximately halfway along its length by brief digestion with trypsin which yields two fragments: heavy meromyosin (HMM, including the double head) and light meromyosin (LMM, the remainder of the tail). Heavy meromyosin can be further cleaved by papain to yield the so-called S_1 and S_2 subunits, the single head lobes and the tail sections of HMM, respectively. When the ionic strength of a solution of LMM is lowered, large, smooth, needlelike aggregates are formed. If this is done slowly, by dialysis, a 430 Å periodicity becomes apparent in electron micrographs after negative staining (Huxley, 1963). Open lattice structures rather like those of tropomyosin (see below) are also observed under these conditions.

2. Actin

In the electron microscope the appearance of filaments of purified actin is very similar to that of thin filaments isolated from muscle fibers (Hanson and Lowy, 1963; Huxley, 1963). Both have a diameter of about 70 Å and consist of beadlike subunits (G-actin monomers) arranged on a two-start helix (Fig. 20). Each helix having a pitch of about 700 Å contains thirteen subunits per turn and the subunits are arranged, axially, midway between those on the other helix. The structure is repeated after 350 Å.

Both types of filaments bind the S_1 subunits of myosin, and the result-

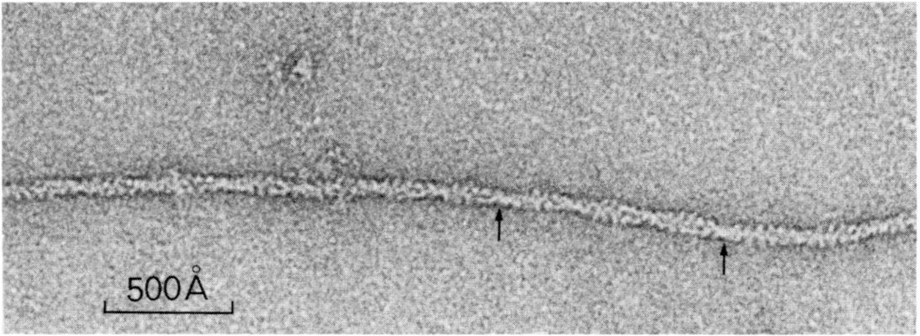

Fig. 20 Image of a negatively stained filament of purified actin showing beadlike subunits arranged on a two-start helical family. Two of the cross-over points of these helices are marked by arrows. Taken from Huxley (1963).

ing "decorated" filaments have a striking polar appearance (Fig. 21), like a helical ribbon of arrowheads pointing the same way (Huxley, 1963). Incompletely disrupted bands of thin filaments still attached to their central transverse Z lines can also be decorated with S_1 subunits. In this instance the arrow pattern points away from the center of the band, i.e., the thin filaments on the two sides of the Z line have opposite polarity. In the muscle fibers, therefore, the polarity of both types of filaments is reversed at each successive overlap of thin and thick filaments. Thus all of the attractive forces between filaments will be cooperative and will tend to reduce the distance between neighboring Z lines, thus contracting the fiber.

Electron micrographs have been used to reconstruct three-dimensional models of filaments of pure actin, natural thin filaments, and thin filaments decorated with S_1 (Moore *et al.*, 1970). Images of straight and well-preserved individual filaments of pure actin were rarely found. However, in the presence of Mg^{2+}, the filaments associate regularly side by side to form paracrystals, and uniform regions of these, only one filament thick, were chosen as a basis for reconstruction. Because it was not clear where to draw the boundary to isolate a single filament from such an array, the Fourier transform of a single filament was computationally reconstructed from the transform of a uniform ribbon of four filaments. It was not necessary to apply this procedure to the extracted and decorated thin filaments because they yielded electron micrographs of well-preserved straight stretches.

A model of the reconstructed actin filament is shown in Fig. 22. The subunits measure about 55 Å axially, 35 Å radially, and 50 Å tangentially; the outermost radius is 40 Å. The striking difference between this reconstruction and the model of the reconstructed thin filament (Fig. 23) is

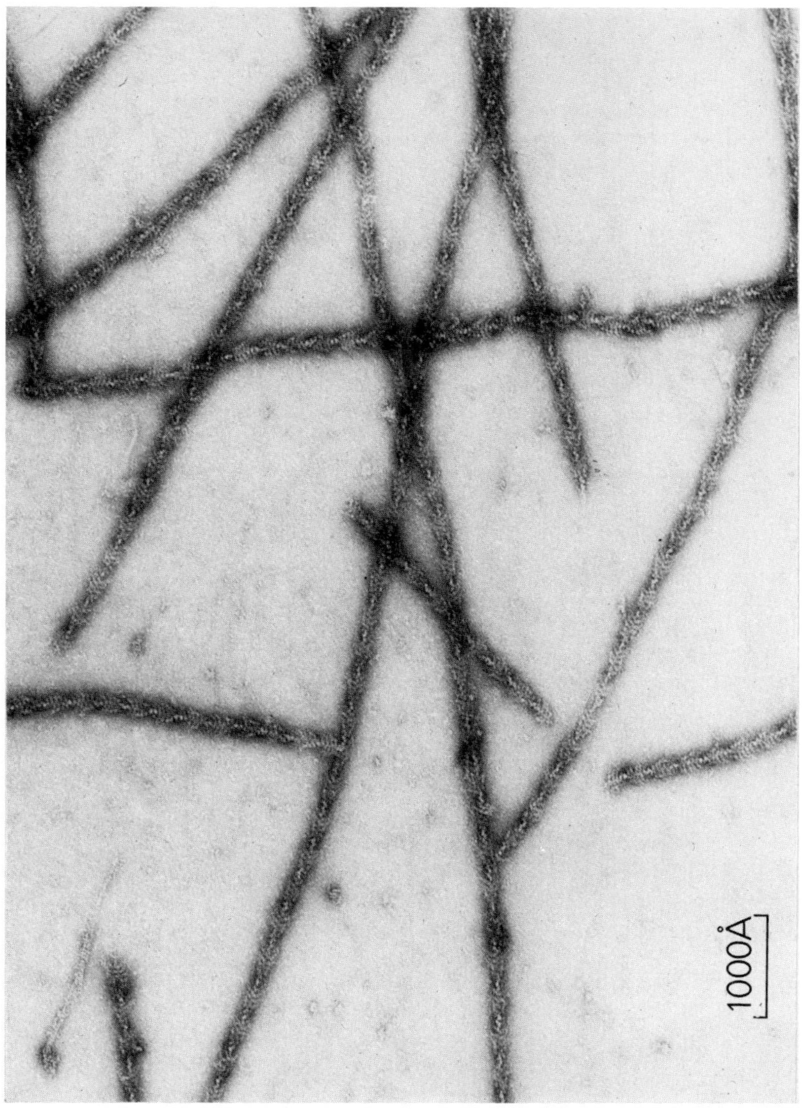

Fig. 21 Thin filaments of muscle fibers "decorated" with the S_1 subfragments of myosin showing the characteristic polar appearance. The individual subunits often have a slightly curved appearance. Taken from Huxley (1963).

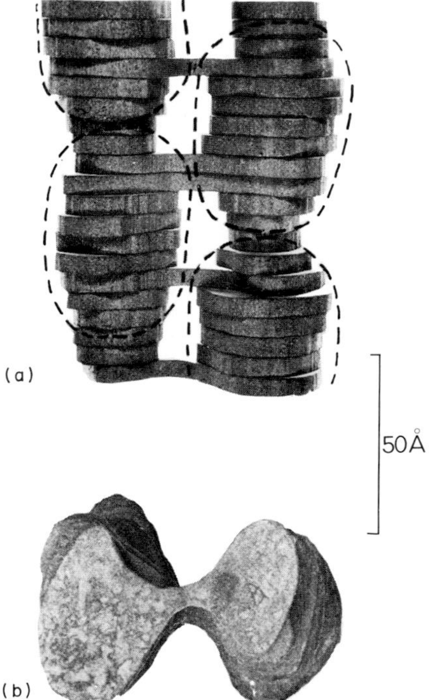

Fig. 22 Model of part of a filament of purified actin reconstructed computationally from electron microscope images (Moore *et al.*, 1970) seen (a) from the side and (b) end-on. The dashed lines in (a) are the probable boundaries of the actin monomers.

the extra density in the latter between the large masses which correspond in position and size to the subunits of actin. Moore *et al.* (1970) suggested that this extra density corresponded to the tropomyosin and troponin present in thin filaments. This interpretation was confirmed by Spudich *et al.* (1972) who reconstructed filaments of the actin–tropomyosin–troponin complex. The tropomyosin strands were found to lie off the center line in each of the long-pitch grooves of the actin 2-start helix, indicating that each tropomyosin molecule apparently contacts only one of the two actin chains.

The reconstructed decorated thin filament is shown in Fig. 24, with the probable positions of the actin and S_1 subunits marked. The S_1 subunits are curved and their long axes are tilted and skewed relative to the axis of the filament. A model indicating how this arrangement gives rise to the arrowhead appearance of decorated filaments is shown in Fig. 25.

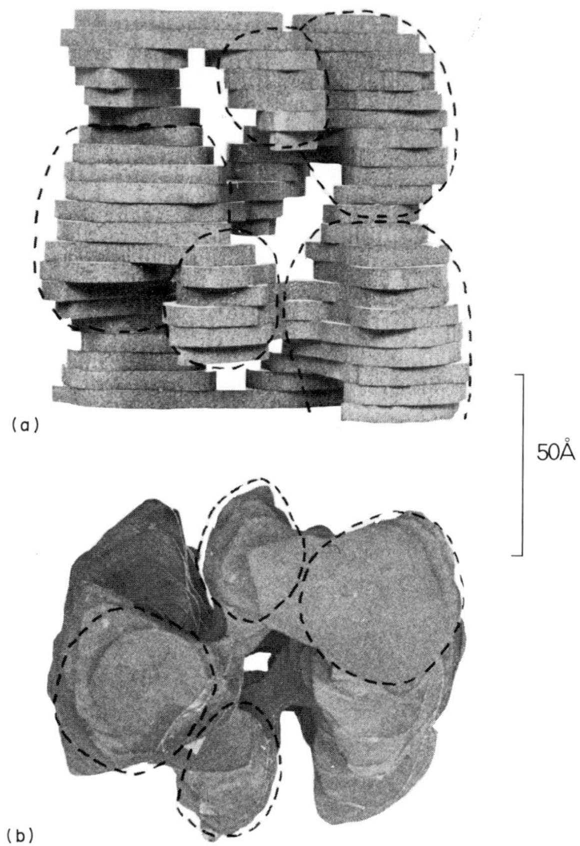

Fig. 23 Model of a reconstruction of part of a thin actin filament (Moore *et al.*, 1970) seen (a) from the side and (b) end-on. The dashed lines are the boundaries between the actin monomers and extra material not present in the reconstructed actin filament.

3. *Tropomyosin*

Tropomyosin is an α-protein having a molecular weight of about 70,000. The results of X-ray diffraction and optical rotatory dispersion studies (Astbury *et al.*, 1948; Cohen and Szent-Györgyi, 1957) indicate that the molecule is about 450 Å in length and has a two chain, α-helical, coiled-coil structure.

Precipitation of tropomyosin under various conditions yields a wide variety of polymorphic forms. In the presence of divalent cations at neutral or slightly alkaline pH, paracrystalline tactoids are formed. These

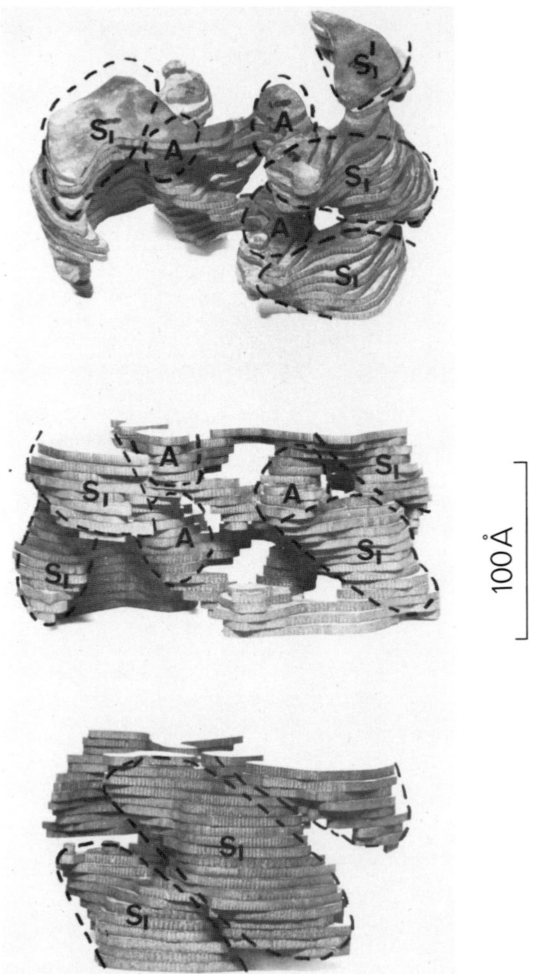

Fig. 24 Three views of a reconstructed model of part of a thin actin filament decorated with the myosin subfragment S_1 (Moore *et al.*, 1970). The dashed lines are the boundaries between the actin monomers (A) and the S_1 subunits. The latter are slightly curved and their long axes are tilted relative to the axis of the filament and also skewed. The model is inverted in the last view.

tactoids show axial periodicities close to 400 Å when negatively stained, the exact values depending upon the nature of the cations (Caspar *et al.*, 1969).

At pH values close to the isoelectric point, i.e., pH 5.1 (Caspar *et al.*, 1969) or after dialysis against ammonium sulfate (Bailey, 1948), tropo-

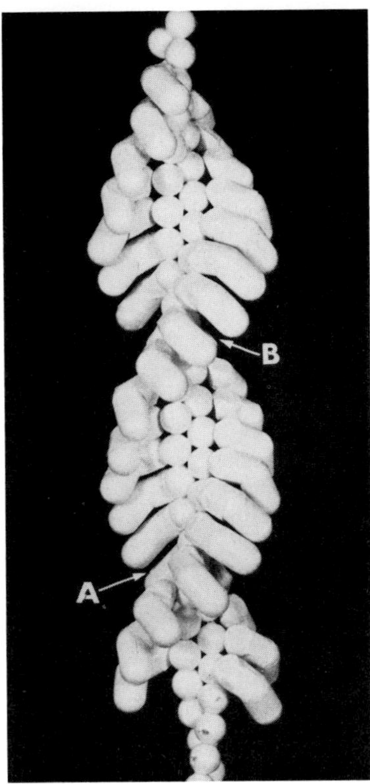

Fig. 25 Model of an actin filament decorated with the myosin subfragment S_1 (Moore *et al.*, 1970). The points A and B are successive "cross-over" points in the actin two-start helix. Between these points the projected view has the striking polar appearance evident in the electron microscope images (Fig. 21).

myosin forms true three-dimensional crystals. Sections of these crystals were examined by Hodge (1959), who found a very open network structure showing 200 Å periodicities and sometimes a 400 Å square net. Huxley (1963) examined fragmented crystals suitable for negative staining and was able to see that the network was formed by thin wavy double filaments about 30 to 40 Å wide.

A combined X-ray diffraction and electron microscope study of the tropomyosin crystal was made by Caspar *et al.* (1969). Although the crystals contain only 4.5% protein and are therefore fragile, prone to disorder, and weakly diffracting, X-ray patterns were obtained extending to spacings of 20 Å. The crystals are orthorhombic, the space group is $P2_12_12$, and the cell dimensions are a = 126 Å, b = 243 Å, and c = 295 Å. An X-ray diagram taken in the direction of the a axis is shown in Fig.

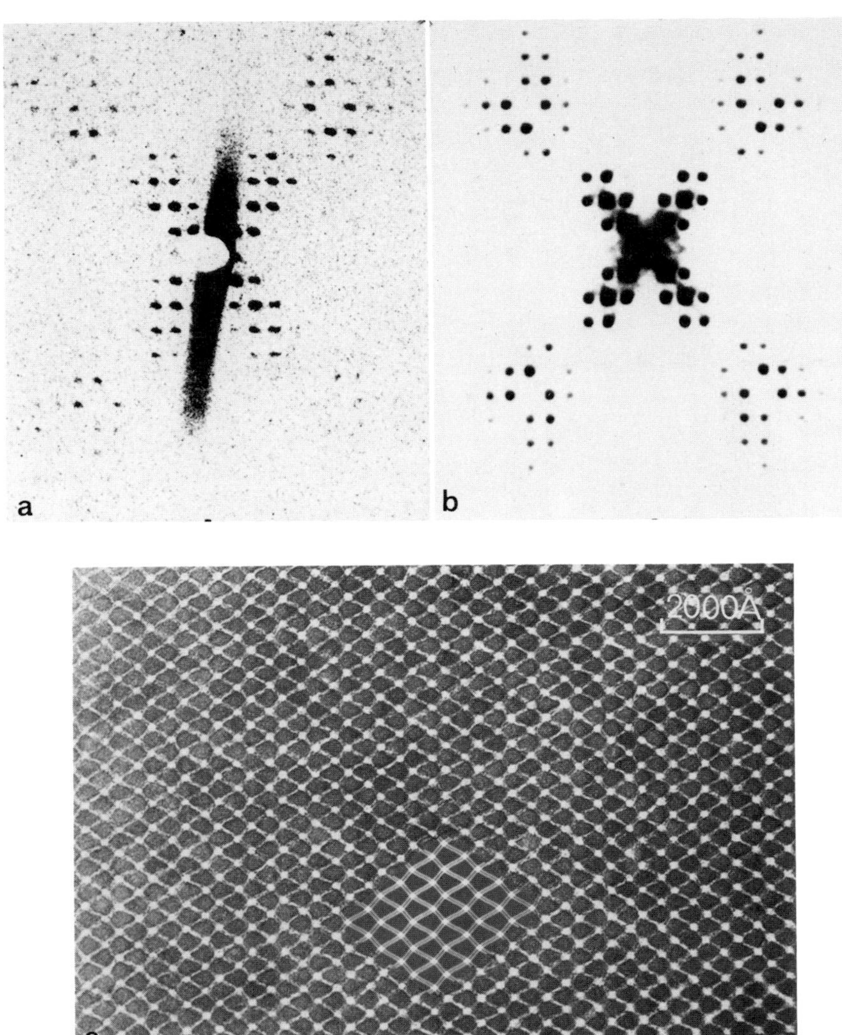

Fig. 26 Ultrastructure of tropomyosin (Caspar *et al.*, 1969). (a) X-Ray diffraction pattern along the *a* axis of a tropomyosin crystal. (b) Optical transform of a continuous filament model derived from the electron density map and shown as the insert in (c). (c) Electron micrograph of a negatively stained thin crystal of tropomyosin. The wavy filament structure was used as a basis for phasing the X-ray pattern in (a), enabling the electron density map (shown superposed) to be calculated.

26a. An electron micrograph of a negatively stained thin crystal seen in the same direction is given in Fig. 26c. The optical diffraction pattern from the micrograph did not correspond in intensity distribution to the X-ray diagram, probably because of the disorder of the filaments between the cross-over points. The micrograph did, however, suggest a trial structure consisting of a network of sinusoidally bent pairs of filaments. This model was used to phase the X-ray reflections, and the refined projection obtained in this way is shown superposed in Fig. 26c. Its optical transform (Fig. 26b) is very similar to the X-ray diagram.

All of the polymorphic forms of tropomyosin show periodicities close to 400 Å, which can be taken as the minimum length of the molecule. (In the crystal this is the distance along the filaments between lattice points.) A 400 Å periodicity has also been observed in bands of thin filaments in muscle fibers (Page and Huxley, 1963), and the presence of tropomyosin in these bands has been confirmed by fluorescent antibody staining (Pepe, 1966). Furthermore, the protein troponin, which binds specifically to tropomyosin, is distributed along the thin filaments with a 400 Å periodicity (Ohtsuki *et al.*, 1967). This periodicity has also been correlated in the X-ray diffraction pattern of muscle (Huxley and Brown, 1967) with a meridional reflection having a spacing of 385 Å. It thus seems likely that when the tropomyosin molecules are bound to thin filaments, they are associated with each other, end-to-end, in a way closely related to their association in tactoids and crystals.

The specific binding of troponin to tropomyosin has been demonstrated by Nonomura *et al.* (1968) and by Hitchcock *et al.* (1973). Electron micrographs of tactoids show bands of extra density, indicating that troponin is attached approximately one-third of the way along the tropomyosin molecule. Electron micrographs of negatively stained crystals of the tropomyosin–troponin complex have a pattern similar to that of tropomyosin (Fig. 26c), but contain additional density on the filaments halfway between the cross-over points (Cohen *et al.*, 1971). Each of these thickenings in the filaments can be identified with pairs of troponin molecules, and their location is consistent with that determined from the tactoids.

4. *Paramyosin*

Paramyosin is an α-protein in the muscles of mollusks and other groups of invertebrate animals. Especially large amounts are present in the large filaments of the unstriated molluscan "catch" muscles. Hydrodynamic and light-scattering data indicate that paramyosin has a molecular weight of approximately 220,000 and that the molecule is rod-shaped with a length of about 1350 Å (Lowey *et al.*, 1963). Optical rotatory dispersion measurements suggest that paramyosin is a fully α-helical molecule (Cohen and Szent-Györgyi, 1957, 1960) and X-ray diffraction patterns (Bear, 1944; Cohen and Szent-Györgyi, 1957, 1960) indicate that it has a two-chain, coiled-coil structure (Cohen and Holmes, 1963).

Paramyosin is soluble in neutral salts or at acid pH. Upon reducing the ionic strength or raising the pH, it precipitates in the form of fibers which do not exhibit the surface net patterns visible in electron micrographs of native filaments but do show similar axial periodicities (Hodge, 1952; Hanson *et al.*, 1957).

Cohen *et al.* (1971) found that, in order to reproduce structures similar to the large native filaments, it was necessary to disperse paramyosin thoroughly in solutions of potassium thiocyanate or urea (sufficiently concentrated to disperse but not denature the protein) and then to precipitate it by adding divalent cations. Paramyosin from a variety of sources yielded a limited number of paracrystalline forms which showed relatively simple but striking band patterns when negatively stained. All of these paracrystalline forms showed axial periodicities in the range of 725 ± 25 Å.

Electron micrographs of one of the most common forms, called P1, are given in Fig. 27a,b. Images of negatively stained P1 paracrystals contain light regions about 550 Å long, separated by shorter dark regions. This pattern was interpreted as arising from bands of molecules aligned in register into which stain has penetrated (the dark regions) except where neighboring bands overlap and exclude the stain (the light regions). This interpretation was confirmed by the appearance of a complete period without overlap at the ends of the paracrystals. Further confirmation was derived from images of positively stained P1 paracrystals. In these images the previously light and dark regions are reversed in relative contrast and the regions interpreted as overlap bind more stain than the regions of nonoverlap. Within the light regions of images of negatively stained P1 paracrystals there is an asymmetric banding pattern which indicates that the structure is polar. Assuming that the molecules are parallel to the axis of the paracrystals, their length must be given by the overlap, 550 Å, plus an integral number of axial repeat periods, $n \times 725$ Å. To be consistent with the hydrodynamic and light-scattering data, the value of n must be 1, and the molecular length 1275 Å.

Other paracrystals exhibited more complex polar banding patterns due to the staggering of P1 arrays by two-fifths of the axial period (Fig.

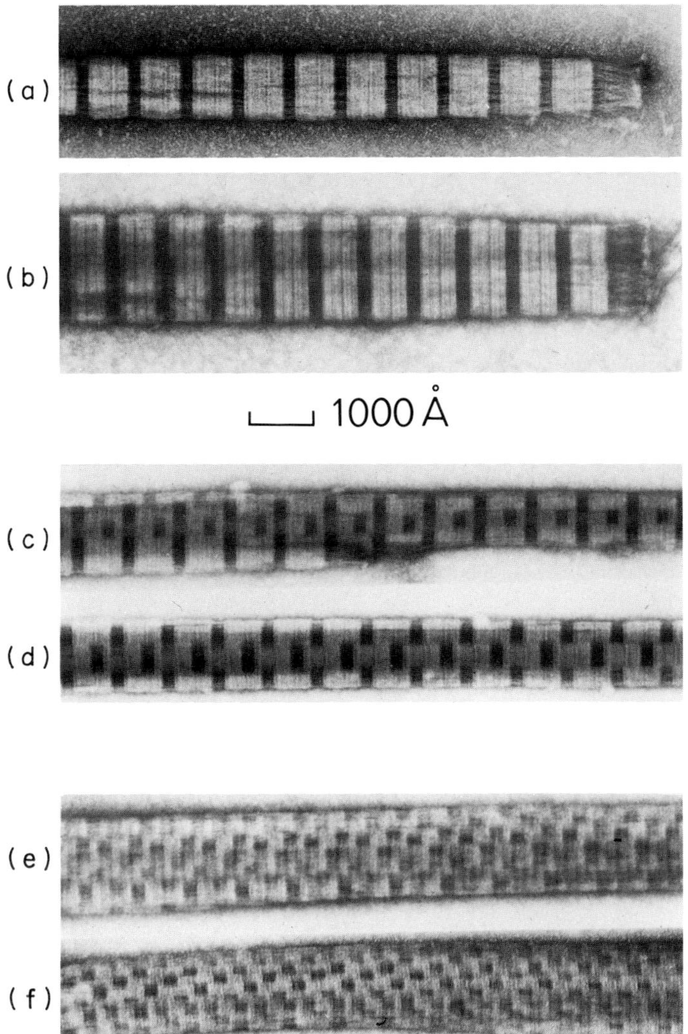

Fig. 27 Electron micrographs of negatively stained paracrystalline aggregates of paramyosin (Cohen *et al.*, 1971). (a, b) P1 forms showing a simple banding pattern of periodicity about 725 Å. The asymmetric banding pattern indicates that the structure is polar. The lengths of the end fringes on the right are approximately equal to the repeat period. (c, d) Forms arising from the staggering of P1 arrays within one paracrystal by two-fifths of the period. (e, f) Forms in which the two-fifths staggering occurs in a more regular way, leading to a surface net pattern similar to that seen in native filaments.

27c,d). This staggering can occur in a regular way to form a net pattern (Fig. 27e,f) similar to the pattern observed in native filaments from which actomyosin has been removed (Szent-Györgyi *et al.*, 1971).

Many paracrystals were observed which showed the P1 type of structure or its variation at each end, but with the polarities reversed, merging in the center to form a nonpolar pattern. This bipolarity is important since it is an essential feature of the sliding filament mechanism of muscle contraction.

C. Viruses and Virus Components

Because of their size, viruses have been convenient specimens for study since the early days of electron microscopy. Until the introduction of the negative staining technique, however, little detail was interpretable beyond size and shape, although images of shadowed specimens did reveal helical grooves in the tails of the T-even bacteriophage (Williams and Fraser, 1956) and the icosahedral shape of the tipula iridescent virus (Williams and Smith, 1958) suggested a symmetrical substructure. From images of negatively stained particles it has been possible to determine the surface structure of a number of viruses, virus components, and associated particles. General rules outlining the ways in which large biologic structures are built from a large number of identical protein molecules (see Section IV,A and Ch. 5) have been derived from these studies.

1. Helical Structures

***a.* Tobacco Mosaic Virus and the Various States of the Virus Protein.** Tobacco mosaic virus (TMV) is a rod-shaped molecule having a length of 3000 Å (Williams and Steere, 1951) and a diameter of 180 Å. The molecule contains 49 identical subunits per three turns of a helix having a pitch of 23 Å. The helical parameters of TMV were determined by X-ray diffraction (Watson, 1954; Franklin and Holmes, 1958) before the substructure of the virus was seen in electron micrographs (Finch, 1964), although in one of the earliest applications of negative staining, Huxley (1957) had shown the presence of a central cylindrical hole. The helical periodicities are most clearly seen in images of negatively stained particles which have been minimally exposed to the electron beam (Williams and Fisher, 1970). Using this technique, the handedness of the structure, which is not revealed by X-ray diffraction, has been determined from images of particles tilted about an axis perpendicular to the electron beam (Finch, 1972). When the particle axis is tilted through an angle of about 3°, the basic helix of subunits on one side of the particle is approximately parallel to the direction of view, giving this side a deeply serrated appearance in contrast to the smoother edge of the other side of the particle (Fig. 28). From the sense of this asymmetry for a known direction of tilt, the basic helix of TMV was found to be right-handed.

When TMV is dissociated in acid or alkali, the protein can be isolated from the nucleic acid. The minimally aggregated form of the protein,

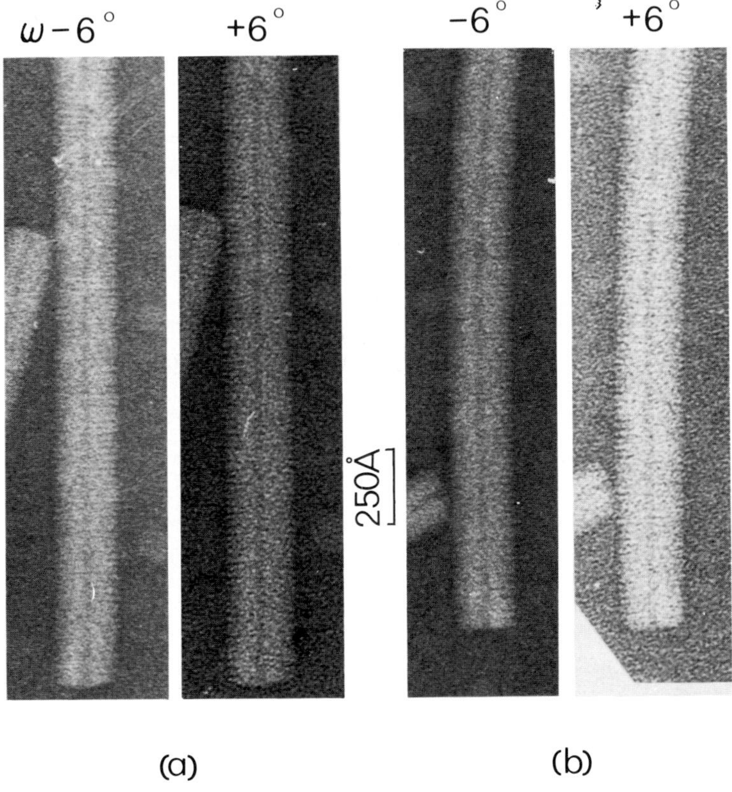

Fig. 28 Pairs of electron micrographs of two particles of tobacco mosaic virus having their axes tilted approximately −6° and +6° out of the plane perpendicular to the electron beam. (The tilt angle, ω, is taken as positive if the top of the particle is tilted toward the observer.) In each case the right half of the image is more serrated than the left for $\omega = -6°$ and the asymmetry is reversed for $\omega = +6°$. The striations are most clearly seen by looking at the images from the side of the page at a glancing angle. The striations occur as a result of the path of the helix on which the virus subunits lie approaching the direction of view, and from the sense of the asymmetry it follows that this helix is right-handed. Taken from Finch (1972).

called "A-protein," has a sedimentation constant of about 4 S and occurs in alkaline solution at low ionic strength and temperature (Schramm, 1947). As the temperature or the ionic strength is raised, larger aggregates having sedimentation constants of about 8 S, 20 S, 30 S, and higher are formed.

In the electron microscope, the aggregates above 20 S appear as short rods with a diameter about the same as that of the intact virus (Schramm and Zillig, 1955). Electron micrographs of preparations containing mainly 20 S aggregates embedded in stain over holes in a carbon film showed circular end-on images and a number of patterns consisting of two parallel lines equal in length to the virus diameter. From these images it was concluded that the 20 S aggregate is a stack of two rings of subunits (Durham and Finch, 1972). Similarly, the 30 S aggregate gave four-line images, indicating that this aggregate is a stack of four rings of subunits.

When these short rods sit flat on the carbon film, are uniformly stained, and are not too close to other particles, 17 protuberances can often be seen around their periphery (Finch *et al.*, 1966). This seventeenfold rotational symmetry was confirmed by Crowther and Amos (1971a), who analyzed these micrographs computationally for rotational harmonics (see Section IV,E, and Figs. 7 and 8). It can therefore be concluded that these aggregates are stacks of rings of 17 subunits. Since there are $16\frac{1}{3}$ subunits per turn of the virus helix, the curvature per subunit in the helix and in the rings is virtually the same, thus allowing the same tangential bonding between subunits to occur in both.

Since the rotationally filtered images (Crowther and Amos, 1971a) all have a skew appearance and both senses of skewness are observed in various images, the stacks of rings must be polar, i.e., all the subunits must point the same way along the axis of the rod. The stacks of rings are not, however, a simple variation of the structure of the virus. The fact that their unit of growth is the 20 S pair of rings indicates that this "disk" is due to a particularly stable association. The pairing of rings into disks is clearly evident in images of longer stacks (Fig. 29). Although the shorter stacks can disaggregate reversibly, for reasons that are not clear at present, these longer stacked disks are not easily disrupted. They are, however, sufficiently long and well-ordered for their images to be analyzed and processed, thus enabling reconstruction of the stacked disk particle (Finch and Klug, 1971). A model of part of a reconstructed rod of stacked disks is shown in Fig. 30. The subunits in adjacent rings skew the same way, confirming the polar nature of the rod. Since the mean angular displacement between rings is very similar to that between turns of the helix, the stacked disks have a structure which can be described as a perturbed variation of that of the helix.

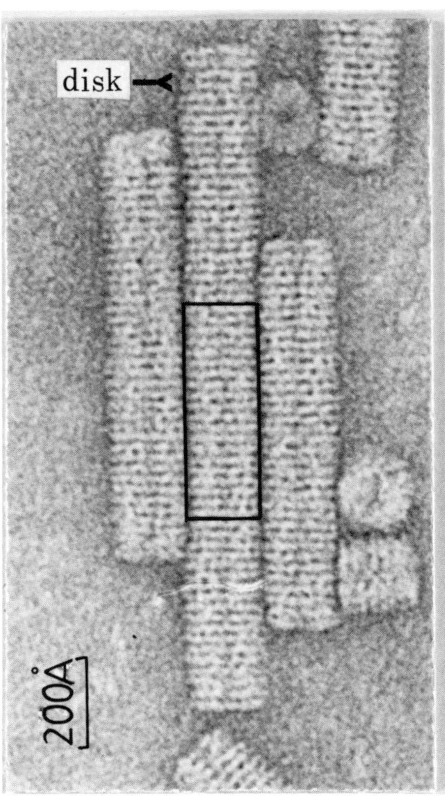

Fig. 29 Images of stacks of disks of tobacco mosaic virus protein. The boxed region was used in the three-dimensional reconstruction of the stacks shown in Fig. 30.

If a solution of TMV protein in the form of 4 S or any of the larger reversible aggregates is acidified to pH 5, the protein assumes a helical structure identical to that of the protein in the intact virus. The various sequences of events have been followed by electron microscopy. In the case of the 4 S protein, polymerization to the helical structure occurs within seconds of acidification. With 20 S disks, the process is slower, as shown in Fig. 31 (Durham *et al.*, 1971). A few seconds after the pH is dropped, short rods with an imperfect, nicked-helical structure are observed. These rods gradually grow longer and anneal to form perfect helical particles after about 20 hr. A schematic diagram of the probable mechanism is shown in Fig. 32. Acidification introduces a dislocation into each disk, converting it into two turns of helix rather like a lock-washer. These lock-washers then stack in random azimuth to form a nicked-helical structure which subsequently anneals into the ordered structure. A similar

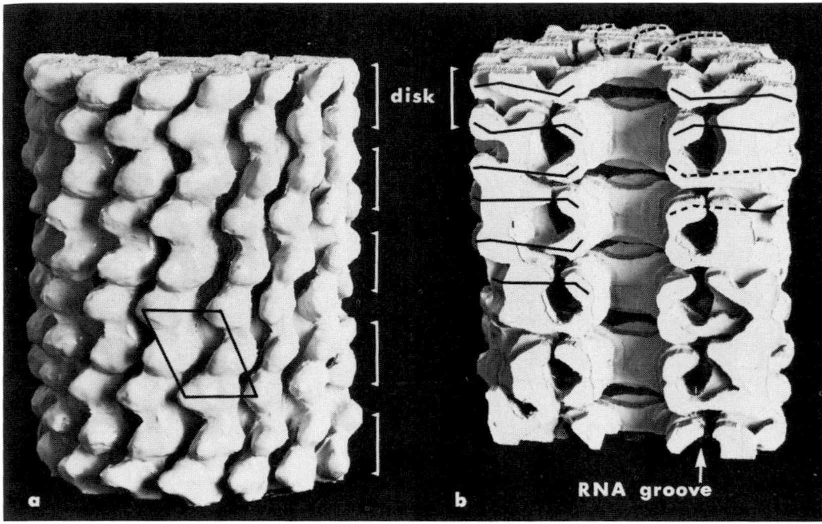

Fig. 30 Model of a three-dimensional reconstruction of a stacked disk rod of tobacco mosaic virus protein. (a) Outer surface showing the marked skewing of the subunits (Fig. 8) and the displacement of the outer ends of subunits which link a pair of rings into a disk. The parallelogram indicates the unit cell which contains two crystallographically nonequivalent subunits. (b) Section through the center showing the groove which in the virus structure is occupied by RNA. The black lines indicate the orientation of the main body of the subunits.

but slower process was observed with 30 S aggregates which are composed of two disks (Durham and Finch, 1972).

In the helical structure two carboxyl groups per subunit titrate with a p*K* near 7.0, whereas in other structures their p*K* is much lower (Caspar, 1963). This difference in p*K* implies that the transition from a disk to a helix must involve the uptake of two protons per subunit. The protons can be thought of as allosteric effectors which convert the relaxed association of subunits in the disks into the tense association in the helix, where the carboxyl groups are in unusual environments. It has been found that the disks play a key role in the nucleation and elongation of TMV (Butler and Klug, 1971) probably by the following mechanism. The ribonucleic acid chain has an end sequence which interacts favorably with the subunits of a disk. This interaction counteracts the effects of the abnormal carboxyl groups and allows the disk to adopt the helical structure. Additional disks can then be added to the stack and are also converted to a helical structure as the nucleic acid chain becomes bound into them, until the whole chain is embedded in the complete helical virus particle.

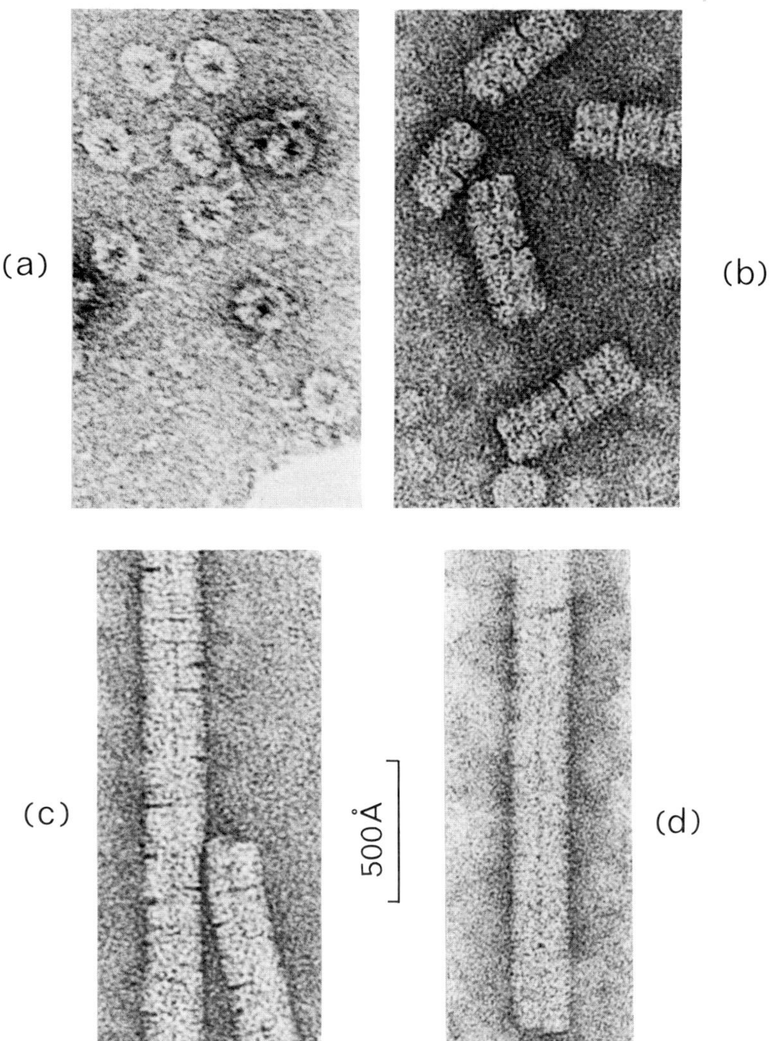

Fig. 31 The effect of lowering the pH on a solution of the 20 S disk aggregate of tobacco mosaic virus protein (Durham *et al.*, 1971). (a) Starting material mainly in the form of 20 S disks at pH 7.0. (b) Short rods with an imperfect helical structure formed 5 sec after dilution in acetate buffer at pH 5.0. (c) Longer and more regular rods formed after 15 min. (d) Long, almost perfect helical rods formed after 18 hr.

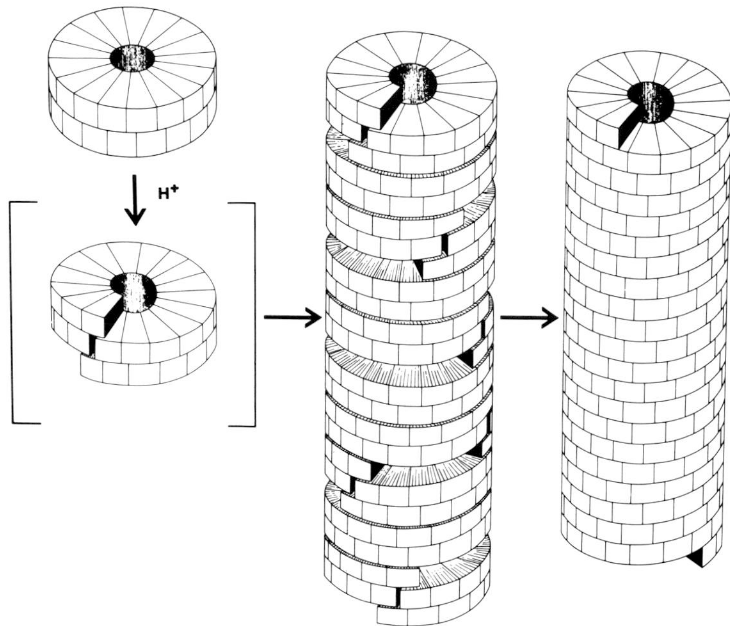

Fig. 32 Schematic representation of the process by which the disk aggregates of tobacco mosaic virus protein are converted into helical rods upon rapid acidification.

***b.* Tail Sheath of Bacteriophage T4.** Intact particles of bacteriophage T4 have a large polyhedral head to which a rod-shaped tail about 900 Å long and about 200 Å in diameter is attached. The outer part of the tail is a sheath around a slender cylindrical core (Fig. 33). This sheath contracts during infection or can be made to contract under various conditions (Brenner *et al.*, 1959). Both forms of the tail have been extensively studied by electron microscopy as have two other large helical protein aggregates which are generated by various defective mutants of T4. One of these aggregates has a structure closely related to that of a contracted sheath (polysheath), and the other is a tubular variant of the head structure (polyhead).

The arrangement of subunits in the contracted sheath has been deduced by Moody (1967a, 1971) from electron micrographs of negatively stained specimens. Since the contracted sheaths are squat, they often sit on end on the substrate when they are free from the remainder of the virus particle. The images of these end-on views consist of twelve spirals which are always of the same sense. Moody (1967a) interpreted this appearance as arising from long-pitch helices of subunits which were

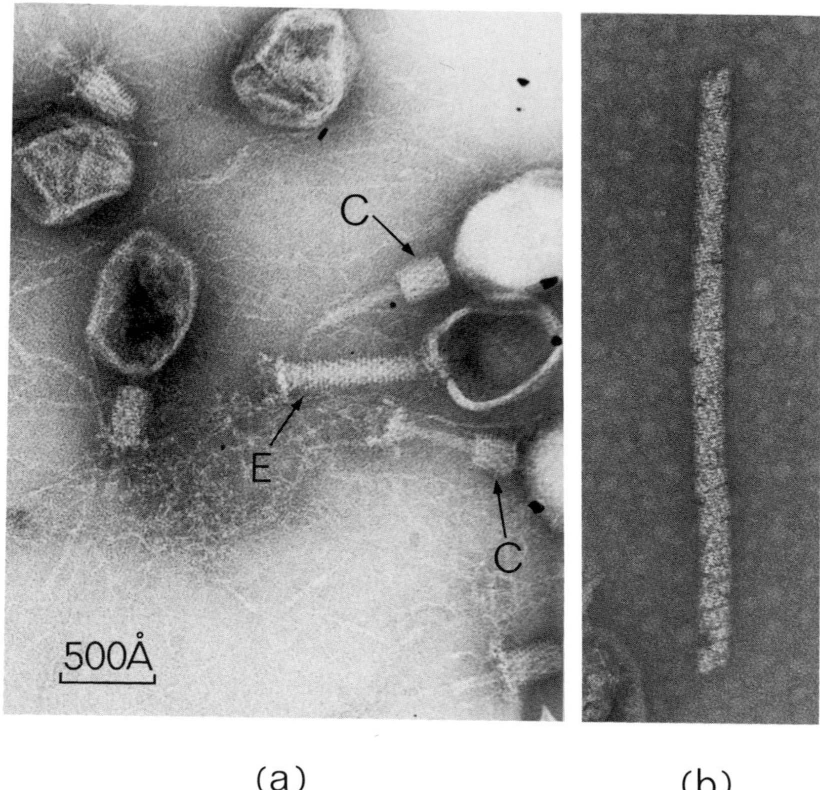

Fig. 33 Electron micrographs of bacteriophage T4. (a) Field showing extended (E) and contracted (C) tail sheaths of bacteriophage T4. (b) A polysheath, a bacteriophage product resembling a contracted sheath in diameter and surface detail.

compressed into a conical shape by the elastic contraction of the stain at the end of the particle furthest from the substrate. On this basis, the sense of the spirals indicated that the corresponding long-pitch helices were right-handed.

The contracted sheaths are too short to give useful optical diffraction patterns. Moody (1967a) therefore continued the structural analysis on polysheaths, which apparently have the same structure as the contracted sheaths; they contain the same long-pitch helices and in addition a family of short-pitch helices which were shown from images of tilted particles to be left-handed. From the positions in the optical transforms of the diffraction spots corresponding to this short-pitch family of helices, they were identified as a six-start family. (This identification was unusually unequivocal since the subunits of polysheaths are confined within a nar-

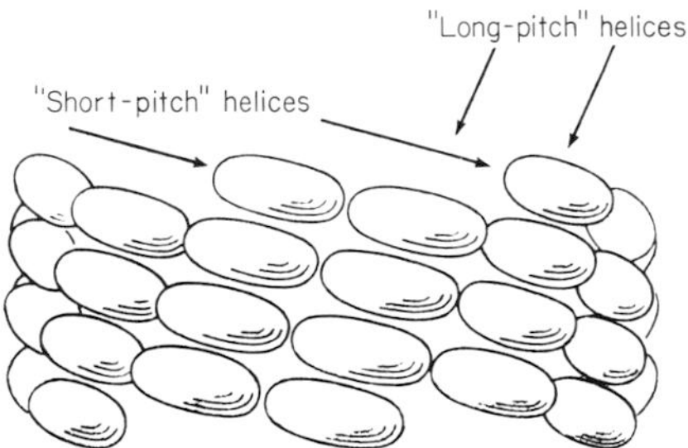

Fig. 34 The arrangement of the outer subunits in a contracted tail sheath of bacteriophage T4 (Moody, 1967a).

row radial band.) The surface lattice was thus established, and a diagram of the subunit arrangement is shown in Fig. 34.

The arrangement of subunits in the extended sheaths has been deduced by optical diffraction (Krimm and Anderson, 1967; DeRosier and Klug, 1968) and by direct measurement of interhelical spacings in images of negatively stained particles (Moody, 1967b). The extended sheaths con-

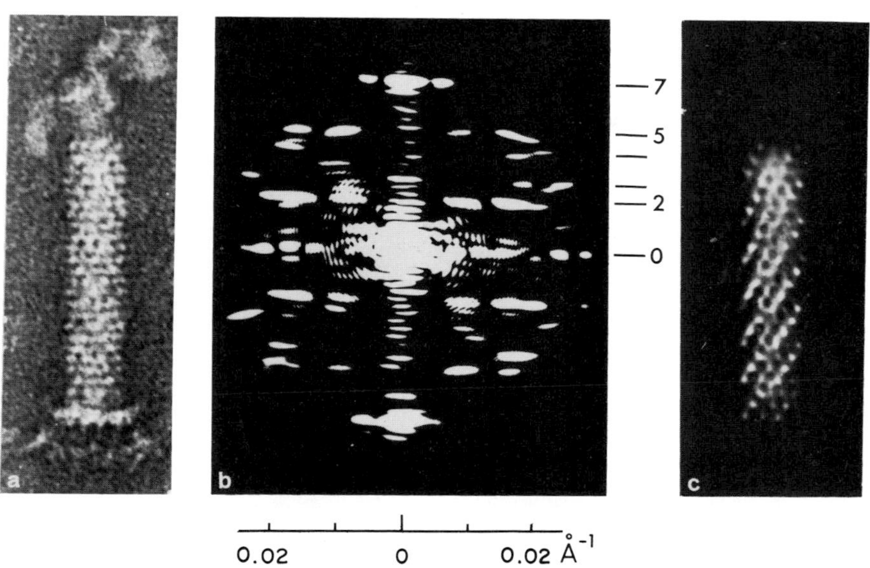

Fig. 35 Optical filtering of an image of an extended tail of bacteriophage T4 (DeRosier and Klug, 1968). (a) Original image of the negatively stained tail sheath. (b) Optical diffraction pattern derived from (a). (c) Optically filtered image of one side of the sheath in (a). The two sets of oblique striations running at slightly different angles across the image correspond to features on the same helical family but at distinctly different radii.

sist of a stack of twenty-four annuli, each with sixfold symmetry, the difference in azimuth between adjacent annuli being about 17°. The optical diffraction patterns from the images (Fig. 35b) are complicated by the doubling of spots on some layer lines due to features at two distinct radii. In the optically filtered image (Fig. 35c), these features are seen as two distinct sets of oblique striations.

The three-dimensional reconstruction technique has been applied to the images of extended sheaths (DeRosier and Klug, 1968), using data to a resolution of 35 Å. The resulting structure (Fig. 36) has a central hole about 30 Å in diameter surrounded by a tube of fairly uniform density with a diameter of about 60 Å, which corresponds to the core within the tail sheath. Six helical tunnels (Fig. 36b) mark the boundary

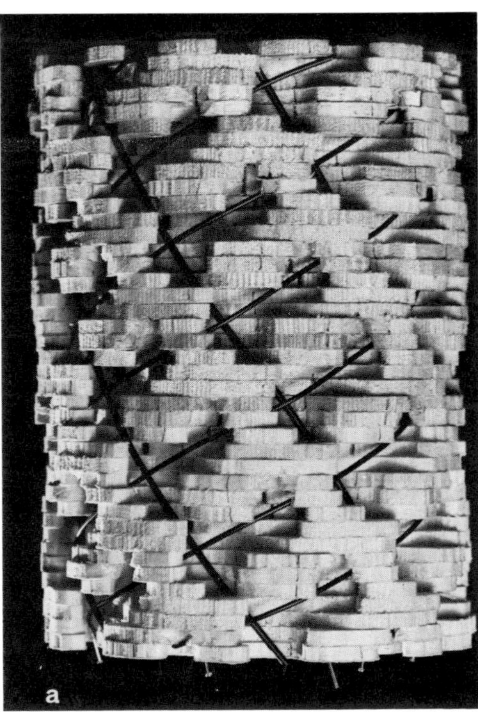

Fig. 36 Model of a reconstructed part of the extended tail sheath of bacteriophage T4 (DeRosier and Klug, 1968). (a) The outside of the structure showing the deep surface grooves indicated by the black wires, which also mark out the unit cell of the structure and probably indicate the approximate boundaries of the intricately shaped subunits. (b) A tilted view with part of the model removed. The fine black wires follow the paths of helical tunnels which lie between the sheath and the tail core at a radius of about 65 Å.

between the tail core and the sheath. The outer surface consists of a complex set of protuberances which extend to an outermost radius of 120 Å and are separated by deep grooves. These grooves and the inner helical tunnels are the two radially distinct features visible in the optically filtered image (Fig. 35c).

c. **Polyheads Associated with Bacteriophage T4.** In the electron microscope the tubular variants of the head of bacteriophage T4 appear in a variety of forms which differ in surface detail, surface lattice, and diameter. Since it is known that changes occur in the head protein during the maturation of the bacteriophage, it was hoped that the various polyhead forms might relate to various states of the head protein. Because of the large areas of their images (the tubes are in the region of 1000 Å in diameter and of indefinite length), they are ideal objects for study by optical diffraction and filtering. The first optical diffraction study showed that the surface lattice of the polyheads was based on a P6 planar hexagonal net which was rolled up into a cylinder (Finch *et al.*, 1964). More recently, a systematic study of the various types of polyheads has been made by DeRosier and Klug (1972) and Yanagida *et al.* (1972), and the results of this study are described below.

Four main classes of polyheads have been observed. The most common type (Fig. 37a) has a diameter of about 750 Å and is referred to as "coarse" because of the relatively large-scale detail which can be seen in the images. Optically filtered images show that the structural units are hexagonally clustered into distinct morphological units arranged on a near-hexagonal lattice (Fig. 37b). The second class, a wider version of the coarse tube, originates as an outer covering of the coarse tubes and is separated from them after lysis of the bacterium (Kellenberger *et al.*, 1968; Yanagida *et al.*, 1970). The morphological units are still distinct but are less regularly hexagonal, and those in adjacent rows have different orientations.

Images of the third class (Fig. 37c) have a very smooth appearance with virtually no substructure visible, rather like the images of the empty heads of the virus particle. Optical diffraction, however, shows the presence of a surface lattice similar to that of the first class. Filtered images suggest that the reason for the smooth appearance is the even distribution of density on the lattice. The hexamers of structural units are more open, fit into each other in a uniform way, and contain an extra unit in each of their centers (Fig. 37d).

Even less detail is visible in images of the last class. These polyheads have a mottled appearance (Fig. 37e) and retain some cylindrical shape, in contrast to the other classes, which are usually flattened against the

Fig. 37 Optically filtered images of polyheads associated with bacteriophage T4 (Yanagida *et al.*, 1972). (a) Image of a "coarse" tube having a diameter of 750 Å, in which periodicities are clearly evident. (b) Optically filtered image of one side of the tube in (a) showing hexamer morphological units. (c) Image of "aged-fine" type of polyhead having a diameter of 800 Å, in which little regular detail is apparent. (d) Filtered image of one side of (c). The hexamers are more open than in (b), leading to a more even distribution of density. (e) Image of "fine-mottled" type of polyhead having a diameter of 750 Å, in which no regular detail can be seen. (f) Filtered image of one side of (e). The presence of extra material within and beween the hexamers of subunits leads to a very uniform, smooth appearance and makes the hexamers almost indistinguishable.

substrate. Filtered images of these "fine-mottled" polyheads show a distribution of matter somewhat similar to that of the smooth class with additional density bridging the "6 + 1" groups of units (Fig. 37f). Antisera prepared against the virus heads become attached to this class but not to the others, indicating that these polyheads bear the closest structural relationship to the heads of mature phage.

Although the conditions for the transition are not known, the coarse tubes can become transformed into fine tubes. Yanagida *et al.* (1972) have suggested that the structural features of this transformation, notably the change in the morphological unit from a compact hexamer to an open 6 + 1 unit, may be correlated with the cleavage of head proteins which occurs during head assembly (Laemmli, 1970).

***d.* Tubular Particles Associated with the Tumor Viruses.** Electron micrographs of negatively stained preparations of polyoma and the papilloma viruses have shown that they are isometric particles with diameters of about 500 Å and large morphological units distributed over their surfaces (Wildy *et al.*, 1960; Williams *et al.*, 1960). Analysis of the particle images (Klug and Finch, 1965; Finch and Klug, 1965) has established that the surface structure of these virus particles is based on the $T = 7$ icosahedral surface lattice (see Section IV,A), the large morphological units being hexamer and pentamer clusters of structure units around the vertices of this lattice.

Unfractionated preparations of these viruses usually contain tubular structures which seem to be very similar to the virus particles in surface detail, again indicating the presence of large morphological units (Williams *et al.*, 1960; Breedis *et al.*, 1962; Finch and Klug, 1965). These tubes appear to be empty and, judging from their density, are composed of protein only (Breedis *et al.*, 1962). The various types of tubes have been classified by Kiselev and Klug (1969), who used optical diffraction and optical filtering to analyze the electron microscope images.

As judged by diameter, the tubes fall into two categories: wide tubes with a diameter of about 500 Å (similar to that of the virus particles) and narrow tubes with a diameter of about 300 Å. Optical diffraction patterns from images of the wide tubes show approximate local hexagonal symmetry. Filtered images show an arrangement rather like that of the coarse polyheads described in the previous section, i.e., distinct, hexagonal morphological units are arranged in an approximate hexagonal lattice.

The images of narrow tubes give more complex diffraction patterns which indicate that there are two related types, one called "zero-start," with a threefold rotational symmetry axis, and a "one-start" helical version of the same lattice. Filtered images from these narrow tubes again show

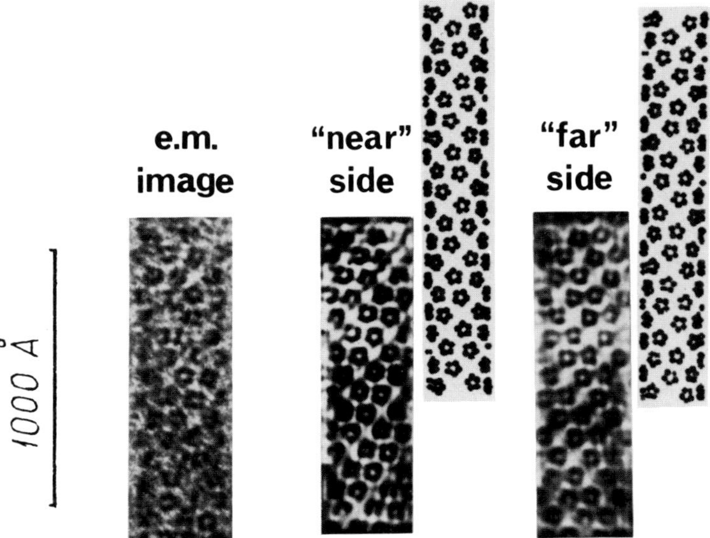

Fig. 38 Electron micrograph and optically filtered images of a narrow tubular structure associated with human wart virus (Kiselev and Klug, 1969). The protein is reproduced black. The filtered images show morphological units which are pentagonal in shape and are arranged on a lattice based on the closest possible packing of pentagons. Computer-generated projections of a model structure are shown alongside the filtered images.

distinct morphological units, but in this case the units are pentagonal in shape (Fig. 38). The local arrangement of the pentamers is based on the planar tessellation shown in Fig. 39, in which pentagons are packed together as closely as possible, each touching three others. Cylindrical surface lattices based on this tessellation can be formed by rolling up the planar arrangement into a cylinder, thus preserving the local pattern along the joining line. It is possible to account for both of the surface lattices of the pentamer tubes in this way.

The two types of pentamer tubes are variants of the same structure, but the coexistence of these with the tubes of hexamers raises the possibility that the two types of morphological units in the virus particle, the pentamers and the hexamers, are built from different protein structure units. The arrangements in the two types of tubes are not, however, inconsistent with the possibility that both are built from the same type of subunit with the same *local* bonding pattern. On this basis, one would expect one type of tube to be thermodynamically the most favorable, although it is possible that once formed, the less stable tube may be difficult to unlock.

2. Isometric Structures

a. **Turnip Yellow Mosaic Virus.** The isometric particles of turnip yellow mosaic virus (TYMV) are about 300 Å in diameter. Electron microscope images of negatively stained particles showed that their surface consists of 32 large morphological units located on the radial lines through the 12 vertices and the centers of the 20 faces of an icosahedron (Huxley and Zubay, 1960b; Nixon and Gibbs, 1960). X-Ray diffraction from single crystals of the nucleic acid-free, viruslike protein shells (top component) showed that most, if not all, of the protein was arranged with strict 532 symmetry (Klug and Finch, 1960). On the basis of the quasi-equivalence theory (see Section IV,A), the arrangement of morphological units on the virus particle corresponded to the clustering of 180 structure units in hexamers and pentamers at the vertices of the $T = 3$ icosahedral surface lattice (Fig. 4d). Finch and Klug (1966) therefore studied many examples of the various types of images of negatively stained virus particles and top component particles for evidence of a consistent substructure within the 32 morphological units. Their results showed that the 5- and 6-coordinated morphological units were in fact composed of rings of 5 and 6 subunits and that the angular relationship between them was consistent with the theory. The nonuniform appearance of individual morphological units near the centers of images suggested that the structure units leaned toward each other at an outer radius. A model of the surface detail, constructed on the basis of these results, is shown in Fig. 40. In order to test the model, simulated negative stain images were produced from it by coating with Plaster of Paris (the "negative stain") and radiographing it with X-rays of appropriate wavelength (Caspar, 1966). The correlation between the resulting simulated image and the corresponding electron microscope image is also shown in Fig. 40.

The three-dimensional reconstruction technique has been applied to TYMV by Mellema and Amos (1972). Data from images of five different views were used to determine the Fourier transform to a resolution of 40 Å. The model of the virus particle obtained by Fourier synthesis from the transform is shown in Fig. 41; it is comfortingly similar to the model obtained by direct inspection of the images.

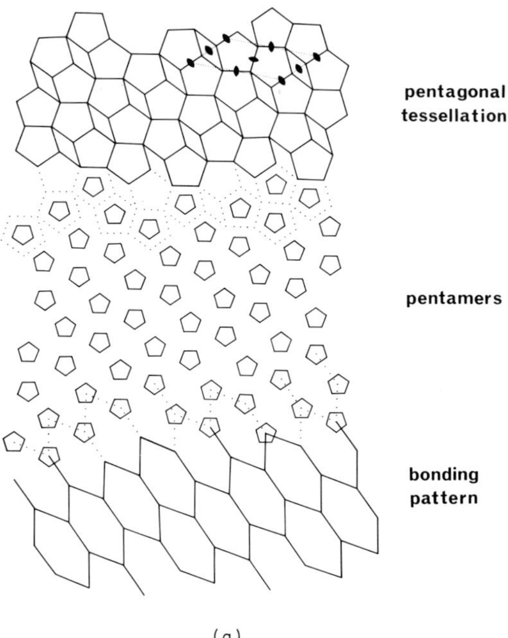
pentagonal
tessellation
pentamers
bonding
pattern

(a)

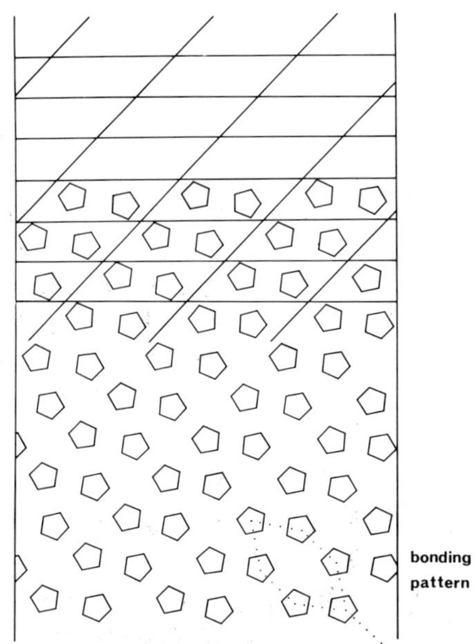
Radial projection of
zero-start pentamer tubes
bonding
pattern

(b)

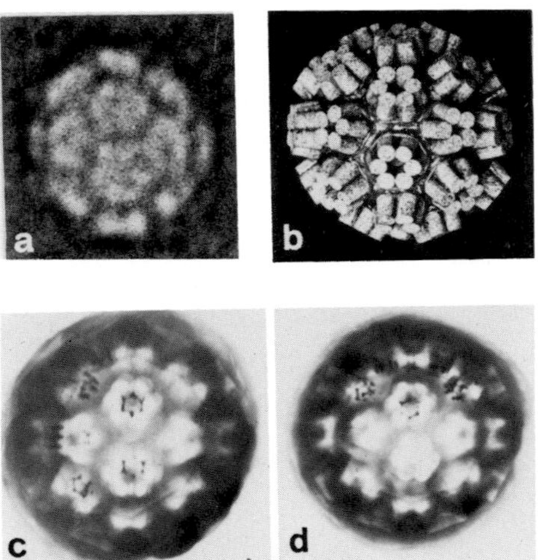

Fig. 40 Turnip yellow mosaic virus. (a) Image of a negatively stained virus particle seen in a direction close to a twofold axis of symmetry. (b) Model of the virus structure deduced by inspection of the electron microscope images (Finch and Klug, 1966). (c, d) Simulated images of "negatively stained" particles obtained by X-radiography of the model in (b) coated with Plaster of Paris (the "negative stain"). Taken from Caspar (1966).

***b.* Other Small Isometric Viruses.** Following the work on TYMV, the structures of a number of other small isometric viruses were determined from electron micrographs. In each case, the protein shells are built from large numbers of identical structure units bonded together quasi-equivalently. Several viruses, e.g., broad bean mottle (Finch and Klug, 1967) and cucumber mosaic (Finch *et al.*, 1967), have been found to have structures similar to that of TYMV, the units being arranged in hexamers and pentamers around the vertices of the $T = 3$ icosahedral surface lattice.

Different types of images have been obtained with two viruses, turnip crinkle and tomato bushy stunt. The patterns are characteristic of the clustering of subunits in dimers on the same $T = 3$ surface lattice (Fig. 4c), with some additional detail in the centers of the rings of five dimers (Finch *et al.*, 1970). Extra density is also found in the three-dimensional

---

Fig. 39 (a) Top, Schematic representation of pentagonal tessellation: the closest regular planar packing of pentagons based on the unit cell indicated at the top right. Middle, An ideal lattice of pentagons based on the above tessellation. Bottom, The "bonds" across the twofold axes connecting a pentagon to its neighbors. (b) Ideal cylindrical surface lattice based on that in (a) with helical parameters chosen to be as close as possible to those of the narrow tubes of the type shown in Fig. 38.

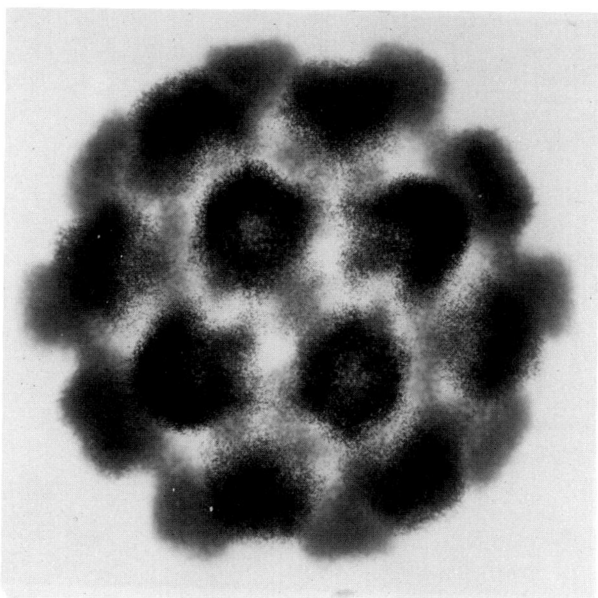

Fig. 41 Turnip yellow mosaic virus reconstructed by computation to a resolution of 40 Å from electron microscope images (Mellema and Amos, 1972) showing hexamer and pentamer morphological units on the $T = 3$ icosahedral surface lattice (cf. Fig. 4d).

reconstruction of tomato bushy stunt virus, shown in Fig. 42 (Crowther *et al.*, 1970). It may be that this density arises from the shape of the structure units and is revealed in these positions and not at the centers of the rings of six dimers because of the slight differences between the two environments. It is also possible that a minor protein component present in the virus is located in these positions (Butler, 1970).

Like the images of TYMV, those of polyoma and the papilloma viruses show large morphological units indicative of the clustering of structure units into hexamers and pentamers at the vertices of an icosahedral surface lattice. The diameter of the virus particles, about 550 Å, is about twice that of TYMV and thus the lattice must be correspondingly larger. In regions where the particles were completely embedded in stain, clear morphological units were never observed over an entire image. This suggested that the surface lattice was skew, since for symmetric lattices, the vertices (and thus the hexamer–pentamer morphological units) on the near and far sides of the lattice will always superpose in views parallel to the twofold axes of the particles. To determine a skew lattice from single images, one must find some which show the structure of only

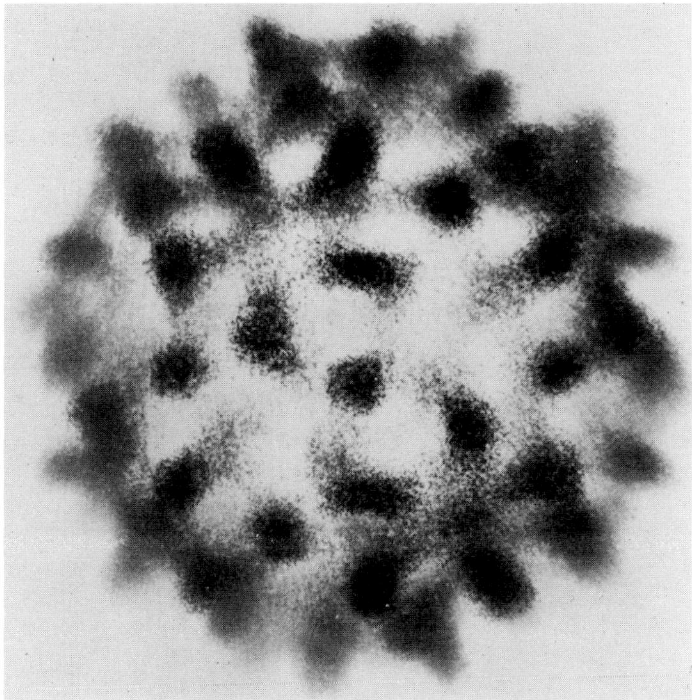

Fig. 42 Tomato bushy stunt virus reconstructed by computation to a resolution of 25 Å from electron microscope images (Crowther *et al.*, 1970) showing the clustering of structure units into 90 dimers on the $T = 3$ icosahedral surface lattice (Fig. 4c).

one side of the particle, where the stain has not appreciably covered the other side. A few one-sided images of this type were found in polyoma and in human and rabbit papilloma viruses. In the resulting patterns, the arrangement between two clearly defined five-coordinated morphological units enabled the surface lattices to be determined as $T = 7d$, $7d$, and $7l$, respectively, d and l referring to the *dextro* and *levo* versions of the $T = 7$ lattice (Klug, 1965; Klug and Finch, 1965; Finch and Klug, 1965). The model structures derived from the above (Fig. 5) were tested by correlating the different images produced as the particles were tilted in the electron microscope with superposition patterns in a gallery computed from the models (Klug and Finch, 1968).

The three-dimensional reconstruction of human papilloma virus shown in Fig. 43 was computed by Crowther and Amos (1971b). The morphological units are hollow and slightly conical with a diameter of about

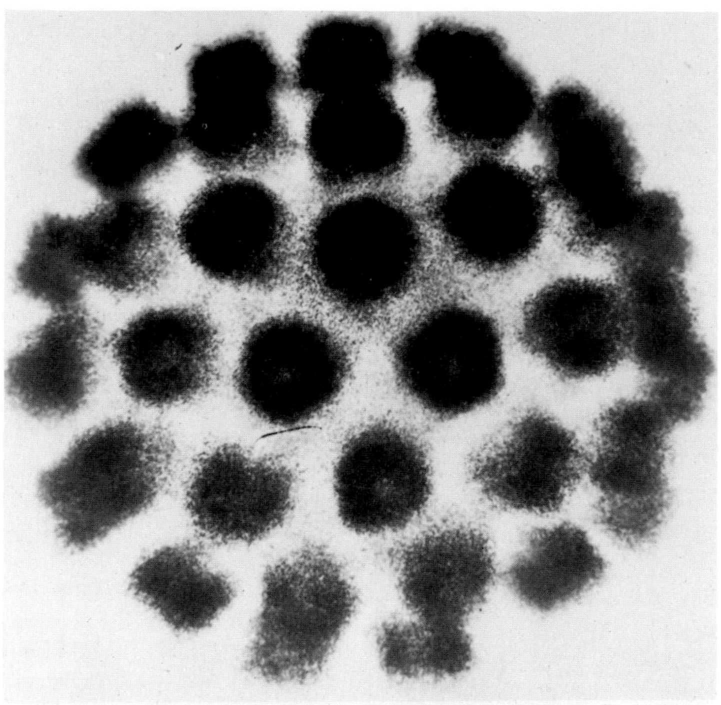

Fig. 43 Computational reconstruction of human papilloma virus to a resolution of 37 Å (Crowther and Amos, 1971b). The 72 morphological units are hollow and slightly conical, and the arrangement corresponds to the hexamer and pentamer clustering of structure units on the $T = 7d$ icosahedral surface lattice.

90 Å, and they extend between radii of about 210 Å and 280 Å. A comparison of the projections of the reconstructed particle with the corresponding electron microscope images is given in Fig. 44. There is very good agreement between the two despite the difficulty of choosing the exact level of density in the reconstructed particle which differentiates stain from protein.

c. **Adenovirus.** In one of the earliest applications of the negative staining technique, Horne *et al.* (1959) established that the particles of adenovirus are composed of 252 large morphological units arranged on the surface of an icosahedron as shown in Fig. 45; the virus particles are about 800 Å in diameter. On the basis of the quasi-equivalence theory (Caspar and Klug, 1962; also see Section IV,A), one would interpret this

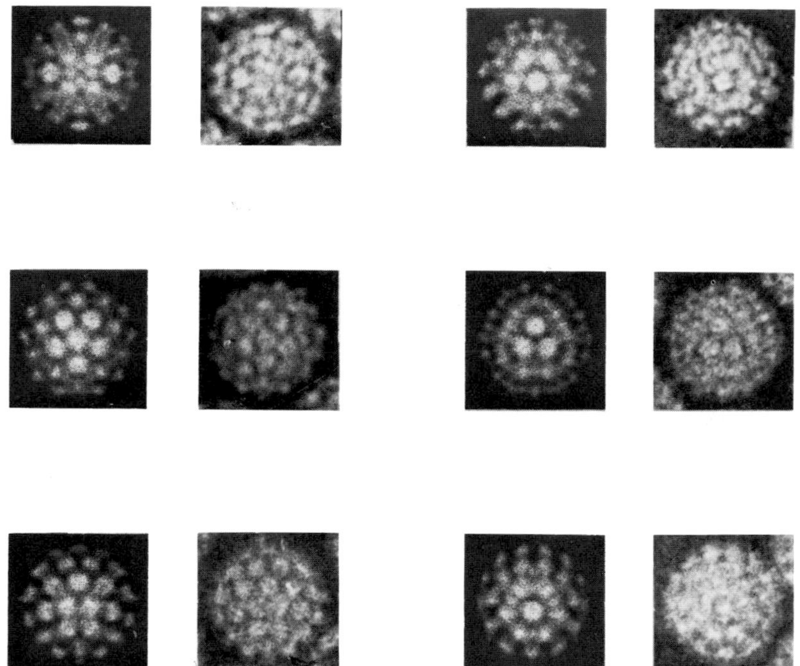

Fig. 44 Comparison of the projections of the reconstructed particle of human papilloma virus (left member of each pair) with electron microscope images of virus particles seen in the corresponding orientations (right member of each pair). Taken from Crowther and Amos (1971b).

surface pattern as arising from 1500 structure units arranged on a $T = 25$ icosahedral surface lattice, the structure units being clustered into unresolved hexamers and pentamers to yield the large morphological units observed. (The black units at the vertices of the icosahedron in Fig. 45 are the twelve pentamers or pentons, and the remaining units are the hexamers or hexons.) However, it soon became evident that this was too simple a picture of the virus structure.

The first indication of this was the observation by Valentine and Pereira (1965) of a radical distinction between the pentons and hexons. They found that the pentons consist of a base unit and a protruding fine fiber about 200 Å long, with a globular end. Chemical studies also indicated that the virus particles contain at least nine different types of protein subunits (Maizel *et al.*, 1968). The major protein component was correlated with the 240 hexons, but there was some controversy over the substructure that this implied within each hexon. Maizel *et al.* (1968) concluded that the hexons were built from three identical subunits having

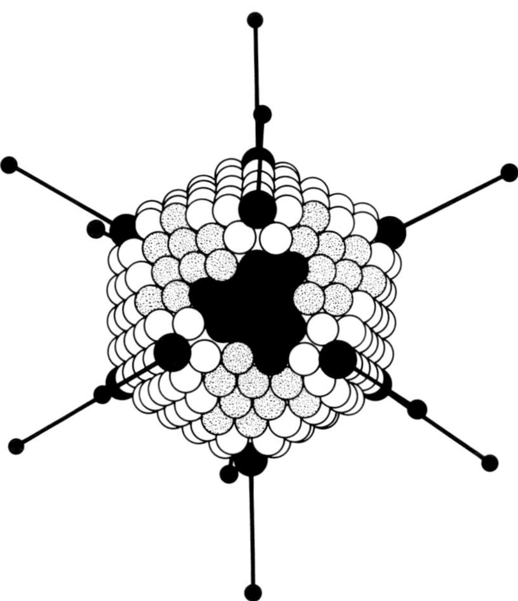

Fig. 45 Diagram of adenovirus showing the different types of morphological units. The pentons (black) are surrounded by the perapentonal hexons (white). Between the perapentonal hexons are hexons (shaded) which remain associated in groups of nine when the virus is mildly disrupted. Taken from Crowther and Franklin (1972).

molecular weights of 120,000, but other workers (Pettersson, 1970; Franklin *et al.*, 1971a) suggested that these were dimers of the true subunit. Crystals of hexons have been studied by X-ray diffraction (Cornick *et al.*, 1971; Franklin *et al.*, 1971a) and the results of these studies indicate that the hexons have at least threefold symmetry. Attempts to find sixfold symmetry in the X-ray data proved unsuccessful (Franklin *et al.*, 1971b). Electron micrographs of isolated hexons have not helped to settle this controversy because it is impossible to tell whether these images have threefold or sixfold symmetry. However, Crowther and Franklin (1972) have analyzed images of aggregates of hexons and concluded that each has threefold symmetry.

The aggregate studied by Crowther and Franklin (1972) is a group of nine hexons around a threefold axis of the particle (shaded in the drawing of the virus structure in Fig. 45). Images showing well-preserved threefold symmetry were selected by using the rotational harmonic analysis technique (Crowther and Amos, 1971a), and these images were then threefold rotationally filtered (Fig. 46). The inner hexons of the filtered images all display threefold symmetry and so do the outer

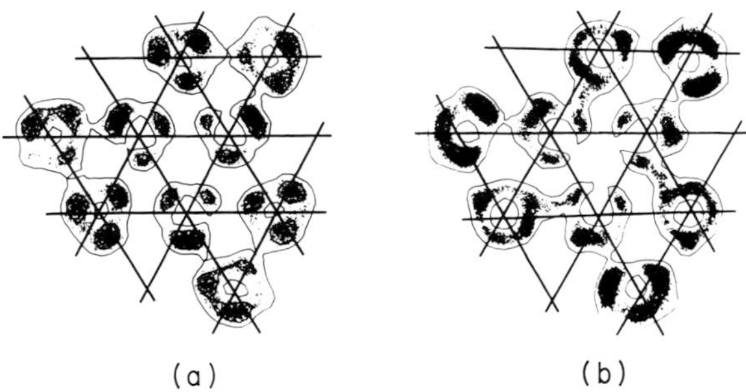

Fig. 46 Threefold rotationally filtered images of two of the aggregates of nine hexons from adenovirus, plotted as combined density and contour maps (Crowther and Franklin, 1972). The inner hexons in both images show clear threefold symmetry, the density distribution in both being similar. The outer hexons, especially in (a), also show approximate threefold symmetry, with the overall density arrangement following the symmetry of the superposed P3 net.

hexons in the best-preserved aggregates. The distribution of density in the three hexons not related by the central threefold axis does, however, appear to be related by a *local* threefold axis, so that the overall density distribution follows the symmetry of a P3 planar lattice, shown superposed in Fig. 46. Approximately the same density arrangement was found in several filtered images. It was thus concluded that in projection the hexons have threefold symmetry only. There remains the possibility that the hexons have 32 symmetry, but this seems unlikely for the following reason. The aggregate of nine hexons has a sense of handedness made apparent by the location of the outermost three hexons. Only images with one sense of handedness are observed in electron micrographs, indicating a pronounced polarity between the top and bottom of the aggregates. If the hexons had 32 symmetry, their top and bottom surfaces would be identical, and an aggregate of hexons would be unlikely to show such striking polarity.

Thus, in spite of its appearance at low resolution, the outer protein shell of adenovirus does not follow the same design principle as the small, simple, icosahedral viruses. However, more detailed studies will probably reveal underlying similarities.

D. Other Proteins

1. *Hemocyanin from Gastropods*

Hemocyanins are copper-containing respiratory proteins which are widely distributed in invertebrate animals. The largest known types, the gastropod hemocyanins, have a molecular weight of 7 to 8 million, and appear in the electron microscope as short cylindrical rods about 360 Å in diameter (van Bruggen *et al.*, 1962; Fernández-Morán *et al.*, 1966). Hemocyanins undergo reversible dissociation controlled by ionic strength and pH. The first stage results in half-molecules about a plane perpendicular to the cylindrical axis. In side views, the images are striated, suggesting that the molecules consist of six rings of subunits (Fig. 47). The images of end-on views show a very dense outer ring as well as some weaker inner density which is variable. Both types of images have been analyzed by Mellema and Klug (1972) with the following results.

The end-on views of whole and half-molecules embedded in stain over holes in carbon film (to reduce nonuniform stain distribution) were analyzed for rotational symmetry, then rotationally filtered following the computational procedure of Crowther and Amos (1971a). In both cases, the strong rotational harmonics are multiples of five. The rotationally filtered images of the half-particles show a strong fivefold modulation about halfway out to the edge of the particle, whereas the outer part is dominated by tenfold symmetry. A similar result was obtained with whole

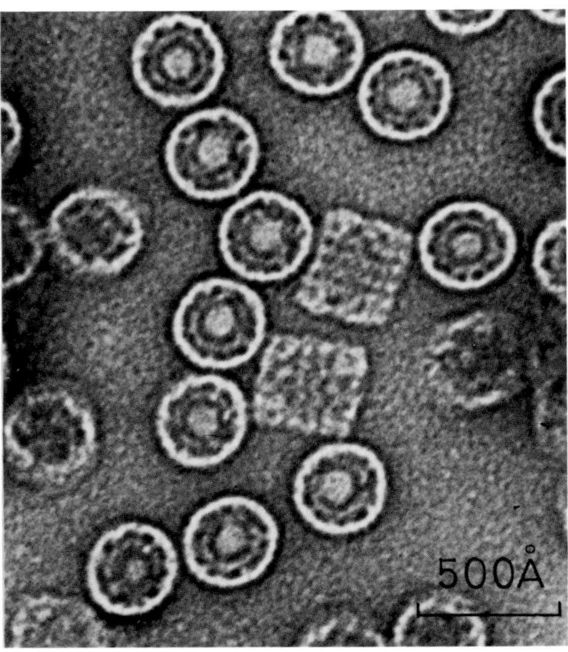

Fig. 47 Electron micrograph of negatively stained hemocyanin from *Busycon canaliculatum* showing side and end-on views. Taken from Mellema and Klug (1972).

molecules, except that the inner fivefold symmetry was greatly reduced. This reduction was later explained by the observation that the fivefold components at the two ends of the molecule are not in register with each other when seen end-on.

Because of the relatively short length of the molecules, detailed analyses of the side views were carried out on end-to-end linear polymers to increase the signal-to-noise ratio and include enough side views to reconstruct the particle to a resolution of 50 Å. Arrays in which the end-to-end polymers were in contact side-to-side were made from *Kelletia* hemocyanin (Fig. 48). There was no regular relationship between adjacent polymers, but along the polymers, adjacent molecules differed

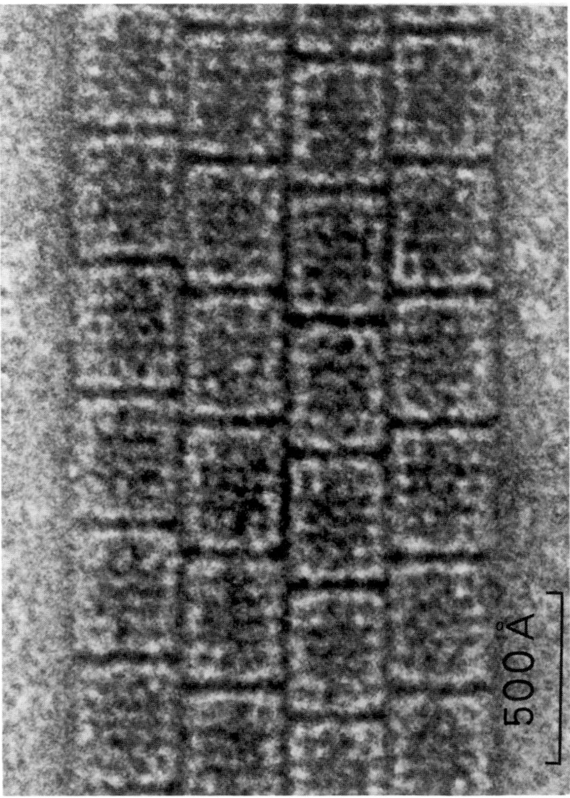

Fig. 48 Electron micrograph of laterally aggregated, linear polymers of *Kelletia* hemocyanin. Inner polymers were used for image processing because these were well preserved and so they have a smaller width than the outer rows which are flattened. Taken from Mellema and Klug (1972).

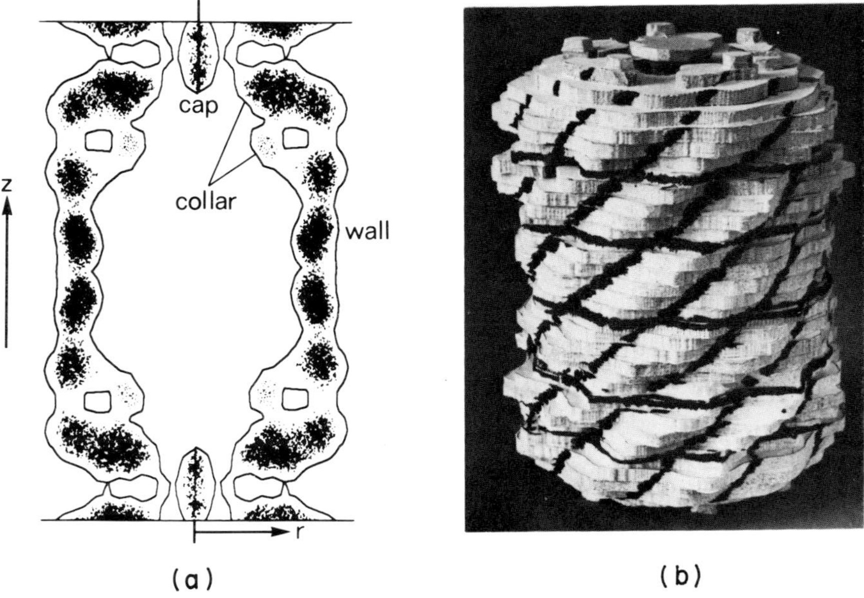

Fig. 49 Structure of *Kelletia* hemocyanin. (a) Cylindrically averaged density distribution in *Kelletia* hemocyanin. A contour line is shown to indicate the boundary of the molecule, and high densities within this are stippled. (b) A model of the computationally reconstructed hemocyanin molecule with the morphological units of the wall outlined. Taken from Mellema and Klug (1972).

regularly in orientation by 120°. That the molecules in the inner rows of these arrays were well preserved was shown by tilting the specimen in the microscope over an angular range of 72°. The images obtained by tilting showed little change in diameter, and the strong off-meridional intensity in optical diffraction patterns did not change in position. Polymer lengths of six molecules were chosen for processing. The phases in the computed Fourier transforms of these lengths showed that, to the resolution of the micrographs, the two halves of the molecules were related by twofold axes through the equatorial plane.

The model of the hemocyanin structure shown in Fig. 49 was constructed from a Fourier density map synthesized from the average of the Fourier transforms of the four best lengths of polymer, and further averaged by including the twofold axes of the molecule. The outer wall of the molecule appears to have tenfold symmetry and to consist of 60 morphological units arranged on an approximately regular surface lattice. As shown in Fig. 50, the surface lattices of adjacent molecules in a polymer are in approximate register, as if the hemocyanin molecules are

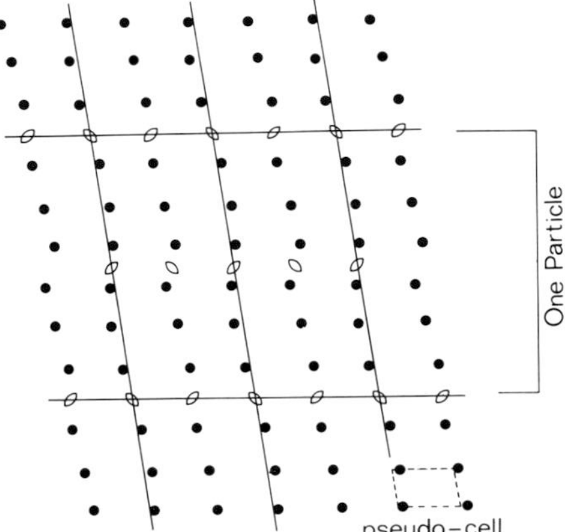

Fig. 50 The surface lattice of the linear polymers of *Kelletia* hemocyanin. The black dots mark the positions of the morphological units and lie approximately on a smaller pseudolattice, which extends across neighboring molecules.

formed by gathering six rings of a continuous stacked ring structure. This view was confirmed by analysis of polymers of *Busycon* hemocyanin, in which the surface lattice continues perfectly across adjacent molecules (Mellema and Klug, 1972). It thus appears that the morphological units are all identical, although in *Kelletia* they are in quasi-equivalent positions. Furthermore they must be dimers, since the molecules contain radial twofold axes. There are thus 120 structure units making up the wall, very close to the number of oxygen molecules bound by hemocyanin as calculated from the molecular weight and copper content.

There are two other components in the reconstructed particle, a central cap at each end, and an inner collar with fivefold symmetry attached to the top and bottom rings of the wall. These components appear to be responsible for the gathering of the six rings into a molecule since in the *Busycon* polymers where they are missing there is no such gathering. Mellema and Klug (1972) suggest that these components may play some crucial role in the assembly of the molecule.

2. *Ferritin*

Electron micrographs of negatively stained particles of the iron-containing protein ferritin have shown that the protein is isometric, has a diameter of about 100 Å and a dense electron-absorbing core (the iron-containing micelle) about 55 Å in diameter (Farrant, 1954). The iron-free protein, apoferritin, has a molecular weight of 480,000 (Richter,

1959; Harrison, 1960) and is built from subunits having a molecular weight of about 25,000 (Hofmann and Harrison, 1963). X-Ray diffraction studies have indicated that the particles must have at least twofold symmetry, probably have 52 symmetry, and certainly have a strong tendency toward pseudo-532 symmetry. From these results it was suggested that the twenty subunits are arranged at the vertices of a pentagonal dodecahedron (Harrison, 1963).

Several attempts have been made to see regular substructure in the images of negatively stained particles. Some selected images obtained by Easterbrook (1970) from unfractionated preparations of ferritin do appear to show a cagelike arrangement of density along the edges of a pentagonal dodecahedron. However, one feels the need for the application of a more objective method of image analysis for extracting details as fine as this from the electron micrographs.

VI. RECENT DEVELOPMENTS AND POSSIBLE FUTURE NEEDS

Although the technique of negative staining was originally introduced as a convenient and effective way of providing sufficient contrast in images of biologic specimens, it has proved the most successful method to date of preserving structural detail in proteins. Electron diffraction from negatively stained crystals of catalase has shown reflections extending to spacings of 8 Å (Glaeser, 1971), and images obtained by the minimal exposure technique of Williams and Fisher (1970) probably show detail extending to this resolution. However, the irradiation required to obtain such an image causes so much damage that the detail in subsequent images is limited to 20 Å or worse. One of the main processes contributing to this damage is the crystallization of staining salts under the action of the electron beam (Zingsheim and Bachmann, 1971). Unwin (1972) has shown that it is possible to inhibit the crystallization of uranyl acetate and formate by mixing them with the corresponding aluminium salt. Eliminating this source of loss of detail could be very useful when a series of images of the same field is required (e.g., in tilting experiments).

For some investigations, images of negatively stained specimens are unsatisfactory. For example, the fine details around the edges of large molecules can be lost in the depths of stain in these regions. The effects of positive staining, achieved by attaching heavy atom markers to specific parts of a molecule, are also counteracted in the image by the presence of negative stain. Moreover, since the images of negatively stained specimens correspond to the distribution of stain rather than to the specimen, in regions where stain cannot penetrate, no detail is seen. In order to reduce the necessity of staining, the problem of contrast has been tackled in a number of other ways, leading to the development of dark field and phase contrast techniques (see, for example, Kleinschmidt, 1971; Unwin, 1971) and the use of very thin, uniform substrate films. The exploitation of these techniques in the study of proteins has, however, been frustrated by the lack of a way to preserve unstained specimens during microscopy.

Structural disruption occurs in two stages: first as a result of dehydration as the specimen is subjected to the high vacuum in the microscope, and second as a result of irradiation with the electron beam. By maintaining a local relative humidity greater than 93% around specimens of catalase crystals in a special stage, Matricardi *et al.* (1972) have shown by electron diffraction that the crystals can remain ordered to 2 Å during the exposure required to obtain the diffraction pattern, but not appreciably longer than this before catastrophic beam damage occurs. In theory, therefore, it would be possible to solve a protein *crystal* structure by this technique to this resolution: the phases of the reflections could be determined by computing the Fourier transform of an image taken with a comparable short exposure. As with X-ray diffraction, one would need diffraction patterns and/or images in many different directions, and the problem of growing crystals which are sufficiently thin in each of these directions but sufficiently extended at right angles could well be insurmountable. There are other problems in extending the technique to isolated molecules, the main one being that the short exposures possible supply only enough electrons to record local detail to a resolution of about 100 Å and hence it is only by averaging over a large number of identical images in a crystal that detail to 2 Å resolution can be obtained. Probably the most suitable objects for study in this way would be helical arrangements of molecules, since their images reveal many different views of each subunit in sufficient numbers to yield high resolution data by averaging.

One way of reducing the damage caused by the electron beam is by using high energy electrons. High voltage electron microscopy has been very successful in the investigation of thick biologic specimens (see the review by Cosslett, 1971). The accelerating voltage used for most conventional work ranges from 80 to 100 kV. By increasing this voltage, the inelastic interaction of electrons with the specimen, which is responsible for damage, is reduced so that at 1 MeV the specimen "lifetime" could well be doubled (Glaeser, 1971). However, the elastic interaction used to form the image is also reduced as the electron energy increases, resulting in reduced contrast. This reduced contrast together with the reduced sensitivity of photographic emulsions and the instrumental problems arising from using high voltages make the probable advantages of this technique marginal in the investigation of isolated molecules and thin aggregates.

From the above discussion it is evident that for high resolution work, the problem of beam damage makes it imperative to extract the maximum data from the interaction of electrons with a specimen. Conventional images and diffraction patterns only utilize the elastically scattered electrons. However, additional information from the inelastically scattered electrons has been obtained in the high resolution scanning electron microscope developed by Crewe (see, for example, Crewe, 1971). In this instrument, the specimen is scanned by a fine electron beam. Because of the different scattering geometries of the elastically and inelastically scattered electrons and the energy loss of the latter, both of these, as well as the remaining beam of noninteracting electrons, can be monitored separately during the scan (Ne, Ni, and No, respectively). From combinations of the signals Ne and No, normal and dark field images can be displayed to a resolution of 5 Å. The full potential of the extra data in the signal Ni remains to be explored. Crewe (1971) has shown that the ratio Ne/Ni is a measure of the atomic number Z of the atom being scanned; it would be proportional to Z in the absence of a substrate film and other atoms. At present, single atoms of $Z > 70$ should be detectable on a very thin carbon film, but with future instruments operating at higher accelerating voltages, one would expect to be able to detect smaller atoms and also perhaps obtain useful information from the energy spectrum of Ni. In this way it may be possible to locate, say, individual metal atoms attached to proteins.

Although the full potentials of the above techniques remain to be explored, it is unlikely that electron microscopy will ever rival X-ray diffraction in the determination of protein structures at the atomic level. The main advantage of electron microscopy in the structural investigation of individual proteins lies in the rapid production of images with limited resolution from very small amounts of material. For obtaining fine detail, the most rewarding procedure at present is the combination of negative staining and minimal exposure to the electron beam, and the most amenable specimens are those with high symmetry, either inherent in the case of closed molecular aggregates, or impressed in extended arrays of molecules with helical or crystalline symmetry. With the available methods of image analysis, the reconstruction of particles to a resolution of 10 Å is feasible in the most favorable cases.

REFERENCES

Amos, L., and Klug, A. (1972). *Proc. Reg. Conf. (Eur.) Electron Microsc., 5th, 1972* p. 580.
Anderson, T. F. (1951). *Trans. N. Y. Acad. Sci.* [2] **13,** 130.
Astbury, W. T., Reed, W. T., and Spark, L. C. (1948). *Biochem. J.* **43,** 282.
Bailey, K. (1948). *Biochem. J.* **43,** 271.
Barynin, V. V., and Vainshtein, B. K. (1972). *Sov. Phys.—Crystallogr.* **16,** 653.
Bear, R. S. (1944). *J. Amer. Chem. Soc.* **66,** 2043.
Bonsignore, A., Lorenzoni, I., Cancedda, R., Cosulich, M. E., and De Flora, A. (1971). *Biochem. Biophys. Res. Commun.* **42,** 159.
Breedis, C., Berwick, L., and Anderson, T. F. (1962). *Virology* **17,** 84.
Brenner, S., and Horne, R. W. (1959). *Biochim. Biophys. Acta* **34,** 103.
Brenner, S., Streisinger, G., Horne, R. W., Champe, S. P., Barnet, L., Benzer, S., and Rees, M. W. (1959). *J. Mol. Biol.* **1,** 281.
Butler, P. J. G. (1970). *J. Mol. Biol.* **52,** 589.
Butler, P. J. G., and Klug, A. (1971). *Nature (London), New Biol.* **229,** 47.
Caspar, D. L. D. (1963). *Advan. Protein Chem.* **18,** 37.
Caspar, D. L. D. (1966). *J. Mol. Biol.* **15,** 365.
Caspar, D. L. D., and Klug, A. (1962). *Cold Spring Harbor Symp. Quant. Biol.* **27,** 1.
Caspar, D. L. D., Cohen, C., and Longley, W. (1969). *J. Mol. Biol.* **41,** 87.
Cassman, M., and Schachman, H. K. (1971). *Biochemistry* **10,** 1015.
Cohen, C., and Holmes, K. C. (1963). *J. Mol. Biol.* **6,** 423.
Cohen, C., and Szent-Györgyi, A. G. (1957). *J. Amer. Chem. Soc.* **79,** 248.
Cohen, C., and Szent-Györgyi, A. G. (1960). *Proc. Int. Congr. Biochem., 4th, 1958* Vol. 8, p. 108.
Cohen, C., Szent-Györgyi, A. G., and Kendrick-Jones, J. (1971). *J. Mol. Biol.* **56,** 223.
Cohen, P., and Rosemeyer, M. A. (1969). *Eur. J. Biochem.* **8,** 1 and 8.
Cohlberg, J. A., Pigiet, U. P., and Schachman, H. K. (1972). *Biochemistry* **11,** 3396.
Cohn, M. (1957). *Bacteriol. Rev.* **21,** 240.
Cornick, G., Sigler, P. B., and Ginsberg, H. S. (1971). *J. Mol. Biol.* **57,** 397.
Cosslett, V. E. (1971). *Phil. Trans. Roy. Soc. London, Ser. B* **261,** 35.
Crewe, A. V. (1971). *Phil. Trans. Roy. Soc. London, Ser. B* **261,** 61.
Crowther, R. A., and Amos, L. A. (1971a). *J. Mol. Biol.* **60,** 123.
Crowther, R. A., and Amos, L. A. (1971b). *Cold Spring Harbor Symp. Quant. Biol.* **36,** 489.
Crowther, R. A., and Franklin, R. M. (1972). *J. Mol. Biol.* **68,** 181.
Crowther, R. A., Amos, L. A., Finch, J. T., DeRosier, D. J., and Klug, A. (1970). *Nature (London)* **226,** 421.
DeRosier, D. J., and Klug, A. (1968). *Nature (London)* **217,** 130.
DeRosier, D. J., and Klug, A. (1972). *J. Mol. Biol.* **65,** 469.
DeRosier, D. J., and Moore, P. B. (1970). *J. Mol. Biol.* **52,** 355.
DeRosier, D. J., and Oliver, R. M. (1971). *Cold Spring Harbor Symp. Quant. Biol.* **36,** 199.
DeRosier, D. J., Oliver, R. M., and Reed, L. J. (1971). *Proc. Nat. Acad. Sci. U. S.* **68,** 1135.

Durham, A. C. H., and Finch, J. T. (1972). *J. Mol. Biol.* **67**, 307.
Durham, A. C. H., Finch, J. T., and Klug, A. (1971). *Nature (London), New Biol.* **229**, 37.
Eagles, P. A. M., and Johnson, L. N. (1972). *J. Mol. Biol.* **64**, 693.
Eagles, P. A. M., Johnson, L. N., Joynson, M. A., McMurray, C. H., and Gutfreund, H. (1969). *J. Mol. Biol.* **45**, 533.
Easterbrook, K. B. (1970). *J. Ultrastruct. Res.* **33**, 442.
Eisenberg, H., and Reisler, E. (1970). *Biopolymers* **9**, 113.
Eisenberg, H., and Tomkins, G. M. (1968). *J. Mol. Biol.* **31**, 37.
Erickson, H. P., and Klug, A. (1971). *Phil. Trans. Roy. Soc. London, Ser. B* **261**, 105.
Farrant, J. L. (1954). *Biochim. Biophys. Acta* **13**, 569.
Fernández-Morán, H., van Bruggen, E. F. J., and Ohtsuki, M. (1966). *J. Mol. Biol.* **16**, 191.
Finch, J. T. (1964). *J. Mol. Biol.* **8**, 872.
Finch, J. T. (1972). *J. Mol. Biol.* **66**, 291.
Finch, J. T., and Klug, A. (1965). *J. Mol. Biol.* **13**, 1.
Finch, J. T., and Klug, A. (1966). *J. Mol. Biol.* **15**, 344.
Finch, J. T., and Klug, A. (1967). *J. Mol. Biol.* **24**, 289.
Finch, J. T., and Klug, A. (1971). *Phil. Trans. Roy. Soc. London, Ser. B* **261**, 211.
Finch, J. T., Klug, A., and Stretton, A. O. W. (1964). *J. Mol. Biol.* **10**, 570.
Finch, J. T., Leberman, R., Chang, Y.-S., and Klug, A. (1966). *Nature (London)* **212**, 349.
Finch, J. T., Klug, A., and van Regenmortel, M. H. V. (1967). *J. Mol. Biol.* **24**, 303.
Finch, J. T., Klug, A., and Leberman, R. (1970). *J. Mol. Biol.* **50**, 215.
Fischer, E. H., and Krebs, E. G. (1966). *Fed. Proc., Fed. Amer. Soc. Exp. Biol.* **25**, 1511.
Franklin, R. E., and Holmes, K. C. (1958). *Acta Crystallogr.* **11**, 213.
Franklin, R. M., Pettersson, U., Åkervall, K., Strandberg, B., and Philipson, L. (1971a). *J. Mol. Biol.* **57**, 383.
Franklin, R. M., Harrison, S. C., Pettersson, U., Philipson, L., Brändén, C. I., and Werner, P.-E. (1971b). *Cold Spring Harbor Symp. Quant. Biol.* **36**, 503.
Fraser, R. D. B., and MacRae, T. P. (1973). "Conformation in Fibrous Proteins." Academic Press, New York.
Gerhart, J. C., and Holoubek, H. (1967). *J. Biol. Chem.* **242**, 2886.
Glaeser, R. M. (1971). *J. Ultrastruct. Res.* **36**, 466.
Green, N. M. (1969). *Advan. Immunol.* **11**, 1.
Green, N. M., Valentine, R. C., Wrigley, N. G., Ahmad, F., Jacobson, B., and Wood, H. G. (1972). *J. Biol. Chem.* **247**, 6284.
Gurskaya, G. V., Lovanova, G. M., and Vainshtein, B. R. (1972). *Sov. Phys.—Crystallogr.* **16**, 662.
Hall, C. E. (1955). *J. Biophys. Biochem. Cytol.* **1**, 1.
Hanson, J., and Lowy, J. (1963). *J. Mol. Biol.* **6**, 46.
Hanson, J., Lowy, J., Huxley, H. E., Bailey, K., Kay, C. M., and Rüegg, J. C. (1957). *Nature (London)* **180**, 1134.
Harrison, P. M. (1960). *Acta Crystallogr.* **13**, 1060.
Harrison, P. M. (1963). *J. Mol. Biol.* **6**, 404.
Hitchcock, S., Huxley, H. E., and Szent-Györgyi, A. G. (1973). *J. Mol. Biol.* **80**, 825.
Hodge, A. J. (1952). *Proc. Nat. Acad. Sci. U. S.* **38**, 850.
Hodge, A. J. (1959). *Rev. Mod. Phys.* **31**, 409.
Hofmann, T., and Harrison, P. M. (1963). *J. Mol. Biol.* **6**, 256.

Horne, R. W., and Greville, G. D. (1963). *J. Mol. Biol.* **6,** 506.
Horne, R. W., Brenner, S., Waterson, A. P., and Wildy, P. (1959). *J. Mol. Biol.* **1,** 84.
Huxley, H. E. (1957). *Electron Microsc., Proc. Stockholm Conf., 1956* p. 260.
Huxley, H. E. (1963). *J. Mol. Biol.* **7,** 281.
Huxley, H. E., and Brown, W. (1967). *J. Mol. Biol.* **30,** 383.
Huxley, H. E., and Zubay, G. (1960a). *J. Mol. Biol.* **2,** 10.
Huxley, H. E., and Zubay, G. (1960b). *J. Mol. Biol.* **2,** 189.
International Tables for X-Ray Crystallography. (1952). Vol. 1, Kynoch Press, Birmingham, England.
Josephs, R. (1971). *J. Mol. Biol.* **55,** 147.
Josephs, R., and Borisy, G. (1972). *J. Mol. Biol.* **65,** 127.
Karlsson, U., Koorajian, S., Zabin, I., Sjöstrand, F. S., and Miller, A. (1964). *J. Ultrastruct. Res.* **10,** 45.
Kellenberger, E., Eiserling, F. A., and Boy de la Tour, E. (1968). *J. Ultrastruct. Res.* **21,** 335.
Kiselev, N. A., and Klug, A. (1969). *J. Mol. Biol.* **40,** 155.
Kiselev, N. A., Spitzberg, C. L., and Vainshtein, B. K. (1967). *J. Mol. Biol.* **25,** 433.
Kiselev, N. A., DeRosier, D. J., and Klug, A. (1968). *J. Mol. Biol.* **35,** 561.
Kiselev, N. A., Lerner, F. Ya., and Livanova, N. B. (1971). *J. Mol. Biol.* **62,** 537.
Kleinschmidt, A. K. (1971). *Phil. Trans. Roy. Soc. London, Ser. B* **261,** 143.
Klug, A. (1965). *J. Mol. Biol.* **11,** 424.
Klug, A. (1968). *Symp. Int. Soc. Cell Biol.* **6,** 1.
Klug, A., and Berger, J. E. (1964). *J. Mol. Biol.* **10,** 565.
Klug, A., and DeRosier, D. J. (1966). *Nature (London)* **212,** 29.
Klug, A., and Finch, J. T. (1960). *J. Mol. Biol.* **2,** 201.
Klug, A., and Finch, J. T. (1965). *J. Mol. Biol.* **11,** 403.
Klug, A., and Finch, J. T. (1968). *J. Mol. Biol.* **31,** 1.
Krimm, S., and Anderson, T. F. (1967). *J. Mol. Biol.* **27,** 197.
Laemmli, U. K. (1970). *Nature (London)* **227,** 680.
Longley, W. (1967). *J. Mol. Biol.* **30,** 323.
Lowey, S., Kucera, J., and Holtzer, A. (1963). *J. Mol. Biol.* **7,** 234.
Madsen, N. B., and Cori, C. F. (1954). *Biochim. Biophys. Acta* **15,** 516.
Madsen, N. B., and Cori, C. F. (1956). *J. Biol. Chem.* **223,** 1055.
Maizel, J. V., White, D. O., and Scharff, M. D. (1968). *Virology* **36,** 115 and 126.
Marchesi, S. L., Steers, E., Jr., and Shifrin, S. (1969). *Biochim. Biophys. Acta* **181,** 20.
Markham, R., Frey, S., and Hills, G. J. (1963). *Virology* **20,** 88.
Markham, R., Hitchborn, J. H., Hills, G. J., and Frey, S. (1964). *Virology* **22,** 342.
Matricardi, V. R., Moretz, R. C., and Parsons, D. F. (1972). *Science* **177,** 268.
Meighen, E. A., Pigiet, V., and Schachman, H. K. (1970). *Proc. Nat. Acad. Sci. U. S.* **65,** 234.
Mellema, J. E., and Amos, L. A. (1972). *J. Mol. Biol.* **72,** 819.
Mellema, J. E., and Klug, A. (1972). *Nature (London)* **239,** 146.
Mellema, J. E., van Bruggen, E. F. J., and Gruber, M. (1968). *J. Mol. Biol.* **31,** 75.
Monod, J., Wyman, J., and Changeux, J.-P. (1965). *J. Mol. Biol.* **12,** 88.
Moody, M. F. (1967a). *J. Mol. Biol.* **25,** 167.
Moody, M. F. (1967b). *J. Mol. Biol.* **25,** 201.
Moody, M. F. (1971). *Phil. Trans. Roy. Soc. London, Ser. B* **261,** 181.
Moore, P. B., Huxley, H. E., and DeRosier, D. J. (1970). *J. Mol. Biol.* **50,** 279.
Morino, Y., and Snell, E. E. (1967a). *J. Biol. Chem.* **242,** 5591.

Morino, Y., and Snell, E. E. (1967b). *J. Biol. Chem.* **242**, 5602.
Moss, J., and Lane, M. D. (1971). *Advan. Enzymol. Relat. Areas Mol. Biol.* **35**, 395.
Munn, E. A. (1971). *Experientia* **27**, 170.
Nixon, H. L., and Gibbs, A. J. (1960). *J. Mol. Biol.* **2**, 197.
Nonomura, Y., Drabikowski, W., and Ebashi, S. (1968). *J. Biochem. (Tokyo)* **64**, 419.
Ohtsuki, I., Masaki, T., Nomura, T., and Ebashi, S. (1967). *J. Biochem. (Tokyo)* **61**, 817.
Olsen, J. A., and Anfinsen, C. B. (1952). *J. Biol. Chem.* **197**, 67.
Page, S. G., and Huxley, H. E. (1963). *J. Cell Biol.* **19**, 369.
Penhoet, E., Kochman, M., Valentine, R., and Rutter, W. J. (1967). *Biochemistry* **6**, 2940.
Pepe, F. (1966). *J. Cell Biol.* **28**, 505.
Pettersson, U. (1970). Ph.D. Thesis, Uppsala University, Uppsala, Sweden.
Pettit, F. H., Hamilton, L., Munk, P., Namihira, G., Eley, M. H., Willms, C. R., and Reed, L. J. (1973). *J. Biol. Chem.* **248**, 5282.
Reisler, E., Pouyet, J., and Eisenberg, H. (1970). *Biochemistry* **9**, 3095.
Rice, R. V. (1961). *Biochim. Biophys. Acta* **52**, 602.
Rice, R. V. (1964). *In* "Biochemistry of Muscle Contraction" (J. Gergely, ed.), p. 41. Churchill, London.
Richards, K. E., and Williams, R. C. (1972). *Biochemistry* **11**, 3393.
Richter, G. W. (1959). *J. Biophys. Biochem. Cytol.* **6**, 531.
Rosenbusch, J. P., and Weber, K. (1971). *J. Biol. Chem.* **246**, 1644.
Samejima, T., and Yang, J. T. (1963). *J. Biol. Chem.* **238**, 3256.
Schramm, G. (1947). *Z. Naturforsch. B* **2**, 249.
Schramm, G., and Zillig, W. (1955). *Z. Naturforsch. B* **10**, 493.
Scrutton, M. C., and Utter, M. F. (1965). *J. Biol. Chem.* **240**, 1.
Seery, V. L., Fischer, E. H., and Teller, D. C. (1967). *Biochemistry* **6**, 3315.
Shapiro, B. M., and Ginsburg, A. (1968). *Biochemistry* **7**, 2153.
Shapiro, B. M., and Stadtman, E. R. (1967). *J. Biol. Chem.* **242**, 5069.
Slayter, H., and Lowey, S. (1967). *Proc. Nat. Acad. Sci. U. S.* **58**, 1611.
Spitzberg, C. L. (1966). *Biophysics* **11**, 766.
Spudich, J. A., Huxley, H. E., and Finch, J. T. (1972). *J. Mol. Biol.* **72**, 619.
Steere, R. L. (1957). *J. Biophys. Biochem. Cytol.* **3**, 45.
Strausbauch, P. H., and Fischer, E. H. (1970). *Biochemistry* **9**, 226.
Sund, H., and Weber, K. (1963). *Biochem. Z.* **337**, 24.
Sund, H., Weber, K., and Mölbert, E. (1967). *Eur. J. Biochem.* **1**, 400.
Sund, H., Pilz, I., and Herbst, M. (1969). *Eur. J. Biochem.* **7**, 517.
Szent-Györgyi, A. G., Cohen, C., and Kendrick-Jones, J. (1971). *J. Mol. Biol.* **56**, 239.
Tanford, C., and Lovrien, R. (1962). *J. Amer. Chem. Soc.* **75**, 179.
To, C. M. (1971). *J. Mol. Biol.* **59**, 215.
Traub, W., and Piez, K. A. (1971). *Advan. Protein Chem.* **25**, 243.
Unwin, P. N. T. (1971). *Phil. Trans. Roy. Soc. London, Ser. B* **261**, 95.
Unwin, P. N. T. (1972). *Proc. Reg. Conf. (Eur.) Electron Microsc., 5th, 1972* p. 232.
Vainshtein, B. K., Barynin, V. V., Gurskaya, G. V., and Nikitin, V. Ya. (1968). *Sov. Phys.—Crystallogr.* **12**, 750.
Valentine, R. C. (1968). *Pre-Congr. Abstr. Reg. Conf. (Eur.) Electron Microsc., 4th, 1968* Vol. 2, p. 3.

Valentine, R. C. (1969). *In* "Symmetry and Function of Biological Systems at the Macromolecular Level" (A. Engstrom and B. Strandberg, eds.), Nobel Symp. No. 11, p. 165. Almqvist & Wiksell, Stockholm.
Valentine, R. C., and Chignell, D. A. (1968). *Nature (London)* **218**, 950.
Valentine, R. C., and Pereira, H. G. (1965). *J. Mol. Biol.* **13**, 13.
Valentine, R. C., Wrigley, N. G., Scrutton, M. C., Irias, J. J., and Utter, M. F. (1966). *Biochemistry* **5**, 3111.
Valentine, R. C., Shapiro, B. M., and Stadtman, E. R. (1968). *Biochemistry* **7**, 2143.
van Bruggen, E. F. J., Wiebenga, E. H., and Gruber, M. (1962). *J. Mol. Biol.* **4**, 1.
Watson, J. D. (1954). *Biochim. Biophys. Acta* **13**, 10.
Weber, K. (1968). *Nature (London)* **218**, 1116.
Weber, K., Sund, H., and Wallenfels, K. (1963). *Angew. Chem., Int. Ed. Engl.* **2**, 481.
Wildy, P., Stoker, M. G. P., Macpherson, I. A., and Horne, R. W. (1960). *Virology* **11**, 444.
Wiley, D. C., and Lipscomb, W. N. (1968). *Nature (London)* **218**, 1119.
Wiley, D. C., Evans, D. R., Warren, S. G., McMurray, C. H., Edwards, B. F. P., Franks, W. A., and Lipscomb, W. N. (1971). *Cold Spring Harbor Symp. Quant. Biol.* **36**, 285.
Williams, R. C. (1953). *Exp. Cell Res.* **4**, 188.
Williams, R. C., and Fisher, H. W. (1970). *J. Mol. Biol.* **52**, 121.
Williams, R. C., and Fraser, D. (1956). *Virology* **2**, 289.
Williams, R. C., and Smith, K. M. (1958). *Biochim. Biophys. Acta* **28**, 464.
Williams, R. C., and Steere, R. L. (1951). *J. Amer. Chem. Soc.* **73**, 2057.
Williams, R. C., Kass, S. J., and Knight, C. A. (1960). *Virology* **12**, 48.
Woolfolk, C. A., Shapiro, B. M., and Stadtman, E. R. (1966). *Arch. Biochem. Biophys.* **116**, 177.
Wrigley, N. G., Heather, J. V., Bonsignore, A., and De Flora, A. (1972). *J. Mol. Biol.* **68**, 483.
Yanagida, M., Boy de la Tour, E., Allf-Steinberger, C., and Kellenberger, E. (1970). *J. Mol. Biol.* **50**, 35.
Yanagida, M., DeRosier, D. J., and Klug, A. (1972). *J. Mol. Biol.* **65**, 489.
Young, M. R., Tolbert, B., Valentine, R. C., and Utter, M. F. (1971). Quoted in Moss and Lane (1971).
Zingsheim, H. P., and Bachmann, L. (1971). *Kolloid-Z. Z. Polym.* **246**, 561.
Zipser, D. (1963). *J. Mol. Biol.* **7**, 113.

Author Index

Numbers in italics refer to the pages on which the complete references are listed.

E

F

H

L

M

N

Q

R

T

U

V

Y

Subject Index

B

C

D

E

F

G

N

Q

R

U

V

W

X

Z

A 5
B 6
C 7
D 8
E 9
F 0
G 1
H 2
I 3
J 4